CAMBRIDGE TRACTS IN MATHEMATICS



**236 Dilation and Model Theory for Pairs of Commuting Contraction Operators**

# CAMBRIDGE TRACTS IN MATHEMATICS



A complete list of books in the series can be found at www.cambridge.org/mathematics.
Recent titles include the following:

203. A Primer on the Dirichlet Space. By O. El-Fallah, K. Kellay, J. Mashreghi, and T. Ransford
204. Group Cohomology and Algebraic Cycles. By B. Totaro
205. Ridge Functions. By A. Pinkus
206. Probability on Real Lie Algebras. By U. Franz and N. Privault
207. Auxiliary Polynomials in Number Theory. By D. Masser
208. Representations of Elementary Abelian p-Groups and Vector Bundles. By D. J. Benson
209. Non-homogeneous Random Walks. By M. Menshikov, S. Popov, and A. Wade
210. Fourier Integrals in Classical Analysis (Second Edition). By C. D. Sogge
211. Eigenvalues, Multiplicities and Graphs. By C. R. JOHNSON and C. M. Saiago
212. Applications of Diophantine Approximation to Integral Points and Transcendence. By P. Corvaja and U. Zannier
213. Variations on a Theme of Borel. By S. Weinberger
214. The Mathieu Groups. By A. A. Ivanov
215. Slenderness I: Foundations. By R. Dimitric
216. Justification Logic. By S. Artemov and M. Fitting
217. Defocusing Nonlinear Schrödinger Equations. By B. Dodson
218. The Random Matrix Theory of the Classical Compact Groups. By E. S. Meckes
219. Operator Analysis. By J. Agler, J. E. McCarthy, and N. J. Young
220. Lectures on Contact 3-Manifolds, Holomorphic Curves and Intersection Theory. By C. Wendl
221. Matrix Positivity. By C. R. Johnson, R. L. Smith, and M. J. Tsatsomeros
222. Assouad Dimension and Fractal Geometry. By J. M. Fraser
223. Coarse Geometry of Topological Groups. By C. Rosendal
224. Attractors of Hamiltonian Nonlinear Partial Differential Equations. By A. Komech and E. Kopylova
225. Noncommutative Function-Theoretic Operator Function and Applications. By J. A. Ball and V. Bolotnikov
226. The Mordell Conjecture. By A. Moriwaki, H. Ikoma, and S. Kawaguchi
227. Transcendence and Linear Relations of 1-Periods. By A. Huber and G. Wüstholz
228. Point-Counting and the Zilber–Pink Conjecture. By J. Pila
229. Large Deviations for Markov Chains. By A. D. De Acosta
230. Fractional Sobolev Spaces and Inequalities. By D. E. Edmunds and W. D. Evans
231. Families of Varieties of General Type. By J. Kollár
232. The Art of Working with the Mathieu Group M24. By R. T. Curtis
233. Category and Measure. By N. H. Bingham and A. J. Ostaszewski
234. The Theory of Countable Borel Equivalence Relations. By A. S. Kechris
235. Weak Solutions to Gradient Flows in Metric Measure Spaces. By W. Górny and J. M. Mazón
236. Dilation and Model Theory for Pairs of Commuting Contraction Operators. By J. A. Ball and H. Sau
237. Birational Geometry and Manin's Conjecture. By S. Tanimoto

# Dilation and Model Theory for Pairs of Commuting Contraction Operators

JOSEPH A. BALL
*Virginia Tech, Blacksburg*

HARIPADA SAU
*Indian Institute of Science Education and Research Pune*

CAMBRIDGE
UNIVERSITY PRESS

CAMBRIDGE
UNIVERSITY PRESS

Shaftesbury Road, Cambridge CB2 8EA, United Kingdom

One Liberty Plaza, 20th Floor, New York, NY 10006, USA

477 Williamstown Road, Port Melbourne, VIC 3207, Australia

314–321, 3rd Floor, Plot 3, Splendor Forum, Jasola District Centre, New Delhi – 110025, India

Cambridge University Press is part of Cambridge University Press & Assessment, a department of the University of Cambridge.

We share the University's mission to contribute to society through the pursuit of education, learning and research at the highest international levels of excellence.

www.cambridge.org
Information on this title: www.cambridge.org/9781009687218

DOI: 10.1017/9781009687225



When citing this work, please include a reference to the DOI 10.1017/9781009687225

First published 2026

*A catalogue record for this publication is available from the British Library*

*A Cataloging-in-Publication data record for this book is available from the Library of Congress*

ISBN 978-1-009-68721-8 Hardback

To all the teachers, collaborators, and friends from
whom we have learned mathematics

# Contents

# Preface

One of the pinnacles of twentieth-century analysis is what is now referred to as the von Neumann inequality. Soon after von Neumann, Sz.-Nagy published his (power) dilation theorem that gave a quick proof of the mighty von Neumann inequality. The dilation theorem of Sz.-Nagy evolved into a spectacularly successful dilation and model theory for contractive Hilbert space operators pioneered by Sz.-Nagy and Foias, the contribution of which has been extraordinary in Hilbert space operator theory and in other fields of mathematics and engineering. The rational dilation problem for a general compact set in several complex variables is a multivariable formulation of the classical dilation theory by Arveson. In Arveson's language, Sz.-Nagy's dilation theorem becomes the rational dilation problem for the disk. This book is about the rational dilation problem for the bidisk. The pioneering work for the bidisk was done by Tsuyosi Ando in 1963. Ando's proof can be seen as a bivariate adaptation of Schäffer's proof of the Sz.-Nagy dilation theorem. Meanwhile, the attention of contemporary experts was soon diverted by the surprising result of Parrott that the tridisk does not enjoy a similar dilation theorem. As a result, until now, there has not been a rigorous follow-up investigation giving a bivariate analogue (to the extent possible) of the Sz.-Nagy–Foias model theory and unitary classification for a single contraction operator $T$ for the setting where the contraction operator $T$ is replaced by a commuting contractive operator pair $(T_1, T_2)$. The goal of this book is to fill in this gap in Hilbert space operator theory by developing a parallel dilation and model theory for commuting contractive pairs. In addition, we give two new proofs of Ando's dilation theorem, which arguably are better motivated and provide additional insight. This book should prove accessible and helpful for graduate students aspiring to study multivariate operator theory. A further chapterwise elaboration of the book is provided in Chapter 1.

This book would not have been possible without the excellent works [33] and [48]. The connecting link between these two papers is the elementary fact first observed in the unpublished preprint [83] by the second author HS that if $(T_1, T_2)$ is a commuting contractive pair, then the triple $(T_1, T_2, T_1T_2)$ has the closure of the tetrablock, which is a bounded domain in the three-dimensional complex space, as a spectral set. One can then apply the operator theory of the tetrablock, a program that Tirthankar Bhattacharyya initiated in [33], to the special case $(T_1, T_2, T_1T_2)$. While this approach leads to verification of some of the results found here for the commuting contractive-pair setting, there are other results that are not found via this route. However, the approach through reduction to the tetrablock setting was the seed of the ideas in [83], which grew to be the present monograph. Let us also note the remarkable fact that while the dilation theory for the tetrablock still remains to be understood (see [20]), the existing theory is strong enough to shed a glittering light on the bivariate dilation theory, which has been a subject of interest since the mid twentieth century.

The second author HS would like to thank the organizers of the "Conference on Functional Analysis" held at IIT Bombay in 2017, where Srijan Sarkar, an author of the article [48], presented some ideas that proved seminal for some of the main concepts presented here. The second author HS is also grateful for all the mathematics he has learned from his PhD supervisor Tirthankar Bhattacharyya and his postdoc mentors, especially the first author JAB and Bata Krishna Das. He also thanks the Science and Engineering Research Board (SERB) and the Department of Science and Technology (DST) for financial support. Both authors thank Google Meet and Dropbox for the opportunity to have many productive meetings to push the book along, while physically separated by the vast geographical divide between India and Virginia.

# 1
# Introduction

The starting point for many future developments in nonselfadjoint operator theory was the Sz.-Nagy dilation theorem from 1953 [86]: *If $T$ is a contraction operator on a Hilbert space $\mathcal{H}$, then there is a unitary operator $\mathcal{U}$ on a larger Hilbert space $\widetilde{\mathcal{K}} \supset \mathcal{H}$ such that $T^n = P_{\mathcal{H}}\mathcal{U}^n|_{\mathcal{H}}$ for all $n = 0, 1, 2, \ldots$.* While the original proofs were more existential than constructive, there followed more concrete constructive proofs (e.g., the Schäffer-matrix construction from [84] to be discussed in Section 2.2), which evolved into a detailed geometric picture of the dilation space (see [92, Chapter II]). Analysis of how the original Hilbert space $\mathcal{H}$ fits into the dilation space $\widetilde{\mathcal{K}}$ and how the abstract contraction operator $T$ can be modeled as a compressed multiplication operator led to the discovery of the characteristic function $\Theta_T$ of the contraction operator $T$. Every isometry operator extends to a unitary operator. For this reason, it is enough to work with the equivalent notion of isometric *lifts* $V$ rather than unitary dilations $\mathcal{U}$ of $T$: We say that an operator $V$ on $\mathcal{K} \supset \mathcal{H}$ is a *lift* of $T$ acting on $\mathcal{H}$ if $\mathcal{H}$ is invariant for $V^*$ and $V^*|_{\mathcal{H}} = T^*$. More generally, if $\Pi \colon \mathcal{H} \to \mathcal{K}$ is an isometric embedding of $\mathcal{H}$ into $\mathcal{K}$ and $V^*\Pi = \Pi T^*$, then we shall say that $(\Pi, V)$ on $\mathcal{K}$ is a lift of $T$. For a contraction operator $T$ acting on a Hilbert space $\mathcal{H}$, the following notations of the defect operators

$$D_T = (I - T^*T)^{\frac{1}{2}}, \quad D_{T^*} = (I - TT^*)^{\frac{1}{2}}$$

and defect spaces

$$\mathcal{D}_T = \overline{\operatorname{Ran}}\, D_T, \quad \mathcal{D}_{T^*} = \overline{\operatorname{Ran}}\, D_{T^*},$$

will be used throughout. The characteristic function of $T$ is given by

$$\Theta_T(z) = \left(-T^* + zD_{T^*}(I - zT^*)^{-1}D_T\right)\Big|_{\mathcal{D}_T} : \mathcal{D}_T \to \mathcal{D}_{T^*}.$$

Consider the pointwise-defect operator of $\Theta_T$:

$$\Delta_{\Theta_T}(\zeta) = (I - \Theta_T(\zeta)^*\Theta_T(\zeta))^{\frac{1}{2}} \quad \text{for } \zeta \in \mathbb{T}.$$

For a Hilbert space $\mathcal{F}$, $H^2(\mathcal{F})$ denotes the $\mathcal{F}$-valued *Hardy space*

$$H^2(\mathcal{F}) := \left\{ f : \mathbb{D} \to \mathcal{F} : f(z) = \sum_{n=0}^{\infty} z^n f_n \text{ and } \sum_{n=0}^{\infty} \|f_n\|^2_{\mathcal{F}} < \infty \right\}.$$

Define Hilbert function spaces

$$\mathcal{K}_{\Theta_T} := \begin{bmatrix} H^2(\mathcal{D}_{T^*}) \\ \overline{\Delta_{\Theta_T} L^2(\mathcal{D}_T)} \end{bmatrix}, \quad \mathcal{H}_{\Theta_T} := \mathcal{K}_{\Theta_T} \ominus \begin{bmatrix} \Theta_T \\ \Delta_{\Theta_T} \end{bmatrix} H^2(\mathcal{D}_T)$$

and define operators $V_{\Theta_T}$ on $\mathcal{K}_{\Theta_T}$ and $T_{\Theta_T}$ on $\mathcal{H}_{\Theta_T}$ by

$$V_{\Theta_T} = \begin{bmatrix} M_z^{\mathcal{D}_{T^*}} & 0 \\ 0 & M_\zeta|_{\overline{\Delta_{\Theta_T} L^2(\mathcal{D}_T)}} \end{bmatrix} \quad \text{and} \quad T_{\Theta_T} = P_{\mathcal{H}_{\Theta_T}} V_{\Theta_T}|_{\mathcal{H}_{\Theta_T}}. \tag{1.1}$$

Then there is an isometric embedding $\Pi_{\text{NF}} : \mathcal{H} \to \mathcal{K}_{\Theta_T}$ such that

$$\operatorname{Ran} \Pi_{\text{NF}} = \mathcal{H}_{\Theta_T} \quad \text{and} \quad V^*_{\Theta_T} \Pi_{\text{NF}} = \Pi_{\text{NF}} T^*,$$

thereby showing that $(\Pi_{\text{NF}}, V_{\Theta_T})$ is an isometric lift of $T$. A direct consequence of this lifting is the realization that the abstract contraction operator $T$ is now unitarily equivalent to the compressed multiplication operator $T_{\Theta_T}$ as in (1.1) via the unitary operator $\Pi_{\text{NF}} : \mathcal{H} \to \mathcal{H}_{\Theta_T} = \operatorname{Ran} \Pi_{\text{NF}}$. The most elegant part of the theory of Sz.-Nagy–Foias is perhaps the fact that there is a converse. The notation $\mathcal{B}(\mathcal{D}, \mathcal{D}_*)$ stands for the space of bounded operators from the Hilbert space $\mathcal{D}$ into another Hilbert space $\mathcal{D}_*$ with the convention that $\mathcal{B}(\mathcal{D}, \mathcal{D}_*) = \mathcal{B}(\mathcal{D})$ when $\mathcal{D} = \mathcal{D}_*$. We shall also use $(\mathcal{D}, \mathcal{D}_*, \Theta)$ for a $\mathcal{B}(\mathcal{D}, \mathcal{D}_*)$-valued bounded analytic function on the unit disk $\mathbb{D}$ with the convention that $(\mathcal{D}, \mathcal{D}_*, \Theta) = (\mathcal{D}, \Theta)$ when $\mathcal{D} = \mathcal{D}_*$. A bounded analytic function $(\mathcal{D}, \mathcal{D}_*, \Theta)$ is (i) a *contractive* if $\|\Theta(z)\| \le 1$ for each $z \in \mathbb{D}$ or (ii) *purely contractive* if, in addition, $\|\Theta(0)d\| < \|d\|$ for $0 \neq d \in \mathcal{D}$. It is known (see, for example, [92, Section 2, Chapter V]) that for any bounded analytic function $(\mathcal{D}, \mathcal{D}_*, \Theta)$, the *radial limit*

$$\Theta(\zeta) := \lim_{r \to 1-} \Theta(r\zeta)$$

exists for every unimodular $\zeta$ except for a set of arc-length measure zero. The convergence above is in the strong operator topology on $\mathcal{B}(\mathcal{D}, \mathcal{D}_*)$. With the radial limit in hand, define $\Delta_\Theta(\zeta)$, $\mathcal{K}_\Theta$, $\mathcal{H}_\Theta$, $V_\Theta$, $T_\Theta$ as above with $(\mathcal{D}, \mathcal{D}_*, \Theta)$ in place of $(\mathcal{D}_T, \mathcal{D}_{T^*}, \Theta_T)$. Then, quite remarkably, $V_\Theta$ *is an isometric lift of the contraction operator* $T_\Theta$ *and, moreover, the characteristic* *function* $(\mathcal{D}_{T_\Theta}, \mathcal{D}_{T_\Theta^*}, \Theta_{T_\Theta})$ *of* $T_\Theta$ *coincides with* $(\mathcal{D}, \mathcal{D}_*, \Theta)$ in the sense that there are unitary operators $u \colon \mathcal{D} \to \mathcal{D}_{T_\Theta}$, $u_* \colon \mathcal{D}_* \to \mathcal{D}_{T_\Theta^*}$ so that

$$\Theta_{T_\Theta}(z)u = u_*\Theta(z) \quad \text{for all } z \in \mathbb{D}.$$

The Ando dilation theorem [10], coming 10 years later, provides a two-variable analogue of the Sz.-Nagy dilation theorem: *Given a commuting pair of contraction operators* $(T_1, T_2)$ *on a Hilbert space* $\mathcal{H}$*, there is a commuting pair of unitary operators* $(\mathcal{U}_1, \mathcal{U}_2)$ *on a larger Hilbert space* $\widetilde{\mathcal{K}} \supset \mathcal{H}$ *so that, for all* $n, m \geq 0$, $T_1^n T_2^m = P_{\mathcal{H}} \mathcal{U}_1^n \mathcal{U}_2^m|_{\mathcal{H}}$*, or equivalently, there is a commuting isometric pair* $(V_1, V_2)$ *acting on a Hilbert space* $\mathcal{K}$ *and an isometric embedding* $\Pi \colon \mathcal{H} \to \mathcal{K}$ *so that* $(V_1^*, V_2^*)\Pi = \Pi(T_1^*, T_2^*)$. The proof in [10] is an expanded version of the Schäffer-matrix construction for the single-operator case. However, the proof failed to shed much light on the geometry of the dilation space. This shortcoming in the original proof of Ando is perhaps expected owing to the fact that, unlike the one-variable scenario, isometric lifts of a commuting contractive pair do not enjoy uniqueness up to unitary equivalence. We will elaborate on this later with concrete examples in various places in the book; see Example 4.1.4 for an elementary example demonstrating nonuniqueness. Perhaps as a consequence of the lack of uniqueness, there has been essentially no follow-up to Ando's result in the direction of a Sz.-Nagy–Foias-type model theory for a commuting pair of contraction operators as there was in the single-operator setting, although there have now been some preliminary results in this direction (see [5, 24, 48]).

The goal of this monograph is to develop a complete analogue of the Sz.-Nagy–Foias dilation theory and operator model theory for the commuting contractive pair setting. This proceeds in several steps. We shall now describe the key developments chapterwise.

**Chapter 2** Following the present Introduction, Chapter 2 reviews the dilation/model theory for a single contraction operator from four points of view: (i) the coordinate-free geometric picture as found in Chapter II of the classic book [92], (ii) the Douglas model theory as found in [54], (iii) the Schäffer model theory [84], and (iv) the Sz.-Nagy–Foias model theory as found in Chapter VI of [92]. This chapter also presents certain canonical decomposition theorems for an isometry (Theorem 2.1.4), a general contraction (Theorem 2.3.3), and the isometric embedding $\Pi$ involved in an isometric lifting $(\Pi, V)$ of a contractive operator $T$ (Theorems 2.3.5 and 2.3.6).

**Chapter 3** It is hard to overestimate the role of isometries in the development of the main results here. Indeed, at the heart of the main results lies a concrete model theory for commuting isometries developed by Berger, Coburn, and Lebow in [31]. It is instructive to describe the model briefly here: *An operator pair* $(V_1, V_2)$ *is a commuting isometric pair if and only if there is a tuple* $(\mathcal{F}, \mathcal{H}_u, P, U, W_1, W_2)$ *consisting of an orthogonal projection* $P$*, a unitary* $U$ *acting on a Hilbert space* $\mathcal{F}$*, and a commuting unitary pair* $(W_1, W_2)$ *acting on* $\mathcal{H}_u$ *such that* $(V_1, V_2)$ *is unitarily equivalent to*

$$\left(\begin{bmatrix} M_{U^*P^\perp+zU^*P} & 0 \\ 0 & W_1 \end{bmatrix}, \begin{bmatrix} M_{PU+zP^\perp U} & 0 \\ 0 & W_2 \end{bmatrix}\right) \quad \text{on} \quad \begin{bmatrix} H^2(\mathcal{F}) \\ \mathcal{H}_u \end{bmatrix}. \tag{1.2}$$

The pair $(S_1, S_2) = (M_{U^*P^\perp+zU^*P}, M_{PU+zP^\perp U})$ appearing in the model above is called the *pure part*, and the commuting unitary part $(W_1, W_2)$ is called the *unitary part* for the commuting isometric pair $(V_1, V_2)$. Note that the pure part has the property that the product $S_1S_2 = S_2S_1$ is equal to the shift operator $M_z^{\mathcal{F}}$.

In stark contrast to the pair case, a projection/unitary operator pair $(P, U)$ on some coefficient Hilbert space $\mathcal{F}$ falls well short of describing up to unitary equivalence commuting isometric tuples of length $d \geq 3$. Indeed, one requires $d$ orthogonal projection/unitary operator pairs $(P_j, U_j)$ $(j = 1, \dots, d)$ acting on a Hilbert coefficient space $\mathcal{F}$ satisfying a stringent set of additional conditions to describe a $d$-tuple of commuting isometries $(V_1, \dots, V_d)$ as

$$\left(M_{U_1^*P_1^\perp+zU_1^*P_1}, M_{U_2^*P_2^\perp+zU_2^*P_2}, \dots, M_{U_d^*P_d^\perp+zU_d^*P_d}\right) \quad \text{on } H^2(\mathcal{F}) \tag{1.3}$$

or, equivalently, as

$$\left(M_{P_1U_1+zP_1^\perp U_1}, M_{P_2U_2+zP_2^\perp U_2}, \dots, M_{P_dU_d+zP_d^\perp U_d}\right) \quad \text{on } H^2(\mathcal{F}). \tag{1.4}$$

Here we have disregarded the commuting unitary tuple that will appear in the model unless the tuple is such that $(V_1V_2\cdots V_d)^{*n} \to 0$ in the strong operator topology. The stringent condition comes *only* with the requirement that we want the model (1.3) (or (1.4)) to commute; while for any projection/unitary pair $(P, U)$ the model pair (1.2) is commuting, the model tuples as in (1.3) or in (1.4) are not commuting without additional assumptions on the projections and unitary operators involved. This stark difference between the model theory for the pair case and the cases of tuples of higher length explains the mysterious failure of the rational dilation for tridisks onwards (see [74]). We further elaborate on this in Remark 5.7.3.

Chapter 3 is divided into three parts. In the recent paper [23], the authors give a thorough treatment of the Berger–Coburn–Lebow (BCL) model theory for commuting isometries. The first part of Chapter 3 recalls with proofs some of the results from [23] that are used here. In the second part, we delve into the model theory developed by Bercovici, Douglas, and Foias in the series of papers [28–30]. Many new results surface. A key component in the Bercovici–Douglas–Foias (BDF) model theory is an operator-valued contractive analytic function $\Theta\colon \mathbb{D} \to \mathcal{B}(\mathcal{F},\mathcal{F})$ for a Hilbert space $\mathcal{F}$ associated with a commuting isometric pair $(V_1, V_2)$, called the *characteristic function* $\Theta_{V_1,V_2}$ of $(V_1, V_2)$. A commuting isometric pair $(V_1, V_2)$ is called $j$-c.n.u., for $j = 1, 2$, if there is no joint $(V_1, V_2)$-reducing subspace on which $V_j$ is unitary (where c.n.u. means completely nonunitary). It is shown in [30] that a $j$-c.n.u. commuting isometric pair is uniquely determined by its characteristic function. For the case $j = 1$ and $\Theta = \Theta_{V_1,V_2}$, the commuting pair $(M_z^{\mathcal{F}}, M_\Theta)$ is not an isometric pair on $H^2(\mathcal{F})$ but it lifts in a canonical way to the commuting isometric pair

$$\left(\begin{bmatrix} M_z^{\mathcal{F}} & 0 \\ 0 & I_{H^2} \otimes M_\zeta \end{bmatrix}, \begin{bmatrix} M_\Theta & 0 \\ \Delta_\Theta|_{H^2(\mathcal{F})} & M_z^{\overline{\Delta_\Theta L^2(\mathcal{F})}} \end{bmatrix}\right) \quad \text{on} \quad \begin{bmatrix} H^2(\mathcal{F}) \\ H^2(\overline{\Delta_\Theta L^2(\mathcal{F})}) \end{bmatrix}, \tag{1.5}$$

which is 1-c.n.u., and conversely, given any 1-c.n.u. commuting isometric pair $(V_1, V_2)$, $(V_1, V_2)$ is unitarily equivalent to the 1-c.n.u. model pair (1.5) with $\Theta = \Theta_{V_1,V_2}$. A commuting isometric pair is 1&2-c.n.u. if it is both 1- and 2-c.n.u. We show in Theorem 3.2.12 that a commuting isometric pair is 1&2-c.n.u. if and only if the corresponding contractive analytic function $\Theta$ is c.n.u., that is, the contractive operator $\Theta(0)$ is a completely nonunitary contraction. This is obtained by first analyzing the jointly reducing subspaces $\mathcal{M}$ for a pair of the form $(M_z, M_\Theta)$ (here $\Theta$ is not necessarily inner but is contractive-valued) where the restriction $M_\Theta|_\mathcal{M}$ is unitary – see Theorem 3.2.9. This leads to the introduction of what we call a refined BDF-model for a commuting isometric pair – see Definition 3.2.14. The refined functional model (see (3.76)) of a commuting isometric pair consists of four components: 1&2-c.n.u, unitary/shift, shift/unitary, and unitary/unitary commuting isometric pairs – also see Corollary 3.2.6. All this amounts to a new proof of the Wold-type decomposition result of Gaşper and Gaşper [63] and [76] – see Remark 3.2.7. The third part of the chapter is devoted to the interplay between the two model theories. Of particular importance is the alternative approach to the well-known realization formula for contractive analytic functions discussed in Section 3.3.1.

**Chapter 4** As is clear from the exposition, the prime object of interest in this monograph is a commuting contractive operator pair $(T_1, T_2)$. The commutativity of the pair ensures the validity of the operator identity

$$I - T_2^*T_2 + T_2^*(I - T_1^*T_1)T_2 = I - T_1^*T_1 + T_1^*(I - T_2^*T_2)T_1.$$

This identity is used in almost every crucial step of the analysis done here. This operator identity was first observed by Ando and has played a key role in his proof. This warrants a discussion of the original proof of Ando. In Chapter 4 we discuss Ando's proof. Given a commuting contractive pair $(T_1, T_2)$ acting on $\mathcal{H}$, a commuting isometric pair $(V_1, V_2)$ acting on $\mathcal{K}$ is said to be an *Ando isometric lift* of $(T_1, T_2)$ via an isometric embedding $\Pi\colon \mathcal{H} \to \mathcal{K}$ if

$$\Pi T_j^* = V_j^* \Pi \quad \text{for } j = 1, 2.$$

The isometric lift is said to be *minimal* if, in addition, $\mathcal{K} = \bigvee_{n_1, n_2 \geq 0} V_1^{n_1} V_2^{n_2} \Pi \mathcal{H}$. Interestingly, while a minimal isomeric lift of a given commuting contractive pair is not unique (up to a natural notion of unitary equivalence of lifts), the unitary part in the BCL decomposition of every minimal isometric lift of a commuting contractive pair is unique up to unitary equivalence. This is the content of Theorem 4.3.1. In Theorem 4.5.1 we present a class of commuting contractive pairs for which any two minimal isometric lifts are unique. Chapter 4 also discusses the celebrated Sz.-Nagy–Foias commutant lifting theorem and adds additional insight into the equivalence of Ando lifting and Sz.-Nagy–Foias commutant lifting (as first observed by Parrott [74]).

**Chapter 5** The attentive reader may have noticed that the model isometric pair (1.2) has the property that the product of the entries is of the form

$$\begin{bmatrix} M_z^{\mathcal{F}} & 0 \\ 0 & W_1 W_2 \end{bmatrix} \quad \text{on} \quad \begin{bmatrix} H^2(\mathcal{F}) \\ \mathcal{H}_u \end{bmatrix}.$$

Since $W_1 W_2$ is a unitary, this is exactly in the Wold decomposition form of a single isometry (see Theorem 2.1.4). Indeed, the first step to obtaining the BCL-model for a commuting isometric tuple $(V_1, V_2, \dots, V_d)$ is to apply the Wold decomposition to the product isometry $V = V_1 V_2 \cdots V_d$. Our strategy, in a nutshell, is to follow the same route: Start with a commuting contractive pair $(T_1, T_2)$, apply the classical Sz.-Nagy–Foias dilation/model theory for the product contraction $T = T_1 T_2 = T_2 T_1$, analyze what the Sz.-Nagy–Foias model versions of $T_1, T_2$ must look like in order to commute and have a product equal to the Sz.-Nagy–Foias model operator on $T$ on $\mathcal{H}_{\Theta_T}$. As with the BCL-model theory, this comes with challenges, as we shall see during the course of the proofs.

Chapter 5 introduces what we refer to as an *Ando tuple*, which plays a key role in the development of the main results. An Ando tuple of a commuting contractive pair $(T_1, T_2)$ is a quadruple $(\mathcal{F}, \Lambda, P, U)$ consisting of a Hilbert (coefficient) space $\mathcal{F}$, an isometric embedding $\Lambda\colon \mathcal{D}_{T_1 T_2} \to \mathcal{F}$, a projection $P$, and a unitary $U$ in $\mathcal{B}(\mathcal{F})$. It is shown in this chapter and in Chapter 6 that not only do Ando tuples describe every minimal Ando isometric lift of a commuting contractive pair (see Theorem 5.2.1) but also they govern the uniqueness of minimal isometric lifts (see Theorem 5.7.1). For the development of this chapter, we need what we call Type I Ando tuples. These are Ando tuples satisfying a set of conditions – see Definition 5.1.5. One of the main results of this chapter is that if $(V_1, V_2)$ acting on $\mathcal{K}$ is a minimal isometric lift of $(T_1, T_2)$ acting on $\mathcal{H}$ via an isometric embedding $\Pi\colon \mathcal{H} \to \mathcal{K}$, then there is a Type I Ando tuple $(\mathcal{F}_*, \Lambda_*, P_*, U_*)$ of $(T_1^*, T_2^*)$ so that, after a unitary change of coordinates

$$\mathcal{K} = \begin{bmatrix} H^2(\mathcal{F}_*) \\ \mathcal{Q}_{T^*} \end{bmatrix}, \quad \Pi = \begin{bmatrix} \Lambda_* D_{T^*}(I - zT^*)^{-1} \\ \lim T^n T^{*n} \end{bmatrix} : \mathcal{H} \to \begin{bmatrix} H^2(\mathcal{F}_*) \\ \mathcal{Q}_{T^*} \end{bmatrix},$$

and $(V_1, V_2)$ is in the BCL-model form (1.2). Here, $T$ is the product contraction $T_1T_2$, the limit $\lim T^n T^{*n}$ is in the strong operator topology on $\mathcal{B}(\mathcal{H})$, and $(W_1, W_2)$ is a commuting unitary pair acting on $\mathcal{Q}_{T^*}$ canonically constructed out of $(T_1, T_2)$ in Section 4.4. In this way, we arrive at a second proof of the fact that the unitary part of every minimal isometric lift is unique, a result that is proved in Chapter 4 in a more elementary way. Further in-depth analysis reveals the existence of special Ando tuples (see Section 5.3) forming a subclass of Type I Ando tuples, which, in turn, leads to a new proof of Ando's dilation theorem. This new proof is discussed in complete detail in Section 5.4.

Let us note that $\Pi$ as above is an "inflation" of the classical isometric embedding $h \mapsto \begin{bmatrix} D_{T^*}(I-zT^*)^{-1}h \\ \lim T^n T^{*n} h \end{bmatrix}$ from $\mathcal{H}$ into $\begin{bmatrix} H^2(\mathcal{D}_{T^*}) \\ \mathcal{Q}_{T^*} \end{bmatrix}$ found by Douglas [54] that does the job in the lifting of the contraction $T$ to an isometry in the Wold decomposition form $\begin{bmatrix} M_z & 0 \\ 0 & W \end{bmatrix}$ acting on $\begin{bmatrix} H^2(\mathcal{D}_{T^*}) \\ \mathcal{Q}_{T^*} \end{bmatrix}$. It is interesting to note here that $W$ is a unitary operator on $\mathcal{Q}_{T^*}$ so that when $T = T_1T_2 = T_2T_1$, we have $W = W_1W_2 = W_2W_1$ where $(W_1, W_2)$ is the commuting unitary pair canonically constructed in Section 4.4.

**Chapter 6** This chapter can be considered to be parallel to Chapter 5, but with the objective being a joint Schäffer-type model for Ando isometric lifts of a commuting contractive pair. A point of distinction with the model in Chapter 5 is that the isometric embedding $\Pi\colon \mathcal{H} \to \mathcal{K}$ into the model dilation space $\mathcal{K}$ is just the inclusion map $h \mapsto \begin{bmatrix} h \\ 0 \end{bmatrix}$. Imposing a BCL-model for the commuting isometric pair $(V_1|_{\mathcal{K}\ominus\mathcal{H}}, V_2|_{\mathcal{K}\ominus\mathcal{H}})$ leads to the embedding of an Ando tuple for $(T_1, T_2)$ in the structure of the space $\mathcal{K}$ which ensures that $(V_1, V_2)$ enjoys lifting property with respect to $(T_1, T_2)$. However, what is more elusive is the commutativity of the lift $(V_1, V_2)$ and the isometric property of the individual operators $V_1$ and $V_2$, leading to auxiliary equations that must be satisfied by the components of the Ando tuple. This analysis then leads to the definition of a Type II Ando tuple and to a second proof of the Ando lifting theorem contingent on the existence of Type II Ando tuples. Again it turns out that special Ando tuples for $(T_1, T_2)$ are automatically also of Type II, so we arrive at a genuine second proof of the Ando lifting theorem.

We also establish an intriguing connection between the Schäffer model for a pair of coisometries $(V_1^*, V_2^*)$ and the BCL1-model for the isometric pair $(V_1, V_2)$. There is also a notion of a weak Type II Ando tuple where the product $V = V_1V_2$ does not fit well with the product $T = T_1T_2$; we show that in a variety of applications it is more convenient to work with the stronger notion of the Type II Ando tuple.

**Chapter 7** Note that if $(V_1, V_2)$ is an Ando lift of the commuting contractive pair $(T_1, T_2)$, then the product isometry $V = V_1V_2$ is a Sz.-Nagy–Foias lift of the product contraction $T = T_1T_2$. However, in general one cannot expect that $V$ is a minimal Sz.-Nagy–Foias lift of $T$, even if $(V_1, V_2)$ is a minimal Ando lift of $(T_1, T_2)$. This chapter characterizes the situation where this phenomenon does happen, in which case that Ando lift is said to be *strongly minimal*. Essentially this case corresponds to the situation where the isometric embedding operator $\Lambda\colon \mathcal{D}_{T^*} \to \mathcal{F}$ in the Ando tuple associated with the Ando lift is unitary. This leads to the definition of *strongly minimal Ando tuples*.

This chapter introduces what we refer to as the *fundamental operators* $F_1, F_2$ (also called the *fundamental-operator pair*) for a commuting contractive pair $(T_1, T_2)$. Three distinct proofs are given for the existence of the fundamental operators. It is shown that the fundamental operators $F_1, F_2$ are, in general, noncommuting but always contractive operators and that they are uniquely determined by the commuting contractive pair $(T_1, T_2)$. The fundamental operators play a significant role in the development of this and later chapters where we develop the model theory and construct the unitary invariants for commuting contractive pairs. It is shown in the proof of Theorem 7.2.3 that if $(\mathcal{F}, \Lambda, P, U)$ is an Ando tuple (Type I or Type II) for $(T_1, T_2)$, then the fundamental operators can be obtained via the formula

$$(F_1, F_2) = \Lambda^*(P^\perp U, U^*P)\Lambda.$$

This is striking in view of the fact that, while the fundamental operators $F_1, F_2$ are uniquely determined by $(T_1, T_2)$, the Ando tuples are not. A commuting contractive pair can have many distinct (i.e., distinct unitary-equivalence classes of) Ando tuples, a fact that is mentioned in various places throughout the book; see, for example, the discussion in Sections 5.3 and 6.4. A main result of this chapter (see Theorem 7.3.4) is that the existence of a strongly minimal Ando lift of a commuting contractive pair $(T_1, T_2)$, the existence of a strongly minimal Type I Ando tuple, and the existence of a strongly minimal Type II Ando tuple are all mutually equivalent, and furthermore each of these is also equivalent to the fundamental-operator pair $(F_1, F_2)$ of $(T_1, T_2)$ satisfying the system of equations

$$F_1F_2 = 0 = F_2F_1 \quad \text{and} \quad F_1^*F_1 + F_2F_2^* = I = F_1F_1^* + F_2^*F_2,$$

as well as equivalent to $(F_1, F_2)$ being of the form $(P^\perp U, U^*P)$ for some projection $P$ and a unitary $U$.

**Chapters 8 and 9** The goal of these chapters is to produce an exact analogue of the classical Sz.-Nagy–Foias model theory for contractive operators in the bivariate setting. Given a commuting contractive pair $(T_1, T_2)$ on $\mathcal{H}$ with Ando lift $(\Pi, V_1, V_2)$ on $\mathcal{K}$, one can always restrict to the subspace

$$\mathcal{K}_0 = \bigvee_{n_1, n_2 \geq 0} V_1^{n_1} V_2^{n_2} \operatorname{Ran} \Pi$$

to get a minimal Ando lift. The subspace

$$\mathcal{K}_{00} = \bigvee_{n \geq 0} V_1^n V_2^n \operatorname{Ran} \Pi$$

is strictly smaller than $\mathcal{K}_0$ unless the minimal lift $(\Pi, V_1|_{\mathcal{K}_0}, V_2|_{\mathcal{K}_0})$ is actually *strongly minimal* (as studied in Chapter 7). Note that $(\Pi, V = V_1V_2|_{\mathcal{K}_{00}})$ is a copy of the (unique up to lifting unitary equivalence) minimal Sz.-Nagy–Foias lift of the product contraction $T = T_1T_2$. Even though the compression $(\Pi, P_{\mathcal{K}_{00}}V_1|_{\mathcal{K}_{00}}, P_{\mathcal{K}_{00}}V_2|_{\mathcal{K}_{00}})$ is no longer an Ando lift of $(T_1, T_2)$, it is the case that $(\Pi, V_1V_2|_{\mathcal{K}_{00}})$ is a minimal isometric lift of $T = T_1T_2$ (and therefore unitarily equivalent to the Sz.-Nagy–Foias model lift of $T = T_1T_2$). Furthermore, there is an axiomatic algebraic characterization of lifts of the commuting pair $(T_1, T_2)$ arising in this way (called *pseudo-commutative contractive lifts* of $(T_1, T_2)$). Moreover, these compressions have explicit expressions in terms of the uniquely determined fundamental-operator pair $(G_1, G_2)$ for $(T_1^*, T_2^*)$, which was introduced in Chapter 7. In-depth analysis of pseudo-commuting contractive lifts leads to our Sz.-Nagy–Foias-like model for the commuting contractive pair $(T_1, T_2)$ factoring the operator $T$ ($T = T_1T_2$), extending the model of Das–Sarkar–Sarkar in [48] beyond the case where $T$ is such that $T^{*n} \to 0$ strongly. In Chapter 9 we furthermore extend all the features of the Sz.-Nagy–Foias model to this bivariate setting, namely: Given a commuting contractive pair $(T_1, T_2)$, which is *c.n.u.* (an extension of the notion of *completely nonunitary* to the setting of the commuting contractive pair), one can associate a *characteristic triple*

$$\Xi_{T_1,T_2} = (\Theta_T, (G_1, G_2), (W_1, W_2)),$$

where $\Theta_T$ is the Sz.-Nagy–Foias characteristic function of $T = T_1T_2$, $(G_1, G_2)$ is the aforementioned fundamental-operator pair for $(T_1^*, T_2^*)$, and $(W_1, W_2)$ is commuting unitary/unitary factorization of the operator $M_\zeta \otimes I_{\overline{\Delta_{\Theta_T} L^2(\mathcal{D}_T)}}$ acting on $\overline{\Delta_{\Theta_T} L^2(\mathcal{D}_T)}$ (where $\zeta$ is the coordinate variable on the unit circle). Based on this characteristic triple (which replaces the *characteristic function* $\Theta_T$ alone in the single-variable setting), one can build a functional-model space $\mathcal{K}_{\Xi_{T_1,T_2}}$ and functional-model commuting contraction operators $(T_{\Xi,1}, T_{\Xi,2})$ so that $(T_1, T_2) \underset{u}{\cong} (T_{\Xi,1}, T_{\Xi,2})$. Conversely, one can define the notion of an *admissible triple* $\Xi := (\Theta, (G_1, G_2), (W_1 W_2))$ together with a model space $\mathcal{K}_\Xi$ with an analogous model c.n.u. commuting contractive pair $(T_{\Xi,1}, T_{\Xi,2})$ so that the characteristic triple of $(T_{\Xi,1}, T_{\Xi,2})$ coincides (in the appropriate sense) with the originally given admissible triple $\Xi$. Working out more examples and applications of this bivariate model should be a stomping ground for further work.

**Chapter A** The concluding chapter is an Appendix that seeks to understand how the detailed results of the previous chapters fit in with the broader current mathematical landscape.

In Section A.1 we explain how the Ando lifting problem fits in as a particular instance of the Arveson $C^*$-algebra dilation framework, which can be formulated with an arbitrary compact subset $X$ of $\mathbb{C}^d$ playing the role of the bidisk $\overline{\mathbb{D}}^2$ in the Ando lifting problem. In addition to the bidisk setting, which is the main thrust of this monograph, here we mention additional specific examples: the *symmetrized bidisk* $\overline{\mathbb{D}}_s^2$, the *tetrablock* $\mathbb{E} \subset \mathbb{C}^3$, and the *polydisk* $\overline{\mathbb{D}}^d \subset \mathbb{C}^d$. We note that the original work on the symmetrized bidisk by Agler–Young [6–8], and on the tetrablock by Abouhajar–White–Young [1], was inspired by the connections with the $\mu$-synthesis problem in robust control theory (see [57, 58]). Within the Arveson framework, we identify an additional distinctive feature, namely the identification of a *companion contraction* as an additional piece of structure that fits with the symmetrized-bidisk, tetrablock, and polydisk settings, and serves to identify additional points of commonality for these three particular settings.

In Section A.2 we identify some remaining open problems concerning commuting contractive pairs that should be a fertile area for future research.

Let us mention that the recent monograph of Bhat–Bhattacharyya [32] is an excellent exposition of the dilation theory concerning four domains (the disk, the bidisk, the Euclidean multivariable unit ball, and the symmetrized bidisk), one of which (the bidisk) is the main focus of the present volume. Both books present an overview of the classical disk case, but with complementary emphases and choice of topics. Our main focus is the second domain (the bidisk) for which case we present new, more complete dilation results generalizing some of the results in [32]. In our speculative Section A.1 in the Appendix, we provide another point of view (i.e., the companion-contraction setting) into which a number of other settings fit, one of which is the symmetrized bidisk. The Euclidean unit ball setting, on the other hand, relies more on operator-algebra techniques that are not part of the toolkit of this book. In summary, one can argue that the book [32] is equally comprehensive and informative, but has a different slant and choice of emphases, making it largely complementary to the present volume but nevertheless an independently valuable contribution to the community.

# 2

# Models for Unitary Dilations and Isometric Lifts of a Contraction Operator

In this chapter, we review the classical dilation and model theory for single contraction operators developed by Schäffer, Douglas, Sz.-Nagy, and Foias. We first begin by reviewing the geometric structure of unitary dilation and isometric lift of a single contraction operator in the coordinate-free form. The term "coordinate-free" means that the setting is in the context of abstract Hilbert space, independent of any more concrete (e.g., functional) representation of the Hilbert space. The analysis enables us to find an *explicit* unitary identification between Douglas and Sz.-Nagy–Foias isometric lift spaces, which proves crucial in the later part of the book.

**Notation 2.0.1** In general we shall use the following notation.

1. Given two Hilbert spaces $\mathcal{E}, \mathcal{F}$ we let $\mathcal{B}(\mathcal{E}, \mathcal{F})$ denote the space of bounded linear operators from $\mathcal{E}$ to $\mathcal{F}$; when $\mathcal{E} = \mathcal{F}$ we write simply $\mathcal{B}(\mathcal{E})$ instead of $\mathcal{B}(\mathcal{E}, \mathcal{E})$.
2. For a (coefficient) Hilbert space $\mathcal{F}$, we write $H^2(\mathcal{F})$ for the *Hardy space* over the unit disk $\mathbb{D}$ centred at the origin in the complex plane $\mathbb{C}$:
$$H^2(\mathcal{F}) := \left\{ f\colon \mathbb{D} \to \mathcal{F} \colon f(z) = \sum_{n=0}^{\infty} z^n f_n \text{ and } \sum_{n=0}^{\infty} \|f_n\|^2_{\mathcal{F}} < \infty \right\}.$$
When $\mathcal{F} = \mathbb{C}$, we write simply $H^2$ rather than $H^2(\mathbb{C})$.
3. We shall use the notation $\mathbf{ev}_{0,\mathcal{F}}$ for the operator of *evaluation at* 0 acting from $H^2(\mathcal{F})$ to $\mathcal{F}$ given by
$$\mathbf{ev}_{0,\mathcal{F}}\colon f(z) \mapsto f(0),$$
with adjoint $\mathbf{ev}^*_{0,\mathcal{F}}\colon \mathcal{F} \to H^2(\mathcal{F})$ given by the identification of a vector $f \in \mathcal{F}$ with the constant function $f \in H^2(\mathcal{F})$ having constant value $f(z) = f$ for $z \in \mathbb{D}$:

$$\mathbf{ev}_{0,\mathcal{F}}^{*}\colon f \in \mathcal{F} \mapsto f(z) = f \in H^2(\mathcal{F}).$$

If the coefficient space $\mathcal{F}$ is clear from the context, we shall often drop the $\mathcal{F}$ from the notation.

4. For $\mathcal{F}$ any coefficient Hilbert space, we shall write $L^2(\mathcal{F})$ for the Hilbert-space tensor product $L^2(\mathcal{F}) = L^2 \otimes \mathcal{F}$, where $L^2$ is the space of scalar-valued measurable functions $f$ on the unit circle that are modulus-squared integrable with respect to normalized Lebesgue measure on the unit circle $\mathbb{T}$: $\int_{\mathbb{T}} |f(\zeta)|^2 \frac{|d\zeta|}{2\pi} < \infty$. One can also view this space as the space of strongly measurable $\mathcal{F}$-valued functions on the unit circle having Fourier expansion on the unit circle $f(\zeta) = \sum_{n=-\infty}^{\infty} f_n \zeta^n$ with $\sum_{n=-\infty}^{\infty} \|f_n\|_{\mathcal{F}}^2 < \infty$.
5. For $\mathcal{X}$ a closed subspace of an ambient Hilbert space $\mathcal{H}$, we use the notation $P_{\mathcal{X}}$ for the orthogonal projection operator from $\mathcal{H}$ to $\mathcal{X}$.
6. Following [92], given two coefficient Hilbert spaces $\mathcal{E}, \mathcal{E}_*$, we use the notation $(\mathcal{E}, \mathcal{E}_*, \Theta)$ for a *contractive analytic function*, that is, an analytic function $\Theta$ on the unit disk $\mathbb{D}$ having values equal to contractive operators $\Theta(z)$ from $\mathcal{E}$ to $\mathcal{E}_*$ for $z \in \mathbb{D}$. When $\mathcal{E} = \mathcal{E}_*$ we compress the notation to $(\mathcal{E}, \Theta)$.

## 2.1 The Coordinate-Free Version

Let $T$ be a contraction operator on a Hilbert space $\mathcal{H}$. We say that a pair $(\widetilde{\Pi}, \mathcal{U})$ is a *unitary dilation* of $T$ if (i) $\widetilde{\Pi}$ is an isometric embedding of $\mathcal{H}$ into a Hilbert space $\widetilde{\mathcal{K}}$, and (ii) $\mathcal{U}$ is a unitary operator on $\widetilde{\mathcal{K}}$ such that

$$\widetilde{\Pi}^* \mathcal{U}^n \widetilde{\Pi} = T^n \quad \text{for } n \geq 0,$$

or, equivalently in local form, for all $h, h' \in \mathcal{H}$ and integers $n, m$ we have

$$\left\langle \mathcal{U}^n \widetilde{\Pi} h, \mathcal{U}^m \widetilde{\Pi} h' \right\rangle = \begin{cases} \langle T^{n-m} h, h' \rangle & \text{if } n \geq m, \\ \langle h, T^{m-n} h' \rangle & \text{if } n < m. \end{cases}$$

We say that $(\widetilde{\Pi}, \mathcal{U})$ is a *minimal unitary dilation* of $T$ if, in addition, the smallest $\mathcal{U}$-reducing subspace of $\widetilde{\mathcal{K}}$ containing $\widetilde{\Pi}\mathcal{H}$ is all of $\widetilde{\mathcal{K}}$. We say that a pair $(\Pi, V)$ is an *isometric dilation* of the contraction operator $T$ if (i) $\Pi$ is an isometric embedding of $\mathcal{H}$ into the Hilbert space $\mathcal{K}$, and (ii) $V$ is an isometric operator on $\mathcal{K}$ such that the dilation property holds:

$$\Pi^* V^n \Pi = T^n \quad \text{for } n = 0, 1, 2, \ldots,$$

or, in local form, for all $h, h' \in \mathcal{H}$ and $n = 1, 2, 3, \ldots$, we have

$$\langle V^n \Pi h, \Pi h' \rangle = \langle T^n h, h' \rangle.$$

We say that $(\Pi, V)$ is a *minimal isometric dilation* if the smallest $V$-invariant subspace of $\mathcal{K}$ containing $\mathcal{H}$ is all of $\mathcal{K}$.

In the case when $(\Pi, V)$ is a minimal isometric dilation of $T$, then $(\Pi, V)$ is an *isometric lift* (after isometric embedding) of $T$, meaning that $V^*$ is a *coisometric extension* (after isometric embedding) of $T^*$, that is, $V^*\Pi = \Pi T^*$ (see [92, Theorem I.4.1]). In general, any isometric lift is also an isometric dilation, as can be seen from the following computation:

$$\begin{aligned} V^*\Pi = \Pi T^* &\Rightarrow V^{*n}\Pi = \Pi T^{*n} \Rightarrow \Pi^* V^n = T^n \Pi^* \quad \text{for } n \in \mathbb{Z}_+ \\ &\Rightarrow \Pi^* V^n \Pi = T^n \Pi^* \Pi = T^n \quad \text{for } n \in \mathbb{Z}_+. \end{aligned}$$

In this introductory section, we assume that isometric lifts $(\Pi, V)$ and unitary dilations $(\widetilde{\Pi}, \mathcal{U})$ for a contraction operator $T$ exist; we discuss various ways for actually constructing isometric lifts in succeeding sections. Without loss of generality, we may identify $\mathcal{H}$ with $\Pi\mathcal{H} \subset \mathcal{K}$ and $\widetilde{\mathcal{K}}$ with $\overline{\bigcup_{n=0}^{\infty} \mathcal{U}^{*n}\mathcal{K}}$ so that after these identifications $\mathcal{H} \subset \mathcal{K} \subset \widetilde{\mathcal{K}}$ and $\Pi : \mathcal{H} \to \mathcal{K}$ and $\widetilde{\Pi} = \iota_{\mathcal{K} \to \widetilde{\mathcal{K}}} \Pi$ are isometric inclusion maps. The book of Sz.-Nagy–Foias (with additional authors Bercovici and Kerchy for the updated version [92]) lays out the Hilbert space geometric structure of a minimal unitary dilation $\mathcal{U}$ on $\widetilde{\mathcal{K}}$ of a contraction operator $T$ on $\mathcal{H} \subset \widetilde{\mathcal{K}}$ with embedded minimal isometric dilation $V$ on $\mathcal{K}$ (so $\mathcal{H} \subset \mathcal{K} \subset \widetilde{\mathcal{K}}$). We next present a mild extension of this picture to the nonminimal case with the inclusion of some additional properties that will be useful for us in the sequel.

We first need some preliminaries concerning notation and terminology. For a contractive operator $T$, the *defect operator* $D_T$ and the *defect space* $\mathcal{D}_T$ are defined as follows:

$$D_T = (I - T^*T)^{\frac{1}{2}} \quad \text{and} \quad \mathcal{D}_T = \overline{\operatorname{Ran}} \, D_T.$$

If $V$ is an isometry on $\mathcal{K}$ and $\mathcal{L}$ is a subspace of $\mathcal{K}$, we say that $\mathcal{L}$ is *wandering* for $V$ if

$$V^n \mathcal{L} \perp \mathcal{L} \quad \text{for } n = 1, 2, \ldots. \tag{2.1}$$

In this case, we can form the (internal) orthogonal direct sum:

$$M_+(\mathcal{L}) := \bigoplus_{n=0}^{\infty} V^n \mathcal{L}$$

(where it is assumed that the operator $V$ is understood from the context). If $V = \mathcal{U}$ is unitary, then in fact the wandering-subspace property (2.1) continues to hold for $n = -1, -2, \ldots$ and we may form the (again internal) orthogonal direct-sum subspace:

$$M(\mathcal{L}) = \bigoplus_{n=-\infty}^{\infty} \mathcal{U}^n \mathcal{L} = \mathcal{M}_-(\mathcal{L}) \oplus \mathcal{M}_+(\mathcal{L}),$$

where we set $\mathcal{M}_-(\mathcal{L}) := \bigoplus_{n=-1}^{-\infty} \mathcal{U}^n \mathcal{L}$.

**Proposition 2.1.1** *Suppose that $T$ is a contraction operator on $\mathcal{H}$ with isometric lift $V$ on $\mathcal{K} \supset \mathcal{H}$ and unitary dilation $\mathcal{U}$ on $\widetilde{\mathcal{K}} \supset \mathcal{K} \supset \mathcal{H}$ (not necessarily minimal) such that also $\mathcal{U}|_{\mathcal{K}} = V$.*

(i) *Then the subspaces*

$$\mathcal{L}_* = \overline{\mathcal{U}(\mathcal{U}^* - T^*)\mathcal{H}} = \overline{(I - \mathcal{U}T^*)\mathcal{H}}, \quad \mathcal{L} = \overline{(\mathcal{U} - T)\mathcal{H}}$$

*are contained in $\mathcal{K}$ and wandering for $V$ (and hence also contained in $\widetilde{\mathcal{K}}$ and wandering for $\mathcal{U}$), respectively. Moreover, if we set*

$$\mathcal{K}_{\min} := \mathcal{H} \oplus M_+(\mathcal{L}) \subset \mathcal{K}, \quad \widetilde{\mathcal{K}}_{\min} := M_-(\mathcal{L}_*) \oplus \mathcal{H} \oplus M_+(\mathcal{L}) \subset \widetilde{\mathcal{K}}, \tag{2.2}$$

*then $\mathcal{K}_{\min}$ is reducing for $V$, $\widetilde{\mathcal{K}}_{\min}$ is reducing for $\mathcal{U}$, and $V|_{\mathcal{K}_{\min}}$ is the essentially unique* minimal isometric lift *for $T$ and $\mathcal{U}|_{\widetilde{\mathcal{K}}_{\min}}$ is the essentially unique* minimal unitary dilation *for $T$.*

(ii) *Define $\widetilde{\mathcal{R}} = \widetilde{\mathcal{K}} \ominus M(\mathcal{L}_*)$ so that we have an alternative orthogonal decomposition for $\widetilde{\mathcal{K}}$:*

$$\widetilde{\mathcal{K}} = M(\mathcal{L}_*) \oplus \widetilde{\mathcal{R}}. \tag{2.3}$$

*Similarly set $\mathcal{R} = \widetilde{\mathcal{K}}_{\min} \ominus M(\mathcal{L}_*)$ so that*

$$\widetilde{\mathcal{K}}_{\min} = M(\mathcal{L}_*) \oplus \mathcal{R}. \tag{2.4}$$

*Thus both $\mathcal{R}$ and $\widetilde{\mathcal{R}}$ are $\mathcal{U}$-reducing with $\mathcal{R} \subset \widetilde{\mathcal{R}}$, and there is a $\mathcal{U}$-invariant subspace $\mathcal{R}_1$ nested between $\mathcal{R}$ and $\widetilde{\mathcal{R}}$, that is,*

$$\mathcal{R} \subset \mathcal{R}_1 \subset \widetilde{\mathcal{R}}, \tag{2.5}$$

*so that*

$$\mathcal{K}_{\min} = M_+(\mathcal{L}_*) \oplus \mathcal{R}, \quad \mathcal{K} = M_+(\mathcal{L}_*) \oplus \mathcal{R}_1. \tag{2.6}$$

*Then also*

*(a) $\mathcal{U}$ is a minimal unitary dilation for $T$ if and only if $\mathcal{R} = \widetilde{\mathcal{R}}$,*
*(b) $V$ is a minimal isometric lift of $T$ if and only if $\mathcal{R} = \mathcal{R}_1$, and*
*(c) $\mathcal{U}$ being a minimal unitary dilation of $T$ implies that $V$ is a minimal isometric lift of $T$.*

(iii) *The maps* $\iota_*\colon \mathcal{L}_* \to \mathcal{D}_{T^*}$ *and* $\iota\colon \mathcal{L} \to \mathcal{D}_T$ *given densely by*

$$\iota_*\colon (I - VT^*)h \mapsto D_{T^*}h, \quad \iota\colon (V - T)h \mapsto D_T h$$

*for* $h \in \mathcal{H}$ *extend to define unitary identification maps from* $\mathcal{L}_*$ *onto* $\mathcal{D}_{T^*}$ *and from* $\mathcal{L}$ *onto* $\mathcal{D}_T$ *respectively.*

(iv) *The projection* $P_{\mathcal{L}_*}$ *onto the wandering subspace* $\mathcal{L}_*$ *restricted to* $\mathcal{H}$ *is given by*

$$P_{\mathcal{L}_*}|_{\mathcal{H}} = (I - VT^*)|_{\mathcal{H}}. \tag{2.7}$$

(v) *For* $h \in \mathcal{H}$*, the orthogonal projection* $P_{M_+(\mathcal{L}_*)}h$ *of* $h$ *onto* $M_+(\mathcal{L}_*)$ *is given by*

$$P_{M_+(\mathcal{L}_*)}h = \sum_{n=0}^{\infty} V^n(I - VT^*)T^{*n}h \tag{2.8}$$

*with*

$$\|P_{M_+(\mathcal{L}_*)}h\|^2 = \sum_{n=0}^{\infty} \|D_{T^*}T^{*n}h\|^2. \tag{2.9}$$

(vi) *The operator* $Q_{T^*}$ *defined as follows (where SOT means strong operator topology)*

$$Q_{T^*}^2 = \text{SOT-lim}_{n\to\infty} T^nT^{*n} \tag{2.10}$$

*exists, that is,* $Q_{T^*}^2 \succeq 0$*, and satisfies*

$$\lim_{n\to\infty} \|(T^nT^{*n} - Q_{T^*}^2)h\| = 0 \quad \text{for all } h \in \mathcal{H},$$

*and furthermore for all* $h \in \mathcal{H}$ *we have*

$$\|P_{\mathcal{R}}h\|^2 = \langle Q_{T^*}^2 h, h\rangle. \tag{2.11}$$

(vii) $P_{\mathcal{R}}\mathcal{H}$ *is invariant under* $V^*$ *and* $V^*|_{P_{\mathcal{R}}\mathcal{H}}$ *extends by continuity to an isometry on* $\overline{P_{\mathcal{R}}\mathcal{H}}$.

(viii) *(a) The linear manifold*

$$\mathcal{D}_{\mathcal{R}} := \bigcup_{n=0}^{\infty} \mathcal{U}^n P_{\mathcal{R}}\mathcal{H} \subset \mathcal{R} \tag{2.12}$$

*is dense in* $\mathcal{R}$*. Hence* $\mathcal{U}$ *on* $\widetilde{\mathcal{K}}$ *is a minimal unitary dilation for* $T$ *(i.e.,* $\mathcal{R} = \widetilde{\mathcal{R}}$*) if and only if* $\mathcal{D}_{\mathcal{R}}$ *is dense in* $\widetilde{\mathcal{R}}$.

*(b) Furthermore the condition*

$$M(\mathcal{L}_*) + M(\mathcal{L}) \text{ is dense in } \widetilde{\mathcal{K}} \tag{2.13}$$

*is sufficient for the minimality of* $\mathcal{U}$ *as a unitary dilation of* $T$.

(ix) *Suppose that $T$ is completely nonunitary, that is, there is no nonzero reducing subspace of $T$ where it is unitary. Then condition* (2.13) *is both necessary and sufficient for minimality of $\mathcal{U}$ as a unitary dilation of $T$. In general, we have the following density condition:*

$$\overline{(I - P_{M(\mathcal{L}_*)})M(\mathcal{L})} = \overline{P_{\mathcal{R}}M(\mathcal{L})} = \mathcal{R}. \tag{2.14}$$

*Proof* We first consider item (i). We can adapt Theorem II.1.1 in [92] to the situation where we do not assume that $\mathcal{U}$ on $\widetilde{\mathcal{K}}$ is a minimal unitary dilation for $T$ to see that $\mathcal{L}$ and $\mathcal{L}_*$ are wandering subspaces for $\mathcal{U}$. Another straightforward adaptation shows the equality of the two orthogonal decompositions

$$\mathcal{L}_* \oplus \mathcal{U}\mathcal{H} = \mathcal{H} \oplus \mathcal{L} \tag{2.15}$$

(see identity (1.7) on page 61 of [92] for details). From the definition $\mathcal{L} = \overline{(\mathcal{U} - T)\mathcal{H}} \subset \overline{(\mathcal{U}\mathcal{H} + \mathcal{H})}$ together with the fact that $V = \mathcal{U}|_{\mathcal{K}}$ and $\mathcal{H} \subset \mathcal{K}$, it is clear that $\mathcal{L} \subset \mathcal{K}$, and hence $\mathcal{L}$ can also be viewed as wandering for $V$. From the identity (2.15), it is also clear that $\mathcal{L}_* \subset \mathcal{K}$; as $\mathcal{L}_*$ is wandering for $\mathcal{U}$, we see that we can also view $\mathcal{L}_*$ as wandering for $V$. Also shown in part (ii) of [92, Theorem II.1.1] are the orthogonality relations

$$M_-(\mathcal{L}_*) \perp \mathcal{H}, \quad M_-(\mathcal{L}_*) \perp M_+(\mathcal{L}), \quad M_+(\mathcal{L}) \perp \mathcal{H}$$

and hence we have the orthogonal decomposition

$$\widetilde{\mathcal{K}}_{\min} := M_-(\mathcal{L}_*) \oplus \mathcal{H} \oplus M_+(\mathcal{L}) \subset \widetilde{\mathcal{K}}.$$

Furthermore, owing to the equality of spaces (2.15), we see that $\widetilde{\mathcal{K}}_{\min}$ is reducing for $\mathcal{U}$. From the definitions, we also see that any reducing subspace for $\mathcal{U}$ containing $\mathcal{H}$ must contain $\widetilde{\mathcal{K}}_{\min}$, so $\mathcal{U}|_{\widetilde{\mathcal{K}}_{\min}}$ is the essentially unique *minimal unitary dilation* for $T$. Similarly we see that $\mathcal{K}_{\min} := \mathcal{H} \oplus M_+(\mathcal{L}) \subset \mathcal{K}$ is invariant for $V$, and from the definitions we see that any $V$-invariant subspace of $\mathcal{K}$ containing $\mathcal{H}$ must contain $\mathcal{K}_{\min}$. It then follows that $V|_{\mathcal{K}_{\min}}$ is a copy of the minimal isometric lift (or equivalently, minimal isometric dilation) of $T$, and all parts of item (i) now follow.

As for item (ii), by assumption $V = \mathcal{U}|_{\mathcal{K}}$ is an isometric lift of $T$, that is, $\mathcal{H}$ is invariant for $V^*$ and $V^*|_{\mathcal{H}} = T^*$. We are also assuming that $\mathcal{U}$ is an extension of $V$, hence for $n \in \mathbb{Z}_+$ we have

$$\mathcal{U}^n|_{\mathcal{K}} = V^n \Rightarrow V^{*n} = P_{\mathcal{K}}\mathcal{U}^{*n}|_{\mathcal{K}}.$$

In particular

$$V^{*n}|_{\mathcal{H}} = P_{\mathcal{K}}\mathcal{U}^{*n}|_{\mathcal{H}}.$$

But from the definitions, we see that

$$\bigvee_{n=0}^{\infty} \mathcal{U}^{*n}\mathcal{H} = M_{-}(\mathcal{L}_*) \oplus \mathcal{H}.$$

Hence we have

$$\mathcal{H} = \bigvee_{n=0}^{\infty} V^{*n}\mathcal{H} = \bigvee_{n=0}^{\infty} P_{\mathcal{K}}\mathcal{U}^{*n}\mathcal{H} = P_{\mathcal{K}}(M_{-}(\mathcal{L}_*) \oplus \mathcal{H}),$$

forcing us to conclude that

$$\mathcal{K} \subset (M_{-}(\mathcal{L}_*))^{\perp} = M_{+}(\mathcal{L}_*) \oplus \widetilde{\mathcal{R}},$$

where we set

$$\widetilde{\mathcal{R}} := \widetilde{\mathcal{K}} \ominus M(\mathcal{L}_*). \tag{2.16}$$

But we know that $M_{+}(\mathcal{L}_*) \subset \mathcal{K}$, so we can now write

$$\mathcal{K} = M_{+}(\mathcal{L}_*) \oplus \mathcal{R}_1, \tag{2.17}$$

where

$$\mathcal{R}_1 \subset \widetilde{\mathcal{R}}. \tag{2.18}$$

As we have already seen that $\mathcal{K} \subset M_{-}(\mathcal{L}_*)^{\perp}$, we also see from (2.17) that

$$\mathcal{R}_1 \subset M(\mathcal{L}_*)^{\perp}. \tag{2.19}$$

As $\mathcal{L}_*$ is wandering for $\mathcal{U}$, we know that $M(\mathcal{L}_*)$ is reducing for $\mathcal{U}$, so also in particular is invariant for $\mathcal{U}^*$. Thus, for $k_+ \in M_{+}(\mathcal{L}_*)$ we have

$$\begin{aligned} V^* k_+ &= P_{\mathcal{K}}\mathcal{U}^* k_+ = P_{M_+(\mathcal{L}_*)\oplus\mathcal{R}_1}\mathcal{U}^* k_+ \quad \text{(by (2.17))} \\ &= P_{M_+(\mathcal{L}_*)}\mathcal{U}^* k_+ \quad \text{(by (2.19))}, \end{aligned}$$

that is, $V^*|_{M_+(\mathcal{L}_*)}$ is the backward unilateral shift with wandering subspace $\mathcal{L}_*$. Recall that $\mathcal{L}_* \subset \mathcal{K}$ as a consequence of (2.15). Furthermore, as $V = \mathcal{U}|_{\mathcal{K}}$ it follows that $M_+(\mathcal{L}_*) = \bigvee_{N=0}^{\infty} \mathcal{U}^n \mathcal{L}_* \subset \mathcal{K}$. Thus $M_+(\mathcal{L}_*)$ is in the domain of $V$ and $V|_{M_+(\mathcal{L}_*)} = \mathcal{U}|_{M_+(\mathcal{L}_*)}$ is the forward unilateral shift with wandering subspace $\mathcal{L}_*$, that is, $M_+(\mathcal{L}_*)$ is reducing for $V$ and $V|_{M_+(\mathcal{L}_*)}$ is the forward unilateral shift with wandering subspace $\mathcal{L}_*$. Then $\mathcal{R}_1 = \mathcal{K} \ominus M_+(\mathcal{L}_*)$ is also reducing for $V$. As $V = \mathcal{U}|_{\mathcal{K}}$, it now follows that $\mathcal{R}_1$ is invariant (but not necessarily reducing) for $\mathcal{U}$ as well.

From the definition of $\mathcal{L}_*$ and the fact that

$$\mathcal{K}_{\min} = \bigvee_{n=0}^{\infty} V^n\mathcal{H} = \bigvee_{n=0}^{\infty} \mathcal{U}^n\mathcal{H},$$

we see that $M_+(\mathcal{L}_*) \subset \mathcal{K}_{\min}$. Let us define the subspace $\mathcal{R}$ by

$$\mathcal{R} := \mathcal{K}_{\min} \ominus M_+(\mathcal{L}_*)$$

so that

$$\mathcal{K}_{\min} = M_+(\mathcal{L}_*) \oplus \mathcal{R}.$$

As it must also be the case that $V$ extends the minimal contractive lift $V|_{\mathcal{K}_{\min}}$ where $\mathcal{K}_{\min} = M_+(\mathcal{L}_*) \oplus \mathcal{R}$, we must also have

$$\mathcal{K} = M_+(\mathcal{L}_*) \oplus \mathcal{R}_1 \supset \mathcal{K}_{\min} := M_+(\mathcal{L}_*) \oplus \mathcal{R}.$$

This then gives us the additional containment $\mathcal{R} \subset \mathcal{R}_1$. Combining this with (2.18) gives us the nesting property (2.5).

Note next that minimality of $\mathcal{U}$ as a unitary dilation of $T$ amounts to the equality $\widetilde{\mathcal{K}}_{\min} = \widetilde{\mathcal{K}}$, which in turn from the formulas (2.3) and (2.6) amounts to $\mathcal{R} = \widetilde{\mathcal{R}}$, while minimality of $V$ as an isometric lift of $T$ by (2.4) and (2.6) amounts to the equality $\mathcal{R} = \mathcal{R}_1$, and items (i) and (ii) follow. Finally, item (iii) then follows from the nesting property (2.5).

Item (iii) also follows from the elementary computations as in the proof of part (i) of Theorem II.1.1 in [92].

To handle item (iv), note that $V|_{\mathcal{K}_{\min}}$ is an isometry where, by (2.2) and (2.6), the space $\mathcal{K}_{\min}$ is given by either one of the two orthogonal decompositions in

$$\mathcal{K}_{\min} = \mathcal{H} \oplus M_+(\mathcal{L}) = M_+(\mathcal{L}_*) \oplus \mathcal{R}. \tag{2.20}$$

The first decomposition exhibits $\mathcal{H}$ as a subspace of $\mathcal{K}_{\min}$ while the second decomposition exhibits the *Wold decomposition* for the isometry $V|_{\mathcal{K}_{\min}}$ with $V|_{\mathcal{K}_{\min}}$ having wandering subspace $\mathcal{L}_*$ (see Theorem 2.1.4 for a discussion of Wold decompositions in general). From the general properties of such isometries, it follows that

$$P_{\mathcal{L}_*}|_{\mathcal{K}_{\min}} = (I - VV^*)|_{\mathcal{K}_{\min}}.$$

When we apply this formula to an element $h \in \mathcal{H} \subset \mathcal{K}_{\min}$ and use that $\mathcal{H}$ is invariant for $V^*$ with $V^*|_{\mathcal{H}} = T^*$, we get that

$$P_{\mathcal{L}_*} h = (I - VV^*)h = (I - VT^*)h$$

as claimed.

We next verify item (v). By definition $M_+(\mathcal{L}_*) = \bigoplus_{N=0}^{\infty} V^n \mathcal{L}_*$, and by elementary properties of isometries, since $\mathcal{L}_*$ is wandering for $V$ we have

$$P_{V^n \mathcal{L}_*}|_{\mathcal{K}_{\min}} = V^n P_{\mathcal{L}_*} V^{*n}|_{\mathcal{K}_{\min}} = V^n (I - VV^*) V^{*n}|_{\mathcal{K}_{\min}}$$

and hence

$$P_{M_+(\mathcal{L}_*)}|_{\mathcal{K}_{\min}} = \sum_{n=0}^{\infty} V^n(I - VV^*)V^{*n}|_{\mathcal{K}_{\min}},$$

where the sum converges strongly. Applying this identity to $h \in \mathcal{H}$ and using that $V^*|_{\mathcal{H}} = T^*$ then gives us the formula (2.8). The formula (2.9) then follows from (2.8) combined with the isometric identification $\iota_*$ in item (iii).

To handle item (vi), note that SOT-$\lim_{n\to\infty} T^n T^{*n}$ exists since $T^n T^{*n}$ is a monotonically decreasing sequence of positive-semidefinite operators as $T^*$ is a contraction, and hence we may define a positive-semidefinite operator $Q_{T^*}$ on $\mathcal{H}$ by

$$Q_{T^*}^2 = \text{SOT-}\lim_{n\to\infty} T^n T^{*n}. \tag{2.21}$$

By combining the first parts of (2.2) and (2.6) we see that

$$\mathcal{K}_{\min} = \mathcal{H} \oplus M_+(\mathcal{L}) = M_+(\mathcal{L}_*) \oplus \mathcal{R}, \tag{2.22}$$

from which we see that

$$\mathcal{H} \subset M_+(\mathcal{L}_*) \oplus \mathcal{R}$$

and hence for $h \in \mathcal{H}$ we have

$$\|P_{\mathcal{R}} h\|^2 = \|h\|^2 - \|P_{M_+(\mathcal{L}_*)} h\|^2.$$

Thus

$$\begin{aligned}
\|P_{\mathcal{R}} h\|^2 &= \|h\|^2 - \|P_{M_+(\mathcal{L}_*)} h\|^2 = \|h\|^2 - \sum_{n=0}^{\infty} \|D_{T^*} T^{*n} h\|^2 \quad \text{(by (2.9))} \\
&= \|h\|^2 - \lim_{N\to\infty} \sum_{n=0}^{N} \langle T^n (I - TT^*) T^{*n} h, h \rangle \\
&= \lim_{N\to\infty} \left\langle T^{N+1} T^{*(N+1)} h, h \right\rangle = \left\langle Q_{T^*}^2 h, h \right\rangle,
\end{aligned}$$

and item (vi) follows.

As for (vii), note that by (2.6) we see that

$$\mathcal{R} \subset M_+(\mathcal{L}_*) \oplus \mathcal{R} = \mathcal{K}_{\min} \subset \mathcal{K}$$

and hence $\mathcal{R}$ is indeed a subspace of the space $\mathcal{K}$, which is the domain of $V$ (and of $V^*$). Making use of the additional observations that $\mathcal{U}|_{\mathcal{K}_{\min}} = V|_{\mathcal{K}_{\min}}$, that $\mathcal{R}$ is reducing for $\mathcal{U}$, and that $\mathcal{R} \subset \mathcal{K}_{\min}$ and hence $P_{\mathcal{R}} = P_{\mathcal{K}_{\min}} P_{\mathcal{R}} = P_{\mathcal{R}} P_{\mathcal{K}_{\min}}$, we compute that, for $h \in \mathcal{H}$,

$$\begin{aligned}
V^* P_{\mathcal{R}} h &= P_{\mathcal{K}_{\min}} \mathcal{U}^* P_{\mathcal{R}} h = P_{\mathcal{K}_{\min}} P_{\mathcal{R}} \mathcal{U}^* h = P_{\mathcal{R}} P_{\mathcal{K}_{\min}} \mathcal{U}^* h = P_{\mathcal{R}} P_{\mathcal{K}_{\min}} V^* h \\
&= P_{\mathcal{R}} T^* h \in P_{\mathcal{R}} \mathcal{H}
\end{aligned}$$

and we conclude that $\overline{P_{\mathcal{R}}\mathcal{H}}$ is indeed invariant for $V^*$. Furthermore, since $\mathcal{R}$ is reducing for $\mathcal{U}$ and again $\mathcal{R} \subset \mathcal{K}_{\min}$, we have, for $P_{\mathcal{R}}h \in P_{\mathcal{R}}\mathcal{H}$,

$$\|V^* P_{\mathcal{R}}h\| = \|P_{\mathcal{K}_{\min}}\mathcal{U}^* P_{\mathcal{R}}h\| = \|\mathcal{U}^* P_{\mathcal{R}}h\| = \|P_{\mathcal{R}}h\|,$$

and we see that $V^*|_{P_{\mathcal{R}}\mathcal{H}}$ extends to an isometry on $\overline{P_{\mathcal{R}}\mathcal{H}}$.

We next discuss item (viii)(a). Let the linear manifold $\mathcal{D}_{\mathcal{R}}$ be given by (2.12). As $\mathcal{R}$ is reducing (in particular, invariant) for $\mathcal{U}$, it is clear that $\mathcal{D}_{\mathcal{R}} \subset \mathcal{R}$. Suppose now that $k \in \mathcal{R}$ is orthogonal to $\mathcal{D}_{\mathcal{R}}$. Thus for $h \in \mathcal{H}$ and $n = 0, 1, 2, \ldots$, we have

$$0 = \langle k, \mathcal{U}^n P_{\mathcal{R}}h\rangle = \langle k, P_{\mathcal{R}}\mathcal{U}^n h\rangle = \langle k, V^n h\rangle,$$

that is, $k \in \mathcal{R}$ is such that $k \perp \bigcup_{n=0}^{\infty} V^n\mathcal{H}$. But

$$\bigvee_{n=0}^{\infty} V^n\mathcal{H} = \mathcal{K}_{\min} = M_+(\mathcal{L}_*) \oplus \mathcal{R}$$

by the first part of (2.6). Thus $k \in \mathcal{R}$ is orthogonal to $M_+(\mathcal{L}_*) \oplus \mathcal{R}$. In particular, $k$ is orthogonal to itself, forcing $k = 0$. Thus the orthogonal complement of $\mathcal{D}_{\mathcal{R}}$ inside $\mathcal{R}$ is $\{0\}$, that is, $\mathcal{D}_{\mathcal{R}}$ is dense in $\mathcal{R}$. As already observed, the minimality of $\mathcal{U}$ as a unitary dilation of $T$ is equivalent to $\mathcal{R}$ being equal to all of $\widetilde{\mathcal{R}}$. Thus we get the density of $\mathcal{D}_{\mathcal{R}}$ inside the larger space $\widetilde{\mathcal{R}}$ as a characterization of the minimality of $\mathcal{U}$.

As for item (viii)(b), we need to show that $M(\mathcal{L}) \bigvee M(\mathcal{L}_*)$ being dense in $\widetilde{\mathcal{K}}$ implies that $\mathcal{U}$ is minimal as a unitary dilation of $T$. The minimality of $\mathcal{U}$ in this sense for our setting here just means that $\widetilde{\mathcal{K}}_{\min} = \widetilde{\mathcal{K}}$, that is,

$$\widetilde{\mathcal{R}}_0 := \widetilde{\mathcal{K}} \ominus \widetilde{\mathcal{K}}_{\min} = \{0\}.$$

From (2.3) and (2.4) we see that

$$\widetilde{\mathcal{R}}_0 := \widetilde{\mathcal{K}} \ominus \widetilde{\mathcal{K}}_{\min} = \widetilde{\mathcal{R}} \ominus \mathcal{R}, \tag{2.23}$$

where $\mathcal{R}$ and $\widetilde{\mathcal{R}}$ are as in item (ii). From (2.3) we see that $\widetilde{\mathcal{R}}$, and hence also $\widetilde{\mathcal{R}}_0$, are orthogonal to $M(\mathcal{L}_*)$.

We claim that $\widetilde{\mathcal{R}}_0$ is also orthogonal to $M(\mathcal{L})$. Toward this end, let us use the two decompositions (2.20) to massage the orthogonal decomposition (2.3) as follows:

$$\begin{aligned}\widetilde{\mathcal{K}} &= M(\mathcal{L}_*) \oplus \widetilde{\mathcal{R}} = M_-(\mathcal{L}_*) \oplus (M_+(\mathcal{L}_*) \oplus \mathcal{R}) \oplus \widetilde{\mathcal{R}}_0\\ &= M_-(\mathcal{L}_*) \oplus \mathcal{K}_{\min} \oplus \widetilde{\mathcal{R}}_0 = M_-(\mathcal{L}_*) \oplus \big(\mathcal{H} \oplus M_+(\mathcal{L})\big) \oplus \widetilde{\mathcal{R}}_0,\end{aligned}$$

from which we see that $\widetilde{\mathcal{R}}_0$ is orthogonal to $M_+(\mathcal{L})$ as well. However, $\widetilde{\mathcal{R}}_0$ is $\mathcal{U}$ reducing and the smallest $\mathcal{U}$-reducing subspace containing $M_+(\mathcal{L})$ is $M(\mathcal{L})$.

Thus it follows that $\mathcal{R}_0$ is orthogonal to $M(\mathcal{L})$ as well. Invoking now the assumption that $M(\mathcal{L}_*) \bigvee M(\mathcal{L})$ is dense in $\widetilde{\mathcal{K}}$ leads us to the conclusion that $\widetilde{\mathcal{R}}_0 = \{0\}$, that is, that (2.23) holds as desired.

To handle item (ix), suppose now that $T$ is a completely nonunitary contraction operator and that minimality holds, so $\widetilde{\mathcal{K}} = \widetilde{\mathcal{K}}_{\text{min}}$. Suppose that $k \in \widetilde{\mathcal{K}}_{\text{min}}$ is orthogonal to $M(\mathcal{L}_*) \bigvee M(\mathcal{L})$. From the decomposition for $\widetilde{\mathcal{K}}_{\text{min}}$ in (2.2), we then see that we must have $k = h$ with $h \in \mathcal{H}$. As $h$ is also orthogonal to $M_-(\mathcal{L})$, we also have, for all $h' \in \mathcal{H}$ and $n = 1, 2, 3, \ldots,$

$$\begin{aligned} 0 = \langle h, \mathcal{U}^{*n}(\mathcal{U} - T)h' \rangle &= \left\langle \mathcal{U}^{n-1}h, h' \right\rangle - \langle \mathcal{U}^n h, Th' \rangle \\ &= \left\langle T^{n-1}h, h' \right\rangle - \langle T^n h, Th' \rangle = \left\langle (I - T^*T)T^{n-1}h, h' \right\rangle, \end{aligned}$$

from which we conclude that $(I - T^*T)T^{n-1}h = 0$ for $n = 1, 2, 3, \ldots$. This in turn gives

$$\|h\|^2 = \|Th\|^2 = \|T^2 h\|^2 = \cdots = \|T^n h\|^2 = \cdots,$$

or $h$ is in the isometric subspace for $T$ (the largest invariant subspace $\mathcal{H}_{\text{i}}$ for $T$ such that $T|_{\mathcal{H}_{\text{i}}}$ is isometric). Similarly, $k \in \mathcal{H}$ and $k \perp M_+(\mathcal{L}_*)$ implies that $k \in \mathcal{H}_{\text{ci}}$, where $\mathcal{H}_{\text{ci}}$ is the largest $T^*$-invariant subspace such that $T^*|_{\mathcal{H}_{\text{ci}}}$ is isometric (where the subscript “ci” is used to denote coisometry). Putting all this together means that $h \in \mathcal{H}_{\text{i}} \cap \mathcal{H}_{\text{ci}} =: \mathcal{H}_{\text{u}}$, where $\mathcal{H}_{\text{u}}$ is the largest $T$-reducing subspace such that $T|_{\mathcal{H}_{\text{u}}}$ is unitary. The hypothesis that $T$ is completely nonunitary amounts to saying that $\mathcal{H}_{\text{u}} = \{0\}$. Thus finally $k = 0$ and we conclude $M(\mathcal{L}) + M(\mathcal{L}_*)$ is dense in $\widetilde{\mathcal{K}}$. It follows that criterion (2.13) is also necessary for minimality in case $T$ is completely nonunitary.

Finally the validity of (2.14) for the case where $M(\mathcal{L}_*) + M(\mathcal{L})$ is dense in $\widetilde{\mathcal{K}}$ is part of Theorem II.2.1 in [92]. This concludes the proof of the last remaining item of Proposition 2.1.1. □

Let us now note the following more concise version of articles (viii)(b) and (ix) in Proposition 2.1.1.

**Corollary 2.1.2** *Suppose that the contraction operator $T$ on $\mathcal{H}$ with isometric lift $V$ on $\mathcal{K} \supset \mathcal{H}$ and unitary dilation $\mathcal{U}$ on $\widetilde{\mathcal{K}} \supset \mathcal{K}$ is as in Proposition 2.1.1. Then the orthogonal complement of $M(\mathcal{L}_*) \bigvee M(\mathcal{L})$ is given by the formula*

$$\widetilde{\mathcal{K}} \ominus \left( M(\mathcal{L}_*) \bigvee M(\mathcal{L}) \right) = \mathcal{H}_u \oplus \widetilde{\mathcal{R}}_0, \tag{2.24}$$

*where*

(i) *$\mathcal{H}_u \subset \mathcal{H}$ is the largest reducing subspace for $T$ on which $T$ is unitary, and*
(ii) *$\widetilde{\mathcal{R}}_0$ is as in* (2.23)*: $\widetilde{\mathcal{R}}_0 := \widetilde{\mathcal{K}} \ominus \widetilde{\mathcal{K}}_{\text{min}} = \widetilde{\mathcal{R}} \ominus \mathcal{R}$.*

*Proof* A computation in the proof of item (viii)(b) in Proposition 2.1.1 shows that it is always the case that $\widetilde{\mathcal{R}}_0 \subset \widetilde{\mathcal{K}} \ominus \left(M(\mathcal{L}_*) \bigvee M(\mathcal{L})\right)$. As

$$M(\mathcal{L}_*) \bigvee M(\mathcal{L}) \subset \widetilde{\mathcal{K}}_{\min} = \widetilde{\mathcal{K}} \ominus \widetilde{\mathcal{R}}_0,$$

it suffices next to characterize the subspace

$$\widetilde{\mathcal{K}}_{\min} \ominus \left(M(\mathcal{L}_*) \bigvee M(\mathcal{L})\right).$$

Hence we are in the minimal case handled in item (ix). There it is shown that if $k \in \mathcal{K}_{\min}$ is orthogonal to $M(\mathcal{L}_*) \bigvee M(\mathcal{L})$, then $k \in \mathcal{H}_u$:

$$\widetilde{\mathcal{K}}_{\min} \ominus \left(M(\mathcal{L}_*) \bigvee M(\mathcal{L})\right) \subset \mathcal{H}_u.$$

But the reverse containment

$$\mathcal{H}_u \subset \widetilde{\mathcal{K}}_{\min} \ominus \left(M(\mathcal{L}_*) \bigvee M(\mathcal{L})\right)$$

follows by simply reversing the steps of the argument for the original containment, and (2.24) follows. □

We note below a useful uniqueness result for minimal isometric lifts/unitary dilations of a contraction operator $T$.

**Theorem 2.1.3** *(1) Suppose that $T$ is a contraction operator on a Hilbert space $\mathcal{H}$ and that $(\mathcal{K}, \Pi, V)$ and $(\mathcal{K}', \Pi', V')$ are two minimal isometric lifts of $T$. Then $(\mathcal{K}, \Pi, V)$ and $(\mathcal{K}', \Pi', V')$ are unitarily equivalent as isometric lifts of $T$, that is, there is a unitary operator $U \colon \mathcal{K} \to \mathcal{K}'$ such that*

$$U\Pi = \Pi', \quad UV = V'U, \tag{2.25}$$

*and furthermore $U$ so specified is unique.*

*(2) A similar statement holds for unitary dilations: If $T$ is a contraction operator on a Hilbert space $\mathcal{H}$ and $\left(\widetilde{\mathcal{K}}, \widetilde{\Pi}, \mathcal{U}\right)$ and $\left(\widetilde{\mathcal{K}}', \widetilde{\Pi}', \mathcal{U}'\right)$ are two minimal unitary dilations for $T$, then $(\mathcal{K}', \Pi', \mathcal{U})$ and $(\mathcal{K}', \Pi', \mathcal{U}')$ are unitarily equivalent as unitary dilations of $T$, that is, there is a unitary operator $\mathcal{U} \colon \mathcal{K} \to \mathcal{K}'$ such that*

$$U\widetilde{\Pi} = \widetilde{\Pi}', \quad U\mathcal{U} = \mathcal{U}'U,$$

*and furthermore $U$ so specified is unique.*

*Proof* As the proof of (2) is similar, we discuss here only (1) in detail. This is an adaptation of Theorem I.4.1 in [92] where we do not assume that the isometric embedding operators $\Pi \colon \mathcal{H} \to \mathcal{K}$ and $\Pi' \colon \mathcal{H} \to \mathcal{K}'$ are inclusion maps. The minimality of $(\mathcal{K}, \Pi, V)$ and $(\mathcal{K}', \Pi', V')$ as isometric lifts of $T$ forces the spaces $\mathcal{K}$ and $\mathcal{K}'$ to be given by

$$\mathcal{K} = \bigvee\{V^n \Pi \mathcal{H} : n = 0, 1, 2, \dots\}, \quad \mathcal{K}' = \bigvee\{V'^n \Pi' \mathcal{H} : n = 0, 1, 2, \dots\}.$$

If $(\mathcal{K}, \Pi, V)$ is any minimal isometric lift, one can check that

$$\left\langle V^n \Pi h, V^m \Pi \widetilde{h} \right\rangle_{\mathcal{K}} = \begin{cases} \left\langle V^{n-m} \Pi h, \Pi \widetilde{h} \right\rangle_{\mathcal{K}} = \left\langle T^{n-m} h, \widetilde{h} \right\rangle_{\mathcal{H}} & \text{if } n \geq m \geq 0, \\ \left\langle h, V^{m-n} \widetilde{h} \right\rangle_{\mathcal{K}} = \left\langle h, T^{m-n} \widetilde{h} \right\rangle_{\mathcal{H}} & \text{if } m \geq n \geq 0, \end{cases}$$

and thus $\left\langle V^n h, V^m \widetilde{h} \right\rangle_{\mathcal{K}}$ does not depend on the choice of minimal isometric lift. Thus, if $(\mathcal{K}, \Pi, V)$ and $(\mathcal{K}', \Pi', V')$ are two minimal isometric lifts, the map defined densely by

$$U : \sum_{n=0}^{N} V^n \Pi h \mapsto \sum_{n=0}^{N} V'^n \Pi' h$$

extends by continuity to a unitary map from $\mathcal{K}$ to $\mathcal{K}'$ implementing a unitary equivalence between the minimal isometric lifts $(\mathcal{K}, \Pi, V)$ and $(\mathcal{K}', \Pi', V')$. Furthermore, from the defining intertwining conditions for unitary equivalence, we see that any such unitary equivalence must be of this form. □

Isometric operators form the building blocks of Hilbert space operator theory. The Wold decomposition is a fundamental structural decomposition for a general isometry and also turns out to be a fundamental stepping stone to the model theory for abstract Hilbert space operators. Let us now recall this decomposition. A *shift operator* is a special type of isometric operator $V$ such that $\bigcap_{n \geq 0} \operatorname{Ran} V^n = \{0\}$. The term *shift* will be clear from the decomposition theorem presented below. A shift operator is also referred to as a *pure isometry*; we shall use both the terms interchangeably.

**Theorem 2.1.4** (Wold decomposition; see [95]) *If $V$ is an isometric operator on a Hilbert space $\mathcal{K}$, then $\mathcal{K}$ has an internal direct sum decomposition of the form*

$$\mathcal{K} = \mathcal{K}_s \oplus \mathcal{K}_u,$$

*which is reducing for $V$ such that $S := V|_{\mathcal{K}_s}$ is a* shift operator *and $W := V|_{\mathcal{K}_u}$ is unitary.*

*Proof* Set $D_{V^*} = (I - VV^*)^{\frac{1}{2}}$, the *defect operator* for $V^*$, and $\mathcal{D}_{V^*} = \overline{\operatorname{Ran}}\, D_{V^*}$, the associated *defect space*. Since $V$ is an isometry, $D_{V^*} = (I - VV^*)$ is just the orthogonal projection onto $(\operatorname{Ran} V)^\perp$, that is, using item 5 in Notation 2.0.1,

$$D_{V^*} = I - VV^* = P_{(\operatorname{Ran} V)^\perp}$$

and $\mathcal{D}_{V^*} = \operatorname{Ran} D_{V^*} = (\operatorname{Ran} V)^\perp$. More generally, we have

$$V^N V^{*N} = P_{\operatorname{Ran} V^N}, \quad I - V^N V^{*N} = P_{(\operatorname{Ran} V^N)^\perp}.$$

Note next that for $d, d' \in \mathcal{D}_{V^*}$ and $n > m$ in $\mathbb{Z}_+$ we have $\langle V^n d, V^m d' \rangle = \langle V^{n-m} d, d' \rangle = 0$ since we have seen that $\mathcal{D}_{V^*} \perp \operatorname{Ran} V$, and hence we have

$$V^n \mathcal{D}_{V^*} \perp V^m \mathcal{D}_{V^*} \quad \text{for } n \neq m \text{ in } \mathbb{Z}_+$$

and in terms of projections we have

$$P_{V^n \mathcal{D}_{V^*}} = V^n D_{V^*} V^{*n}.$$

Then from the identity

$$\begin{aligned} \sum_{n=0}^{N} P_{V^n \mathcal{D}_{V^*}} &= \sum_{n=0}^{N} V^n D_{V^*} V^{*n} = \sum_{n=0}^{N} \left( V^n V^{*n} - V^{n+1} V^{*n+1} \right) \\ &= \sum_{n=0}^{N} (P_{\operatorname{Ran} V^n} - P_{\operatorname{Ran} V^{n+1}}) = I - P_{\operatorname{Ran} V^{N+1}} \\ &= P_{(\operatorname{Ran} V^{N+1})^\perp}, \end{aligned}$$

we see that, for each $N = 0, 1, 2, \ldots$, the space $\mathcal{K}$ has the orthogonal decomposition

$$\mathcal{K} = \left( \bigoplus_{n=0}^{N} V^n \mathcal{D}_{V^*} \right) \oplus \operatorname{Ran} V^{N+1}.$$

Letting $N \to \infty$ then gives us

$$\mathcal{K} = \left( \bigoplus_{n=0}^{\infty} V^n \mathcal{D}_{V^*} \right) \oplus \bigcap_{N \geq 0} \operatorname{Ran} V^N.$$

It is now a simple exercise to check that setting

$$\mathcal{K}_s = \bigoplus_{n=0}^{\infty} V^n \mathcal{D}_{V^*}, \quad \mathcal{K}_u = \bigcap_{N \geq 0} \operatorname{Ran} V^N = \operatorname{Ran} Q_{V^*}$$

gives the claimed Wold decomposition. Moreover, we also see that

$$P_{\mathcal{K}_s} = \sum_{n=0}^{\infty} V^n D_{V^*} V^{*n}, \quad P_{\mathcal{K}_u} = \lim_{n \to \infty} V^n V^{*n},$$

where the convergence of the infinite series for $P_{\mathcal{K}_s}$ and the existence of the limit for $P_{\mathcal{K}_u}$ are both in the strong operator topology. □

Note that this decomposition,

$$\mathcal{K} = \left( \bigoplus_{n=0}^{\infty} V^n \mathcal{D}_{V^*} \right) \oplus \bigcap_{N \geq 0} \operatorname{Ran} V^N, \tag{2.26}$$

amounts to a coordinate-free version of the Wold decomposition for the isometry $V$. By construction it follows that $V$ is a forward shift operator on the first component and a unitary operator on the second.

Let us convert this decomposition to a more functional form. We introduce a unitary Fourier representation operator

$$\tau_{\mathrm{F}} \colon \mathcal{K} \to \begin{bmatrix} H^2(\mathcal{D}_{V^*}) \\ \mathcal{K}_u \end{bmatrix}$$

given by

$$\tau_{\mathrm{F}} \colon h \mapsto \begin{bmatrix} \sum_{n=0}^{\infty} (D_{V^*} V^{*n} h) z^n \\ \text{SOT-}\lim_{n\to\infty} V^n V^{*n} h \end{bmatrix}. \tag{2.27}$$

Note that the first component $\sum_{n=0}^{\infty} (D_{V^*} V^{*n} h) z^n$ for the projection onto the shift part amounts to formally replacing the operator $V^n$ with the multiplier $z^n$ in the expansion $P_{\mathcal{K}_s} = \sum_{n=0}^{\infty} V^n D_{V^*} V^{*n}$ for the projection operator $P_{\mathcal{K}_s} \colon \mathcal{K} \to \mathcal{K}_s$. Then one easily checks that $\tau_{\mathrm{F}}$ has the intertwining property

$$\tau_{\mathrm{F}} V = \left( M_z^{\mathcal{D}_{V^*}} \oplus W \right) \tau_{\mathrm{F}},$$

where $M_z^{\mathcal{D}_{V^*}}$ is the forward shift on $H^2(\mathcal{D}_{V^*})$ and where $W = V|_{\mathcal{K}_u}$ is a unitary.

We close this section with a remark.

**Remark 2.1.5** Suppose that we are in the situation of part (ix) of Proposition 2.1.1 where $T$ *is completely nonunitary and* $V$ *is a minimal isometric lift of* $T$. Then the first decomposition in (2.6) exhibits the Wold decomposition for the isometry $V$ on $\mathcal{K}$. In the case when $\mathcal{K}_u$ is zero, then the operator $V$ itself is a shift operator. In the case when the residual space $\mathcal{R} = \mathcal{H}_u$ is zero, then the minimal isometric dilation $V$ for $T$ is itself a shift operator. This turns out also to be equivalent to: $T$ *is pure*, that is, powers of $T^*$ tend strongly to zero. Indeed, if $\mathcal{R} = \{0\}$, then $\mathcal{H} \subset M_+(\mathcal{L}_*)$ and powers of $V^*$ tend to zero strongly. As $T^*$ is the restriction of $V^*$ to $\mathcal{H} \subset \mathcal{K}$, it follows that $\|T^{*n} h\| = \|V^* h\| \to 0$ as $n \to \infty$ for each $h \in \mathcal{H}$. Conversely, if it is the case that $\mathcal{R} \neq 0$, part (viii)(a) of Proposition 2.1.1 implies that there is an $h \in \mathcal{H}$ such that $P_{\mathcal{R}} h \neq 0$. Then for all $n = 0, 1, 2, \ldots$ it is the case that

$$\|T^{*n} h\| = \|V^{*n} h\| \geq \|P_{\mathcal{R}} V^{*n} h\| = \|V^{*n} P_{\mathcal{R}} h\| = \|P_{\mathcal{R}} h\| > 0,$$

implying that $T$ is not pure.

We shall see that, despite the lack of uniqueness of two minimal isometric lifts of a commuting contractive pair (see Examples 4.1.4, 5.5.2, and 5.5.3), an appropriate analogue of the phenomenon above involving the residual space of the minimal isometric lift space and the original contraction operator, extends to the two-variable case. See Theorem 4.3.1 and Remark 4.3.4 for an elaborate discussion.

## 2.2 Functional Models for Isometric Lifts of a Contraction Operator

In this section, we describe three distinct models of the minimal isometric lift of a single contraction operator. We shall assume throughout this section that $T$ is a completely nonunitary contraction operator on $\mathcal{H}$ (i.e., there is no $T$-reducing subspace of $\mathcal{H}$ on which $T$ is unitary). This condition, although not essential, is convenient for us as it saves us from carrying around the unitary summand of $T$ in every minimal isometric lift of $T$. The ideas presented here are basic tools for developing the bivariate context to follow in succeeding chapters.

### 2.2.1 Schäffer Functional Model

Let us write $\mathcal{K}_S$ for the *Schäffer isometric lift space*

$$\mathcal{K}_S = \begin{bmatrix} \mathcal{H} \\ H^2(\mathcal{D}_T) \end{bmatrix} \tag{2.28}$$

(where $H^2(\mathcal{D}_T)$ is as in Notation 2.0.1 with $\mathcal{F} = \mathcal{D}_T$), and $\Pi_S$ for the isometric embedding operator

$$\Pi_S = \left[ \begin{smallmatrix} I_{\mathcal{H}} \\ 0 \end{smallmatrix} \right] : \mathcal{H} \to \mathcal{K}_S, \tag{2.29}$$

and let $V_S$ on $\mathcal{K}_S$ be the operator given by

$$V_S \colon \begin{bmatrix} h \\ f \end{bmatrix} \mapsto \begin{bmatrix} T & 0 \\ (\mathbf{ev}_0^{\mathcal{D}_T})^* D_T & M_z^{\mathcal{D}_T} \end{bmatrix} \begin{bmatrix} h \\ f \end{bmatrix} = \begin{bmatrix} Th \\ D_T h + z f(z) \end{bmatrix}, \tag{2.30}$$

where the "evaluation at 0" map $\mathbf{ev}_0^{\mathcal{D}_T} : H^2(\mathcal{D}_T) \to \mathcal{D}_T$ is as in Notation 2.0.1. In the following discussion we drop the $\mathcal{D}_T$ from the notation as the coefficient space here is clear from the context. Then one can check that $V_S$ is isometric on $\mathcal{K}_S$ and that $\Pi_S T^* = V_S^* \Pi_S$ (due to the block lower-triangular form in the matrix representation of $V_S$), and hence $(\Pi_S, V_S)$ is an isometric lift of $T$. Let us formally give this a name.

**Definition 2.2.1** If $\mathcal{K}_S$, $\Pi_S$, and $V_S$ are given as in (2.28), (2.29), and (2.30), then $(\Pi_S, V_S)$ is an isometric lift of $T$ on $\mathcal{K}_S$ and we shall say that $(\Pi_S, V_S)$ is the *Schäffer model isometric lift* of $T$.

We next analyze how the Schäffer model isometric lift can be embedded into the Sz.-Nagy–Foias coordinate-free geometric picture of the minimal lift for $T$. To compute the space $\mathcal{L} = \overline{(V\Pi_S - \Pi_S T)\mathcal{H}}$ in these coordinates, note that, for $h \in \mathcal{H}$, we have

$$(V_S \Pi_S - \Pi_S T)h = \begin{bmatrix} Th \\ \mathbf{ev}_0^* D_T h \end{bmatrix} - \begin{bmatrix} Th \\ 0 \end{bmatrix} = \begin{bmatrix} 0 \\ \mathbf{ev}_0^* D_T h \end{bmatrix}$$

and hence

$$V_S^n \begin{bmatrix} 0 \\ \mathbf{ev}_0^* D_T h \end{bmatrix} = \begin{bmatrix} 0 \\ z^n D_T h \end{bmatrix}$$

and then, from the definitions, we see that

$$\mathcal{H} \oplus M_+(\mathcal{L}) = \begin{bmatrix} \mathcal{H} \\ H^2(\mathcal{D}_T) \end{bmatrix} = \mathcal{K}_S, \tag{2.31}$$

thereby verifying the minimality of the Schäffer lift.

Let us next presume that this minimal isometric lift is presented in geometric coordinate-free form as in Section 2.1, so we have $\mathcal{K} = \mathcal{H} \oplus M_+(\mathcal{L})$ as in the first part of (2.6). Define $U_S : \mathcal{K} \to \mathcal{K}_S$ by

$$U_S = \begin{bmatrix} P_{\mathcal{H}} \\ \iota P_{M_+(\mathcal{L})} \end{bmatrix},$$

where

$$\iota \colon \bigoplus_{n=0}^{\infty} V^n \ell_n \mapsto \sum_{n=0}^{\infty} \iota \ell_n z^n \tag{2.32}$$

for $\bigoplus V^n \ell_n \in M_+(\mathcal{L})$, with $\iota$ given by item (iii) of Proposition 2.1.1. Then one can check that the intertwining

$$\Pi_S = U_S \, \iota_{\mathcal{H} \to \mathcal{K}}$$

holds. Furthermore, the calculation

$$\begin{aligned} U_S V \left( h + \sum_{n=0}^{\infty} V^n \ell_n \right) &= U_S \left( Th + (V - T)h + \sum_{n=0}^{\infty} V^{n+1} \ell_n \right) \\ &= \begin{bmatrix} Th \\ D_T h + M_z \cdot \sum_{n=0}^{\infty} (\iota \ell_n) z^n \end{bmatrix} = \begin{bmatrix} T & 0 \\ D_T & M_z \end{bmatrix} U_S \left( h + \sum_{n=0}^{\infty} V^n z^n \right) \end{aligned}$$

shows that

$$U_S V = V_S U_S. \tag{2.33}$$

We conclude that $U_S$ implements a unitary equivalence between the minimal coordinate-free isometric lift $(\iota_{\mathcal{H}\to\mathcal{K}}, V)$ and the Schäffer isometric lift $(\Pi_S, V_S)$.

Note that the Schäffer isometric-lift space $\mathcal{K}_S = \begin{bmatrix} \mathcal{H} \\ H^2(\mathcal{D}_T) \end{bmatrix}$ has its first component $\mathcal{H}$ equal to the original abstract Hilbert space while the second component $H^2(\mathcal{D}_T)$ is a functional Hilbert space, so strictly speaking the Schäffer model is only a semifunctional model. The original Schäffer model as presented in [92, Section I.5.1] has a purely matricial form as the second component $H^2(\mathcal{D}_T)$ is written in matricial form as $\ell^2_{\mathbb{Z}^+}(\mathcal{D}_T)$ rather than in the functional form $H^2(\mathcal{D}_T)$. Also, the construction extends to a matricial model for the minimal unitary dilation $\mathcal{U}$ of $T$. Note that the presentation of the Schäffer model space as

$$\mathcal{K}_S = \begin{bmatrix} \mathcal{H} \\ H^2(\mathcal{D}_T) \end{bmatrix} = \begin{bmatrix} \mathcal{H} \\ 0 \end{bmatrix} \oplus \begin{bmatrix} 0 \\ H^2(\mathcal{D}_T) \end{bmatrix} = U_S\mathcal{H} \oplus U_S M_+(\mathcal{L})$$

corresponds to the decomposition for $\mathcal{K}_{\min}$ in (2.2).

### 2.2.2 Douglas Functional Model

For $T$ a contraction operator on $\mathcal{H}$, a fundamental map for the construction of the Douglas model is the map

$$\mathcal{O}_{D_{T^*},T^*} : \mathcal{H} \to H^2(\mathcal{D}_{T^*}),$$

given by[1]

$$\mathcal{O}_{D_{T^*},T^*} : h \mapsto \sum_{n=0}^{\infty} D_{T^*} T^{*n} h\, z^n = D_{T^*}(I - zT^*)^{-1} h \tag{2.34}$$

as well as the *evaluation at* 0 map $\mathbf{ev}_0 \colon H^2 \to \mathbb{C}$, and more generally $\mathbf{ev}_{0,\mathcal{F}}\colon H^2(\mathcal{F})] \to \mathcal{F}$ and its adjoint $\mathbf{ev}^*_{0,\mathcal{F}}\colon \mathcal{F} \to H^2(\mathcal{F})$ as introduced in Notation 2.0.1.

[1] The notation $\mathcal{O}_{D_{T^*},T^*}$ is suggested by the fact that the operator $\mathcal{O}_{D_{T^*},T^*}$ can be viewed as the *frequence-domain observability operator* for the discrete-time state/output linear system

$$\left\{ \begin{array}{lcl} x(t+1) & = & T^* x(t), \\ y(t) & = & D_{T^*} x(t), \end{array} \right. \quad t = 0, 1, 2, \ldots$$

($x(t)$ equal to the *state* at time $t$, $y(t)$ equal to the output at time $t$) since running the system with initial condition $x(0)$ to produce an output string $\{y(t)\}_{n\in\mathbb{Z}_+}$ results in the $Z$-transform $\widehat{y}(z) := \sum_{n=0}^{\infty} y(n)\, z^n$ of the output string $\{y(n)\}_{n\in\mathbb{Z}^+}$ being given by

$$\widehat{y}(z) = \mathcal{O}_{D_{T^*},T^*} x_0.$$

We now focus on a general contraction operator $T$ acting on the Hilbert space $\mathcal{H}$. Since $T^*$ is then also a contraction, the easy computation

$$\begin{aligned}\|\mathcal{O}_{D_{T^*},T^*}x\|^2_{H^2(\mathcal{D}_{T^*})} &= \sum_{n=0}^{\infty}\|D_{T^*}T^{*n}x\|^2 = \sum_{n=0}^{\infty}\langle T^n(I-TT^*)T^{*n}x,x\rangle \\ &= \|x\|^2 - \|Q_{T^*}x\|^2 \quad (\text{where } Q_{T^*} \succeq 0 \text{ is as in (2.21)})\end{aligned}$$

shows that $\mathcal{O}_{D_{T^*},T^*}$ is contractive as an operator from $\mathcal{H}$ into the Hardy space $H^2(\mathcal{D}_{T^*})$, and is an isometry exactly when $Q_{T^*} = 0$.

We now note that $TQ^2_{T^*}T^* = Q^2_{T^*}$; hence the formula

$$X^*Q_{T^*}h = Q_{T^*}T^*h \tag{2.35}$$

defines an isometry $X^*$ on $\operatorname{Ran} Q_{T^*}$ that extends by continuity to an isometry (still denoted as $X^*$) on $\overline{\operatorname{Ran}}\, Q_{T^*}$. If not already unitary, this operator has a minimal unitary extension on a space $\mathcal{Q}_{T^*} \supset \overline{\operatorname{Ran}}\, Q_{T^*}$ that we denote by $W_D^*$. A dense subspace of $\mathcal{Q}_{T^*}$ is

$$\bigcup_{n=0}^{\infty} W_D^n \operatorname{Ran} Q_{T^*} \text{ dense in } \mathcal{Q}_{T^*} \tag{2.36}$$

and then the extension $W_D^*$ is given densely by

$$\begin{aligned} W_D^*W_D^nQ_{T^*}h &= W_D^{n-1}X^*Q_{T^*}h = W_D^{n-1}Q_T^*T^*h \quad \text{for } n \geq 1, \\ W_D^*Q_{T^*}h &= X^*Q_{T^*}h = Q_{T^*}T^*h. \end{aligned} \tag{2.37}$$

Let us introduce the Hilbert space

$$\mathcal{K}_D := \begin{bmatrix} H^2(\mathcal{D}_{T^*}) \\ \mathcal{Q}_{T^*} \end{bmatrix} \tag{2.38}$$

and define an isometric operator $V_D$ on $\mathcal{K}_D$ by

$$V_D = \begin{bmatrix} M_z & 0 \\ 0 & W_D \end{bmatrix}. \tag{2.39}$$

There is a canonical isometric embedding operator $\Pi_D \colon \mathcal{H} \to \mathcal{K}_D$ given by

$$\Pi_D \colon h \mapsto \begin{bmatrix} \mathcal{O}_{D_{T^*},T^*} \\ Q_{T^*} \end{bmatrix} h. \tag{2.40}$$

Furthermore, we have the intertwining relation

$$\Pi_D T^* = (V_D)^* \Pi_D. \tag{2.41}$$

Let us give all this a formal name.

**Definition 2.2.2** Define the space $\mathcal{K}_D$, the operator $\Pi_D$ and the operator $V_D$ as in (2.38), (2.40), and (2.39). Then $(\Pi_D, V_D)$ is a lift of $T$ on the space $\mathcal{K}_D$ that we shall refer to as the *Douglas model isometric lift* of $T$.

It is easy to compute the wandering subspace $\mathcal{L}_* = \overline{\operatorname{Ran}}\,(\Pi_D - V_D\Pi_D T^*)$ in the coordinates of the Douglas model. Indeed, for $h \in \mathcal{H}$ we have

$$(\Pi_D - V_D\Pi_D T^*)h = (I - V_D V_D^*)\Pi_D h = \begin{bmatrix} \mathbf{ev}^*_{0,\mathcal{D}_{T^*}} D_{T^*} h \\ 0 \end{bmatrix}$$

from which we see that $\mathcal{L}_* = \begin{bmatrix} \mathbf{ev}^*_{0,\mathcal{D}_{T^*}} \mathcal{D}_{T^*} \\ 0 \end{bmatrix}$. It then follows that

$$M_+(\mathcal{L}_*) = \bigoplus_{n=0}^{\infty} V^n \mathcal{L}_* = \begin{bmatrix} H^2(\mathcal{D}_{T^*}) \\ 0 \end{bmatrix}$$

and that the natural decomposition for the Douglas model space

$$\mathcal{K}_D := \begin{bmatrix} H^2(\mathcal{D}_{T^*}) \\ \mathcal{Q}_{T^*} \end{bmatrix} = \begin{bmatrix} H^2(\mathcal{D}_{T^*}) \\ 0 \end{bmatrix} \oplus \begin{bmatrix} 0 \\ \mathcal{Q}_{T^*} \end{bmatrix}$$

corresponds to the decomposition for $\mathcal{K}_{\min} = M_+(\mathcal{L}_*) \oplus \mathcal{R}$ appearing in (2.6).

To make the identification of the Douglas model with its geometric coordinate-free version complete, we wish to define a unitary identification map $U_D$ from the coordinate-free lift space

$$\mathcal{K}_{\min} = M_+(\mathcal{L}_*) \oplus \mathcal{R}$$

discussed in Section 2.1 to the Douglas model space

$$\mathcal{K}_D = \begin{bmatrix} H^2(\mathcal{D}_{T^*}) \\ \mathcal{Q}_{T^*} \end{bmatrix}.$$

The identification of $M_+(\mathcal{L}_*)$ with $H^2(\mathcal{D}_{T^*})$ is straightforward. Let $\iota_*$ be the extension of the identification of the wandering subspace $\mathcal{L}_*$ with $\mathcal{D}_{T^*}$ given as in Proposition 2.1.1 (iii) to an identification $\iota_*$ of $M_+(\mathcal{L}_*)$ with $H^2(\mathcal{D}_{T^*})$:

$$\iota_* \colon \bigoplus_{n=0}^{\infty} V^n \ell_{*n} \mapsto \sum_{n=0}^{\infty} \iota_* \ell_{*n}\, z^n. \tag{2.42}$$

Define $\omega_D \colon \mathcal{R} \to \mathcal{R}_D$ first on the linear submanifold $P_{\mathcal{R}}\mathcal{H} \subset \mathcal{R}$ by

$$\omega_D \colon P_{\mathcal{R}} h \mapsto Q_{T^*} h \in \mathcal{Q}_{T^*}. \tag{2.43}$$

By Proposition 2.1.1 (vi) (see (2.11)), we see that $\omega_D|_{P_{\mathcal{R}}\mathcal{H}}$ is isometric. Then there is no problem to extend $\omega_D$ to a well-defined isometry from $\overline{P_{\mathcal{R}}\mathcal{H}}$ into $\mathcal{Q}_{T^*}$. Recall that we have seen in Proposition 2.1.1 (viii) that $\cup_{n=0}^{\infty} V^n P_{\mathcal{R}}\mathcal{H}$ is dense in $\mathcal{R}$ and in the construction of $W_D$ that $W_D^n P_{\mathcal{R}}\mathcal{H}$ is dense in $Q_{T^*}$. Then it is a routine matter to check that the formula

$$\omega_D \colon V^n(P_{\mathcal{R}}h) \mapsto W_D^n Q_{T^*}h \tag{2.44}$$

extends to a well-defined unitary identification map $\omega_D$ from $\mathcal{R}$ onto $Q_{T^*}$. We may then define the unitary identification map $U_D$ by

$$U_D \colon k \mapsto \begin{bmatrix} \iota_* P_{M_+(\mathcal{L}_*)}k \\ \omega_D P_{\mathcal{R}} k \end{bmatrix}$$

for $k \in \mathcal{K}$ with the property that $U_D = I_{\mathcal{K}}$ if the isometric lift $V$ is given in Douglas model coordinates. Furthermore one can check the intertwinings:

$$\Pi_D = U_D|_{\mathcal{H}}, \quad U_D V = V_D U_D. \tag{2.45}$$

Thus $(\Pi_D, V_D)$ and $(\iota_{\mathcal{H}\to\mathcal{K}}, V)$ are unitarily equivalent as isometric lifts of $T$.

As $(\iota_{\mathcal{H}\to\mathcal{K}}, V)$ is minimal, it follows that $(U_D, V_D)$ is minimal as well. One can also see this fact as a consequence of Lemma 1 in Douglas's paper [54]. For completeness, we include here a simple direct proof of the context of the notation that we are using here.

**Proposition 2.2.3** *The Douglas model isometric lift* $(\Pi_D, V_D)$ *((2.39) and (2.40)) of a contraction operator $T$ is minimal.*

*Proof* Verification of the minimality of $(\Pi_D, V_D)$ amounts to showing that $\mathcal{K}_D = \bigvee_{n=0,1,2,\dots} V_D^n \operatorname{Ran}\Pi_D$, that is,

$$\mathcal{K}_{\min} := \bigvee_{n=0,1,2,\dots} \begin{bmatrix} M_z^n & 0 \\ 0 & W_D^n \end{bmatrix} \begin{bmatrix} \mathcal{O}_{D_{T^*},T^*} \\ Q_{T^*} \end{bmatrix} \mathcal{H} = \begin{bmatrix} H^2(\mathcal{D}_{T^*}) \\ Q_{T^*} \end{bmatrix}. \tag{2.46}$$

To see this, note first that for each $n \in \mathbb{Z}_+$ we have in particular that

$$\begin{bmatrix} M_z^n & 0 \\ 0 & W_D^n \end{bmatrix} \begin{bmatrix} \mathcal{O}_{D_{T^*},T^*} \\ Q_{T^*} \end{bmatrix} T^{*n}h \in \mathcal{K}_{\min},$$

where

$$\begin{bmatrix} M_z^n & 0 \\ 0 & W_D^n \end{bmatrix} \begin{bmatrix} \mathcal{O}_{D_{T^*},T^*} \\ Q_{T^*} \end{bmatrix} T^{*n}h = \begin{bmatrix} M_z^n & 0 \\ 0 & W_D^n \end{bmatrix} \begin{bmatrix} M_z^{*n} & 0 \\ 0 & W_D^{*n} \end{bmatrix} \begin{bmatrix} \mathcal{O}_{D_{T^*},T^*} \\ Q_{T^*} \end{bmatrix} h$$

$$= \begin{bmatrix} M_z^n M_z^{*n} & 0 \\ 0 & W_D^n W_D^{*n} \end{bmatrix} \begin{bmatrix} \mathcal{O}_{D_{T^*},T^*} \\ Q_{T^*} \end{bmatrix} h = \begin{bmatrix} M_z^n M_z^{*n} \mathcal{O}_{D_{T^*},T^*}h \\ Q_{T^*}h \end{bmatrix}, \tag{2.47}$$

where

$$\lim_{n\to\infty} \|M_z^n M_z^{*n} \mathcal{O}_{D_{T^*},T^*}h\| = \|M_z^{*n} \mathcal{O}_{D_{T^*},T^*}h\| = 0$$

since $M_z$ is a shift operator. Combining this observation with (2.47) leads us to

$$\lim_{n\to\infty} \begin{bmatrix} M_z^n & 0 \\ 0 & W_D^n \end{bmatrix} \begin{bmatrix} \mathcal{O}_{D_{T^*},T^*} \\ Q_{T^*} \end{bmatrix} T^{*n} h = \begin{bmatrix} 0 \\ Q_{T^*} h \end{bmatrix} \in \mathcal{K}_{\min}.$$

Subtracting this element off from $\begin{bmatrix} \mathcal{O}_{D_{T^*},T^*} \\ Q_{T^*} \end{bmatrix} h \in \mathcal{K}_{\min}$, we conclude that

$$\begin{bmatrix} \mathcal{O}_{D_{T^*},T^*} h \\ 0 \end{bmatrix} h \in \mathcal{K}_{\min}.$$

But then we also have, for all $h \in \mathcal{H}$,

$$\begin{bmatrix} (\mathcal{O}_{D_{T^*},T^*} - M_z \mathcal{O}_{D_{T^*},T^*} T^*) h \\ 0 \end{bmatrix} \in \mathcal{K}_{\min},$$

where

$$\begin{aligned} ((\mathcal{O}_{D_{T^*},T^*} - M_z \mathcal{O}_{D_{T^*},T^*} T^*) h)(z) &= \Big( D_{T^*}(I - zT^*)^{-1} - z D_{T^*}(I - zT^*)^{-1} T^* \Big) h \\ &= \mathbf{ev}^*_{0, \mathcal{D}_{T^*}} D_{T^*} h. \end{aligned} \tag{2.48}$$

Thus $\begin{bmatrix} \mathbf{ev}^*_{0,\mathcal{D}_{T^*}} D_{T^*} h \\ 0 \end{bmatrix} \in \mathcal{K}_{\min}$ and hence also

$$\begin{aligned} \bigvee_{n=0,1,2,\ldots} \begin{bmatrix} M_z^n & 0 \\ 0 & W_D^n \end{bmatrix} \begin{bmatrix} \mathbf{ev}^*_{0,\mathcal{D}_{T^*}} D_{T^*} \mathcal{H} \\ 0 \end{bmatrix} &= \begin{bmatrix} \bigvee_{n=0,1,2,\ldots} M_z^n \mathbf{ev}^*_{0,\mathcal{D}_{T^*}} \mathcal{D}_{T^*} \\ 0 \end{bmatrix} \\ &= \begin{bmatrix} H^2(\mathcal{D}_{T^*}) \\ 0 \end{bmatrix} \subset \mathcal{K}_{\min}. \end{aligned}$$

Finally let us recall that $\cup_{n \ge 0} W_D^n Q_{T^*} \mathcal{H}$ is dense in $\mathcal{Q}_{T^*}$ and hence

$$\bigvee_{n=0,1,2,\ldots} \begin{bmatrix} M_z^n & 0 \\ 0 & W_D^n \end{bmatrix} \begin{bmatrix} 0 \\ Q_{T^*} \mathcal{H} \end{bmatrix} = \begin{bmatrix} 0 \\ \mathcal{Q}_{T^*} \end{bmatrix} \subset \mathcal{K}_{\min}.$$

Putting all the pieces together then gives us

$$\mathcal{K}_D := \begin{bmatrix} H^2(\mathcal{D}_{T^*}) \\ \mathcal{Q}_{T^*} \end{bmatrix} \subset \mathcal{K}_{\min} \subset \mathcal{K}_D$$

and (2.46) follows as desired. □

**Remark 2.2.4** As the Douglas isometric-lift space $\mathcal{K}_D = \begin{bmatrix} H^2(\mathcal{D}_{T^*}) \\ \mathcal{Q}_{T^*} \end{bmatrix}$ has its first component $H^2(\mathcal{D}_{T^*})$ equal to a functional Hilbert space but its second component $\mathcal{Q}_{T^*}$ equal to an abstract (nonfunctional) Hilbert space, so strictly speaking one can think of the Douglas model as really being only a semifunctional model. However, the space $\mathcal{Q}_{T^*}$ comes equipped with a unitary operator $W_D$. By the spectral theorem for unitary operators, one can convert this space to a direct-integral $L^2$-space $\mathcal{Q}_{T^*} \cong \bigoplus \int_{\mathbb{T}} \mathcal{H}_\zeta \, \mathrm{d}\nu(\zeta)$ for a scalar spectral measure

$\nu$ and multiplicity function $m(\zeta) = \dim \mathcal{H}_\zeta$, with $W_D$ then given as the multiplication operator $W_D = M_\zeta \colon f(\zeta) \mapsto \zeta \cdot f(\zeta)$ (where $\zeta \mapsto f(\zeta) \in \mathcal{H}_\zeta$ is a measurable square-integrable cross-section of $\{\mathcal{H}_\zeta\}_{\zeta \in \mathbb{T}}$). This becomes more precise when we make connections with the Sz.-Nagy–Foias model in Section 2.2.3.

**Remark 2.2.5** While the Douglas model is mostly explicitly constructed in terms of the operator $T$, one could ask for a more explicit construction of the residual part $Q_{T^*}$, where the unitary part $W_D$ of the minimal isometric lift $V_D = M_z \oplus W_D$ is defined. There is a later construction due to Durszt [59], which, while arguably more difficult to work with, gives such an explicit construction, with the first component exactly the same as in the Douglas construction (after interchanging the contraction operator $T$ with its adjoint $T^*$), while the second component is a bilateral unitary shift operator with a more complicated but completely explicit formula for the map from $\mathcal{H}$ into this $\ell^2_{\mathbb{Z}}$-space.

### 2.2.3 Sz.-Nagy–Foias Functional Model

As is standard (see [92, Chapter VI]), we do not use the Sz.-Nagy–Foias model to prove that unitary dilations of a contraction operator exist. Rather we assume that a minimal unitary dilation has been constructed explicitly by some other method (e.g., via constructing the Schäffer or Douglas model), which we then convert to the coordinate-free geometric form as discussed in Section 2.1. Here all subspace decompositions are internal direct sums and all subspaces appearing are explicit in terms of the original contraction operator $T$, the minimal isometric lift $V$, and/or the minimal unitary dilation $\mathcal{U}$. The exercise then is to convert this coordinate-free presentation to a unitarily equivalent functional form.

We therefore assume that we are given a c.n.u. contraction operator $T$ together with its minimal unitary dilation $\mathcal{U}$ on $\widetilde{\mathcal{K}}_{\min}$ and minimal isometric lift $V$ on $\mathcal{K}_{\min}$ where

$$\mathcal{L} = \overline{(V - T)\mathcal{H}}, \quad \mathcal{L}_* = \overline{(I - \mathcal{U}T^*)\mathcal{H}}$$

are wandering subspaces for $\mathcal{U}$,

$$\begin{aligned} \widetilde{\mathcal{K}}_{\min} &= M(\mathcal{L}_*) \oplus \mathcal{R} = M_-(\mathcal{L}_*) \oplus \mathcal{H} \oplus M_+(\mathcal{L}), \\ \mathcal{K}_{\min} &= M_+(\mathcal{L}_*) \oplus \mathcal{R} = \mathcal{H} \oplus M_+(\mathcal{L}), \end{aligned}$$

as in Proposition 2.1.1. Define the operator $\boldsymbol{\Theta} \colon M(\mathcal{L}) \to M(\mathcal{L}_*)$ as the restricted projection

$$\boldsymbol{\Theta} = P_{M(\mathcal{L}_*)}|_{M(\mathcal{L})}.$$

Let $\iota_* \colon M(\mathcal{L}_*) \to L^2(\mathcal{D}_{T^*})$ and $\iota \colon M(\mathcal{L}) \to L^2(\mathcal{D}_T)$ be the Fourier representation operators

$$\iota_* \colon \sum_{n=-\infty}^{\infty} \mathcal{U}^n \ell_{*n} \mapsto \sum_{n=-\infty}^{\infty} \ell_{*n} e^{int}, \quad \iota \colon \sum_{n=-\infty}^{\infty} \mathcal{U}^n \ell_n \mapsto \sum_{n=-\infty}^{\infty} \ell_n e^{int}, \tag{2.49}$$

that is, the operator $\iota_*$ in (2.42) extended in the natural way to a unitary identification of the bilateral-shift space $M(\mathcal{L})$ to the $L^2$-space $L^2(\mathcal{D}_{T^*})$ and similarly for the operator $\iota$ in (2.32). Then it is easily checked that

$$\boldsymbol{\Theta}\left(\mathcal{U}|_{M(\mathcal{L})}\right) = \left(\mathcal{U}|_{M(\mathcal{L}_*)}\right)\boldsymbol{\Theta}.$$

Let $\widehat{\boldsymbol{\Theta}} = \iota_* \boldsymbol{\Theta} \iota \colon L^2(\mathcal{D}_T) \to L^2(\mathcal{D}_{T^*})$. Then the previous intertwining relation becomes the function-space intertwining

$$\widehat{\boldsymbol{\Theta}} M_\zeta = M_\zeta \widehat{\boldsymbol{\Theta}}.$$

By a standard result (see, e.g., [92, Lemma V.3.1]), it follows that $\widehat{\boldsymbol{\Theta}}$ is a multiplication operator

$$\widehat{\boldsymbol{\Theta}} \colon h(\zeta) \mapsto \Theta(\zeta) \cdot h(\zeta)$$

for a measurable $\mathcal{B}(\mathcal{D}_T, \mathcal{D}_{T^*})$-valued function $\zeta \mapsto \Theta(\zeta)$ (notation as in Notation 2.0.1 with $\mathcal{E} = \mathcal{D}_T$ and $\mathcal{F} = \mathcal{D}_{T^*}$).

As $\boldsymbol{\Theta}$ is a restricted projection, it follows that $\|\boldsymbol{\Theta}\| \le 1$, and also $\|M_\Theta\| \le 1$ as an operator from $L^2(\mathcal{D}_T)$ to $L^2(\mathcal{D}_{T^*})$, from which it follows that $\|\Theta(\zeta)\| \le 1$ for almost all $\zeta$ in the unit circle. Furthermore, from the second decomposition in (2.4), we see that $M_\Theta$ maps $H^2(\mathcal{D}_T)$ into $H^2(\mathcal{D}_{T^*})$; thus in fact $\Theta$ is a contractive $H^\infty$-function with values in $\mathcal{B}(\mathcal{D}_T, \mathcal{D}_{T^*})$, known as the *Sz.-Nagy–Foias characteristic function* of $T$, given by the explicit formula

$$\Theta(z) = \Theta_T(z) = -T + zD_{T^*}(I_{\mathcal{H}} - zT^*)^{-1} D_T|_{\mathcal{D}_T} \colon \mathcal{D}_T \to \mathcal{D}_{T^*} \tag{2.50}$$

(see [92, Proposition VI.2.2]).

Suppose next that $k \in \mathcal{R}$ has the form $k = P_{\mathcal{R}}\ell$ for some $\ell \in M(\mathcal{L})$. Then

$$\|k\|^2 = \|P_{\mathcal{R}}\ell\|^2 = \|\ell\|^2 - \|\boldsymbol{\Theta}\ell\|^2 = \|\iota\ell\|^2 - \|\Theta_T \cdot \iota\ell\|^2 = \|\Delta_{\Theta_T} \cdot \iota\ell\|^2,$$

where we let $\Delta_{\Theta_T}$ be the $\mathcal{D}_T$-valued operator function on the unit circle $\mathbb{T}$ given by

$$\Delta_{\Theta_T}(\zeta) := (I - \Theta_T(\zeta)^* \Theta_T(\zeta))^{1/2}$$

(the pointwise defect operator of $\Theta_T$: $\Delta_{\Theta_T}(\zeta) = D_{\Theta_T(\zeta)}$). As we are assuming that $T$ is c.n.u., by Proposition 2.1.1 (ix) (see condition (2.14)) we know that the space $(I - P_{M(\mathcal{L}_*)})M(\mathcal{L}) = P_{\mathcal{R}} M(\mathcal{L})$ is dense in $\mathcal{R}$. Hence we can define a unitary map $\omega_{\mathrm{NF}}$ from $\mathcal{R}$ to $\overline{\Delta_{\Theta_T} L^2(\mathcal{D}_T)}$ densely defined on $P_{\mathcal{R}} M(\mathcal{L})$ by

$$\omega_{\mathrm{NF}} \colon P_{\mathcal{R}}\ell \mapsto \Delta_{\Theta_T} \cdot \iota\ell. \tag{2.51}$$

From this formula, we can see the validity of the intertwining relation

$$\omega_{\rm NF}(V|_{\mathcal{R}}) = M_{\zeta}^{\mathcal{D}_T}|_{\overline{\Delta_{\Theta_T} L^2(\mathcal{D}_T)}}\, \omega_{\rm NF}. \tag{2.52}$$

Let us introduce the Sz.-Nagy–Foias functional-model lift space (which depends only on the characteristic function $\Theta_T$) $\mathcal{K}_{\Theta_T}$ by

$$\mathcal{K}_{\Theta_T} := \begin{bmatrix} H^2(\mathcal{D}_{T^*}) \\ \overline{\Delta_{\Theta_T} L^2(\mathcal{D}_T)} \end{bmatrix}. \tag{2.53}$$

We next define a unitary identification map $U_{\rm NF}$ from $\mathcal{K} = M_+(\mathcal{L}_*) \oplus \mathcal{R}$ to $\mathcal{K}_{\Theta_T}$ by

$$U_{\rm NF} k = \begin{bmatrix} \iota_* P_{M_+(\mathcal{L}_*)} k \\ \omega_{\rm NF} P_{\mathcal{R}} k \end{bmatrix}. \tag{2.54}$$

Since $\iota_*$ is unitary from $M_+(\mathcal{L}_*)$ to $H^2(\mathcal{D}_{T^*})$, $\omega_{\rm NF}$ is unitary from $\mathcal{R}$ to $\overline{\Delta_{\Theta_T} L^2(\mathcal{D}_T)}$, and $\mathcal{K}$ has the internal orthogonal decomposition $\mathcal{K} = M_+(\mathcal{L}_*) \oplus \mathcal{R}$, we see that $U_{\rm NF}$ so defined is unitary from $\mathcal{K}$ onto $\mathcal{K}_{\Theta_T}$. Observing the intertwining relation $M_z \iota_* |_{M_+(\mathcal{L}_*)} = \iota V|_{M_+(\mathcal{L}_*)}$ and recalling (2.52), we arrive at the intertwining relation

$$U_{\rm NF} V = V_{\rm NF} U_{\rm NF}, \tag{2.55}$$

where we set $V_{\rm NF}$ equal to the isometric operator on $\mathcal{K}_{\Theta_T}$ given by

$$V_{\rm NF} = \begin{bmatrix} M_z^{\mathcal{D}_T} & 0 \\ 0 & M_\zeta|_{\overline{\Delta_{\Theta_T} L^2(\mathcal{D}_T)}} \end{bmatrix}. \tag{2.56}$$

Define the isometric embedding $\Pi_{\rm NF}$ of $\mathcal{H}$ into $\mathcal{K}_{\Theta_T}$ as

$$\Pi_{\rm NF} = U_{\rm NF}\, \iota_{\mathcal{H} \to \mathcal{K}}. \tag{2.57}$$

More explicitly, for $h \in \mathcal{H}$ we have

$$\begin{aligned} \iota_* P_{M_+(\mathcal{L}_*)} h &= \iota_* \left( \sum_{n=0}^{\infty} V^n P_{\mathcal{L}_*} T^{*n} h \right) \\ &= \sum_{n=0}^{\infty} z^n D_{T^*} T^{*n} h = D_{T^*}(I - zT^*)^{-1} h \end{aligned}$$

while, from (2.44) with $n = 0$, we see that $P_{\mathcal{R}} h = \omega_D^* Q_{T^*} h$ from which we get $\omega_{\rm NF} P_{\mathcal{R}} h = \omega_{\rm NF} \omega_D^* h$, and hence the action of $\Pi_{\rm NF}$ can be given more explicitly as

$$\Pi_{\rm NF} h = \begin{bmatrix} \iota_* P_{M_+(\mathcal{L}_*)} h \\ \omega_{\rm NF} P_{\mathcal{R}} h \end{bmatrix} = \begin{bmatrix} D_{T^*}(I - zT^*)^{-1} h \\ \omega_{\rm NF} \omega_D^* Q_{T^*} h \end{bmatrix} =: \begin{bmatrix} D_{T^*}(I - zT^*)^{-1} h \\ \omega_{\rm NF,D} Q_{T^*} h \end{bmatrix}. \tag{2.58}$$

The space $U_{\rm NF}M_+(\mathcal{L})$ can be identified explicitly as follows: For $\ell \in M_+(\mathcal{L})$,

$$U_{\rm NF}\ell = \begin{bmatrix} \Theta_T \cdot \iota\ell \\ \omega_{\rm NF}P_{\mathcal{R}}\ell \end{bmatrix} = \begin{bmatrix} \Theta_T \cdot \iota\ell \\ \Delta_{\Theta_T} \cdot \iota\ell \end{bmatrix},$$

and hence the space $\mathcal{H}_{\Theta_T} := \Pi_{\rm NF}\mathcal{H} = U_{\rm NF}\mathcal{H}$ (which turns out to depend only on the characteristic function $\Theta_T$) is given by

$$\mathcal{H}_{\Theta_T} = \begin{bmatrix} H^2(\mathcal{D}_{T^*}) \\ \hline \Delta_{\Theta_T}L^2(\mathcal{D}_T) \end{bmatrix} \ominus \begin{bmatrix} \Theta_T \\ \Delta_{\Theta_T} \end{bmatrix} \cdot H^2(\mathcal{D}_T). \tag{2.59}$$

Note that the subspace $U_{\rm NF}M_+(\mathcal{L}) = \left[\begin{smallmatrix} \Theta_T \\ \Delta_{\Theta_T} \end{smallmatrix}\right] \cdot H^2(\mathcal{D}_T)$ is invariant for $V_{\rm NF}$ and hence $\mathcal{H}_{\Theta_{\rm T}}$ is invariant for $V_{\rm NF}^*$. Rewrite (2.55) in the form

$$U_{\rm NF}V^* = V_{\rm NF}^*U_{\rm NF}$$

and restrict this identity to $\mathcal{H}$ to arrive at

$$\Pi_{\rm NF}T^* = V_{\rm NF}^*\Pi_{\rm NF}.$$

This suggests the following formal definition.

**Definition 2.2.6** Let the space $\mathcal{K}_{\Theta_T}$ and the operators $\Pi_{\rm NF} \colon \mathcal{H} \to \mathcal{K}_{\Theta_T}$ and $V_{\rm NF}$ on $\mathcal{K}_{\Theta_T}$ be given as in (2.53), (2.58), and (2.56). Then $(\Pi_{\rm NF}, V_{\rm NF})$ is a minimal isometric lift of $T$ on $\mathcal{K}_{\Theta_T}$ that we shall refer to as the *Sz.-Nagy–Foias functional-model isometric lift of* $T$.

Let us next note that the operator $U_{\rm NF} \colon \mathcal{K} \to \mathcal{K}_{\Theta_T}$ implements a unitary equivalence of the Sz.-Nagy–Foias isometric lift $(\Pi_{\rm NF}, V_{\rm NF})$ with the coordinate-free minimal isometric lift $(\iota_{\mathcal{H}\to\mathcal{K}}, V)$, and hence the Sz.-Nagy–Foias functional-model isometric lift $(\Pi_{\rm NF}, V_{\rm NF})$ is also minimal.

We note that both components of the Sz.-Nagy–Foias isometric-lift space $\mathcal{K}_{\Theta_T} = \left[\begin{smallmatrix} H^2(\mathcal{D}_{T^*}) \\ \hline \Delta_{\Theta_T}L^2(\mathcal{D}_T) \end{smallmatrix}\right]$ are functional Hilbert spaces (albeit with the first component $H^2(\mathcal{D}_{T^*})$ consisting of holomorphic functions and with the second component consisting of $L^2$-measurable functions). For this reason it makes sense to say that $\mathcal{K}_{\Theta_T}$ is a *functional model space* and that $V_{\rm NF}$ is the *Sz.-Nagy–Foias functional-model isometric lift* of $T$.

**Remark 2.2.7** The isometric embedding operator $\Pi_{\rm NF}$ for the Sz.-Nagy–Foias model (the analogue of the operator $\Pi_S$ (2.29) for the Schäffer model and of the operator $\Pi_D$ (2.40) for the Douglas model) is the embedding operator

$$\Pi_{\rm NF} \colon \mathcal{H} \to \mathcal{K}_{\Theta_T} = \begin{bmatrix} H^2(\mathcal{D}_{T^*}) \\ \hline \Delta_{\Theta_T}L^2(\mathcal{D}_T) \end{bmatrix}$$

with range given by

$$\operatorname{Ran}\Pi_{\mathrm{NF}} = \mathcal{H}_{\Theta_T} := \begin{bmatrix} H^2(\mathcal{D}_{T^*}) \\ \overline{\Delta_{\Theta_T} L^2(\mathcal{D}_T)} \end{bmatrix} \ominus \begin{bmatrix} \Theta_T \\ \Delta_{\Theta_T} \end{bmatrix} H^2(\mathcal{D}_T)$$

and

$$\Pi_{\mathrm{NF}} T^* = V_{\mathrm{NF}}^* \Pi_{\mathrm{NF}}. \tag{2.60}$$

Let us write $\Pi_{\mathrm{NF},0}$ for the operator $\Pi_{\mathrm{NF}}$ where it is considered to have codomain equal to its range $\mathcal{H}_{\Theta_T}$ rather than all of $\mathcal{K}_{\Theta_T}$. Then $\Pi_{\mathrm{NF},0} \colon \mathcal{H} \to \mathcal{H}_{\Theta_T}$ is unitary and from (2.60) one can see that

$$T = \Pi_{\mathrm{NF},0}^* (P_{\mathcal{H}_{\Theta_T}} V|_{\mathcal{H}_{\Theta_T}}) \Pi_{\mathrm{NF},0},$$

thereby showing that $T$ is unitarily equivalent to its functional-model operator $T_{\Theta_T}$:

$$T \underset{u}{\cong} T_{\Theta_T} := P_{\mathcal{H}_{\Theta_T}} \begin{bmatrix} M_z^{\mathcal{D}_{T^*}} & 0 \\ 0 & M_\zeta^{\mathcal{D}_T}|_{\overline{\Delta_{\Theta_T} L^2(\mathcal{D}_T)}} \end{bmatrix}\Bigg|_{\mathcal{H}_{\Theta_T}}.$$

It turns out that the characteristic operator function $\Theta_T$ (2.50) is a *complete unitary invariant* for a c.n.u. contraction operator $T$ in the sense that two c.n.u. contraction operators $T$ and $T'$ are unitarily equivalent if and only if $\Theta_T$ and $\Theta_{T'}$ *coincide* (i.e., they are the same after a unitary change of basis on the input coefficient Hilbert space and on the output coefficient Hilbert space).

But there is more. Let us first note that any characteristic function $\Theta_T$ (written as $(\mathcal{D}_T, \mathcal{D}_{T^*}, \Theta_T)$ when we wish to emphasize that the values of $\Theta_T$ are operators from $\mathcal{D}_T$ to $\mathcal{D}_{T^*}$) is a *contractive analytic function* in the terminology of [92], meaning that $\Theta_T$ is an analytic function on the unit disk with contractive operator values. We shall have need of special contractive analytic functions throughout this book. We put them together in the following definition.

**Definition 2.2.8** Let $(\mathcal{E}, \mathcal{E}_*, \Theta)$ be a contractive analytic function, that is, $\Theta \colon \mathbb{D} \to \mathcal{B}(\mathcal{E}, \mathcal{E}_*)$ is analytic and $\|\Theta(z)\| \le 1$ for all $z \in \mathbb{D}$.

(i) We say that $\Theta$ is *purely contractive*, if the operator $\Theta(0)$ is such that

$$\|\Theta(0)x\| < \|d\| \quad \text{for } 0 \ne x \in \mathcal{E}.$$

(ii) When $\mathcal{E} = \mathcal{E}_*$, we say that $\Theta$ is *completely nonunitary*, or c.n.u. for short, if the operator $\Theta(0)$ is completely nonunitary, that is, there is no nontrivial reducing subspace where $\Theta(0)$ is unitary.

(iii) When $\mathcal{E} = \mathcal{E}_*$, we say that $\Theta$ is *pure* if the operator $\Theta(0)$ is pure, that is, $\Theta(0)^{*n} \to 0$ strongly.

Note that the notions of *c.n.u.* and of *pure* are operator-theoretic and can be applied equally well to operators that are not of the form $\Theta(0)$ for some contractive operator function $\Theta$; for example, we can speak of the multiplication operator $M_\Theta$ (acting on $H^2(\mathcal{F})$) being *c.n.u.* or *pure*. The notion of *purely contractive* is not usually used in the context where the operator $\Theta(0)$ is replaced by a general contraction operator $T$: such operators come up but then are said to have the property of *failing to be norm-attaining*.

We note that notion (i) in Definition 2.2.8 comes up in the context of the Sz.-Nagy–Foias characteristic function for a Hilbert space contraction operator being discussed here, while the two notions (ii) and (iii) come up in the context of the Becovici–Douglas–Foias characteristic function for a commuting isometric operator pair $(V_1, V_2)$, to be discussed in Section 3.2.

It is worthwhile to note that when $\mathcal{E} = \mathcal{E}_*$ is finite dimensional, then all the three classes (i), (ii), and (iii) coincide. This is because if a contractive square matrix $A$ satisfies any of the three conditions in items (i), (ii), and (iii) that $\Theta(0)$ satisfies, $A$ will also satisfy the other two. The proof is routine; interested readers can see [49, Lemma 2.1]. Let us also mention that in this finite-dimensional situation, purity of $M_\Theta$ is the same as purity of $\Theta(0)$ (see [28, Lemma 3.4]) and that $M_\Theta$ being c.n.u. is the same as $\Theta(0)$ being c.n.u. (this also is a consequence of the analysis surrounding Lemma 2.1 in [49]). However, these equivalences fail in the general infinite-dimensional setting (see [23] and [30]).

The notion of *purely contractive* (item (i) in Definition 2.2.8) was introduced by Sz.-Nagy and Foias in connection with characterizing which contractive operator functions can arise as the characteristic operator function for a c.n.u. contraction operator $T$, where the appropriate notion of *equivalence* for two nonsquare contractive operator functions $(\mathcal{E}, \mathcal{E}_*, \Theta)$ and $(\mathcal{E}', \mathcal{E}'_*, \Theta')$ is the existence of two unitary operators $\tau\colon \mathcal{E} \to \mathcal{E}'$ and $\tau_*\colon \mathcal{E}_* \to \mathcal{E}'_*$ so that $\tau_*\Theta(z) = \Theta'(z)\tau$ for all $z \in \mathbb{D}$. This is in contrast with the notion appropriate in the context of square Bercovici–Douglas–Foias characteristic functions $(\mathcal{F}, \Theta)$ for a commuting isometric pair $(V_1, V_2)$, to be discussed in Section 3.3.1, which is that two such square contractive operator functions $(\mathcal{F}, \Theta)$ and $(\mathcal{F}', \Theta')$ are unitarily equivalent if there is a single unitary operator $\tau\colon \mathcal{F} \to \mathcal{F}'$ so that $\tau\Theta(z) = \Theta'(z)\tau$ for all $z \in \mathbb{D}$.

It works out that the *coincidence envelope* of the characteristic functions consists exactly of those contractive analytic functions $(\mathcal{E}, \mathcal{E}_*, \Theta)$, which are also purely contractive and the model-theory point of view can be reversed: Given any purely contractive analytic function $(\mathcal{E}, \mathcal{E}_*, \Theta)$, one can form the space $\mathcal{H}_\Theta$ exactly as in (2.59) but with $(\mathcal{E}, \mathcal{E}_*, \Theta)$ in place of $(\mathcal{D}_T, \mathcal{D}_{T^*}, \Theta_T)$:

$$\mathcal{H}_\Theta := \begin{bmatrix} H^2(\mathcal{E}_*) \\ \overline{\Delta_\Theta L^2(\mathcal{E})} \end{bmatrix} \ominus \begin{bmatrix} \Theta \\ \Delta_\Theta \end{bmatrix} H^2(\mathcal{E}), \tag{2.61}$$

where $\Delta_\Theta(\zeta) = (I_\mathcal{D} - \Theta(\zeta)^*\Theta(\zeta))^{1/2}$ for $\zeta \in \mathbb{T}$. Define the *model operator* associated with the purely contractive analytic function $(\mathcal{D}, \mathcal{D}_*, \Theta)$:

$$T_\Theta := P_{\mathcal{H}_\Theta} \begin{bmatrix} M_z^{\mathcal{E}_*} & 0 \\ 0 & M_\zeta^{\mathcal{E}}|_{\overline{\Delta_\Theta L^2(\mathcal{E})}} \end{bmatrix}\Bigg|_{\mathcal{H}_\Theta}. \tag{2.62}$$

Then it can be shown that the model operator $T_\Theta$ is a c.n.u. contraction operator on $\mathcal{H}_\Theta$, and its characteristic function $(\mathcal{D}_{T_\Theta}, \mathcal{D}_{T_\Theta^*}, \Theta_{T_\Theta})$ coincides with the original pure contractive analytic function $(\mathcal{D}, \mathcal{D}_*, \Theta)$. Consequently, the study of abstract c.n.u. contraction operators $T$ on a Hilbert space $\mathcal{H}$ is equivalent to the study of concrete functional-model operators of the form $T_\Theta$ associated with a purely contractive analytic function $\Theta$. Furthermore, the restriction that $T$ be c.n.u. is not really a restriction because any contraction operator can be decomposed as $T_{\rm cnu} \oplus T_u$, where $T_{\rm cnu}$ is c.n.u. and $T_u$ is unitary, where unitaries are essentially well understood via spectral theory. In Chapter 9 we present an extension of all these ideas to the setting of a commuting contractive pair $(T_1, T_2)$.

**Remark 2.2.9** A drawback of the Sz.-Nagy–Foias approach versus that of Douglas or Schäffer is that the formula $\Pi_{\rm NF}$ lacks explicitness in terms of the original contraction operator $T$. The best formula that we have been able to obtain for such an operator $\Pi_{\rm NF}$ is

$$\Pi_{\rm NF} = \begin{bmatrix} \mathcal{O}_{D_{T^*}, T^*} \\ \omega_{\rm NF} P_\mathcal{R}|_\mathcal{H} \end{bmatrix}, \tag{2.63}$$

where $\omega_{\rm NF}$ is given by (2.51). By referring back to the construction of the Douglas model, one can see that an equivalent formula is

$$\Pi_{\rm NF} = \begin{bmatrix} \mathcal{O}_{D_{T^*}, T^*} \\ \omega_{\rm NF}\omega_D^* Q_{T^*} \end{bmatrix}, \tag{2.64}$$

where $Q_{T^*}$ is as in (2.21) and $\omega_D$ is as in (2.44). To make sense of either formula, one must have already constructed some isometric lift (or unitary dilation) for $T$ from which one can then convert to the coordinate-free geometric version (as in Proposition 2.1.1), where one can make use of the operators $\mathcal{U}$ and $V$ to make sense of all the key subspaces $\mathcal{L}$, $\mathcal{L}_*$, $M(\mathcal{L})$, $M(\mathcal{L}_*)$, $\mathcal{R}$ and the associated projection operators $P_\mathcal{L}$, $P_{\mathcal{L}_*}$, $P_\mathcal{R}$, etc.

**Remark 2.2.10** Since the isometric lifts $(\Pi_D, V_D)$ and $(\Pi_{\text{NF}}, V_{\text{NF}})$ are minimal, they are unitarily equivalent (as lifts) by Theorem 2.1.3. We find the explicit unitary involved in the equivalence of the minimal isometric lifts $(\Pi_D, V_D)$ and $(\Pi_{\text{NF}}, V_{\text{NF}})$ as follows. Let $\omega_D$ and $\omega_{\text{NF}}$ be the unitaries defined in (2.44) and (2.51), respectively. Let us introduce the notation $\omega_{\text{NF,D}}$ for the unitary operator

$$\omega_{\text{NF,D}} := \omega_{\text{NF}}\omega_D^* \colon \mathcal{Q}_{T^*} \to \overline{\Delta_{\Theta_T} L^2(\mathcal{D}_T)}. \tag{2.65}$$

Let us define another operator by

$$U_{\text{NF,D}} := \begin{bmatrix} I_{H^2(\mathcal{D}_{T^*})} & 0 \\ 0 & \omega_{\text{NF,D}} \end{bmatrix} : \mathcal{K}_D := \begin{bmatrix} H^2(\mathcal{D}_{T^*}) \\ \mathcal{Q}_{T^*} \end{bmatrix} \to \mathcal{K}_{\Theta_T} := \begin{bmatrix} H^2(\mathcal{D}_{T^*}) \\ \overline{\Delta_{\Theta_T} L^2(\mathcal{D}_T)} \end{bmatrix}. \tag{2.66}$$

Clearly $U_{\text{NF,D}}$ is unitary. To show that $U_{\text{NF,D}}$ establishes a unitary equivalence from the Douglas lift $(\Pi_D, V_D)$ on $\mathcal{K}_D$ to the Sz.-Nagy–Foias lift $(\Pi_{\text{NF}}, V_{\text{NF}})$ on $\mathcal{K}_{\Theta_T}$, we must verify the equations

$$U_{\text{NF,D}} V_D = V_{\text{NF}} U_{\text{NF,D}}, \quad U_{\text{NF,D}} \Pi_D = \Pi_{\text{NF}},$$

or, in more detail,

$$\begin{bmatrix} I_{H^2(\mathcal{D}_{T^*})} & 0 \\ 0 & \omega_{\text{NF,D}} \end{bmatrix} \begin{bmatrix} M_z^{\mathcal{D}_{T^*}} & 0 \\ 0 & W_D \end{bmatrix} = \begin{bmatrix} M_z^{\mathcal{D}_{T^*}} & 0 \\ 0 & M_\zeta^{\mathcal{D}_T}|_{\overline{\Delta_{\Theta_T} L^2(\mathcal{D}_T)}} \end{bmatrix} \begin{bmatrix} I_{H^2(\mathcal{D}_{T^*})} & 0 \\ 0 & \omega_{\text{NF,D}} \end{bmatrix},$$

which in turn is equivalent to the trivial identity $M_z^{\mathcal{D}_{T^*}} = M_z^{\mathcal{D}_{T^*}}$ along with

$$\omega_{\text{NF,D}} W_D = \left( M_\zeta^{\mathcal{D}_T}|_{\overline{\Delta_{\Theta_T} L^2(\mathcal{D}_T)}} \right) \omega_{\text{NF,D}}. \tag{2.67}$$

Owing to the definition (2.65), this last identity is a consequence of the two identities

$$\omega_D^* W_D = (V|_{\mathcal{R}})\omega_D^*, \quad \omega_{\text{NF}}(V|_{\mathcal{R}}) = \left( M_\zeta^{\mathcal{D}_T}|_{\overline{\Delta_{\Theta_T} L^2(\mathcal{D}_T)}} \right) \omega_{\text{NF}},$$

which follow as immediate consequences of (2.43), (2.44), and (2.52). This unitary identification will be used in what follows.

## 2.3 More on Isometric Lifts of a Contraction

In this section we exhibit an interplay between any minimal isometric lift and a minimal isometric lift of a contraction operator. This will be used later in what follows. We start by recording a lemma, the first part of which is well known (see, e.g., [90, page 227]) and the second part is an easy corollary of the first part. We include short proofs of both results for completeness.

**Lemma 2.3.1** (i) *The only bounded linear operator intertwining a unitary operator with a shift operator is the zero operator, that is, if $\mathcal{K}$ and $\mathcal{K}'$ are Hilbert spaces, $U$ is a unitary operator on $\mathcal{K}$, $S$ is a shift operator on $\mathcal{K}'$, and $\Gamma\colon \mathcal{K} \to \mathcal{K}'$ is a bounded linear operator such that $\Gamma U = S\Gamma$, then $\Gamma = 0$.*

(ii) *Suppose that $S$ and $S'$ are shift operators on $\mathcal{K}_1$ and $\mathcal{K}_1'$ respectively, $U$ and $U'$ are unitary operators on $\mathcal{K}_2$ and $\mathcal{K}_2'$ respectively, and $\Gamma = \left[\begin{smallmatrix} \Gamma_{11} & \Gamma_{12} \\ \Gamma_{21} & \omega \end{smallmatrix}\right]$ is a unitary operator from $\mathcal{K}_1 \oplus \mathcal{K}_2$ to $\mathcal{K}_1' \oplus \mathcal{K}_2'$ that intertwines $S \oplus U$ with $S' \oplus U'$:*

$$\begin{bmatrix} \Gamma_{11} & \Gamma_{12} \\ \Gamma_{21} & \omega \end{bmatrix} \begin{bmatrix} S & 0 \\ 0 & U \end{bmatrix} = \begin{bmatrix} S' & 0 \\ 0 & U' \end{bmatrix} \begin{bmatrix} \Gamma_{11} & \Gamma_{12} \\ \Gamma_{21} & \omega \end{bmatrix}. \tag{2.68}$$

*Then $\Gamma$ is block-diagonal, that is, $\Gamma_{12} = 0$ and $\Gamma_{21} = 0$.*

*Proof of (i)* The intertwining condition $\Gamma U = S\Gamma$ implies that $\Gamma U^n = S^n \Gamma$ for all $n = 0, 1, 2, \ldots$. As $U$ is unitary, $\operatorname{Ran} U^n$ is the whole space $\mathcal{K}$ for all $n = 0, 1, 2, \ldots$ and we conclude that $\operatorname{Ran} \Gamma \subset \cap_{n=0}^{\infty} \operatorname{Ran} S^n$. As $S$ is a shift, $\cap_{n=0}^{\infty} \operatorname{Ran} S^n = \{0\}$, and we are forced to conclude that $\Gamma$ is the zero operator.

*Proof of (ii)* From the (1,2)-entry of (2.68) we see that $\Gamma_{12} U = S' \Gamma_{12}$. From part (i) of the lemma we conclude that $\Gamma_{12} = 0$.

As $\Gamma$ is unitary, it follows that (2.68) can be rewritten as

$$\begin{bmatrix} S & 0 \\ 0 & U \end{bmatrix} \begin{bmatrix} \Gamma_{11}^* & \Gamma_{21}^* \\ 0 & \omega^* \end{bmatrix} = \begin{bmatrix} \Gamma_{11}^* & \Gamma_{21}^* \\ 0 & \omega^* \end{bmatrix} \begin{bmatrix} S' & 0 \\ 0 & U' \end{bmatrix}.$$

The (1,2)-entry of this equality gives

$$S\Gamma_{21}^* = \Gamma_{21}^* U'.$$

Again by part (i) of the lemma, we conclude that $\Gamma_{21}^* = 0$, and hence also $\Gamma_{21} = 0$. □

**Remark 2.3.2** Despite some confusion in the literature,[2] it happens that part (i) of Lemma 2.3.1 fails if the hypothesis is changed to: *$U$ is unitary on $\mathcal{K}$, $S$ is a shift on $\mathcal{K}'$ and $X\colon \mathcal{K}' \to \mathcal{K}$ is such that $XS = UX$.* As an example, take $U = M_\zeta$ on $\mathcal{K} = L^2$, $S = M_z$ on $\mathcal{K}' = H^2$, and $X\colon H^2 \to L^2$ equal to the embedding $X\colon f(z) \mapsto f(\zeta)$ of $H^2$ into $L^2$.

[2] The proof of Theorem 4.1 in [80] relies on the incorrect version of Lemma 2.3.1 (i) (the correct result cited there from another paper is misapplied). Nevertheless the result is true but via a different proof.

Just like isometric operators, every Hilbert space contractive operator also enjoys a decomposition as follows. This is often referred to as the *canonical decomposition* for contractions. As is expected, perhaps, this is more general than the Wold decomposition for isometries; the details of this assertion are left as an exercise for interested readers. However, we include a sketch of the proof of the canonical decomposition because not only do we use it throughout the book but we also prove a multivariable analogue of it later (see Theorem 9.2.2).

**Theorem 2.3.3** (See Theorem 3.2 in [92]) *If $T$ is a contraction acting on a Hilbert space $\mathcal{H}$, then there exist subspaces $\mathcal{H}_0$, $\mathcal{H}_1$ of $\mathcal{H}$ such that*

$$\mathcal{H} = \begin{bmatrix} \mathcal{H}_0 \\ \mathcal{H}_1 \end{bmatrix} \quad \text{and} \quad T = \begin{bmatrix} T_u & 0 \\ 0 & T_{\text{cnu}} \end{bmatrix} : \begin{bmatrix} \mathcal{H}_0 \\ \mathcal{H}_1 \end{bmatrix} \to \begin{bmatrix} \mathcal{H}_0 \\ \mathcal{H}_1 \end{bmatrix}, \tag{2.69}$$

*where $T_u$ is a unitary and $T_{\text{cnu}}$ is a completely nonunitary. Moreover, $\mathcal{H}_0$ is given by*

$$\mathcal{H}_0 = \bigcap_{n \in \mathbb{Z}} \operatorname{Ker} D_{T^n} = \{h \in \mathcal{H} \colon \|T^n h\| = \|h\| \quad \text{for all } n \in \mathbb{Z}\}.$$

*Here we use the convention that for a negative integer $n$, $T^n = T^{*-n}$.*

*Proof* We show that $\mathcal{H}_0$ is a reducing subspace where $T$ can be unitary. The fact that $T$ is a unitary on $\mathcal{H}_0$ follows from the fact that for a contraction $T$ and a vector $h$, $\|Th\| = \|h\|$ if and only if $h = T^*Th$. To get $TT^*h = h$, just apply the same fact for the contraction $T^*$.

To see that $\mathcal{H}_0$ is reducing, it is enough to show that it is invariant under $T$ because $\mathcal{H}_0$ is symmetric with respect to $T$ and $T^*$. Suppose $h \in \mathcal{H}$ and denote $Th = h'$. Then, for every $n \geq 1$, we note that

$$\|T^{*n}Th\| = \left\|T^{*(n-1)}T^*Th\right\| = \left\|T^{*(n-1)}h\right\| = \|h\|.$$

Here we again used the fact that $\|Th\| = \|h\|$ implies $T^*Th = h$. The fact that $\|T^nTh\| = \|h\|$ is obvious because $h \in \mathcal{H}_0$.

We finish the proof by showing that $\mathcal{H}_0$ is the largest reducing subspace for $T$ where $T$ is unitary, that is, if $\mathcal{H}_2$ is a reducing subspace for $T$ such that $T|_{\mathcal{H}_2}$ is unitary, then $\mathcal{H}_2$ must be contained in $\mathcal{H}_0$. This is obvious from the way $\mathcal{H}_0$ is defined and the general fact that if $\tau$ is a unitary, then so are $\tau^n$ for any integer $n$. Note that this, in particular, proves that $T|_{\mathcal{H}_1}$ is c.n.u. □

It is natural to wonder if Lemma 2.3.1 extends to the more general setting of the canonical decomposition stated previously, namely whether or not Lemma 2.3.1 remains true if all the instances of "shift" are replaced by "c.n.u." in the statement of Lemma 2.3.1. We address this in Lemma 2.3.4.

**Lemma 2.3.4** (i) *Let $T$ and $U$ be a c.n.u. contraction and a unitary acting on $\mathcal{H}$, respectively. There are nonzero operators $Y_1, Y_2$ such that*

$$Y_1 U = T Y_1 \quad \text{and} \quad U Y_2 = Y_2 T.$$

(ii) *Nevertheless, if*

$$\tau : \begin{bmatrix} \mathcal{H}_0 \\ \mathcal{H}_1 \end{bmatrix} \to \begin{bmatrix} \mathcal{H}'_0 \\ \mathcal{H}'_1 \end{bmatrix}$$

*is a unitary operator such that*

$$\tau \begin{bmatrix} T_u & 0 \\ 0 & T_{\text{cnu}} \end{bmatrix} = \begin{bmatrix} T'_u & 0 \\ 0 & T'_{\text{cnu}} \end{bmatrix} \tau,$$

*where $T_u, T'_u$ are unitary operators and $T_{\text{cnu}}, T'_{\text{cnu}}$ are c.n.u. contractions, then $\tau$ must be of the form*

$$\tau = \begin{bmatrix} \tau_0 & 0 \\ 0 & \tau_1 \end{bmatrix}.$$

*Proof of (i)* Let us take

$$T = \begin{bmatrix} M_z & 0 \\ 0 & M_z^* \end{bmatrix} : \begin{bmatrix} H^2 \\ H^2 \end{bmatrix} \to \begin{bmatrix} H^2 \\ H^2 \end{bmatrix}.$$

To see that $T$ is a c.n.u. contraction, suppose that $\mathcal{M} \neq \{0\}$ is a $T$-reducing proper subspace where $T$ is unitary. Suppose there exists a vector $g \in \mathcal{M}$ so that

$$0 \neq f = P|_{\left[\begin{smallmatrix} 0 \\ H^2 \end{smallmatrix}\right]}(g).$$

On account of $T|_{\mathcal{M}}$ being a unitary, we must have, for every $n \geq 1$,

$$\|f\| = \left\|T^n \left[\begin{smallmatrix} 0 \\ f \end{smallmatrix}\right]\right\| = \|M_z^{*n} f\|.$$

Since $\|M_z^{*n} f\| \to 0$ as $n \to \infty$, the above is not possible unless $f = 0$. Thus

$$P|_{\left[\begin{smallmatrix} 0 \\ H^2 \end{smallmatrix}\right]} \mathcal{M} = \{0\}$$

and, consequently, $\mathcal{M}$ must be a subspace of $\left[\begin{smallmatrix} H^2 \\ 0 \end{smallmatrix}\right]$. This too is not possible, because this would imply that $\mathcal{M}$ is a reducing subspace for $M_z$. A standard result in functional analysis is that the shift $M_z$ acting on $H^2$ has no proper reducing subspace.

To see the existence of a $Y_1$ as stated, take the unitary operator

$$U = \begin{bmatrix} M_\zeta^* & 0 \\ 0 & M_\zeta \end{bmatrix} : \begin{bmatrix} L^2 \\ L^2 \end{bmatrix} \to \begin{bmatrix} L^2 \\ L^2 \end{bmatrix},$$

and with $X\colon H^2 \to L^2$ as the embedding operator $X\colon f(z) \mapsto f(\zeta)$, take

$$Y_1 = \begin{bmatrix} 0 & 0 \\ X^* & 0 \end{bmatrix} : \begin{bmatrix} L^2 \\ L^2 \end{bmatrix} \to \begin{bmatrix} H^2 \\ H^2 \end{bmatrix}.$$

We observed in Remark 2.3.2 that $XM_z = M_\zeta X$, and so $X^* M_\zeta^* = M_z^* X^*$. This and a plain easy matrix computation give $Y_1 U = TY_1$:

$$U = \begin{bmatrix} M_\zeta & 0 \\ 0 & M_\zeta \end{bmatrix} : \begin{bmatrix} L^2 \\ L^2 \end{bmatrix} \to \begin{bmatrix} L^2 \\ L^2 \end{bmatrix}.$$

To see the existence of a nonzero $Y_2$ as stated, take $U$ and $X$ as earlier. Take

$$Y_2 = \begin{bmatrix} 0 & 0 \\ X & 0 \end{bmatrix} : \begin{bmatrix} H^2 \\ H^2 \end{bmatrix} \to \begin{bmatrix} L^2 \\ L^2 \end{bmatrix}.$$

Then it is an easy check that $Y_2 T = UY_2$.
*Proof of (ii)* If we denote

$$T = \begin{bmatrix} T_u & 0 \\ 0 & T_{\text{cnu}} \end{bmatrix} \quad \text{and} \quad T' = \begin{bmatrix} T'_u & 0 \\ 0 & T'_{\text{cnu}} \end{bmatrix},$$

then we conclude from the canonical decomposition of Theorem 2.3.3 that

$$\mathcal{H}_0 = \{h\colon \|T^n h\| = \|h\| \text{ for all } n \in \mathbb{Z}\} \quad \text{and}$$
$$\mathcal{H}'_0 = \{h\colon \|T'^n h\| = \|h\| \text{ for all } n \in \mathbb{Z}\}.$$

We show that both $\tau$ and $\tau^*$ take $\mathcal{H}_0$ into $\mathcal{H}'_0$. This will complete the proof. Since $\tau T = T'\tau$ and $\tau$ is a unitary, we also have $\tau T^* = T'^* \tau$. So pick $h \in \mathcal{H}_0$. We use all the information to compute that, for $n \geq 1$,

$$\|T'^n \tau h\| = \|\tau T^n h\| = \|T^n h\| = \|h\| = \|\tau h\|$$

and

$$\|T'^{*n} \tau h\| = \|\tau T^{*n} h\| = \|T^{*n} h\| = \|h\| = \|\tau h\|.$$

This shows that $\tau$ takes $\mathcal{H}_0$ into $\mathcal{H}'_0$. Similarly, one can show that $\tau^*$ takes $\mathcal{H}_0$ into $\mathcal{H}'_0$. □

Returning to the analysis of isometric lifts, we end this chapter with the following result. This will play a pivotal role in later developments.

**Theorem 2.3.5** *Let the unitary operator $W_D$ and the positive contraction operator $Q_{T^*}$, for a contraction operator $T$, be as discussed in Section 2.2.2. Let $(\mathbf{\Pi}, \mathbf{V})$ acting on $\mathcal{K}$ be any (not necessarily minimal) isometric lift of a* *given contraction operator $T$ acting on $\mathcal{H}$. Let $W$ be the unitary operator arising from the Wold decomposition of* $\mathbf{V}$*, namely (see Theorem 2.1.4)*

$$W = \mathbf{V}|_{\cap_{n\geq 0}\mathbf{V}^n\mathcal{K}} \colon \bigcap_{n\geq 0}\mathbf{V}^n\mathcal{K} \to \bigcap_{n\geq 0}\mathbf{V}^n\mathcal{K}.$$

*Then there are isometric operators $\Lambda_* \colon \mathcal{D}_{T^*} \to \mathcal{D}_{\mathbf{V}^*}$ and $\omega \colon \mathcal{Q}_{T^*} \to \mathcal{Q}_{\mathbf{V}^*}$, such that with the isometry*

$$\widehat{\Gamma} \colon \begin{bmatrix} I_{H^2} \otimes \Lambda_* & 0 \\ 0 & \omega \end{bmatrix} \colon \begin{bmatrix} H^2(\mathcal{D}_{T^*}) \\ \mathcal{Q}_{T^*} \end{bmatrix} \to \begin{bmatrix} H^2(\mathcal{D}_{\mathbf{V}^*}) \\ \mathcal{Q}_{\mathbf{V}^*} \end{bmatrix} \tag{2.70}$$

*we have the intertwining relations*

$$\widehat{\Gamma} \begin{bmatrix} M_z^{\mathcal{D}_{T^*}} & 0 \\ 0 & W_D \end{bmatrix} = \begin{bmatrix} M_z^{\mathcal{D}_{\mathbf{V}^*}} & 0 \\ 0 & W \end{bmatrix} \widehat{\Gamma}, \quad \widehat{\Gamma} \begin{bmatrix} O_{D_{T^*},T^*} \\ Q_{T^*} \end{bmatrix} = \begin{bmatrix} O_{D_{\mathbf{V}^*},\mathbf{V}^*} \\ Q_{\mathbf{V}^*} \end{bmatrix} \mathbf{\Pi} \tag{2.71}$$

*and the range of the isometry $\widehat{\Gamma}$ is given by*

$$\widehat{\Gamma} \begin{bmatrix} H^2(\mathcal{D}_{T^*}) \\ \mathcal{Q}_{T^*} \end{bmatrix} = \bigvee_{n\geq 0} \begin{bmatrix} M_z^{\mathcal{D}_{\mathbf{V}^*}} & 0 \\ 0 & W \end{bmatrix}^n \begin{bmatrix} O_{D_{\mathbf{V}^*},\mathbf{V}^*} \\ Q_{\mathbf{V}^*} \end{bmatrix} \mathbf{\Pi}\mathcal{H}. \tag{2.72}$$

*Moreover,* Ran $\omega$ *is reducing for the unitary operator $W$.*

*Proof* By the Wold decomposition of Theorem 2.1.4, we know that $V$ is unitarily equivalent to

$$\begin{bmatrix} M_z^{\mathcal{D}_{\mathbf{V}^*}} & 0 \\ 0 & W \end{bmatrix} \colon \begin{bmatrix} H^2(\mathcal{D}_{\mathbf{V}^*}) \\ \mathcal{Q}_{\mathbf{V}^*} \end{bmatrix} \to \begin{bmatrix} H^2(\mathcal{D}_{\mathbf{V}^*}) \\ \mathcal{Q}_{\mathbf{V}^*} \end{bmatrix}$$

via the unitary identification map given by

$$\tau_W := \begin{bmatrix} O_{D_{\mathbf{V}^*},\mathbf{V}^*} \\ Q_{\mathbf{V}^*} \end{bmatrix} \colon \mathbf{k} \mapsto \begin{bmatrix} O_{D_{\mathbf{V}^*},\mathbf{V}^*} \\ Q_{\mathbf{V}^*} \end{bmatrix} \mathbf{k}.$$

Define an isometric embedding map $\Pi \colon \mathcal{H} \to \mathcal{K} = \left[\begin{smallmatrix} H^2(\mathcal{D}_{\mathbf{V}^*}) \\ \mathcal{Q}_{\mathbf{V}^*} \end{smallmatrix}\right]$ according to

$$\Pi = \tau_W \mathbf{\Pi} = \begin{bmatrix} O_{D_{\mathbf{V}^*},\mathbf{V}^*} \\ Q_{\mathbf{V}^*} \end{bmatrix} \mathbf{\Pi}. \tag{2.73}$$

Then

$$\left(\Pi, \begin{bmatrix} M_z^{\mathcal{D}_{\mathbf{V}^*}} & 0 \\ 0 & W \end{bmatrix}\right) \quad \text{on} \quad \mathcal{K} = \begin{bmatrix} H^2(\mathcal{D}_{\mathbf{V}^*}) \\ \mathcal{Q}_{\mathbf{V}^*} \end{bmatrix}$$

is also an isometric lift of $T$, which is unitarily equivalent to the abstract lift $(\mathbf{\Pi}, \mathbf{V})$. Note that

$$\left(\Pi, \begin{bmatrix} M_z^{\mathcal{D}_{\mathbf{V}^*}} & 0 \\ 0 & W \end{bmatrix}\Big|_{\mathcal{K}_{00}}\right) \quad \text{on} \quad \mathcal{K}_{00} := \bigvee_{n \geq 0} \begin{bmatrix} M_z^{\mathcal{D}_{\mathbf{V}^*}} & 0 \\ 0 & W \end{bmatrix}^n \operatorname{Ran} \Pi \tag{2.74}$$

is a *minimal* isometric lift for $T$. On the other hand, the Douglas model for the minimal isometric lift of $T$ (see Section 2.2.2) has the form $(\Pi_D, V_D)$, where $\Pi_D$ is the isometric embedding of $\mathcal{H}$ into $\mathcal{K}_D = \left[\begin{smallmatrix} H^2(\mathcal{D}_{T^*}) \\ \mathcal{Q}_{T^*} \end{smallmatrix}\right]$ given by

$$\Pi_D \colon h \mapsto \begin{bmatrix} \mathcal{O}_{D_{T^*},T^*} \\ Q_{T^*} \end{bmatrix} h \quad \text{and} \quad V_D = \begin{bmatrix} M_z^{\mathcal{D}_{T^*}} & 0 \\ 0 & W_D \end{bmatrix} \colon \begin{bmatrix} H^2(\mathcal{D}_{T^*}) \\ \mathcal{Q}_{T^*} \end{bmatrix} \to \begin{bmatrix} H^2(\mathcal{D}_{T^*}) \\ \mathcal{Q}_{T^*} \end{bmatrix}.$$

By the uniqueness statement for such isometric lifts (see item (1) in Theorem 2.1.3), there exists a unique unitary operator $\widehat{\Gamma}_0 \colon \mathcal{K}_D \to \mathcal{K}_{00} \subset \mathcal{K}$ so that

$$\widehat{\Gamma}_0 \begin{bmatrix} M_z^{\mathcal{D}_{T^*}} & 0 \\ 0 & W_D \end{bmatrix} = \left(\begin{bmatrix} M_z^{\mathcal{D}_{\mathbf{V}^*}} & 0 \\ 0 & W \end{bmatrix}\Big|_{\mathcal{K}_{00}}\right) \widehat{\Gamma}_0, \quad \text{and} \tag{2.75}$$

$$\widehat{\Gamma}_0 \begin{bmatrix} \mathcal{O}_{D_{T^*},T^*} \\ Q_{T^*} \end{bmatrix} = \Pi \colon \mathcal{H} \to \mathcal{K}_{00} \subset \mathcal{K}. \tag{2.76}$$

To gain added flexibility, let us view the unitary operator $\widehat{\Gamma}_0$ from $\left[\begin{smallmatrix} H^2(\mathcal{D}_{T^*}) \\ \mathcal{Q}_{T^*} \end{smallmatrix}\right]$ to $\mathcal{K}_{00}$ as actually an isometry, now denoted more simply as $\widehat{\Gamma}$, mapping into $\mathcal{K} = \left[\begin{smallmatrix} H^2(\mathcal{D}_{\mathbf{V}^*}) \\ \mathcal{Q}_{\mathbf{V}^*} \end{smallmatrix}\right]$. Thus, Theorem 2.3.5 will be proved if we can show that $\hat{\Gamma}$ has the form as in (2.70). Let us write out $\widehat{\Gamma}$ as a $2 \times 2$-block operator matrix

$$\widehat{\Gamma} = \begin{bmatrix} \Gamma & \Gamma_{12} \\ \Gamma_{21} & \omega \end{bmatrix} \colon \begin{bmatrix} H^2(\mathcal{D}_{T^*}) \\ \mathcal{Q}_{T^*} \end{bmatrix} \to \begin{bmatrix} H^2(\mathcal{D}_{\mathbf{V}^*}) \\ \mathcal{Q}_{\mathbf{V}^*} \end{bmatrix}. \tag{2.77}$$

Recalling the formula (2.73) for $\Pi$ we may rewrite (2.76) as

$$\widehat{\Gamma} \begin{bmatrix} \mathcal{O}_{D_{T^*},T^*} \\ Q_{T^*} \end{bmatrix} = \begin{bmatrix} \mathcal{O}_{D_{\mathbf{V}^*},\mathbf{V}^*} \\ Q_{\mathbf{V}^*} \end{bmatrix} \mathbf{\Pi}. \tag{2.78}$$

From (2.75) we see that

$$\begin{aligned} \Gamma M_z^{\mathcal{D}_{T^*}} &= M_z^{\mathcal{D}_{\mathbf{V}^*}} \Gamma, \quad \Gamma_{12} W_D = M_z^{\mathcal{D}_{\mathbf{V}^*}} \Gamma_{12}, \\ \Gamma_{21} M_z^{\mathcal{D}_{T^*}} &= W \Gamma_{21}, \quad \omega W_D = W \omega. \end{aligned} \tag{2.79}$$

As $W_D$ is unitary and $M_z^{\mathcal{D}_{\mathbf{V}^*}}$ is a shift, we see from item (i) in Lemma 2.3.1 that the second intertwining condition in (2.79) implies that $\Gamma_{12} = 0$, and hence $\widehat{\Gamma}$ collapses to

$$\widehat{\Gamma} = \begin{bmatrix} \Gamma & 0 \\ \Gamma_{21} & \omega \end{bmatrix}. \tag{2.80}$$

From (2.78) we then pick up the additional relations

$$\Gamma \mathcal{O}_{D_{T^*},T^*} = \mathcal{O}_{D_{\mathbf{V}^*},\mathbf{V}^*}\mathbf{\Pi}, \quad \Gamma_{21}\mathcal{O}_{D_{T^*},T^*} + \omega Q_{T^*} = Q_{\mathbf{V}^*}\mathbf{\Pi}. \tag{2.81}$$

Consider next the following variants of the identity (2.48):

$$\begin{aligned} \mathcal{O}_{D_{T^*},T^*} - M_z^{\mathcal{D}_{T^*}} \mathcal{O}_{D_{T^*},T^*}T^* &= \mathbf{ev}_0^* D_{T^*}, \\ \mathcal{O}_{D_{\mathbf{V}^*},\mathbf{V}^*} - M_z^{\mathcal{D}_{\mathbf{V}^*}} \mathcal{O}_{D_{\mathbf{V}^*},\mathbf{V}^*}\mathbf{V}^* &= \mathbf{ev}_0^* D_{\mathbf{V}^*}. \end{aligned} \tag{2.82}$$

Use the first relation in (2.81) together with the fact that $(\mathbf{\Pi}, \mathbf{V})$ is a lift of $T$ to get

$$\begin{aligned} M_z^{\mathcal{D}_{\mathbf{V}^*}} \left(\Gamma \mathcal{O}_{D_{T^*},T^*}\right) T^* &= M_z^{\mathcal{D}_{\mathbf{V}^*}} \left(\mathcal{O}_{D^*_{\mathbf{V}},\mathbf{V}^*}\mathbf{\Pi}\right) T^* \\ &= M_z^{\mathcal{D}_{\mathbf{V}^*}} \mathcal{O}_{D_{\mathbf{V}^*},\mathbf{V}^*}\mathbf{V}^*\mathbf{\Pi}. \end{aligned} \tag{2.83}$$

Using the second relation in (2.82) together with again (2.81) then enables us to compute

$$\begin{aligned} \Gamma \mathbf{ev}_0^* D_{T^*} &= \Gamma\left(\mathcal{O}_{D_{T^*},T^*} - M_z^{\mathcal{D}_{T^*}} \mathcal{O}_{D_{T^*},T^*}T^*\right) \quad \text{(by (2.82))} \\ &= \left(\mathcal{O}_{D_{\mathbf{V}^*},\mathbf{V}^*} - M_z^{\mathcal{D}_{\mathbf{V}^*}} \mathcal{O}_{D_{\mathbf{V}^*},\mathbf{V}^*}\mathbf{V}^*\right)\mathbf{\Pi} \\ &= \mathbf{ev}_0^* D_{\mathbf{V}^*}\mathbf{\Pi}, \end{aligned} \tag{2.84}$$

that is, the operator $\Gamma\colon H^2(\mathcal{D}_{T^*}) \to H^2(\mathcal{D}_{\mathbf{V}^*})$ maps constant functions in $H^2(\mathcal{D}_{T^*})$ to constant functions in $H^2(\mathcal{D}_{\mathbf{V}^*})$. Since $\Gamma\colon H^2(\mathcal{D}_{T^*}) \to H^2(\mathcal{D}_{\mathbf{V}^*})$ has the shift intertwining property (the first relation in (2.79)), we conclude that there must be an operator $\Lambda_*\colon \mathcal{D}_{T^*} \to \mathcal{D}_{\mathbf{V}^*}$ so that $\Gamma$ is "multiplication by a constant": $\Gamma = I_{H^2} \otimes \Lambda_*$. From the implicit formula (2.84), we see that $\Lambda_*$ must be given by

$$\Lambda_*\colon D_{T^*}h \mapsto D_{\mathbf{V}^*}\mathbf{\Pi}h. \tag{2.85}$$

Let us note next that

$$\begin{aligned} \|D_{T^*}h\|^2 &= \|h\|^2 - \|T^*h\|^2 = \|\mathbf{\Pi}h\|^2 - \|\mathbf{\Pi}T^*h\|^2 = \|\mathbf{\Pi}h\|^2 - \|\mathbf{V}^*\mathbf{\Pi}h\|^2 \\ &= \langle (I - \mathbf{V}\mathbf{V}^*)\mathbf{\Pi}h, \mathbf{\Pi}h\rangle = \|D_{\mathbf{V}^*}\mathbf{\Pi}h\|^2 = \|\Lambda_* D_{T^*}h\|^2, \end{aligned}$$

so $\Lambda_*$ is an isometry. Combining the representation (2.80) with the fact that $\Gamma = I_{H^2} \otimes \Lambda_*$ is an isometry since $\Lambda_*$ is isometric and the fact that $\widehat{\Gamma}$ is also an isometry, we are now finally able to conclude that $\Gamma_{21} = 0$ as well. Thus, we have now reduced $\widehat{\Gamma}$ to the diagonal form

$$\widehat{\Gamma} = \begin{bmatrix} I_{H^2} \otimes \Lambda_* & 0 \\ 0 & \omega \end{bmatrix} : \begin{bmatrix} H^2(\mathcal{D}_{T^*}) \\ Q_{T^*} \end{bmatrix} \to \begin{bmatrix} H^2(\mathcal{D}_{\mathbf{V}^*}) \\ Q_{\mathbf{V}^*} \end{bmatrix}, \tag{2.86}$$

where $\omega$ is an isometry from $Q_{T^*}$ to $Q_{\mathbf{V}^*}$.

Finally, the fourth condition in (2.79), together with its adjoint version $\omega W_D^* = W^*\omega$, implies that $\operatorname{Ran}\omega$ is reducing for $W$. □

In Theorem 2.3.5, we used the Douglas model of isometric lifts of a contractive operator. Had we used the Sz.-Nagy–Foias model instead, the statement of the theorem would be as follows. This version of the theorem will also be used in what follows. We do not prove it here as the proof is very similar to the previous proof.

**Theorem 2.3.6** *Let $(\mathbf{\Pi}, \mathbf{V})$ acting on $\mathcal{K}$ be any (not necessarily minimal) isometric lift of a given contraction operator $T$ acting on $\mathcal{H}$. Set $W$ to be the unitary given by*

$$W = \mathbf{V}|_{\cap_{n\ge0}\mathbf{V}^n\mathcal{K}} \colon \bigcap_{n\ge0} \mathbf{V}^n\mathcal{K} \to \bigcap_{n\ge0} \mathbf{V}^n\mathcal{K}.$$

*Then there are isometric operators $\Lambda_* \colon \mathcal{D}_{T^*} \to \mathcal{D}_{\mathbf{V}^*}$ and $\omega \colon \mathcal{Q}_{T^*} \to \mathcal{Q}_{\mathbf{V}^*}$ such that with*

$$\hat{\Gamma} \colon \begin{bmatrix} I_{H^2}\otimes\Lambda_* & 0 \\ 0 & \omega \end{bmatrix} \colon \begin{bmatrix} H^2(\mathcal{D}_{T^*}) \\ \overline{\Delta_{\Theta_T} L^2(\mathcal{D}_T)} \end{bmatrix} \to \begin{bmatrix} H^2(\mathcal{D}_{\mathbf{V}^*}) \\ \mathcal{Q}_{\mathbf{V}^*} \end{bmatrix}$$

*we have*

$$\begin{aligned} \widehat{\Gamma} \begin{bmatrix} M_z^{\mathcal{D}_{T^*}} & 0 \\ 0 & M_\zeta|_{\overline{\Delta_{\Theta_T} L^2(\mathcal{D}_T)}} \end{bmatrix} &= \begin{bmatrix} M_z^{\mathcal{D}_{\mathbf{V}^*}} & 0 \\ 0 & W \end{bmatrix} \widehat{\Gamma}, \\ \widehat{\Gamma} \begin{bmatrix} \mathcal{O}_{D_{T^*},T^*} \\ \omega_{\mathrm{NF,D}}\mathcal{Q}_{T^*} \end{bmatrix} &= \begin{bmatrix} \mathcal{O}_{D_{\mathbf{V}^*},\mathbf{V}^*} \\ \mathcal{Q}_{\mathbf{V}^*} \end{bmatrix} \mathbf{\Pi} \quad (\omega_{\mathrm{NF,D}} \textit{ as in } (2.65)), \\ \textit{and} \quad \widehat{\Gamma} \begin{bmatrix} H^2(\mathcal{D}_{T^*}) \\ \overline{\Delta_{\Theta_T} L^2(\mathcal{D}_T)} \end{bmatrix} &= \bigvee_{n\ge0} \begin{bmatrix} M_z^{\mathcal{D}_{\mathbf{V}^*}} & 0 \\ 0 & W \end{bmatrix}^n \begin{bmatrix} \mathcal{O}_{D_{\mathbf{V}^*},\mathbf{V}^*} \\ \mathcal{Q}_{\mathbf{V}^*} \end{bmatrix} \mathbf{\Pi}\mathcal{H}. \end{aligned}$$

*Moreover,* $\operatorname{Ran}\omega$ *is reducing for the unitary operator $W$.*

We conclude this chapter with the following result, which will be used later.

**Lemma 2.3.7** *Let $(\Pi, V)$ acting on $\mathcal{K}$ be any isometric lift of $T$ acting on $\mathcal{H}$. Consider the subspace*

$$\mathcal{K}_0 = \bigvee_{n\ge0} V^n\Pi\mathcal{H} \subset \mathcal{K}.$$

*Then $\mathcal{K}_0$ is a reducing subspace for $V$ and $(\Pi, V|_{\mathcal{K}_0})$ is a minimal isometric lift of $T$.*

*Proof* The minimality of the lift $(\Pi, V|_{\mathcal{K}_0})$ will follow from the definition once we prove that $\mathcal{K}_0$ is reducing for $V$. That $\mathcal{K}_0$ is invariant for $V$ is clear from the definition of $\mathcal{K}_0$. Let us note that for $n \geq 1$, and any $h \in \mathcal{H}$,

$$V^*(V^n \Pi h) = V^{n-1} \Pi h \in \mathcal{K}_0$$

and, for $n = 0$,

$$V^* \Pi h = \Pi T^* h \in \mathcal{K}_0.$$

Thus $V^*$ keeps the set of generating elements $\{V^n \Pi h : n \geq 0, h \in \mathcal{H}\}$ of $\mathcal{K}_0$ invariant and therefore $\mathcal{K}_0$ is invariant under $V^*$ as well. □

## 2.4 Notes to Chapter 2

**1. Survey of proofs of the Sz.-Nagy dilation theorem** In addition to the Schäffer model $(\Pi_S, V_S)$ and the Douglas model $(\Pi_D, V_D)$ constructions mentioned previously, let us also mention two earlier less explicit constructions due to Sz.-Nagy (see [92, page 50]) : (i) one based on the Riesz solution of a trigonometric moment problem followed by an application of the Naimark dilation theorem (a positive operator-valued measure can always be dilated to a spectral measure), and (ii) another based on showing that the function

$$n \mapsto \begin{cases} T^n & \text{if } n \geq 0, \\ (T^*)^{-n} & \text{if } n < 0 \end{cases}$$

is a positive-definite operator-valued function on the group $\mathbb{Z}$ of integers whenever $T$ is a Hilbert space contraction operator and then applying another theorem of Naimark guaranteeing that any such positive-definite function is the compression of a unitary representation of the group.

Yet another approach is implicit in the connection of the Sz.-Nagy dilation theorem with Lax–Phillips scattering appearing in the work of Adamjan–Arov [2], which is worked out in detail and extended to the multivariable setting in the work of Ball–Sadosky–Vinnikov [19]. For the single-variable context such as we have here, it suffices to look at the Introduction and part (3) of Theorem 3.2 appearing in [19]. The idea is as follows. Given a c.n.u. contraction operator $T$ on a Hilbert space $\mathcal{H}$, one can embed $T$ into a unitary system matrix

$$\mathbf{U} := \begin{bmatrix} A & B \\ C & D \end{bmatrix} = \begin{bmatrix} T^* & D_T \\ D_{T^*} & -T \end{bmatrix} \Bigg|_{\left[\begin{smallmatrix} \mathcal{H} \\ \mathcal{D}_T \end{smallmatrix}\right]} : \begin{bmatrix} \mathcal{H} \\ \mathcal{D}_T \end{bmatrix} \to \begin{bmatrix} \mathcal{H} \\ \mathcal{D}_{T^*} \end{bmatrix}$$

having transfer function $\Theta_{\mathbf{U}}(z) := D + zC(I_{\mathcal{H}} - zA)^{-1}B$ equal to the Sz.-Nagy–Foias characteristic function $\Theta_T(z)$ (2.50) for $T$. It turns out that the bilateral shift acting on the space of finite-energy system trajectories encodes the minimal unitary dilation $\mathcal{U}$ for $T$, that is, we get an alternative construction (and proof of existence) for a unitary dilation for the contraction operator $T$. One can then use appropriate Fourier representation operators to arrive at the Sz.-Nagy–Foias model directly from this system-theoretic construction.

**2. Epilogue on the de Branges–Rovnyak functional model** We have noted that the Schäffer construction yields only a *semifunctional model* in that only one component of the isometriclift space is a functional Hilbert space. The Douglas model as initially presented is also only a semifunctional model but can be brought to the form of a truly functional model by massaging the second coordinate space with the spectral theory for unitary operators to get an $L^2$-space in the second coordinate (see Remark 2.2.4). The Sz.-Nagy–Foias construction leads to a true functional model since both components of the isometric-lift space are functional Hilbert spaces (although the first component consists of holomorphic functions while the second consists of measurable functions). Let us note here that there is yet another functional model space for c.n.u. contraction operators, namely that of de Branges–Rovnyak (see [51, 52] for the original work and [15, 16, 18, 71, 72, 79] for later treatments) consisting of two (in general) coupled components, each of which is a holomorphic function on the unit disk $\mathbb{D}$, so the study of c.n.u. contraction operators is again reduced to pure holomorphic-function theory rather than a hybrid of holomorphic- and measurable-function theory.

Let us also note that the case where the residual space $\mathcal{R}$ in the minimal dilation space is trivial (discussed in Remark 2.1.5 for the coordinate-free setting) is the case where the Sz.-Nagy–Foias and de Branges–Rovnyak models are equal to each other isometrically. In this case, the model space collapses to have the form $H^2(\mathcal{D}_*) \ominus \Theta \cdot H^2(\mathcal{D})$ for an inner function $(\mathcal{D}, \mathcal{D}_*, \Theta)$.

**3. Functional models for the unitary dilation** Let us mention that this chapter has close overlap with the work of Nikolskii–Vasyunin in [72] but from a different point of view. These authors assume the existence of unitary dilations and then explore how different spectral representations for the unitary dilation can lead to three distinct functional models (Pavlov, de Branges–Rovnyak, Sz.-Nagy–Foias) for the embedded contraction operator $T$ and unitary dilation $\mathcal{U}$.

# 3

# The Berger–Coburn–Lebow and Bercovici–Douglas–Foias Models for Pairs of Commuting Isometries

Berger, Coburn, and Lebow (see [31, Theorem 3.1]) gave a concrete model for a $d$-tuple of commuting isometries (namely, a $d$-tuple of commuting linear-pencil multiplication operators on a vectorial Hardy space) that played a basic role in their investigation of the structure of the $C^*$-algebra generated by commuting isometries and of the Fredholm theory of its elements. Later work of Bercovici–Douglas–Foias [28–30] reviewed the model theory of Berger–Coburn–Lebow (BCL) but also included a particular focus on the pair case ($d = 2$); whereas the starting point in the BCL approach for obtaining a functional model for a commuting isometric pair $(V_1, V_2)$ is to look at the Wold decomposition of the product isometry $V = V_1V_2 = V_2V_1$, the starting point for the Bercovici–Douglas–Foias (BDF) approach is to look at the Wold decomposition for the first isometry $V_1$.

In this chapter, we review both the BCL and BDF approaches and obtain new results on how one can map a BCL-model commuting isometric pair to a unitarily equivalent BDF-model commuting isometric pair and vice versa. We also present a more complete systematic account of the BDF-model, which fleshes out various scattered results of [28, 30]; on the other hand Bercovici, Douglas, and Foias include some probing results concerning the tuple case that go well beyond our results for the pair case. Here we also make contact with some ideas from discrete-time input/state/output linear systems (e.g., realization of a contractive analytic function as the transfer function of a unitary system node) that are important for understanding how to go from a contractive analytic function $(\mathcal{F}, \Theta)$ (an important invariant for the BDF-model theory) to a $2 \times 2$-block unitary operator matrix $\left[\begin{smallmatrix} A & B \\ C & D \end{smallmatrix}\right]$ (an important invariant for the BCL-model theory). However, it is the BCL-model for a commuting isometric pair that will be our main tool for obtaining functional models for Ando lifts $(V_1, V_2)$ of a commuting contractive pair $(T_1, T_2)$.

## 3.1 BCL-Model for Commuting Pairs of Isometries

Recently the authors have given a thorough systematic treatment of the $d = 2$ case [23]. Here we highlight the key features that we shall need to develop our main theme, namely: the structure of Ando lifts of a commuting contractive pair. In Section 3.1.1, we recall the main features of the BCL-model for the pair case that will be relevant for applications to the Ando lifting theory to come. Section 3.1.2 introduces some new invariants (not covered in [23]), namely, *normalized BCL-tuple* (Definition 3.1.10) and *refined BCL-tuple* (Lemma 3.1.14), which are essential in the discussion of the connections between the BCL-model and the BDF-model developed in Section 3.3. Section 3.1.3 reviews some additional material concerning BCL-models appearing in more detail in [23] that will be needed for our approach to the Ando lifting problem in later chapters.

### 3.1.1 Derivation of the BCL-Model

We give here a sketch of the main ideas leading to a proof of the BCL-model for the pair case. Complete details can be found in Section 2 of [23].

As mentioned in the introduction to this chapter, the idea of the BCL-model for $(V_1, V_2)$ is to apply the Wold decomposition (Theorem 2.1.4) to the single product isometry $V = V_1 V_2 = V_2 V_1$; thus the Hilbert space $\mathcal{K}$ has the orthogonal decomposition $\mathcal{K} = \mathcal{K}_s \oplus \mathcal{K}_u$, which splits the operator $V$ as $V = S \oplus W$, where $S$ is a shift operator on $\mathcal{K}_s$ and $W$ is a unitary operator on $\mathcal{K}_u$. Furthermore, by Theorem 2.1.4 we may take $\mathcal{K}_s$ to be $\mathcal{K}_s = H^2(\mathcal{F})$ where $\mathcal{F} = \mathcal{D}_{V^*}$ (the *defect space* for the isometry $V$) and then the shift $S$ can be transformed to have the model form $M_z^{\mathcal{F}}$ acting on the Hardy space $H^2(\mathcal{F})$ with the implementing unitary similarity operator given by the unitary Fourier representation operator (now denoted as $\tau_{\rm BCL}$ rather than $\tau_F$ in this context of the BCL-model) $\tau_{\rm BCL}$ (3.11):

$$\tau_{\rm BCL} = \begin{bmatrix} \sum_{n=0}^{\infty} D_{V^*} V^{*n} z^n \\ \text{SOT-}\lim_{n\to\infty} V^n V^{*n} \end{bmatrix} = \begin{bmatrix} \mathcal{O}_{D_{V^*}, V^*} \\ Q_{V^*}^2 \end{bmatrix} : \mathcal{K} \to \begin{bmatrix} H^2(\mathcal{F}) \\ \mathcal{K}_u \end{bmatrix}.$$

Note that $Q_{V^*}^2$ is just the orthogonal projection of $\mathcal{K}$ onto $\mathcal{K}_u =: Q_{V^*}$.

Then one must show the following.

**Step 1** *$V_1$ and $V_2$ also split with respect to the decomposition $\mathcal{K} = \mathcal{K}_s \oplus \mathcal{K}_u$*

$$V_1 = S_1 \oplus W_1, \quad V_2 = S_2 \oplus W_2,$$

*where $S_1$ and $S_2$ are commuting isometries on $\mathcal{K}_s$ and $W_1$ and $W_2$ are commuting unitaries on $\mathcal{K}_u$* or, in model form, *there are two inner functions $\varphi_1(z)$ and $\varphi_2(z)$ with values in $\mathcal{B}(\mathcal{F}) = \mathcal{B}(\mathcal{D}_{V^*})$ so that*

$$V_1 = M_{\varphi_1} \oplus W_1, \quad V_2 = M_{\varphi_2} \oplus W_2 \tag{3.1}$$

*with* $W_1, W_2$ *commuting unitary operators on* $\mathcal{K}_u$.

To verify (3.1), let us take for $j = 1, 2$,

$$\Gamma^{(j)} = \begin{bmatrix} \Gamma_{11}^{(j)} & \Gamma_{12}^{(j)} \\ \Gamma_{21}^{(j)} & \Gamma_{22}^{(j)} \end{bmatrix}$$

as the $(2 \times 2)$-block matrix representation of the operator $\tau_{\rm BCL} V_j \tau_{\rm BCL}^*$ acting on $\left[\begin{smallmatrix} H^2(\mathcal{F}) \\ \mathcal{K}_u \end{smallmatrix}\right]$. As $V_j V = V V_j$ for $j = 1, 2$, we see that

$$\Gamma^{(j)} \begin{bmatrix} M_z^{\mathcal{F}} & 0 \\ 0 & W \end{bmatrix} = \begin{bmatrix} M_z^{\mathcal{F}} & 0 \\ 0 & W \end{bmatrix} \Gamma^{(j)}$$

for $j = 1, 2$. The fact that $\Gamma_{12}^{(j)} = 0$ as well as $\Gamma_{21}^{(j)} = 0$ can now be seen as a direct consequence of Lemma 2.3.1. The fact that $\Gamma_{11}^{(j)} = M_{\varphi_j}$ for an inner function $\varphi_j$ is a direct consequence of the well-known characterization of the commutant of the shift operator $M_z^{\mathcal{F}}$.

**Step 2** *There exists what we call a* BCL-tuple $(\mathcal{F}, P, U, W_1, W_2)$ (see Definition 3.1.2 below) *so that* $\varphi_1$ *and* $\varphi_2$ *have the form*

$$\varphi_1(z) = U^* P^\perp + z U^* P, \quad \varphi_2(z) = PU + z P^\perp U \quad \text{(BCL1-model)} \tag{3.2}$$

*or, equivalently,*

$$\varphi_1(z) = P^\perp U + z PU, \quad \varphi_2(z) = U^* P + z U^* P^\perp \quad \text{(BCL2-model).} \tag{3.3}$$

To see how to complete this step, note that from the fact that $V_1 = V_2^* V$ we see that $M_{\varphi_1} = M_{\varphi_2}^* M_z^{\mathcal{F}}$; consideration of power-series expansions for $\varphi_1$ and $\varphi_2$ then leads to the conclusion that

$$\varphi_1(z) = E_1 + z E_2^*, \quad \varphi_2(z) = E_2 + z E_1^*$$

for some operators $E_1$ and $E_2$ on $\mathcal{F}$. The fact that $M_{\varphi_1}$ is an isometry then leads to the identity

$$E_1^* E_1 + E_2 E_2^* = I_{\mathcal{F}}, \tag{3.4}$$

while the fact that $M_{\varphi_2}$ is an isometry leads to the parallel identity

$$E_2^* E_2 + E_1 E_1^* = I_{\mathcal{F}}. \tag{3.5}$$

From the fact that $V = V_1 V_2 = V_2 V_1$, we get

$$M_z^{\mathcal{F}} = M_{E_1 + z E_2^*} M_{E_2 + z E_1^*} = M_{E_2 + z E_1^*} M_{E_1 + z E_2^*},$$

which then leads to

$$E_1E_2 = 0 = E_2E_1. \tag{3.6}$$

We may then apply the following general elegant lemma to arrive at (3.2). This lemma first appeared in the unpublished preprint [83, Lemma 14] by the second author, and can also be found in [23, Lemma 2.1] and [32, Lemma 1.14.2].

**Lemma 3.1.1** *Let $\mathcal{F}$ be any Hilbert space and let $E_1, E_2$ be operators on $\mathcal{F}$. Then $E_1, E_2$ are partial isometries of the form*

$$(E_1, E_2) = (U^*P^\perp, PU) \tag{3.7}$$

*for some projection $P$ and unitary $U$ in $\mathcal{B}(\mathcal{F})$ if and only if $E_1$, $E_2$ satisfy relations* (3.4), (3.5), *and* (3.6).

To arrive at (3.3), simply write out the model (3.2) for $(\varphi_1, \varphi_2)$ but with projection/unitary pair $(U^*PU, U^*)$ in place of projection/unitary pair $(P, U)$.

Finally, note that from the decomposition (3.1) and the fact that $V_1$ and $V_2$ are isometries we see that $W_1$ and $W_2$ are isometries. Furthermore, from the fact that $V_1V_2 = V_2V_1 = V$ we see that $W_1W_2 = W_2W_1 = W$, where $W$ is unitary. It now follows that $(W_1, W_2)$ is a commuting unitary pair with product equal to the unitary $W$ on $\mathcal{K}_u = \mathcal{Q}_{V^*}$.

To formulate the BCL-model theorem, the following terminology will be useful.

**Definition 3.1.2** We say that any collection of the form $(\mathcal{F}, P, U, W_1, W_2)$ where

(i) $\mathcal{F}$ is a coefficient Hilbert space,
(ii) $(P, U)$ is a projection/unitary operator pair on $\mathcal{F}$, and
(iii) $(W_1, W_2)$ is a commuting unitary operator pair on some other Hilbert space $\mathcal{K}_u$

is a *BCL-tuple*. Given a BCL-tuple $(\mathcal{F}, P, U, W_1, W_2)$, consider an operator pair $(V_1, V_2)$ of the form

$$(V_1, V_2) = \left( \begin{bmatrix} M_{\varphi_1} & 0 \\ 0 & W_1 \end{bmatrix}, \begin{bmatrix} M_{\varphi_2} & 0 \\ 0 & W_2 \end{bmatrix} \right), \tag{3.8}$$

where $(\varphi_1, \varphi_2)$ is either as in (3.2) or as in (3.3). When $(\varphi_1, \varphi_2)$ is as in (3.2) we say that the pair (3.8) is the BCL1-model associated with the BCL-tuple

$(\mathcal{F}, P, U, W_1, W_2)$. When $(\varphi_1, \varphi_2)$ is as in (3.3), we say that the pair (3.8) is the BCL2-model associated with the same BCL-tuple $(\mathcal{F}, P, U, W_1, W_2)$.

Finally, if it is the case that the BCL1-model associated with the BCL-tuple $(\mathcal{F}, P, U, W_1, W_2)$ is unitarily equivalent to $(V_1, V_2)$, let us say simply that *the BCL-tuple* $\Xi$ *generates a BCL1-model for* $(V_1, V_2)$, and similarly with *BCL2-tuple* in place of the *BCL1-model.*

We shall see in the sequel that the BCL1-model is more convenient for discussing what we call the *Douglas model* for an Ando lift $(V_1, V_2)$ of a given commuting contractive pair $(T_1, T_2)$ while the BCL2-model is more convenient for discussing what we call the *Schäffer model* for an Ando lift $(V_1, V_2)$; hence it is convenient to carry around both models (BCL1 and BCL2) for commuting isometric pairs.

**Remark 3.1.3 Difference in terminology in Definition 3.1.2 versus terminology in [23]** What we call the BCL1-model here is called the BCL-2 model in [23], and vice versa. We decided that this swap of terminologies is necessary here in view of the applications of these models to come in later chapters.

The following theorem gives a precise summary of the preceding observations. Note that the converse direction follows via a direct check.

**Theorem 3.1.4** (See Berger–Coburn–Lebow [31, Section 3], Ball–Sau [23, Theorem 2.4].) *Let* $(V_1, V_2)$ *be a pair of commuting isometries on a Hilbert space* $\mathcal{H}$.

(i) *Then there exists a BCL-tuple* $(\mathcal{F}, P, U, W_1, W_2)$ *generating a BCL1-model*

$$\left(\begin{bmatrix} M_{U^*P^\perp + zU^*P} & 0 \\ 0 & W_1 \end{bmatrix}, \begin{bmatrix} M_{PU+zP^\perp U} & 0 \\ 0 & W_2 \end{bmatrix}\right) \tag{3.9}$$

*for* $(V_1, V_2)$.

(ii) *There exists a BCL-tuple* $(\mathcal{F}, P, U, W_1, W_2)$ *generating a BCL2-model for* $(V_1, V_2)$*:*

$$\left(\begin{bmatrix} M_{P^\perp U+zPU} & 0 \\ 0 & W_1 \end{bmatrix}, \begin{bmatrix} M_{U^*P+zU^*P^\perp} & 0 \\ 0 & W_2 \end{bmatrix}\right). \tag{3.10}$$

*In both cases the implementing unitary similarity* $\tau\colon \mathcal{K} \to \begin{bmatrix} H^2(\mathcal{F}) \\ \mathcal{Q}_{V^*} \end{bmatrix}$ *is given by* $\tau = \tau_{\mathrm{BCL}}$*, where*

$$\tau_{\mathrm{BCL}}\colon h \mapsto \sum_{n=0}^{\infty} z^n D_{V^*} V^{*n} h. \tag{3.11}$$

*Conversely, given any BCL-tuple* $(\mathcal{F}, P, U, W_1, W_2)$*, the associated BCL1-model as well as the associated BCL2-model are commuting isometric pairs.*

### 3.1.2 BCL-Tuple as a Complete Unitary Invariant

We see from Theorem 2.1.4 that any isometry $V$ can be represented as a direct sum $V = S \oplus W$ with $S$ equal to a shift operator (also called a *pure isometry*) and $W$ equal to a unitary operator. The previous discussion suggests the following analogous terminology for the case of the commuting isometric pair.

**Definition 3.1.5** Suppose that $(V_1, V_2)$ is a commuting isometric operator pair on $\mathcal{K}$ with decomposition

$$V_1 = S_1 \oplus W_1, \quad V_2 = S_2 \oplus W_2,$$

induced by the univariate Wold decomposition for the product isometry $V = V_1V_2 = V_2V_1 : V = S \oplus W$ on $\mathcal{K} = \mathcal{K}_s \oplus \mathcal{K}_u$, where $S$ is a shift (i.e., a *pure* isometry) and where $W$ is unitary. Then we shall say that

- $(S_1, S_2) = (V_1|_{\mathcal{K}_s}, V_2|_{\mathcal{K}_s})$ is the *pure part* of the commuting isometric pair $(V_1, V_2)$, and
- $(W_1, W_2) = (V_1|_{\mathcal{K}_u}, V_2|_{\mathcal{K}_u})$ is the *unitary part* of the commuting isometric pair $(V_1, V_2)$.

When the subspace $\mathcal{K}_u$ is absent ($\mathcal{K}_u = \{0\}$), we shall say that $(V_1, V_2)$ is a *pure commuting isometric pair.* When the subspace $\mathcal{K}_s$ is absent ($\mathcal{K}_s = \{0\}$), we shall say that $(V_1, V_2)$ is a *commuting unitary pair.*

While it is the case that $\mathcal{K}_s = \{0\}$ implies that $(V_1, V_2) = (W_1, W_2)$ is a commuting unitary operator pair, it need not be the case (as we shall see in examples to come) that $\mathcal{K}_u = \{0\}$ forces either of $V_1$ or $V_2$ to be a shift operator, despite the fact that $V_1V_2 = V_2V_1 = V$ is a shift operator.

We shall have need in the sequel for the following uniqueness result from [23].

**Theorem 3.1.6** (See [23, Theorem 2.6].) *Suppose that*

$$\Xi = (\mathcal{F}, P, U, W_1, W_2), \quad \Xi' = \left(\mathcal{F}', P', U', W_1', W_2'\right)$$

*are two BCL-tuples and that*

$$(V_1, V_2) \text{ on } \mathcal{K}, \quad \left(V_1', V_2'\right) \text{ on } \mathcal{K}'$$

*are two commuting isometric pairs such that* $\Xi$ *generates a BCL1-model for* $(V_1, V_2)$ *and* $\Xi'$ *generates a BCL1-model for* $\left(V_1', V_2'\right)$*. Then the operator pairs* $(V_1, V_2)$ *and* $\left(V_1', V_2'\right)$ *are jointly unitarily equivalent as operator pairs if and only if the BCL-tuples* $\Xi$ *and* $\Xi'$ *are unitarily equivalent as BCL-tuples in the* *following sense: There exist unitary operators* $\tau\colon \mathcal{F} \to \mathcal{F}'$ *and* $\tau_u\colon \mathcal{K}_u \to \mathcal{K}'_u$ *such that*

$$(\tau P \tau^*, \tau U \tau^*) = (P', U'), \quad (\tau_u W_1 \tau_u^*, \tau_u W_2 \tau_u^*) = (W'_1, W'_2). \tag{3.12}$$

*Moreover, a BCL1-model uniquely determines the associated BCL-tuple in the following sense: If* $(\mathcal{F}, P, U, W_1, W_2)$ *and* $(\mathcal{F}', P', U', W'_1, W'_2)$ *are two BCL-tuples with the same model coefficient spaces* $\mathcal{F} = \mathcal{F}'$ *and the same unitary-part ambient spaces* $\mathcal{K}_u = \mathcal{K}'_u$ *such that the associated model commuting isometric pairs are the same*

$$\left( \begin{bmatrix} M_{U^*P^\perp + zU^*P} & 0 \\ 0 & W_1 \end{bmatrix}, \begin{bmatrix} M_{PU+zP^\perp U} & 0 \\ 0 & W_2 \end{bmatrix} \right) = \left( \begin{bmatrix} M_{U'^*P'^\perp + zU'^*P'} & 0 \\ 0 & W'_1 \end{bmatrix}, \begin{bmatrix} M_{P'U'+zP'^\perp U'} & 0 \\ 0 & W'_2 \end{bmatrix} \right),$$

*then in fact the associated BCL-tuples are identical*

$$(\mathcal{F}, P, U, W_1, W_2) = (\mathcal{F}', P', U', W'_1, W'_2).$$

*Hence unitary equivalence classes of commuting isometric operator pairs* $(V_1, V_2)$ *are in bijective correspondence with unitary equivalence classes of BCL-tuples* $(\mathcal{F}, P, U, W_1, W_2)$ *(in the sense of* (3.12)*).*

*Similar statements hold true with BCL1-model in place of BCL2-model.*

*Proof* We include the proof here for completeness. Since the BCL2-model can be derived from the BCL1-model by the transformation rule

$$\mathfrak{f}\colon (P, U) \to (U^*PU, U^*), \tag{3.13}$$

it suffices to prove the result for BCL1-model.

If there exist unitary operators $\tau\colon \mathcal{F} \to \mathcal{F}'$ and $\tau_u\colon \mathcal{K}_u \to \mathcal{K}'_u$ such that (3.12) holds, then the pairs

$$\left( \begin{bmatrix} M_{U^*P^\perp + zU^*P} & 0 \\ 0 & W_1 \end{bmatrix}, \begin{bmatrix} M_{PU+zP^\perp U} & 0 \\ 0 & W_2 \end{bmatrix} \right) \tag{3.14}$$

and

$$\left( \begin{bmatrix} M_{U'^*P'^\perp + zU'^*P'} & 0 \\ 0 & W'_1 \end{bmatrix}, \begin{bmatrix} M_{P'U'+zP'^\perp U'} & 0 \\ 0 & W'_2 \end{bmatrix} \right) \tag{3.15}$$

are unitarily equivalent via the unitary similarity

$$\begin{bmatrix} I_{H^2} \otimes \tau & 0 \\ 0 & \tau_u \end{bmatrix} : \begin{bmatrix} H^2(\mathcal{F}) \\ \mathcal{K}_u \end{bmatrix} \to \begin{bmatrix} H^2(\mathcal{F}') \\ \mathcal{K}'_u \end{bmatrix}.$$

Then by Theorem 3.1.4 the pairs $(V_1, V_2)$ and $(V'_1, V'_2)$ are unitarily equivalent.

Conversely, suppose that the pairs in (3.14) and (3.15) are unitarily equivalent via the unitary similarity

$$\hat{\tau} = \begin{bmatrix} \tau' & \tau_{12} \\ \tau_{21} & \tau_u \end{bmatrix} : \begin{bmatrix} H^2(\mathcal{F}) \\ \mathcal{K}_u \end{bmatrix} \to \begin{bmatrix} H^2(\mathcal{F}') \\ \mathcal{K}'_u \end{bmatrix}.$$

From the intertwining $\hat{\tau}V_1V_2 = V_1'V_2'\hat{\tau}$, we have

$$\begin{bmatrix} \tau' & \tau_{12} \\ \tau_{21} & \tau_u \end{bmatrix} \begin{bmatrix} M_z^{\mathcal{F}} & 0 \\ 0 & W \end{bmatrix} = \begin{bmatrix} M_z^{\mathcal{F}'} & 0 \\ 0 & W' \end{bmatrix} \begin{bmatrix} \tau' & \tau_{12} \\ \tau_{21} & \tau_u \end{bmatrix}, \tag{3.16}$$

where we write $M_z^{\mathcal{F}}$ for multiplication by $z$ on $H^2(\mathcal{F})$ and $M_z^{\mathcal{F}'}$ for multiplication by $z$ on $H^2(\mathcal{F}')$. By part (ii) of Lemma 2.3.1 we conclude that $\tau_{12} = 0$, $\tau_{21} = 0$ and hence $\hat{\tau}$ collapses to the diagonal form

$$\hat{\tau} = \begin{bmatrix} \tau' & 0 \\ 0 & \tau_u \end{bmatrix}.$$

Therefore the unitary $\tau_u$ intertwines $(W_1, W_2)$ with $\left(W_1', W_2'\right)$ and the unitary $\tau'$ intertwines $M_z^{\mathcal{F}}$ with $M_z^{\mathcal{F}'}$, forcing $\tau$ to have the form $\tau' = I_{H^2} \otimes \tau$ for some unitary $\tau\colon \mathcal{F} \to \mathcal{F}'$. Since $\tau'$ intertwines $M_{U^*(P^\perp + zP)}$ with $M_{U'^*(P'^\perp + zP')}$, we have $\tau U^* P^\perp = U'^* P'^\perp \tau$ and $\tau U^* P = U'^* P' \tau$. Therefore

$$\tau U^* = \tau U^*(P^\perp + P) = U'^*(P'^\perp + P')\tau = U'^*\tau,$$

from which we see that

$$U'^*\tau P = \tau U^* P = U'^* P' \tau \Rightarrow \tau P = P'\tau.$$

In particular, if $\hat{\tau}$ is the identity operator, then $\tau$ and $\tau_u$ are identity operators, implying that $(\mathcal{F}, P, U, W_1, W_2) = \left(\mathcal{F}', P', U', W_1', W_2'\right)$, thereby verifying the last statement. This completes the proof. □

Theorem 3.1.6 suggests that there should be a canonical choice of BCL-tuple generating a BCL1-(or BCL2) model for a given commuting isometric pair. This is indeed the case and is the content of the next result (see also Proposition 7.1 in [30]). Since the unitary part can be handled separately by spectral theory, we assume that the product isometry $V = V_1V_2 = V_2V_1$ is a pure isometry ($V^{*n} \to 0$ strongly as $n \to \infty$).

**Theorem 3.1.7** (See [23, Theorem 2.7].) *Suppose that $(V_1, V_2)$ is a commuting isometric pair on $\mathcal{H}$ such that $V = V_1V_2 = V_2V_1$ is a pure isometry. Then we have the following.*

*(1) The operators*

$$D_{V^*},\ D_{V_1^*},\ D_{V_2^*},\ V_1 D_{V_2^*} V_1^*,\ V_2 D_{V_1^*} V_2,\ D_{V_2^*} V_1^* + V_2 D_{V_1^*},\ D_{V_2^*} V_1^* + V_2 D_{V_1^*}$$

*all have range and cokernel contained in $\mathcal{D}_{V^*} = \operatorname{Ran} D_{V^*}$ and, therefore, when restricted to $\mathcal{D}_{V^*}$, can be viewed as elements of $\mathcal{B}(\mathcal{D}_{V^*})$.*

*(2) The operators $D_{V^*}$, $D_{V_1^*}$, $D_{V_2^*}$, $V_1 D_{V_2^*} V_1^*$, $V_2 D_{V_1^*} V_2^*$ all map $\mathcal{D}_{V^*}$ into itself and thus the restriction of these operators to $\mathcal{D}_{V^*}$ may be considered as elements of $\mathcal{B}(\mathcal{D}_{V^*})$ (bounded linear operators on $\mathcal{D}_{V^*}$). When this is done,*

*all are orthogonal projections on $\mathcal{D}_{V^*}$ (with $D_{V^*}|_{\mathcal{D}_{V^*}} = I_{\mathcal{D}_{V^*}}$) and we have the orthogonal decompositions:*

$$\mathcal{D}_{V^*} = \operatorname{Ran} D_{V_1^*} \oplus \operatorname{Ran} V_1 D_{V_2^*} V_1^*, \quad \mathcal{D}_{V^*} = \operatorname{Ran} V_2 D_{V_1^*} V_2^* \oplus \operatorname{Ran} D_{V_2^*}. \tag{3.17}$$

*(3) When considered as operators on $\mathcal{D}_{V^*}$, the operators*

$$U := \left(V_1 D_{V_2^*} + D_{V_1^*} V_2^*\right)\Big|_{\mathcal{D}_{V^*}}, \quad U_* := \left(D_{V_2^*} V_1^* + V_2 D_{V_1^*}\right)\Big|_{\mathcal{D}_{V^*}}$$

*are unitary.*

*(4) A BCL-tuple generating a BCL1-model for $(V_1, V_2)$ expressed directly in terms of $(V_1, V_2)$ is given by*

$$\left(\mathcal{F}^{(1)}, P^{(1)}, U^{(1)}\right) = \left(\mathcal{D}_{V^*}, V_2 D_{V_1^*} V_2^*\Big|_{\mathcal{D}_{V^*}}, \left(D_{V_2^*} V_1^* + V_2 D_{V_1^*}\right)\Big|_{\mathcal{D}_{V^*}}\right), \tag{3.18}$$

*while a BCL-tuple giving rise to the BCL2-model for $(V_1, V_2)$ is similarly given by*

$$\left(\mathcal{F}^{(2)}, P^{(2)}, U^{(2)}\right) = \left(\mathcal{D}_{V^*}, D_{V_1^*}\Big|_{\mathcal{D}_{V^*}}, \left(V_1 D_{V_2^*} + D_{V_1^*} V_2^*\right)\Big|_{\mathcal{D}_{V^*}}\right). \tag{3.19}$$

*Proof* From the fact that $V_1$ and $V_2$ are isometries, it is easily checked that each of the operators $X$ in statement (2) is a projection (i.e., $X = X^*$ and $X^2 = X$). The two orthogonal decompositions in statement (2) then follow from the general identities

$$I - VV^* = (I - V_1 V_1^*) + V_1 (I - V_2 V_2^*) V_1^* = I - V_2 V_2^* + V_2 (I - V_1 V_1^*) V_2^*. \tag{3.20}$$

These identities also show that all the projection operators have range contained in $\mathcal{D}_{V^*}$.

As for $U^{(1)} = D_{V_2^*} V_1^* + V_2 D_{V_1^*}$, noting that $D_{V_2^*} V_2 = 0$ since $V_2$ is an isometry, we can compute

$$\begin{aligned}
&(U^{(1)})^* U^{(1)} \\
&\quad = \left(V_1 D_{V_2^*} + D_{V_1^*} V_2^*\right)\left(D_{V_2^*} V_1^* + V_2 D_{V_1^*}\right) \\
&\quad = V_1 (I - V_2 V_2^*) V_1^* + (I - V_1 V_1^*) = V_1 V_1^* - VV^* + I - V_1 V_1^* = D_{V^*}
\end{aligned}$$

and

$$U^{(1)} \left(U^{(1)}\right)^* = \left(D_{V_2^*} V_1^* + V_2 D_{V_1^*}\right)\left(V_1 D_{V_2^*} + D_{V_1^*} V_2^*\right) = D_{V_2^*} + V_2 D_{V_1^*} V_2^* = D_{V^*}.$$

These identities not only verify that $U^{(1)}$ has range and cokernel inside $\mathcal{D}_{V^*}$, but furthermore that $U^{(1)}$ is unitary when considered as an operator on $\mathcal{D}_{V^*}$. Similar computations verify the corresponding properties for $U^{(2)} = D_{V_2^*} V_1^* + V_2 D_{V_1^*}$. This completes the verification of statements (1), (2), and (3).

Let us now assume that $V_1, V_2$ are presented in the form of a BCL1-model

$$V_1 = M_{U^{(1)*}(P^{(1)\perp}+zP^{(1)})}, \quad V_2 = M_{(P^{(1)}+zP^{(1)\perp})U^{(1)}}$$

acting on the Hardy space $H^2(\mathcal{F})$. Our next goal is to understand how to recover the operator pair $(P^{(1)}, U^{(1)})$ directly in terms of the operator pair $\mathbf{V} = (V_1, V_2)$. Note first that then

$$\begin{aligned} V_1^* - V_2V^* &= \big(I_{H^2} \otimes U^{(1)*}P^{(1)\perp} + M_z \otimes U^{(1)*}P^{(1)}\big)^* \\ &\quad - \big(I_{H^2} \otimes P^{(1)}U^{(1)} + M_z \otimes P^{(1)\perp}U^{(1)}\big)\big(M_z \otimes I_{\mathcal{D}_{V^*}}\big)^* \\ &= \big(I_{H^2} \otimes P^{(1)\perp}U^{(1)} + M_z^* \otimes P^{(1)}U^{(1)}\big) \\ &\quad - \big(M_z^* \otimes P^{(1)}U^{(1)} + M_zM_z^* \otimes P^{(1)\perp}U^{(1)}\big) \\ &= \big((I_{H^2} - M_zM_z^*) \otimes P^{(1)\perp}U^{(1)}\big) = \mathbf{ev}_{0,\mathcal{F}}^* P^{(1)\perp}U^{(1)}. \end{aligned}$$

If we identify the coefficient space $\mathcal{F}$ with the $V^*$-defect space $\mathcal{D}_{V^*} = \operatorname{Ran} I - VV^*$, then it is convenient to view the operators $P^{(1)}$ and $U^{(1)}$ as operators on $\mathcal{D}_{V^*}$ rather than just on $\mathcal{F}$, and then we have for $n = 1, 2, 3, \ldots$,

$$D_{V^*}V^{*n}(V_1^* - V_2V^*) = 0 \quad \text{and} \quad D_{V^*}(V_1^* - V_2V^*) = P^{(1)\perp}U^{(1)}D_{V^*}.$$

It follows that $V_1^* - V_2V^*$ has range in $\mathcal{D}_{V^*}$, and hence we actually have

$$V_1^* - V_2V^* = P^{(1)\perp}U^{(1)}D_{V^*} = D_{V^*}P^{(1)\perp}U^{(1)}D_{V^*}. \tag{3.21}$$

A similar computation gives us

$$V_2^* - V_1V^* = D_{V^*}U^{(1)*}P^{(1)}D_{V^*}. \tag{3.22}$$

Next note that

$$\begin{aligned} D_{V^*}U^{(1)*}D_{V^*} &= D_{V^*}U^{(1)*}(P^{(1)\perp} + P^{(1)})D_{V^*} = (V_1^* - V_2V^*)^* + (V_2^* - V_1V^*) \\ &= V_1 - V_1V_2V_2^* + V_2 - V_1V_1^*V_2^* = V_1D_{V_2^*} + D_{V_1^*}V_2^*. \end{aligned}$$

As we have already checked that the operator $V_1D_{V_2^*} + D_{V_1^*}V_2^*$ has range and cokernel contained in $\mathcal{D}_{V^*}$, we may cancel the projection $D_{V^*}$ on the left and the right of the left-hand side and deduce that necessarily $U^{(1)*} = (V_1D_{V_2^*} + D_{V_1^*}V_2^*)|_{\mathcal{D}_{V^*}}$. Hence $U^{(1)} = (D_{V_2^*}V_1^* + V_2D_{V_1^*})|_{\mathcal{D}_{V^*}}$ as in (3.18).

To find $P^{(1)}$ we see from (3.22) that

$$\begin{aligned} P^{(1)} &= U^{(1)}\,(V_2^* - V_1V^*) = \left(D_{V_2^*}V_1^* + V_2D_{V_1^*}\right)(V_2^* - V_1V^*) \\ &= \left(D_{V_2^*}V_1^* + V_2D_{V_1^*}\right)(V_2^* - V_1V^*) \\ &= D_{V_2^*}V^* - D_{V_2^*}V^* + V_2D_{V_1^*}V_2^* - V_2D_{V_1^*}V_1V^* \\ &= V_2D_{V_1^*}V_2^* \quad (\text{since } D_{V_1^*}V_1 = 0) \end{aligned}$$

as in (3.18).

Alternatively, one can arrive at the same conclusion by a different route as explained in the following. Since $V = V_1V_2$ is assumed to be a shift operator, the unitary operator $\tau_{\rm BCL}$ as in (3.24) simplifies to the formula

$$\tau_{\rm BCL}h = D_{V^*}(I_{\mathcal{H}} - zV^*)^{-1}h.$$

By a direct computation, one can then establish the following intertwinings:

$$\tau_{\rm BCL}V_1 = M_{U^{(1)*}P^{(1)\perp}+zU^{(1)*}P^{(1)}}\tau_{\rm BCL}, \quad \tau_{\rm BCL}V_2 = M_{P^{(1)}U^{(1)}+zP^{(1)\perp}U^{(1)}}\tau_{\rm BCL},$$

where $(P^{(1)}, U^{(1)})$ are as in (3.18). The readers are referred to [23] for the computations establishing the aforementioned intertwinings.

Parallel analysis can be used to prove the assertion with the BCL2-model. □

To facilitate computation, it is convenient to replace the space $\mathcal{D}_{V^*}$ known to have the internal direct-sum decompositions (3.17) with the external direct sum $\begin{bmatrix} \mathcal{D}_{V_1^*} \\ \mathcal{D}_{V_2^*} \end{bmatrix}$ via either of the identification maps

$$\Phi = \begin{bmatrix} D_{V_1^*} & V_1D_{V_2^*} \end{bmatrix} : \begin{bmatrix} \mathcal{D}_{V_1^*} \\ \mathcal{D}_{V_2^*} \end{bmatrix} \to \mathcal{D}_{V^*}, \quad \Phi_* = \begin{bmatrix} V_2D_{V_1^*} & D_{V_2^*} \end{bmatrix} : \begin{bmatrix} \mathcal{D}_{V_1^*} \\ \mathcal{D}_{V_2^*} \end{bmatrix} \to \mathcal{D}_{V^*}.$$

The same idea appears in [30, Corollary 7.2]. We then have the following result. For the rest of this chapter we shall focus on BCL1-models generated by a given BCL-tuple; of course, parallel results will hold with BCL2-models in place of BCL1-models, but we shall not have need at this point to work out the parallel results explicitly.

**Corollary 3.1.8** (See Corollary 2.8 in [23].) *Given a commuting isometric pair* $\mathbf{V} := (V_1, V_2)$ *acting on* $\mathcal{H}$ *such that* $V := V_1V_2$ *is a shift, an alternative expression for the canonical BCL-tuple generating a BCL1-model for* $(V_1, V_2)$ *is*

$$\begin{aligned} \Xi_{\mathbf{V}} &= \left( \begin{bmatrix} \mathcal{F}_{\mathbf{V}} \\ \mathcal{E}_{\mathbf{V}} \end{bmatrix}, \begin{bmatrix} I_{\mathcal{F}_{\mathbf{V}}} & 0 \\ 0 & 0 \end{bmatrix}, \begin{bmatrix} A_{\mathbf{V}} & B_{\mathbf{V}} \\ C_{\mathbf{V}} & D_{\mathbf{V}} \end{bmatrix} \right) \\ &:= \left( \begin{bmatrix} \mathcal{D}_{V_1^*} \\ \mathcal{D}_{V_2^*} \end{bmatrix}, \begin{bmatrix} I_{\mathcal{D}_{V_1^*}} & 0 \\ 0 & 0 \end{bmatrix}, \begin{bmatrix} D_{V_1^*}V_2|_{\mathcal{D}_{V_1^*}} & D_{V_1^*}|_{\mathcal{D}_{V_2^*}} \\ V_1^*V_2|_{\mathcal{D}_{V_1^*}} & V_1^*|_{\mathcal{D}_{V_2^*}} \end{bmatrix} \right). \end{aligned} \tag{3.23}$$

*Moreover, the unitary operator*

$$\tau'_{\rm BCL} \colon \mathcal{H} \to H^2\left( \begin{bmatrix} \mathcal{D}_{V_1^*} \\ \mathcal{D}_{V_2^*} \end{bmatrix} \right)$$

*implementing the unitary identification between* $(V_1, V_2)$ *and the BCL1-model generated by the BCL-tuple* (3.23) *is given by*

$$h \mapsto \sum_{n=0}^{\infty} z^n \begin{bmatrix} D_{V_1^*} V_2^* \\ D_{V_2^*} \end{bmatrix} V^{*n} h \tag{3.24}$$

*for every* $h \in \mathcal{H}$.

*Proof* We have already noted in item (4) of Theorem 3.1.7 that

$$\left(\mathcal{F}^{(1)}, P^{(1)}, U^{(1)}\right) = \left(\mathcal{D}_{V^*}, V_2 D_{V_1^*} V_2^*, \left(D_{V_2^*} V_1^* + V_2 D_{V_1^*}\right)\Big|_{\mathcal{D}_{V^*}}\right)$$

generates a BCL1-model for $(V_1, V_2)$ with the underlining unitary identification operator being

$$\tau_{\mathrm{BCL}} : h \mapsto D_{V^*}(I_{\mathcal{H}} - zV^*)^{-1} h = \sum_{n=0}^{\infty} z^n D_{V^*} V^{*n} h. \tag{3.25}$$

One can see as a consequence of the identities (3.20) that $\Phi^{(1)}$ acting from $\begin{bmatrix} \mathcal{D}_{V_1^*} \\ \mathcal{D}_{V_2^*} \end{bmatrix}$ to $\mathcal{D}_{V^*}$, given by

$$\Phi^{(1)} = \begin{bmatrix} V_2 D_{V_1^*} & D_{V_2^*} \end{bmatrix} : \begin{bmatrix} \mathcal{D}_{V_1^*} \\ \mathcal{D}_{V_2^*} \end{bmatrix} \to \mathcal{D}_{V^*},$$

is unitary. It is then a matter of again repeatedly using the identities (3.20) to see that

$$\begin{aligned}
\Phi^{(1)*} P^{(1)} \Phi^{(1)} &= \begin{bmatrix} D_{V_1^*} V_2^* \\ D_{V_2^*} \end{bmatrix} V_2 D_{V_1^*} V_2^* \begin{bmatrix} V_2 D_{V_1^*} & D_{V_2^*} \end{bmatrix} \\
&= \begin{bmatrix} I_{\mathcal{D}_{V_1^*}} & 0 \\ 0 & 0 \end{bmatrix} = \begin{bmatrix} I_{\mathcal{F}_V} & 0 \\ 0 & 0 \end{bmatrix}, \\
\Phi^{(1)*} U^{(1)} \Phi^{(1)} &= \begin{bmatrix} D_{V_1^*} V_2^* \\ D_{V_2^*} \end{bmatrix} \left(D_{V_2^*} V_1^* + V_2 D_{V_1^*}\right) \begin{bmatrix} V_2 D_{V_1^*} & D_{V_2^*} \end{bmatrix} \\
&= \begin{bmatrix} D_{V_1^*} V_2 D_{V_1^*} & D_{V_1^*} D_{V_2^*} \\ D_{V_2^*} V_1^* V_2 D_{V_1^*} & D_{V_2^*} V_1^* D_{V_2^*} \end{bmatrix}.
\end{aligned}$$

Let us note on the side that

$$\begin{aligned}
D_{V_2^*} V_1^* V_2 D_{V_1^*} &= (I - V_2 V_2^*) V_1^* V_2 D_{V_1^*} = V_1^* V_2 D_{V_1^*} - V_2 V_1^* V_2^* V_2 D_{V_1^*} \\
&= V_1^* V_2 D_{V_1^*} - V_2 V_1^* D_{V_1^*} = V_1^* V_2 D_{V_1^*}.
\end{aligned}$$

Hence the formula for $\Phi^{(1)*}U^{(1)*}\Phi^{(1)}$ can be completed to

$$\Phi^{(1)*}U^{(1)}\Phi^{(1)} = \begin{bmatrix} D_{V_1^*}V_2D_{V_1^*} & D_{V_1^*}D_{V_2^*} \\ V_1^*V_2D_{V_1^*} & D_{V_2^*}V_1^*D_{V_2^*} \end{bmatrix} = \begin{bmatrix} D_{V_1^*}V_2|_{\mathcal{D}_{V_1^*}} & D_{V_1^*}|_{\mathcal{D}_{V_2^*}} \\ V_1^*V_2|_{\mathcal{D}_{V_1^*}} & V_1^*|_{\mathcal{D}_{V_2^*}} \end{bmatrix}$$
$$=: \begin{bmatrix} A_{\mathbf{V}} & B_{\mathbf{V}} \\ C_{\mathbf{V}} & D_{\mathbf{V}} \end{bmatrix}.$$

We conclude that the BCL-tuple $\Xi_{\mathbf{V}}$ is unitarily equivalent to the BCL-tuple $(\mathcal{F}^{(1]}, P^{(1)}, U^{(1)})$, which in turn is known to generate a BCL1-model for $\mathbf{V} := (V_1, V_2)$. As a consequence of Theorem 3.1.6, it then follows that the BCL-tuple $\Xi_{\mathbf{V}}$ also generates a BCL1-model for $\mathbf{V} := (V_1, V_2)$.

Moreover, let us note that the unitary operator that identifies $(V_1, V_2)$ with the BCL1-model corresponding to the BCL-tuple $\Xi_{\mathbf{V}}$ is given by

$$\tau'_{\mathrm{BCL}}h = \Phi^{(1)^*}\tau_{\mathrm{BCL}}h = \Phi^{(1)^*}\sum_{n=0}^{\infty} z^n D_{V^*}V^{*n}h = \sum_{n=0}^{\infty} z^n \begin{bmatrix} D_{V_1^*}V_2^* \\ D_{V_2^*} \end{bmatrix} V^{*n}h.$$

This completes the proof. □

**Remark 3.1.9** Given a pure commuting isometric pair $(V_1, V_2)$ (so the product isometry $V_1V_2$ is a shift), Corollary 3.1.8 provides an explicit formula directly in terms of $(V_1, V_2)$ for a BCL-tuple $\Xi$ generating a BCL1-model for $(V_1, V_2)$. It is of interest to understand how such a BCL-tuple relates to a BCL-tuple generating a BCL1-model for $(V_2, V_1)$ rather than for $(V_1, V_2)$. This amounts to interchanging the indices $(1, 2) \to (2, 1)$ in the formula for $\Xi_{\mathbf{V}}$ (3.23) to arrive at

$$\Xi'_{\mathbf{V}} := \left( \begin{bmatrix} \mathcal{D}_{V_2^*} \\ \mathcal{D}_{V_1^*} \end{bmatrix}, \begin{bmatrix} I_{\mathcal{D}_{V_2^*}} & 0 \\ 0 & 0 \end{bmatrix}, \begin{bmatrix} D_{V_2^*}V_1|_{\mathcal{D}_{V_2^*}} & D_{V_2^*}|_{\mathcal{D}_{V_1^*}} \\ D_{V_1^*}V_2^*V_1|_{\mathcal{D}_{V_2^*}} & D_{V_1^*}V_2^*|_{\mathcal{D}_{V_1^*}} \end{bmatrix} \right)$$
$$= \left( \begin{bmatrix} \mathcal{E}_{\mathbf{V}} \\ \mathcal{F}_{\mathbf{V}} \end{bmatrix}, \begin{bmatrix} I_{\mathcal{F}_{\mathbf{V}}} & 0 \\ 0 & 0 \end{bmatrix}, \begin{bmatrix} D_{\mathbf{V}}^* & B_{\mathbf{V}}^* \\ C_{\mathbf{V}}^* & A_{\mathbf{V}}^* \end{bmatrix} \right), \tag{3.26}$$

where we use occasional unnecessary projections in the last formula for additional clarity. The conclusion is: *The BCL-tuple* $\Xi_{\mathbf{V}}$ (3.23) *generates a BCL1-model for* $(V_1, V_2)$ *while the BCL-tuple* $\Xi'_{\mathbf{V}}$ (3.26) *generates a BCL1-model for* $(V_2, V_1)$.

Since it will be convenient to work with canonical BCL-tuples having the block $2 \times 2$-matrix form in (3.23), it will be convenient to work with BCL-tuples with entries also having such a $2 \times 2$-matrix decomposition. This leads us to the next definition.

**Definition 3.1.10** We say that a BCL-tuple $(\mathcal{F}, P, U)$ is *normalized* if it has the form

$$(\mathcal{F}, P, U) = \left( \begin{bmatrix} \mathcal{F} \\ \mathcal{E} \end{bmatrix}, \begin{bmatrix} I_{\mathcal{F}} & 0 \\ 0 & 0 \end{bmatrix}, \begin{bmatrix} A & B \\ C & D \end{bmatrix} \right). \tag{3.27}$$

Note that in particular the canonical BCL-tuple (3.23) is a normalized BCL-tuple.

Then an important point concerning normalized BCL-tuples is as follows.

**Lemma 3.1.11** (i) *Any BCL-tuple $(\mathcal{F}, P, U)$ is unitarily equivalent (in the sense of BCL-tuples) to a normalized BCL-tuple as in* (3.27).

(ii) *Let* $\Xi = \left( \left[ \begin{smallmatrix} \mathcal{F} \\ \mathcal{E} \end{smallmatrix} \right], \left[ \begin{smallmatrix} I_{\mathcal{F}} & 0 \\ 0 & 0 \end{smallmatrix} \right], \left[ \begin{smallmatrix} A & B \\ C & D \end{smallmatrix} \right] \right)$ *and* $\Xi' = \left( \left[ \begin{smallmatrix} \mathcal{F}' \\ \mathcal{E}' \end{smallmatrix} \right], \left[ \begin{smallmatrix} I_{\mathcal{F}'} & 0 \\ 0 & 0 \end{smallmatrix} \right], \left[ \begin{smallmatrix} A' & B' \\ C' & D' \end{smallmatrix} \right] \right)$ *be two BCL-tuples in normalized form; then $\Xi$ and $\Xi'$ are unitarily equivalent as BCL-tuples if and only if there are unitary transforms $\tau_{\mathcal{F}} \colon \mathcal{F} \to \mathcal{F}'$ and $\tau_{\mathcal{E}} \colon \mathcal{E} \to \mathcal{E}'$, so that*

$$\begin{bmatrix} \tau_{\mathcal{F}} & 0 \\ 0 & \tau_{\mathcal{E}} \end{bmatrix} \begin{bmatrix} A & B \\ C & D \end{bmatrix} = \begin{bmatrix} A' & B' \\ C' & D' \end{bmatrix} \begin{bmatrix} \tau_{\mathcal{F}} & 0 \\ 0 & \tau_{\mathcal{E}} \end{bmatrix}. \tag{3.28}$$

*Proof of (i)* Let $(\mathcal{F}, P, U)$ be a BCL-tuple. Set $\mathcal{F} = \operatorname{Ran} P$ and $\mathcal{E} = \operatorname{Ker} P$. Then it is an elementary fact that there exists a unitary operator $\alpha \colon \left[ \begin{smallmatrix} \mathcal{F} \\ \mathcal{E} \end{smallmatrix} \right] \to \mathcal{F}$ so that $\alpha \left[ \begin{smallmatrix} I_{\mathcal{F}} & 0 \\ 0 & 0 \end{smallmatrix} \right] \alpha^* = P$. If we set $\left[ \begin{smallmatrix} A & B \\ C & D \end{smallmatrix} \right] = \alpha^* U \alpha$, then the normalized BCL-tuple $\left( \left[ \begin{smallmatrix} \mathcal{F} \\ \mathcal{E} \end{smallmatrix} \right], \left[ \begin{smallmatrix} I_{\mathcal{F}} & 0 \\ 0 & 0 \end{smallmatrix} \right], \left[ \begin{smallmatrix} A & B \\ C & D \end{smallmatrix} \right] \right)$ is unitarily equivalent to the original BCL-tuple $(\mathcal{F}, P, U)$.

*Proof of (ii)* Suppose that $\Xi$ and $\Xi'$ as in the statement of item (ii) are unitarily equivalent as BCL-tuples. Then there is a unitary

$$\tau = \begin{bmatrix} \tau_{11} & \tau_{12} \\ \tau_{21} & \tau_{22} \end{bmatrix} : \begin{bmatrix} \mathcal{F} \\ \mathcal{E} \end{bmatrix} \to \begin{bmatrix} \mathcal{F}' \\ \mathcal{E}' \end{bmatrix}$$

such that

$$\begin{bmatrix} \tau_{11} & \tau_{12} \\ \tau_{21} & \tau_{22} \end{bmatrix} \begin{bmatrix} I_{\mathcal{F}} & 0 \\ 0 & 0 \end{bmatrix} = \begin{bmatrix} I_{\mathcal{F}'} & 0 \\ 0 & 0 \end{bmatrix} \begin{bmatrix} \tau_{11} & \tau_{12} \\ \tau_{21} & \tau_{22} \end{bmatrix}, \tag{3.29}$$

$$\begin{bmatrix} \tau_{11} & \tau_{12} \\ \tau_{21} & \tau_{22} \end{bmatrix} \begin{bmatrix} A & B \\ C & D \end{bmatrix} = \begin{bmatrix} A' & B' \\ C' & D' \end{bmatrix} \begin{bmatrix} \tau_{11} & \tau_{12} \\ \tau_{21} & \tau_{22} \end{bmatrix}. \tag{3.30}$$

From the off-diagonal entries in (3.29) we see that $\tau_{12} = 0$ and $\tau_{21} = 0$, so $\tau$ has the diagonal form $\tau = \left[ \begin{smallmatrix} \tau_{\mathcal{F}} & 0 \\ 0 & \tau_{\mathcal{E}} \end{smallmatrix} \right] : \left[ \begin{smallmatrix} \mathcal{F} \\ \mathcal{E} \end{smallmatrix} \right] \to \left[ \begin{smallmatrix} \mathcal{F}' \\ \mathcal{E}' \end{smallmatrix} \right]$. As $\tau$ is unitary, it follows that $\tau_{\mathcal{F}} \colon \mathcal{F} \to \mathcal{F}'$ and $\tau_{\mathcal{E}} \colon \mathcal{E} \to \widetilde{\mathcal{E}}$ are unitary. Plugging this information in the second equation (3.30) then leaves us with (3.28) as desired.

Conversely, if (3.28) is satisfied, then the equation

$$\begin{bmatrix} \tau_{\mathcal{F}} & 0 \\ 0 & \tau_{\mathcal{E}} \end{bmatrix} \begin{bmatrix} I_{\mathcal{F}} & 0 \\ 0 & 0 \end{bmatrix} = \begin{bmatrix} I_{\mathcal{F}'} & 0 \\ 0 & 0 \end{bmatrix} \begin{bmatrix} \tau_{\mathcal{F}} & 0 \\ 0 & \tau_{\mathcal{E}} \end{bmatrix}$$

reduces to $\tau_{\mathcal{F}} = \tau_{\mathcal{F}}$ and hence is automatic, and it follows that $\Xi$ and $\Xi'$ are unitarily equivalent as BCL-tuples. □

**Example 3.1.12** We can illustrate the formula (3.18) for the case where the BCL-tuple is assumed to be in normalized form (3.27) as follows. For a normalized BCL-tuple $\Xi$ as in (3.27), the BCL1-model commuting isometric pair generated by $\Xi$ is

$$(V_1, V_2) = \left( \begin{bmatrix} M_z \otimes A^* & I_{H^2} \otimes C^* \\ M_z \otimes B^* & I_{H^2} \otimes D^* \end{bmatrix}, \begin{bmatrix} I_{H^2} \otimes A & I_{H^2} \otimes B \\ M_z \otimes C & M_z \otimes D \end{bmatrix} \right) \quad \text{on } H^2\left(\left[\begin{smallmatrix} \mathcal{F} \\ \mathcal{E} \end{smallmatrix}\right]\right). \tag{3.31}$$

Let us identify the coefficient space $\left[\begin{smallmatrix} \mathcal{F} \\ \mathcal{E} \end{smallmatrix}\right]$ with the space of constants Ran $\mathbf{ev}^*_{0,\left[\begin{smallmatrix} \mathcal{F} \\ \mathcal{E} \end{smallmatrix}\right]}$ in $H^2\left(\left[\begin{smallmatrix} \mathcal{F} \\ \mathcal{E} \end{smallmatrix}\right]\right)$. Then straightforward manipulations using formulas (3.31) show that

$$V_1^* = \begin{bmatrix} M_z^* \otimes A & M_z^* \otimes B \\ I_{H^2} \otimes C & I_{H^2} \otimes D \end{bmatrix}, \quad V_2^* = \begin{bmatrix} I_{H^2} \otimes A^* & M_z^* \otimes C^* \\ I_{H^2} \otimes B^* & M_z^* \otimes D^* \end{bmatrix},$$

$$V_1^*|_{\left[\begin{smallmatrix} \mathcal{F} \\ \mathcal{E} \end{smallmatrix}\right]} = \begin{bmatrix} 0 & 0 \\ C & D \end{bmatrix}, \quad V_2^*|_{\left[\begin{smallmatrix} \mathcal{F} \\ \mathcal{E} \end{smallmatrix}\right]} = \begin{bmatrix} A^* & 0 \\ B^* & 0 \end{bmatrix},$$

$$D_{V_1^*}|_{\left[\begin{smallmatrix} \mathcal{F} \\ \mathcal{E} \end{smallmatrix}\right]} := \left( I_{\left[\begin{smallmatrix} \mathcal{F} \\ \mathcal{E} \end{smallmatrix}\right]} - V_1 V_1^*|_{\left[\begin{smallmatrix} \mathcal{F} \\ \mathcal{E} \end{smallmatrix}\right]} \right) = \begin{bmatrix} I_{\mathcal{F}} & 0 \\ 0 & I_{\mathcal{E}} \end{bmatrix} - \begin{bmatrix} C^*C & C^*D \\ D^*C & D^*D \end{bmatrix} = \begin{bmatrix} A^*A & A^*B \\ B^*A & B^*B \end{bmatrix},$$

using that $U$ is unitary for the last step. Similarly we have

$$\begin{aligned} D_{V_2^*}|_{\left[\begin{smallmatrix} \mathcal{F} \\ \mathcal{E} \end{smallmatrix}\right]} &= \left( \begin{bmatrix} I_{\mathcal{F}} & 0 \\ 0 & I_{\mathcal{E}} \end{bmatrix} - \begin{bmatrix} I_{H^2} \otimes A & I_{H^2} \otimes B \\ M_z \otimes C & M_z \otimes D \end{bmatrix} \begin{bmatrix} I_{H^2} \otimes A^* & M_z^* \otimes C^* \\ I_{H^2} \otimes B^* & M_z^* \otimes D^* \end{bmatrix} \Big|_{\left[\begin{smallmatrix} \mathcal{F} \\ \mathcal{E} \end{smallmatrix}\right]} \right) \\ &= \left( \begin{bmatrix} I_{\mathcal{F}} & 0 \\ 0 & I_{\mathcal{E}} \end{bmatrix} - \begin{bmatrix} I_{H^2} \otimes A & I_{H^2} \otimes B \\ M_z \otimes C & M_z \otimes D \end{bmatrix} \begin{bmatrix} I_{H^2} \otimes A^* & 0 \\ I_{H^2} \otimes B^* & 0 \end{bmatrix} \Big|_{\left[\begin{smallmatrix} \mathcal{F} \\ \mathcal{E} \end{smallmatrix}\right]} \right) \\ &= \left( \begin{bmatrix} I_{\mathcal{F}} & 0 \\ 0 & I_{\mathcal{E}} \end{bmatrix} - \begin{bmatrix} I_{H^2} \otimes (AA^* + BB^*) & 0 \\ M_z \otimes (CA^* + DB^*) & 0 \end{bmatrix} \Big|_{\left[\begin{smallmatrix} \mathcal{F} \\ \mathcal{E} \end{smallmatrix}\right]} \right) = \begin{bmatrix} 0 & 0 \\ 0 & I_{\mathcal{E}} \end{bmatrix}, \end{aligned}$$

again using that $\left[\begin{smallmatrix} A & B \\ C & D \end{smallmatrix}\right]$ is unitary for the last step.

Hence

$$
\begin{aligned}
V_2 D_{V_1^*} V_2^* &= \begin{bmatrix} I_{H^2} \otimes A & I_{H^2} \otimes B \\ M_z \otimes C & M_z \otimes D \end{bmatrix} \begin{bmatrix} A^*A & A^*B \\ B^*A & B^*B \end{bmatrix} \begin{bmatrix} A^* & 0 \\ B^* & 0 \end{bmatrix} \\
&= \begin{bmatrix} I_{H^2} \otimes A & I_{H^2} \otimes B \\ M_z \otimes C & M_a \otimes D \end{bmatrix} \begin{bmatrix} A^*AA^* + A^*BB^* & 0 \\ B^*AA^* + B^*BB^* & 0 \end{bmatrix} \\
&= \begin{bmatrix} I_{H^2} \otimes A & I_{H^2} \otimes B \\ M_z \otimes C & M_z \otimes D \end{bmatrix} \begin{bmatrix} A^* & 0 \\ B^* & 0 \end{bmatrix} \\
&= \begin{bmatrix} AA^* + BB^* & 0 \\ M_z \otimes (CA^* + DB^*) & 0 \end{bmatrix} = \begin{bmatrix} I_{\mathcal{F}} & 0 \\ 0 & 0 \end{bmatrix},
\end{aligned}
\tag{3.32}
$$

again using that $U = \left[\begin{smallmatrix} A & B \\ C & D \end{smallmatrix}\right]$ is unitary for the last step. We see next that

$$
D_{V_2^*} V_1^* |_{\left[\begin{smallmatrix} \mathcal{F} \\ \mathcal{E} \end{smallmatrix}\right]} = D_{V_2^*} \begin{bmatrix} 0 & 0 \\ C & D \end{bmatrix} = \begin{bmatrix} 0 & 0 \\ C & D \end{bmatrix}, \tag{3.33}
$$

while

$$
\begin{aligned}
V_2 D_{V_1^*} |_{\left[\begin{smallmatrix} \mathcal{F} \\ \mathcal{E} \end{smallmatrix}\right]} &= \begin{bmatrix} I_{H^2} \otimes A & I_{H^2} \otimes B \\ M_z \otimes C & M_z \otimes D \end{bmatrix} \begin{bmatrix} A^*A & A^*B \\ B^*A & B^*B \end{bmatrix} \\
&= \begin{bmatrix} AA^*A + BB^*A & AA^*B + BB^*B \\ M_z \otimes (CA^*A + DB^*A) & M_z^*(CA^*B + DB^*B) \end{bmatrix} \\
&= \begin{bmatrix} (AA^* + BB^*)A & (AA^* + BB^*)B \\ 0 & 0 \end{bmatrix} = \begin{bmatrix} A & B \\ 0 & 0 \end{bmatrix},
\end{aligned}
\tag{3.34}
$$

again using that $U = \left[\begin{smallmatrix} A & B \\ C & D \end{smallmatrix}\right]$ is unitary for the last step. Combining (3.33) and (3.34) then gives us

$$
(D_{V_2^*} V_1^* + V_2 D_{V_1^*})|_{\left[\begin{smallmatrix} \mathcal{F} \\ \mathcal{E} \end{smallmatrix}\right]} = \begin{bmatrix} A & B \\ C & D \end{bmatrix}. \tag{3.35}
$$

Thus the formulas (3.32) and (3.35) show that we recover formula (3.18). In this way we give an alternative proof of formula (3.18) for recovering the BCL-tuple $\Xi$ (at least up to unitary equivalence) prescribed in normalized form (3.27) directly from the BCL1-model commuting isometric pair $\mathbf{V} = (V_1, V_2)$ via the formula (3.18). Here the unitary equivalence is rather mild: We identified the defect space $\mathcal{D}_{\mathbf{V}^*} = \operatorname{Ran} \mathbf{ev}_{0, \left[\begin{smallmatrix} \mathcal{F} \\ \mathcal{E} \end{smallmatrix}\right]} \subset H^2\left(\left[\begin{smallmatrix} \mathcal{F} \\ \mathcal{E} \end{smallmatrix}\right]\right)$ with the coefficient space $\left[\begin{smallmatrix} \mathcal{F} \\ \mathcal{E} \end{smallmatrix}\right]$.

**Theorem 3.1.13** *Suppose that the normalized BCL-tuple*

$$
\Xi = \left( \begin{bmatrix} \mathcal{F} \\ \mathcal{E} \end{bmatrix}, P = \begin{bmatrix} I_{\mathcal{F}} & 0 \\ 0 & 0 \end{bmatrix}, U = \begin{bmatrix} A & B \\ C & D \end{bmatrix} \text{ acting on } \begin{bmatrix} \mathcal{F} \\ \mathcal{E} \end{bmatrix} \right) \tag{3.36}
$$

*generates a BCL1-model for the commuting isometric pair* $\mathbf{V} = (V_1, V_2)$. *Then the transformed normalized BCL-tuple*

$$\Xi' = \left( \begin{bmatrix} \mathcal{E} \\ \mathcal{F} \end{bmatrix}, P' = \begin{bmatrix} I_{\mathcal{E}} & 0 \\ 0 & 0 \end{bmatrix}, U' = \begin{bmatrix} D^* & B^* \\ C^* & A^* \end{bmatrix} \text{ acting on } \begin{bmatrix} \mathcal{E} \\ \mathcal{F} \end{bmatrix} \right) \tag{3.37}$$

*generates a BCL1-model for the commuting isometric pair* $\mathbf{V}' := (V_2, V_1)$.

*Proof* By hypothesis, $\Xi$ as in (3.36) generates a BCL1-model for $\mathbf{V} = (V_1, V_2)$. By Corollary 3.1.8, the $\mathbf{V}$-canonical BCL-tuple $\Xi_{\mathbf{V}}$ as in (3.23) also generates a BCL1-model for the same $\mathbf{V}$. Hence by Theorem 3.1.6 the BCL-tuples $\Xi$ and $\Xi_{\mathbf{V}}$ are unitarily equivalent. Hence, by Theorem (3.1.6) combined with part (ii) of Lemma 3.1.11 there are unitary operators $\tau_{\mathcal{F}} \colon \mathcal{F} \to \mathcal{F}_{\mathbf{V}}$ and $\tau_{\mathcal{E}} \colon \mathcal{E} \to \mathcal{E}_{\mathbf{V}}$ so that

$$\begin{aligned} \begin{bmatrix} A_{\mathbf{V}} & B_{\mathbf{V}} \\ C_{\mathbf{V}} & D_{\mathbf{V}} \end{bmatrix} &= \begin{bmatrix} \tau_{\mathcal{F}} & 0 \\ 0 & \tau_{\mathcal{E}} \end{bmatrix} \begin{bmatrix} A & B \\ C & D \end{bmatrix} \begin{bmatrix} \tau_{\mathcal{F}}^* & 0 \\ 0 & \tau_{\mathcal{E}}^* \end{bmatrix} \\ &= \begin{bmatrix} \tau_{\mathcal{F}} A \tau_{\mathcal{F}}^* & \tau_{\mathcal{F}} B \tau_{\mathcal{E}}^* \\ \tau_{\mathcal{E}} C \tau_{\mathcal{F}}^* & \tau_{\mathcal{E}} D \tau_{\mathcal{E}}^* \end{bmatrix}. \end{aligned} \tag{3.38}$$

We also know by the observation in Remark 3.1.9 that

$$\Xi'_{\mathbf{V}} := \left( \begin{bmatrix} \mathcal{E}_{\mathbf{V}} \\ \mathcal{F}_{\mathbf{V}} \end{bmatrix}, \begin{bmatrix} I_{\mathcal{E}_{\mathbf{V}}} & 0 \\ 0 & 0 \end{bmatrix}, \begin{bmatrix} D_{\mathbf{V}}^* & B_{\mathbf{V}}^* \\ C_{\mathbf{V}}^* & A_{\mathbf{V}}^* \end{bmatrix} \right)$$

is a BCL-tuple generating a BCL1-model for $(V_2, V_1)$. From (3.38) we see that

$$\begin{bmatrix} D_{\mathbf{V}}^* & B_{\mathbf{V}}^* \\ C_{\mathbf{V}}^* & A_{\mathbf{V}}^* \end{bmatrix} = \begin{bmatrix} \tau_{\mathcal{E}} D^* \tau_{\mathcal{E}}^* & \tau_{\mathcal{E}} B^* \tau_{\mathcal{F}}^* \\ \tau_{\mathcal{F}} C^* \tau_{\mathcal{E}}^* & \tau_{\mathcal{F}} A^* \tau_{\mathcal{F}}^* \end{bmatrix} = \begin{bmatrix} \tau_{\mathcal{E}} & 0 \\ 0 & \tau_{\mathcal{F}} \end{bmatrix} \begin{bmatrix} D^* & B^* \\ C^* & A^* \end{bmatrix} \begin{bmatrix} \tau_{\mathcal{E}}^* & 0 \\ 0 & \tau_{\mathcal{F}}^* \end{bmatrix}$$

and trivially we also have

$$\begin{bmatrix} I_{\mathcal{E}_{\mathbf{V}}} & 0 \\ 0 & 0 \end{bmatrix} = \begin{bmatrix} \tau_{\mathcal{E}} & 0 \\ 0 & \tau_{\mathcal{F}} \end{bmatrix} \begin{bmatrix} I_{\mathcal{E}} & 0 \\ 0 & 0 \end{bmatrix} \begin{bmatrix} \tau_{\mathcal{E}}^* & 0 \\ 0 & \tau_{\mathcal{F}}^* \end{bmatrix},$$

from which we conclude that $\Xi'$ as in (3.37) is unitarily equivalent to $\Xi'_{\mathbf{V},\mathrm{can}}$. We conclude that the transformed BCL-tuple $\Xi'$ (3.37) is unitarily equivalent to the transformed canonical BCL-tuple $\Xi'_{\mathbf{V}}$. Hence again by Theorem 3.1.6, it follows that the BCL-tuple $\Xi'$ also generates a BCL1-model for $(V_2, V_1)$. □

The next lemma extends the result of Lemma 3.1.11 to a more refined setting.

**Lemma 3.1.14** *Let* $\Xi$ *and* $\widetilde{\Xi}$ *be BCL-tuples of the form* (3.27). *Suppose that we perform the Sz.-Nagy–Foias canonical decomposition on the contraction operators* $A$ *and* $D$*:*

$$A \text{ on } \mathcal{F} \underset{\text{u}}{\cong} \begin{bmatrix} A_0 & 0 \\ 0 & A_1 \end{bmatrix} \text{ on } \begin{bmatrix} \mathcal{F}_0 \\ \mathcal{F}_1 \end{bmatrix}, \ A_0 \text{ unitary on } \mathcal{F}_0, \ A_1 \text{ c.n.u. on } \mathcal{F}_1,$$

$$D \text{ on } \mathcal{E} \underset{\text{u}}{\cong} \begin{bmatrix} D_1 & 0 \\ 0 & D_0 \end{bmatrix} \text{ on } \begin{bmatrix} \mathcal{E}_1 \\ \mathcal{E}_0 \end{bmatrix}, \ D_1 \text{ c.n.u. on } \mathcal{E}_1, \ D_0 \text{ unitary on } \mathcal{E}_0.$$

*Since the original operator matrix* $\left[\begin{smallmatrix} A & B \\ C & D \end{smallmatrix}\right]$ *is unitary, this partitioning of* $\mathcal{F}$ *as* $\left[\begin{smallmatrix} \mathcal{F}_0 \\ \mathcal{F}_1 \end{smallmatrix}\right]$ *and of* $\mathcal{E}$ *as* $\left[\begin{smallmatrix} \mathcal{E}_1 \\ \mathcal{E}_0 \end{smallmatrix}\right]$ *induces a finer representation of* $U = \left[\begin{smallmatrix} A & B \\ C & D \end{smallmatrix}\right]$ *of the form*

$$\begin{bmatrix} A & B \\ C & D \end{bmatrix} \underset{\text{u}}{\cong} \begin{bmatrix} A_0 & 0 & 0 & 0 \\ 0 & A_1 & B_1 & 0 \\ 0 & C_1 & D_1 & 0 \\ 0 & 0 & 0 & D_0 \end{bmatrix} \quad \text{acting on} \quad \begin{bmatrix} \mathcal{F}_0 \\ \mathcal{F}_1 \\ \mathcal{E}_1 \\ \mathcal{E}_0 \end{bmatrix}. \tag{3.39}$$

*The projection* $P$ *in the BCL-tuple then assumes the form* $P \underset{\text{u}}{\cong} \left[\begin{smallmatrix} I_{\mathcal{F}_0} & 0 & 0 & 0 \\ 0 & I_{\mathcal{F}_1} & 0 & 0 \\ 0 & 0 & 0 & 0 \\ 0 & 0 & 0 & 0 \end{smallmatrix}\right]$ *leaving us with* $\Xi$ *being unitarily equivalent (as a BCL-tuple) to its refined version*

$$\Xi \underset{\text{u}}{\cong} \Xi_{\text{ref}} := \left( \begin{bmatrix} \mathcal{F}_0 \\ \mathcal{F}_1 \\ \mathcal{E}_1 \\ \mathcal{E}_0 \end{bmatrix}, \begin{bmatrix} I_{\mathcal{F}_0} & 0 & 0 & 0 \\ 0 & I_{\mathcal{F}_1} & 0 & 0 \\ 0 & 0 & 0 & 0 \\ 0 & 0 & 0 & 0 \end{bmatrix}, \begin{bmatrix} A_0 & 0 & 0 & 0 \\ 0 & A_1 & B_1 & 0 \\ 0 & C_1 & D_1 & 0 \\ 0 & 0 & 0 & D_0 \end{bmatrix} \right). \tag{3.40}$$

*Performing the corresponding sequence of operations on* $\widetilde{\Xi}$ *gives us*

$$\widetilde{\Xi} \underset{\text{u}}{\cong} \widetilde{\Xi}_{\text{ref}} := \left( \begin{bmatrix} \widetilde{\mathcal{F}}_0 \\ \widetilde{\mathcal{F}}_1 \\ \widetilde{\mathcal{E}}_1 \\ \widetilde{\mathcal{E}}_0 \end{bmatrix}, \begin{bmatrix} I_{\widetilde{\mathcal{F}}_0} & 0 & 0 & 0 \\ 0 & I_{\widetilde{\mathcal{F}}_1} & 0 & 0 \\ 0 & 0 & 0 & 0 \\ 0 & 0 & 0 & 0 \end{bmatrix}, \begin{bmatrix} \widetilde{A}_0 & 0 & 0 & 0 \\ 0 & \widetilde{A}_1 & \widetilde{B}_1 & 0 \\ 0 & \widetilde{C}_1 & \widetilde{D}_1 & 0 \\ 0 & 0 & 0 & \widetilde{D}_0 \end{bmatrix} \right).$$

*Then* $\Xi_{\text{ref}}$ *is unitarily equivalent to* $\widetilde{\Xi}_{\text{ref}}$ *if and only if there exist unitary operators*

$$\tau_{\mathcal{F}_0} \colon \mathcal{F}_0 \to \widetilde{\mathcal{F}}_0, \quad \tau_{\mathcal{F}_1} \colon \mathcal{F}_1 \to \widetilde{\mathcal{F}}_1, \quad \tau_{\mathcal{E}_1} \colon \mathcal{E}_1 \to \widetilde{\mathcal{E}}_1, \quad \tau_{\mathcal{E}_0} \colon \mathcal{E}_0 \to \widetilde{\mathcal{E}}_0$$

*so that*

$$\begin{bmatrix} \tau_{\mathcal{F}_0} & 0 & 0 & 0 \\ 0 & \tau_{\mathcal{F}_1} & 0 & 0 \\ 0 & 0 & \tau_{\mathcal{E}_1} & 0 \\ 0 & 0 & 0 & \tau_{\mathcal{E}_0} \end{bmatrix} \begin{bmatrix} A_0 & 0 & 0 & 0 \\ 0 & A_1 & B_1 & 0 \\ 0 & C_1 & D_1 & 0 \\ 0 & 0 & 0 & D_0 \end{bmatrix}$$
$$= \begin{bmatrix} \widetilde{A}_0 & 0 & 0 & 0 \\ 0 & \widetilde{A}_1 & \widetilde{B}_1 & 0 \\ 0 & \widetilde{C}_1 & \widetilde{D}_1 & 0 \\ 0 & 0 & 0 & \widetilde{D}_0 \end{bmatrix} \begin{bmatrix} \tau_{\mathcal{F}_0} & 0 & 0 & 0 \\ 0 & \tau_{\mathcal{F}_1} & 0 & 0 \\ 0 & 0 & \tau_{\mathcal{E}_1} & 0 \\ 0 & 0 & 0 & \tau_{\mathcal{E}_0} \end{bmatrix}. \tag{3.41}$$

*Proof* We apply the result of part (ii) of Lemma 3.1.11 to the case where

$$\mathcal{F} = \begin{bmatrix} \mathcal{F}_0 \\ \mathcal{F}_1 \end{bmatrix}, \quad \mathcal{E} = \begin{bmatrix} \mathcal{E}_1 \\ \mathcal{E}_0 \end{bmatrix},$$
$$A = \begin{bmatrix} A_0 & 0 \\ 0 & A_1 \end{bmatrix}, \quad B = \begin{bmatrix} 0 & 0 \\ B_1 & 0 \end{bmatrix}, \quad C = \begin{bmatrix} 0 & C_1 \\ 0 & 0 \end{bmatrix}, \quad D = \begin{bmatrix} D_1 & 0 \\ 0 & D_0 \end{bmatrix}$$

and similarly,

$$\widetilde{\mathcal{F}} = \begin{bmatrix} \widetilde{\mathcal{F}}_0 \\ \widetilde{\mathcal{F}}_1 \end{bmatrix}, \quad \widetilde{\mathcal{E}} = \begin{bmatrix} \widetilde{\mathcal{E}}_0 \\ \widetilde{\mathcal{E}}_1 \end{bmatrix},$$
$$\widetilde{A} = \begin{bmatrix} \widetilde{A}_0 & 0 \\ 0 & \widetilde{A}_1 \end{bmatrix}, \quad \widetilde{B} = \begin{bmatrix} 0 & 0 \\ \widetilde{B}_1 & 0 \end{bmatrix}, \quad \widetilde{C} = \begin{bmatrix} 0 & \widetilde{C}_1 \\ 0 & 0 \end{bmatrix}, \quad \widetilde{D} = \begin{bmatrix} \widetilde{D}_1 & 0 \\ 0 & \widetilde{D}_0 \end{bmatrix},$$

where here we introduce Sz.-Nagy–Foias canonical decompositions for $A$, $D$, $\widetilde{A}$, $\widetilde{D}$. By the result from item (i) we know that $\Xi_{\rm ref} \underset{\rm u}{\cong} \widetilde{\Xi}_{\rm ref}$ if and only if there exist

$$\tau_{\mathcal{F}} = \begin{bmatrix} \tau_{\mathcal{F}}^{(11)} & \tau_{\mathcal{F}}^{(12)} \\ \tau_{\mathcal{F}}^{(21)} & \tau_{\mathcal{F}}^{(22)} \end{bmatrix} : \begin{bmatrix} \mathcal{F}_0 \\ \mathcal{F}_1 \end{bmatrix} \to \begin{bmatrix} \widetilde{\mathcal{F}}_0 \\ \widetilde{\mathcal{F}}_1 \end{bmatrix}, \quad \tau_{\mathcal{E}} = \begin{bmatrix} \tau_{\mathcal{E}}^{(11)} & \tau_{\mathcal{E}}^{(12)} \\ \tau_{\mathcal{E}}^{(21)} & \tau_{\mathcal{E}}^{(22)} \end{bmatrix} : \begin{bmatrix} \mathcal{E}_1 \\ \mathcal{E}_0 \end{bmatrix} \to \begin{bmatrix} \widetilde{\mathcal{E}}_1 \\ \widetilde{\mathcal{E}}_0 \end{bmatrix}$$

so that

$$\begin{bmatrix} \tau_{\mathcal{F}} & 0 \\ 0 & \tau_{\mathcal{E}} \end{bmatrix} \begin{bmatrix} \begin{bmatrix} A_0 & 0 \\ 0 & A_1 \end{bmatrix} & \begin{bmatrix} 0 & 0 \\ B_1 & 0 \end{bmatrix} \\ \begin{bmatrix} 0 & C_1 \\ 0 & 0 \end{bmatrix} & \begin{bmatrix} D_1 & 0 \\ 0 & D_0 \end{bmatrix} \end{bmatrix} = \begin{bmatrix} \begin{bmatrix} \widetilde{A}_0 & 0 \\ 0 & \widetilde{A}_1 \end{bmatrix} & \begin{bmatrix} 0 & 0 \\ \widetilde{B}_1 & 0 \end{bmatrix} \\ \begin{bmatrix} 0 & \widetilde{C}_1 \\ 0 & 0 \end{bmatrix} & \begin{bmatrix} \widetilde{D}_1 & 0 \\ 0 & \widetilde{D}_0 \end{bmatrix} \end{bmatrix} \begin{bmatrix} \tau_{\widetilde{\mathcal{F}}} & 0 \\ 0 & \tau_{\widetilde{\mathcal{E}}} \end{bmatrix}.$$

A particular consequence of this matrix equality is that

$$\tau_{\mathcal{F}} \begin{bmatrix} A_0 & 0 \\ 0 & A_1 \end{bmatrix} = \begin{bmatrix} \widetilde{A}_0 & 0 \\ 0 & \widetilde{A}_1 \end{bmatrix} \tau_{\mathcal{F}},$$

or, in more detail,

$$\begin{bmatrix} \tau_{\mathcal{F}}^{(11)} & \tau_{\mathcal{F}}^{(12)} \\ \tau_{\mathcal{F}}^{(21)} & \tau_{\mathcal{F}}^{(22)} \end{bmatrix} \begin{bmatrix} A_0 & 0 \\ 0 & A_1 \end{bmatrix} = \begin{bmatrix} \widetilde{A}_0 & 0 \\ 0 & \widetilde{A}_1 \end{bmatrix} \begin{bmatrix} \tau_{\widetilde{\mathcal{F}}}^{(11)} & \tau_{\widetilde{\mathcal{F}}}^{(12)} \\ \tau_{\widetilde{\mathcal{F}}}^{(21)} & \tau_{\widetilde{\mathcal{F}}}^{(22)} \end{bmatrix}.$$

As $\left[\begin{smallmatrix} A_0 & 0 \\ 0 & A_1 \end{smallmatrix}\right]$ and $\left[\begin{smallmatrix} \widetilde{A}_0 & 0 \\ 0 & \widetilde{A}_1 \end{smallmatrix}\right]$ are canonical decompositions for $A$ and $\widetilde{A}$ respectively, it follows from item (ii) in Lemma 2.3.4 that $\tau_{\mathcal{F}}$ is diagonal: $\tau_{\mathcal{F}} = \left[\begin{smallmatrix} \tau_{\mathcal{F}_0} & 0 \\ 0 & \tau_{\mathcal{F}_1} \end{smallmatrix}\right]$, where $\tau_{\mathcal{F}_0} \colon \mathcal{F}_0 \to \widetilde{\mathcal{F}}_0$ and $\tau_{\mathcal{F}_1} \colon \mathcal{F}_1 \to \widetilde{\mathcal{F}}_1$ are unitary. A similar analysis implies that $\tau_{\mathcal{E}} = \left[\begin{smallmatrix} \tau_{\mathcal{E}}^{(11)} & \tau_{\mathcal{E}}^{(12)} \\ \tau_{\mathcal{E}}^{(21)} & \tau_{\mathcal{E}}^{(22)} \end{smallmatrix}\right]$ is diagonal, that is, $\tau_{\mathcal{E}} = \left[\begin{smallmatrix} \tau_{\mathcal{E}_1} & 0 \\ 0 & \tau_{\mathcal{E}_0} \end{smallmatrix}\right]$. Putting all the pieces together results in the criterion (3.41). □

**Definition 3.1.15** We shall say that a tuple of spaces and operators of the form (3.40) with $A_0$ and $D_0$ unitary, $A_1$ and $D_1$ c.n.u., and $\left[\begin{smallmatrix} A_1 & B_1 \\ C_1 & D_1 \end{smallmatrix}\right]$ unitary is a *refined BCL-tuple*. From the formula (3.9) we see that the BCL1-model operator pair $(V_1, V_2)$ generated by the refined BCL-tuple $\Xi$ (3.39) is given by

$$V_1 = \begin{bmatrix} M_z \otimes A_0^* & 0 & 0 & 0 \\ 0 & M_z \otimes A_1^* & I_{H^2} \otimes C_1^* & 0 \\ 0 & M_z \otimes B_1^* & I_{H^2} \otimes D_1^* & 0 \\ 0 & 0 & 0 & I_{H^2} \otimes D_0^* \end{bmatrix},$$

$$V_2 = \begin{bmatrix} I_{H^2} \otimes A_0 & 0 & 0 & 0 \\ 0 & I_{H^2} \otimes A_1 & I_{H^2} \otimes B_1 & 0 \\ 0 & M_z \otimes C_1 & M_z \otimes D_1 & 0 \\ 0 & 0 & 0 & M_z \otimes D_0 \end{bmatrix} \quad \text{on } H^2\left(\begin{bmatrix} \mathcal{F}_0 \\ \mathcal{F}_1 \\ \mathcal{E}_1 \\ \mathcal{E}_0 \end{bmatrix}\right). \tag{3.42}$$

**Remark 3.1.16** Let $\Xi_{\text{ref}}$ as in (3.40) be a refined BCL-tuple in the sense of Definition 3.1.15. Let us decompose the ambient space $H^2\left(\left[\begin{smallmatrix} \mathcal{F}_0 \\ \mathcal{F}_1 \\ \mathcal{E}_1 \\ \mathcal{E}_0 \end{smallmatrix}\right]\right)$ as the orthogonal direct sum of three subspaces

$$\mathcal{H}_{\text{su}} = \begin{bmatrix} H^2(\mathcal{F}_0) \\ 0 \\ 0 \\ 0 \end{bmatrix}, \quad \mathcal{H}_{ws} = \begin{bmatrix} 0 \\ H^2(\mathcal{F}_1) \\ H^2(\mathcal{E}_1) \\ 0 \end{bmatrix}, \quad H_{\text{us}} = \begin{bmatrix} 0 \\ 0 \\ 0 \\ H^2(\mathcal{E}_0) \end{bmatrix}. \tag{3.43}$$

From the block-diagonal form of $\varphi_1(z)$ and $\varphi_2(z)$ we see that each of these three subspaces is jointly reducing for $(M_{\varphi_1}, M_{\varphi_2})$. The three direct summands can be described in more detail as follows.

1. $(V_1, V_2)|_{\mathcal{H}_{\text{su}}} = \left(M_z \otimes A_0^*, I_{H^2} \otimes A_0\right)$ on $H^2(\mathcal{F}_0)$ has the form of a commuting shift/unitary pair (with the BCL-property that $V_1 V_2 = M_z \otimes I_{\mathcal{F}_0}$).

2. $(V_1, V_2))|_{\mathcal{H}_{us}} = \big(I_{H^2} \otimes D_0^*, M_z \otimes D_0\big)$ on $H^2(\mathcal{E}_0)$ is a commuting unitary/shift pair (with the BCL-property that $V_1V_2|_{\mathcal{H}_{us}} = M_z \otimes I_{\mathcal{E}_0}$).

3. $(V_1, V_2)\big|_{\mathcal{H}_{ws}} = \left( M_{\left[\begin{smallmatrix} zA_1^* & C_1^* \\ zB_1^* & D_1^* \end{smallmatrix}\right]}, M_{\left[\begin{smallmatrix} A_1 & B_1 \\ zC_1 & zD_1 \end{smallmatrix}\right]} \right)\Bigg|_{\left[\begin{smallmatrix} H^2(\mathcal{F}_1) \\ H^2(\mathcal{E}_1) \end{smallmatrix}\right]}$ can be shown to be what is variously called a *weak bishift* (see [23]), by which we mean that *there is no nontrivial jointly reducing subspace on which either $M_{\varphi_1}$ or $M_{\varphi_2}$ is unitary*. Another term (used in [30] and in Section 3.2.2 later) is simultaneously *1-c.n.u.* and *2-c.n.u.*, where *j-c.n.u.* means that there is no nontrivial jointly reducing subspace where $V_j$ is unitary for $j = 1, 2$. Let us note that verification of this last point relies on the assumption that $A_1$ and $D_1$ are c.n.u.

All of this is done under the assumption that the product isometry $V_1V_2$ is a shift. As we have seen, for the case of a general isometric pair $(V_1, V_2)$ one only needs to add a unitary/unitary direct summand to the three-fold orthogonal direct-sum decomposition (3.43). One can view all this as a bivariate version of the Wold decomposition for the case of commuting isometric pairs $(V_1, V_2)$ rather than for the case of a single isometry $V$.

We will stumble again on the bivariate Wold decomposition in Section 3.2.2 in the context of the BDF-models (see in particular Corollary 3.2.4 and Remark 3.2.7 for more details).

### 3.1.3 Applications and Examples of BCL-Models

We next seek a characterization of the joint reducing subspaces for the pure part (in the sense of Definition 3.1.5) of a commuting isometric pair $(V_1, V_2)$ in terms of the BCL-tuple $(\mathcal{F}, P, U)$. It is convenient first to introduce a definition.

**Definition 3.1.17** Suppose $(\mathcal{F}, P, U)$ is a BCL-tuple (with commuting unitary operators $W_1, W_2$ assumed to be trivial). Suppose that $\mathcal{F}_0$ is a subspace of $\mathcal{F}$ such that

(i) $\mathcal{F}_0$ is invariant for $P$,
(ii) $\mathcal{F}_0$ is reducing for $U$.

Set $P_0 = P|_{\mathcal{F}_0}$ and $U_0 = U|_{\mathcal{F}_0}$. Then we say that the BCL-tuple $(\mathcal{F}_0, P_0, U_0)$ is a *reduced sub-BCL-tuple* of $(\mathcal{F}, P, U)$.

Then we have the following result.

**Theorem 3.1.18** (See [63, Theorem 1.2] and [23, Theorem 3.3].) *Suppose that $(V_1, V_2)$ is the (BCL1- or) BCL2-model commuting isometric pair associated with a BCL-tuple $(\mathcal{F}, P, U)$. Then joint reducing subspaces for $(V_1, V_2)$ are in one-to-one correspondence with reducing sub-Ando tuples $(\mathcal{F}_0, P_0, U_0)$ with associated reducing subspace equal to $H^2(\mathcal{F}_0)$ viewed as a subspace of $H^2(\mathcal{F})$ in the natural way.*

*Proof* Suppose that $\mathcal{M} \subset H^2(\mathcal{F})$ is a joint reducing subspace for $(V_1, V_2)$ given in the BCL1-model form

$$(V_1, V_2) = \left(M_{U^*(P^\perp + zP)}, M_{(P+zP^\perp)U}\right).$$

Then, in particular, $\mathcal{M}$ is reducing for $V = V_1 V_2 = M_z^{\mathcal{F}}$ on $H^2(\mathcal{F})$. The Beurling–Lax theorem characterizes the invariant subspaces $\mathcal{M}$ for the shift operator $M_z^{\mathcal{F}}$ on a vectorial Hardy space $H^2(\mathcal{F})$ as those of the form $\Theta \cdot H^2(\mathcal{E})$ for an inner function $\Theta$ (i.e., an $\mathcal{B}(\mathcal{U}, \mathcal{Y})$-valued function $z \mapsto \Theta(z)$ on the unit disk with radial-limit boundary-value function $\zeta \mapsto \Theta(\zeta)$ having isometric values a.e. on $\mathbb{T}$). If $\mathcal{M}$ is reducing for $M_z$ then both $\mathcal{M}$ and $\mathcal{M}^\perp$ have Beurling–Lax representations

$$\mathcal{M} = \Theta \cdot H^2(\mathcal{E}), \quad \mathcal{M}^\perp = \Psi \cdot H^2(\mathcal{E}')$$

for inner functions $\Theta$ and $\Psi$ with values in $\mathcal{B}(\mathcal{E}, \mathcal{F})$ and $\mathcal{B}(\mathcal{E}', \mathcal{F})$ for appropriate coefficient Hilbert spaces $\mathcal{E}$ and $\mathcal{E}'$ respectively. Furthermore, we have the orthogonal decomposition

$$H^2(\mathcal{F}) = \mathcal{M} \oplus \mathcal{M}^\perp = \Theta H^2(\mathcal{E}) \oplus \Psi H^2(\mathcal{E}'),$$

implying that $\begin{bmatrix} \Theta & \Psi \end{bmatrix}$ is also inner as a function with values in $\mathcal{B}\left(\left[\begin{smallmatrix} \mathcal{E} \\ \mathcal{E}' \end{smallmatrix}\right], \mathcal{F}\right)$. As, in general, the Beurling–Lax representer for a given shift-invariant subspace $\mathcal{M}$ is unique up to a unitary-constant right factor, we see from all this that $\begin{bmatrix} \Theta & \Psi \end{bmatrix}$ is a unitary constant from $\left[\begin{smallmatrix} \mathcal{E} \\ \mathcal{E}' \end{smallmatrix}\right]$ onto $\mathcal{F}$. In particular, we see that $\Theta$ must be equal to a constant $\Theta(z) = \Theta(0)$ isometric embedding of $\mathcal{E}$ onto a subspace $\mathcal{F}_0$ of $\mathcal{F}$ and $\mathcal{M}$ has the form $\mathcal{M} = H^2(\mathcal{F}_0) \subset H^2(\mathcal{F})$ for the subspace $\mathcal{F}_0 = \Theta(0)\mathcal{E}$ of the coefficient space $\mathcal{F}$.

It remains to understand when a subspace of this form is also reducing for $V_1$ and $V_2$. Let $f(z) = f_0$, where $f_0 \in \mathcal{F}_0$. Then $V_1 \colon f(z) \mapsto U^* P^\perp f_0 + z U^* P f_0 \in H^2(\mathcal{F}_0)$ forces $U^* P^\perp f_0 \in \mathcal{F}_0$, $U^* P f_0 \in \mathcal{F}_0$, that is,

(i) invariance of $H^2(\mathcal{F}_0)$ under $V_1$ implies invariance of $\mathcal{F}_0$ under $U^* P^\perp$ and $U^* P$.

Similarly, from the formula $V_1^* \colon f(z) = f_0 \mapsto P^\perp U f_0$ for $f(z) = f_0 \in H^2(\mathcal{F}_0)$, we see that

(ii) invariance of $H^2(\mathcal{F}_0)$ under $V_1^*$ implies invariance of $\mathcal{F}_0$ under $P^\perp U$.

A parallel argument with $V_2$ in place of $V_1$ then gives us

(iii) invariance of $H^2(\mathcal{F}_0)$ under $V_2$ implies invariance of $\mathcal{F}_0$ under $PU$ and $P^\perp U$, and
(iv) invariance of $H^2(\mathcal{F}_0)$ under $V_2^*$ implies invariance of $\mathcal{F}_0$ under $U^*P$.

By summing the two operators in item (i) and in item (iii) respectively, we see that $\mathcal{F}_0$ is invariant under $U^*$ and under $U$. Then from either (ii) or (iv) we see that $\mathcal{F}_0$ is invariant under $P$ or $P^\perp$ (and hence also under $P^\perp = I - P$ or $P = I - P^\perp$). Conversely, if $\mathcal{F}_0$ is invariant under $U$, $U^*$, and $P$ (and hence also $P^\perp$), it is routine to verify by direct computation that $H^2(\mathcal{F}_0)$ is invariant under all of $V_1$, $V_1^*$, $V_2$, $V_2^*$, and hence is jointly reducing for $(V_1, V_2)$. Then restriction of $(V_1, V_2)$ to $H^2(\mathcal{F}_0)$ amounts to the BCL1-model corresponding to the reduced sub-BCL-tuple $(\mathcal{F}_0, P_0, U_0)$ as expected. □

It is well known (see [92, Section I.6]) that an arbitrary family of commuting isometries can always be extended to a family of commuting unitaries. The following result shows that when the family is finite and one of the isometries in the family is the product of the rest of the isometries, then the family can be extended to a family of commuting unitaries with additional structure. We include the short proof here as an illustrative application of the BCL2-model for commutative isometric pairs.

**Lemma 3.1.19** (See [31, Theorem 3.6] and [23, Lemma 3.1].) *Let $(V_1, V_2)$ be a pair of commuting isometries. Then the tuple $\underline{V} = (V_1, V_2, V_1V_2)$ of commuting isometries has a commuting unitary extension $\underline{Y} = (Y_1, Y_2, Y)$ such that $Y = Y_1Y_2$ is the minimal unitary extension of $\underline{V}$.*

*Proof* We can assume without loss of generality that $\mathcal{H} = \begin{bmatrix} H^2(\mathcal{F}) \\ \mathcal{K}_u \end{bmatrix}$ and

$$(V_1, V_2, V_1V_2) = \left( \begin{bmatrix} M_{P^\perp U + zPU} & 0 \\ 0 & W_1 \end{bmatrix}, \begin{bmatrix} M_{U^*P + zU^*P^\perp} & 0 \\ 0 & W_2 \end{bmatrix}, \begin{bmatrix} M_z & 0 \\ 0 & W \end{bmatrix} \right),$$

where $(\mathcal{F}, P, U, W_1, W_2)$ is a BCL-tuple for $(V_1, V_2)$. Now define a pair of operators $Y_1, Y_2$ on $L^2(\mathcal{F}) \oplus \mathcal{K}_u$ by

$$(Y_1, Y_2) := \left( \begin{bmatrix} M_{P^\perp U + \zeta PU} & 0 \\ 0 & W_1 \end{bmatrix}, \begin{bmatrix} M_{U^*P + \zeta U^*P^\perp} & 0 \\ 0 & W_2 \end{bmatrix} \right),$$

where $\zeta$ is the coordinate variable on the unit circle $\mathbb{T}$. Then one can check that $(Y_1, Y_2)$ is a pair of commuting unitaries and that $(Y_1, Y_2)$ is an extension of $(V_1, V_2)$, where $\begin{bmatrix} H^2(\mathcal{F}) \\ \mathcal{K}_u \end{bmatrix}$ is identified as a subspace of $\begin{bmatrix} L^2(\mathcal{F}) \\ \mathcal{K}_u \end{bmatrix}$ via

$$\begin{bmatrix} z^n\xi \\ \eta \end{bmatrix} \mapsto \begin{bmatrix} \zeta^n\xi \\ \eta \end{bmatrix} \quad \text{for all } \xi \in \mathcal{F} \quad \text{and} \quad \eta \in \mathcal{K}_u \quad \text{for } n = 0, 1, 2, \ldots.$$

Then $Y = Y_1 Y_2 = \begin{bmatrix} M_\zeta & 0 \\ 0 & W \end{bmatrix}$ on $\begin{bmatrix} L^2(\mathcal{F}) \\ \mathcal{K}_u \end{bmatrix}$ is clearly the minimal unitary extension of $V = V_1 V_2 = \begin{bmatrix} M_z & 0 \\ 0 & W \end{bmatrix}$ on $\begin{bmatrix} H^2(\mathcal{F}) \\ \mathcal{H}_u \end{bmatrix}$. □

As an illustrative example where a BCL-model for a concretely given commuting isometric pair $(V_1, V_2)$ can be computed explicitly, we offer the following; we shall make use of this example in the sequel.

**Example 3.1.20** (See Example 3.9 in [23].) Consider the commuting pair of shift operators $(V_1, V_2) = (M_{z_1}, M_{z_2})$ acting on the Hardy space over the bidisk

$$H^2_{\mathbb{D}^2} := \left\{ f(z_1, z_2) = \sum_{(m,n)\in\mathbb{Z}^2_+} a_{ij} z_1^i z_2^j \colon \sum_{(m,n)\in\mathbb{Z}^2_+} |a_{ij}|^2 < \infty \right\}.$$

Note that the operators $M_{z_1}$ and $M_{z_2}$ are shifts on $H^2_{\mathbb{D}^2}$ so the $(W_1, W_2)$-component in a BCL-tuple for $(M_{z_1}, M_{z_2})$ is trivial. We shall show: *The commuting isometric pair $(M_{z_1}, M_{z_2})$ on $H^2_{\mathbb{D}^2}$ is unitarily equivalent to the BCL1-model commuting isometric pair* (3.2) *generated by the BCL-tuple*

$$(\mathcal{F}, P, U) = \left(\ell^2_{\mathbb{Z}}, P_{\ell^2_{[1,\infty)}}, \mathbf{S}\right), \tag{3.44}$$

*where $\ell^2_{\mathbb{Z}}$ is the space of absolutely square-summable sequences indexed by the integers $\mathbb{Z}$, $P_{\ell^2_{[1,\infty)}}$ is the orthogonal projection on $\ell^2_{\mathbb{Z}}$ with range equal to the subspace of sequences supported on the subset $\{n \in \mathbb{Z} \colon 1 \le n\}$, and $\mathbf{S}$ is the bilateral shift operator*

$$\mathbf{S} \colon \mathbf{e}_n \mapsto \mathbf{e}_{n+1}$$

*on $\ell^2_{\mathbb{Z}}$ (where $\{\mathbf{e}_n \colon n \in \mathbb{Z}\}$ is the standard orthonormal basis for $\ell^2_{\mathbb{Z}}$).*

We need to compute the wandering subspace for the shift $V_1 V_2 = M_{z_1 z_2}$. Note that Ran $M_{z_1 z_2}$ consists of functions with Taylor coefficients $a_{ij}$ supported on the set $\{(i,j) \in \mathbb{Z}^2_+ \colon i, j \ge 1\}$. Hence $\mathcal{F} = (\text{Ran}\, M_{z_1 z_2})^\perp$ is the subspace

$$\mathcal{F} = \left\{ f(z_1, z_2) = a_{00} + \sum_{i>0} a_{i0} z_1^i + \sum_{j>0} a_{0j} z_2^j \colon |a_{00}|^2 + \sum_{i>0} |a_{i0}|^2 + \sum_{j>0} |a_{0j}|^2 < \infty \right\}. \tag{3.45}$$

It is convenient to identify $\mathcal{F}$ with $\ell^2_{\mathbb{Z}}$ via the map $\tau_{\text{bd}}$ (the subscript bd suggesting *bidisk*) defined on the orthonormal basis of monomials for $\mathcal{F}$ according to the formula

$$\tau_{\text{bd}} \colon z_1^i \mapsto \mathbf{e}_{-i} \text{ for } i \ge 0, \quad \tau_{\text{bd}} \colon z_2^j \mapsto \mathbf{e}_j \text{ for } j \ge 0. \tag{3.46}$$

We wish to extend $\tau_{\text{bd}}$ to a map from all of $H^2_{\mathbb{D}^2}$ to $H^2\left(\ell^2_{\mathbb{Z}}\right) := H^2 \otimes \ell^2_{\mathbb{Z}}$ so that we have the intertwining $\tau_{\text{bd}} M_{z_1 z_2} = M_z \tau_{\text{bd}}$. Thus we require that

$$\tau_{\text{bd}}\left((z_1 z_2)^k z_1^i\right) = z^k \tau_{\text{bd}}\left(z_1^i\right) = z^k \mathbf{e}_{-i}, \quad \tau_{\text{bd}}\left((z_1 z_2)^k z_2^j\right) = z^k \tau_{\text{bd}}\left(z_2^j\right) = z^k \mathbf{e}_j$$

for $i \geq 0$ and $j \geq 0$ or, in a more closed form,

$$\tau_{\rm bd} \colon z_1^i z_2^j \mapsto \begin{cases} \mathbf{e}_{j-i} z^j & \text{for } i \geq j, \\ \mathbf{e}_{j-i} z^i & \text{for } i \leq j. \end{cases} \tag{3.47}$$

As $\tau_{\rm bd}$ so defined is a well-defined bijection from an orthonormal basis for $H^2_{\mathbb{D}^2}$ to an orthonormal basis for $H^2\left(\ell^2_{\mathbb{Z}}\right)$, $\tau_{\rm bd}$ extends to a well-defined unitary map from the scalar-valued Hardy space over the bidisk $H^2_{\mathbb{D}^2}$ onto the $\ell^2_{\mathbb{Z}}$-valued Hardy space over the disk $H^2\left(\ell^2_{\mathbb{Z}}\right)$, which satisfies the intertwining property

$$\tau_{\rm bd} M_{z_1 z_2} = M_z \tau_{\rm bd}.$$

By the construction in the proof of Theorem 3.1.4 we are guaranteed that there is a projection operator $P$ and a unitary operator $U$ on $\mathcal{F} \cong \ell^2_{\mathbb{Z}}$ so that

$$\tau_{\rm bd} M_{z_1} = M_{U^*(P^\perp + zP)} \tau_{\rm bd}, \quad \tau_{\rm bd} M_{z_2} = M_{(P + zP^\perp)U} \tau_{\rm bd}.$$

Once one discovers the candidate, it is a matter of direct checking to see that $P = P_{\ell^2_{[1,\infty)}}$ and $U = \mathbf{S}$ on $\mathcal{F} = \ell^2_{\mathbb{Z}}$ do the job.

## 3.2 The BDF-Model for a Commuting Isometric Operator Pair

This section develops the BDF-model for a commuting isometric pair systematically from first principles, recovering the main results of Bercovici–Douglas–Foias for the pair case as well as some new results. We identify the components of the bivariate Wold decomposition (shift/unitary, 1&2-completely nonunitary (1&2-c.n.u.), unitary/shift, unitary/unitary) for the general commuting isometric pair, and identify a complete unitary invariant and an associated BDF-functional model for a commuting isometric pair of each elementary type in the Wold decomposition.

### 3.2.1 Derivation of the BDF-Model

Let $(V_1, V_2)$ be a commuting isometric operator pair on the Hilbert space $\mathcal{K}$. Our goal now is to obtain a functional model for $(V_1, V_2)$ by embedding the pair into the functional-model Wold decomposition for $V_1$ (rather than for $V = V_1 V_2$ as we did to obtain the BCL-model) to arrive at a new model (in fact the model of Bercovici–Douglas–Foias described in [29]) for $(V_1, V_2)$). The details are as follows.

Let $V_1 = V_{1s} \oplus V_{1u}$ be the Wold decomposition for the isometry $V_1$ acting on the space $\mathcal{K} = \mathcal{K}_{1s} \oplus \mathcal{K}_{1u}$ with $V_{1s} = V_1|_{\mathcal{K}_{1s}}$ a shift operator and $V_{1u} =$

$V_1|_{\mathcal{K}_{1u}}$ equal to a unitary operator. By Theorem 2.1.4 we know that the Fourier representation operator

$$\tau_{\rm BDF} = \begin{bmatrix} \sum_{n=0}^{\infty} D_{V_1^*} V_1^{*n} z^n \\ \text{SOT-lim}_{n\to\infty} V_1^n V_1^{*n} \end{bmatrix} =: \begin{bmatrix} O_{D_{V_1^*}, V_1^*} \\ Q_{V_1^*} \end{bmatrix} \tag{3.48}$$

is unitary from $\mathcal{K}$ to $\begin{bmatrix} H^2(\mathcal{D}_{V_1^*}) \\ \mathcal{Q}_{V_1^*} \end{bmatrix}$ and implements the unitary equivalence between the isometry $V_1$ and its model Wold decomposition acting on $\begin{bmatrix} H^2(\mathcal{D}_{V_1^*}) \\ \mathcal{Q}_{V_1^*} \end{bmatrix}$:

$$\tau_{\rm BDF} V_1 = \begin{bmatrix} M_z^{\mathcal{D}_{V_1^*}} & 0 \\ 0 & W_1 \end{bmatrix} \tau_{\rm BDF}. \tag{3.49}$$

Note that for the case here where $V_1$ is an isometry, the strong-limit operator SOT-$\lim_{n\to\infty} V_1^n V_1^{*n}$ is just the projection $P_{\mathcal{K}_{1u}}$ of $\mathcal{K}$ onto the unitary subspace in its Wold decomposition, and hence $\mathcal{Q}_{V_1^*} = \operatorname{Ran} P_{Q_{V_1^*}}$ is just another expression for the unitary subspace $\mathcal{K}_{1u}$ in the Wold decomposition of $V_1$ and $W_1 := V_1|_{\mathcal{K}_{1u}}$. Let us note that it is easy to compute $\tau_{\rm BDF}^* = \tau_{\rm BDF}^{-1}$:

$$\tau_{\rm BDF}^* \colon \begin{bmatrix} \sum_{n=0}^{\infty} d_n z^n \\ q \end{bmatrix} \mapsto \begin{bmatrix} \sum_{n=0}^{\infty} V_1^n d_n \\ q \end{bmatrix}. \tag{3.50}$$

Thus the functional-model version $V_1' = \tau_{\rm BDF} V_1 \tau_{\rm BDF}^*$ of $V_1$ has the form

$$V_1' = \begin{bmatrix} M_z^{\mathcal{D}_{V_1^*}} & 0 \\ 0 & W_1 \end{bmatrix} \quad \text{acting on} \quad \begin{bmatrix} H^2\left(\mathcal{D}_{V_1^*}\right) \\ \mathcal{Q}_{V_1^*} \end{bmatrix}. \tag{3.51}$$

The transformed version $V_2' = \tau_{\rm BDF} V_2 \tau_{\rm BDF}^*$ of $V_2$ then has the $2 \times 2$-block matrix decomposition

$$V_2' = \begin{bmatrix} A & B \\ C & E \end{bmatrix} \quad \text{acting on} \quad \begin{bmatrix} H^2\left(\mathcal{D}_{V_1^*}\right) \\ \mathcal{Q}_{V_1^*} \end{bmatrix}.$$

From the fact that $V_2'$ commutes with $V_1' = \begin{bmatrix} M_z^{\mathcal{D}_{V_1^*}} & 0 \\ 0 & W_1 \end{bmatrix}$, we get

$$\begin{bmatrix} A M_z^{\mathcal{D}_{V_1^*}} & B W_1 \\ C M_z^{\mathcal{D}_{V_1^*}} & E W_1 \end{bmatrix} = \begin{bmatrix} M_z^{\mathcal{D}_{V_1^*}} A & M_z^{\mathcal{D}_{V_1^*}} B \\ W_1 C & W_1 E \end{bmatrix}. \tag{3.52}$$

From the (1,1)-entry of (3.52), we see that $A$ commutes with $M_z^{\mathcal{D}_{V_1^*}}$; by a well-known characterization of the commutant of the shift operator $M_z^{\mathcal{D}_{V_1^*}}$ (see, e.g., [92]), it follows that $A = M_\Theta$ for a contractive analytic operator function $\left(\mathcal{D}_{V_1^*}, \Theta\right)$:

$$A = M_\Theta \colon f(z) \mapsto \Theta(z) f(z) \quad \text{for } f \in H^2\left(\mathcal{D}_{V_1^*}\right).$$

From the (1,2)-entry of (3.52) we get $BW_1 = M_z^{\mathcal{D}_{V_1^*}} B$; as $W_1$ is unitary and $M_z^{\mathcal{F}}$ is a shift, we see from Lemma 2.3.1 that

$$B = 0,$$

so we are left with

$$V_2' = \begin{bmatrix} M_\Theta & 0 \\ C & E \end{bmatrix} \colon \begin{bmatrix} H^2\left(\mathcal{D}_{V_1^*}\right) \\ Q_{V_1^*} \end{bmatrix} \to \begin{bmatrix} H^2\left(\mathcal{D}_{V_1^*}\right) \\ Q_{V_1^*} \end{bmatrix}.$$

From the formulas (3.48) and (3.50) we see that, for $d(z) = \sum_{n=0}^{\infty} d_n z^n \in H^2\left(\mathcal{D}_{V_1^*}\right)$, we have

$$A \colon \sum_{n=1}^{\infty} d_n z^n \mapsto O_{D_{V_1^*}, V_1^*} V_2 \left( \sum_{n=0}^{\infty} V_1^n d_n \right).$$

To simplify the notation in the sequel, let us write $\mathcal{F}$ in place of $\mathcal{D}_{V_1^*}$ for the coefficient space appearing in all these formulas, and we write $\mathcal{K}_{1u}$ for the Hilbert space $Q_{V_1^*}$ (the unitary space in the Wold decomposition of $V_1$). Eventually we shall interpret all the formulas for the case where $\mathcal{F}$ and $\mathcal{K}_{1u}$ are general ambient Hilbert spaces and all reference to an initial commuting isometric pair $(V_1, V_2)$ has been removed.

We organize the ensuing discussion in a series of incremental steps as follows.

**Step 1** *Assume that*

$$V_2' = \begin{bmatrix} M_\Theta & 0 \\ C & E \end{bmatrix} \colon \begin{bmatrix} H^2(\mathcal{F}_1) \\ \mathcal{K}_{1u} \end{bmatrix} \to \begin{bmatrix} H^2(\mathcal{F}_1) \\ \mathcal{K}_{1u} \end{bmatrix} \tag{3.53}$$

*is an isometry. Then*

(i) *$E$ is an isometry acting on $\mathcal{K}_{1u}$,*
(ii) *$C \colon H^2(\mathcal{F}) \to \mathcal{K}_{1u}$ satisfies*

$$C^* C = I_{H^2(\mathcal{F})} - M_\Theta^* M_\Theta \tag{3.54}$$

*as well as*

$$\operatorname{Ran} C \subset \mathcal{D}_{E^*} = (\operatorname{Ran} E)^\perp. \tag{3.55}$$

Indeed, from the fact that $V_2'$ is an isometry we get

$$\begin{bmatrix} M_\Theta^* & C^* \\ 0 & E^* \end{bmatrix} \begin{bmatrix} M_\Theta & 0 \\ C & E \end{bmatrix} = \begin{bmatrix} M_\Theta^* M_\Theta + C^* C & C^* E \\ E^* C & E^* E \end{bmatrix} = \begin{bmatrix} I_{H^2_{\mathcal{F}}} & 0 \\ 0 & I_{\mathcal{K}_{1u}} \end{bmatrix}.$$

From this identity we see that $E$ is an isometry as well as properties (3.54) and (3.55) of $C$.

**Step 2** *With notation as in the result of Step 1, let us next introduce the model Wold decomposition* $\mathcal{K}_{1u} \cong \left[\begin{smallmatrix} H^2(\mathcal{D}_{E^*}) \\ \mathcal{Q}_{E^*} \end{smallmatrix}\right]$ *for the operator* $E = V_2'|_{\mathcal{K}_{1u}}$*, so then we have*

$$\mathcal{K} \cong \begin{bmatrix} H^2(\mathcal{F}) \\ \mathcal{K}_{1u} \end{bmatrix} \cong \begin{bmatrix} H^2(\mathcal{F}) \\ H^2(\mathcal{D}_{E^*}) \\ \mathcal{Q}_{E^*} \end{bmatrix}$$

*with the commuting isometric pair* $(V_1, V_2)$ *having representation with respect to this decomposition as*

$$(V_1'', V_2'') = \left( \begin{bmatrix} M_z^{\mathcal{F}} & 0 & 0 \\ 0 & I_{H^2} \otimes u & 0 \\ 0 & 0 & Y_1 \end{bmatrix}, \begin{bmatrix} M_\Theta & 0 & 0 \\ C_1 & M_z^{\mathcal{D}_{E^*}} & 0 \\ 0 & 0 & Y_2 \end{bmatrix} \right), \tag{3.56}$$

*where* $Y_1, Y_2$ *are commuting unitary operators acting on* $\mathcal{Q}_{E^*}$*,* $u$ *is a unitary operator acting on the coefficient space* $\mathcal{D}_{E^*}$*, and where* $C_1 \colon H^2(\mathcal{F}) \to \mathcal{D}_{E^*}$ *satisfies the intertwining condition*

$$C_1 M_z^{\mathcal{F}} = u C_1 \tag{3.57}$$

*as well as the factorization property*

$$C_1^* C_1 = I_{H^2\left(\mathcal{D}_{V_1}^*\right)} - M_\Theta^* M_\Theta.$$

To see this, if we work with the representation from Step 1 with the second component $\mathcal{Q}_{V_1^*}$ now split as $\mathcal{Q}_{V_1^*} \cong \left[\begin{smallmatrix} \mathcal{D}_{E^*} \\ \mathcal{Q}_{E^*} \end{smallmatrix}\right]$, we get a more refined decomposition:

$$V_1' = \begin{bmatrix} M_z^{\mathcal{F}} & 0 \\ 0 & W_1 \end{bmatrix} \cong V_1'' = \begin{bmatrix} M_z^{\mathcal{F}} & 0 & 0 \\ 0 & W_{11}^{(1)} & W_{12}^{(1)} \\ 0 & W_{21}^{(1)} & W_{22}^{(1)} \end{bmatrix},$$

$$V_2' = \begin{bmatrix} M_\Theta & 0 \\ C & E \end{bmatrix} \cong V_2'' = \begin{bmatrix} M_\Theta & 0 & 0 \\ C_1 & M_z^{\mathcal{D}_{E^*}} & 0 \\ C_2 & 0 & Y_2 \end{bmatrix},$$

where $Y_2 = E|_{\mathcal{Q}_{E^*}}$, the unitary part of the isometry $E$, is unitary on $\mathcal{Q}_{E^*}$. From property (3.55) we see that $C \cong \left[\begin{smallmatrix} C_1 \\ C_2 \end{smallmatrix}\right]$ actually has the form $\left[\begin{smallmatrix} C_1 \\ 0 \end{smallmatrix}\right]$ (i.e., $C_2 = 0$) and the factorization identity $C^*C = I_{H^2(\mathcal{F})} - M_\Theta^* M_\Theta$ collapses to

$$C_1^* C_1 = I_{H^2(\mathcal{F})} - M_\Theta^* M_\Theta.$$

Then commutativity of $V_1''$ with $V_2''$ means that

$$\begin{bmatrix} W_{11}^{(1)} & W_{12}^{(1)} \\ W_{21}^{(1)} & W_{22}^{(1)} \end{bmatrix} \begin{bmatrix} C_1 \\ 0 \end{bmatrix} = \begin{bmatrix} C_1 M_z^{\mathcal{F}} \\ 0 \end{bmatrix}, \tag{3.58}$$

$$\begin{bmatrix} W_{11}^{(1)} & W_{12}^{(1)} \\ W_{21}^{(1)} & W_{22}^{(1)} \end{bmatrix} \begin{bmatrix} M_z^{\mathcal{D}_{E^*}} & 0 \\ 0 & Y_2 \end{bmatrix} = \begin{bmatrix} M_z^{\mathcal{D}_{E^*}} & 0 \\ 0 & Y_2 \end{bmatrix} \begin{bmatrix} W_{11}^{(1)} & W_{12}^{(1)} \\ W_{21}^{(1)} & W_{22}^{(1)} \end{bmatrix}. \tag{3.59}$$

Equation (3.59) combined with part (ii) of Lemma 2.3.1 then tells us that $W_1 = \begin{bmatrix} W_{11}^{(1)} & W_{12}^{(1)} \\ W_{21}^{(1)} & W_{22}^{(1)} \end{bmatrix}$ is block-diagonal (i.e., $W_{12}^{(1)} = 0$ and $W_{21}^{(1)} = 0$) and we are left with $W_1 = \begin{bmatrix} W_{11}^{(1)} & 0 \\ 0 & W_{22}^{(1)} \end{bmatrix}$. Then the new information in (3.58) is

$$W_{11}^{(1)} C_1 = C_1 M_z^{\mathcal{F}}. \tag{3.60}$$

As $W_1 = \begin{bmatrix} W_{11}^{(1)} & 0 \\ 0 & W_{22}^{(1)} \end{bmatrix}$ is unitary, it follows that $W_{11}^{(1)}$ is unitary on $H^2(\mathcal{D}_{E^*})$ and $W_{22}^{(1)}$ is unitary on $\mathcal{Q}_{E^*}$. As $W_1$ commutes with $\begin{bmatrix} M_z^{\mathcal{D}_{E^*}} & 0 \\ 0 & \mathcal{Q}_{E^*} \end{bmatrix}$, we see that

$$W_{11}^{(1)} M_z^{\mathcal{D}_{E^*}} = M_z^{\mathcal{D}_{E^*}} W_{11}^{(1)}, \quad W_{22}^{(1)} \mathcal{Q}_{E^*} = \mathcal{Q}_{E^*} W_{22}^{(1)}.$$

As $W_{11}^{(1)}$ is unitary and $M_z^{\mathcal{D}_{E^*}}$ is a shift, the first equation then forces $W_{11}^{(1)}$ to have the form $W_{11}^{(1)} = I_{H^2} \otimes u$, where $u$ is a unitary operator on the coefficient space $\mathcal{D}_{E^*}$. We now set

$$Y_1 = W_{22}^{(1)}, \quad Y_2 = W_{22}^{(2)}.$$

With this notation in place, we now see that all assertions of Step 2 have been verified.

**Step 3** *Assume notation as in the assertions of Step 2. Then we have the following.*

(i) *The operator $C_1 \colon H^2(\mathcal{F}) \to \mathcal{D}_{E^*}$ factors as*

$$C_1 = \iota \Delta_\Theta |_{H^2(\mathcal{F})}.$$

*Here $\Delta_\Theta$ is the operator of multiplication by the function*

$$\Delta_\Theta(\zeta) := \left(I_{L^2(\mathcal{F})} - \Theta(\zeta)^* \Theta(\zeta)\right)^{\frac{1}{2}} \text{ acting on } L^2(\mathcal{F})$$

*and $\iota \colon \Delta_\Theta H^2(\mathcal{F}) \to \mathcal{D}_{E^*}$ is an isometry, where we here identify $\mathcal{D}_{E^*}$ with its image under the isometric embedding operator $\mathbf{ev}_{0,\mathcal{D}_{E^*}} \colon \mathcal{D}_{E^*} \to H^2(\mathcal{D}_{E^*})$, that is, the space of constant functions in $H^2(\mathcal{D}_{E^*})$.*

(ii) *The isometry* $\iota\colon \Delta_\Theta H^2(\mathcal{F}) \to \mathcal{D}_{E^*}$ *extends to an isometry* $\hat{\iota}\colon \overline{\Delta_\Theta L^2(\mathcal{F})} \to \mathcal{D}_{E^*}$ *uniquely determined by the validity of the intertwining condition*

$$\hat{\iota} M_{\zeta^{-1}} = u^* \hat{\iota}. \tag{3.61}$$

In detail, note that

$$\overline{\cup_{n=1}^{\infty} M_{\zeta^{-n}} \Delta_\Theta H^2(\mathcal{F})} = \overline{\Delta_\Theta L^2(\mathcal{F})}.$$

Hence we may view $\hat{\iota}$ as an isometric identification map of $\overline{\Delta_\Theta L^2(\mathcal{F})}$ with the subspace $\hat{\iota}\left(\overline{\Delta_\Theta L^2(\mathcal{F})}\right)$ of $\mathcal{Q}_{E^*}$.

(iii) *The commuting isometric operator pair* $(V_1'', V_2'')$ *from Step 2 is unitarily equivalent to the commuting isometric operator pair* $(V_1''', V_2''')$ *acting on the space*

$$\mathcal{K}''' = \begin{bmatrix} H^2(\mathcal{F}) \\ H^2(\overline{\Delta_\Theta L^2(\mathcal{F})}) \\ H^2(\mathcal{E}_0) \\ \mathcal{Q}_{E^*}. \end{bmatrix},$$

*given by*

$$(V_1''', V_2''') = \left( \begin{bmatrix} M_z^{\mathcal{F}} & 0 & 0 & 0 \\ 0 & I_{H^2} \otimes M_\zeta & 0 & 0 \\ 0 & 0 & I_{H^2} \otimes u_{\mathcal{E}_0} & 0 \\ 0 & 0 & 0 & Y_1 \end{bmatrix}, \begin{bmatrix} M_\Theta & 0 & 0 & 0 \\ \Delta_\Theta|_{H^2(\mathcal{F})} & M_z^{\mathcal{D}_{E^*}} & 0 & 0 \\ 0 & 0 & M_z^{\mathcal{E}_0} & 0 \\ 0 & 0 & 0 & Y_2 \end{bmatrix} \right), \tag{3.62}$$

*where* $\mathcal{E}_0$ *is yet another coefficient space and* $u_{\mathcal{E}_0}$ *is a unitary operator on* $\mathcal{E}_0$.

Note that item (i) follows as a consequence of the factorization

$$C_1^* C_1 = I_{H^2(\mathcal{F})} - M_\Theta^* M_\Theta = P_{H^2(\mathcal{F})} \Delta_\Theta^* \Delta_\Theta|_{H^2(\mathcal{F})}.$$

The fact that we can take the range space of $\iota$ to be $\mathcal{D}_{E^*} \cong \operatorname{Ran} \mathbf{ev}_0^* \mathcal{D}_{E^*} \subset H^2(\mathcal{D}_{E^*})$ (rather than all of $H^2(\mathcal{D}_{E^*})$) is a consequence of property (3.55) already noted in Step 1.

As for item (ii), we extend $\iota$ to an operator $\hat{\iota}\colon \overline{\Delta_\Theta L^2(\mathcal{F})} \to \mathcal{D}_{V_1^*}$ while maintaining the intertwining relation $\hat{\iota} M_\zeta = u \hat{\iota}$ as follows. Note that a dense subset of $\overline{\Delta_\Theta L^2(\mathcal{F})}$ is $\bigcup_{n=0}^{\infty} \Delta_\Theta \zeta^{-n} H^2(\mathcal{F})$; hence by an approximation argument it suffices to define $\hat{\iota}$ on $\Delta_\Theta \zeta^{-n} H^2(\mathcal{F})$ for an arbitrary fixed $n \in \mathbb{Z}_+$. Define $\hat{\iota}$ on $\zeta^{-n} \Delta_\Theta f$ (for $f \in H^2(\mathcal{F})$) by

$$\hat{\iota} \zeta^{-n} \Delta_\Theta f = u^{*n} \iota \Delta_\Theta f.$$

Note that then

$$\|\hat{\iota}\zeta^{-n}\Delta_\Theta f\| = \|u^{*n}\iota\Delta_\Theta f\| = \|\iota\Delta_\Theta f\| = \|\Delta_\Theta f\| = \|\zeta^{-n}\Delta_\Theta f\|,$$

so $\hat{\iota}$ is isometric on $\zeta^{-n}\Delta_\Theta H^2(\mathcal{F})$ for each $n \in \mathbb{Z}_+$. Moreover

$$\begin{aligned}\hat{\iota}(\zeta \cdot \zeta^{-n}\Delta_\Theta f) &= \hat{\iota}(\zeta^{-n}\Delta_\Theta(\zeta \cdot f) = u^{*n}\iota(\zeta \cdot \Delta_\Theta f) \quad \text{(by (3.61))}\\ &= u \cdot u^{*n}\iota(\Delta_\Theta f) = u\iota(\zeta^{-n}\Delta_\Theta f),\end{aligned}$$

showing that $\hat{\iota}M_\zeta = u\hat{\iota}$ as desired.

To show that $(V_1'', V_2'')$ (3.56) is unitarily equivalent to $(V_1''', V_2''')$ (3.62) (for some choice of coefficient space $\mathcal{E}_0$) as claimed in item (iii), we proceed as follows. As a first step we note that the isometry $\hat{\iota}\colon \overline{\Delta_\Theta L^2(\mathcal{F})} \to \mathcal{D}_{E^*}$ has the intertwining property

$$\hat{\iota}M_\zeta = u_1\hat{\iota}$$

as well as (3.61). Moreover, we can view $\hat{\iota}$ as a unitary transformation from $\overline{\Delta_\Theta L^2(\mathcal{F})}$ onto its image $\hat{\iota}\left(\overline{\Delta_\Theta L^2(\mathcal{F})}\right) \subset \mathcal{D}_{E^*}$ that furthermore establishes a unitary equivalence between the operators $M_\zeta|_{\overline{\Delta_\Theta L^2(\mathcal{F})}}$ and $u|_{\operatorname{Ran}\hat{\iota}}$. As $\operatorname{Ran}\hat{\iota}$ is reducing for $u$, so also is $\mathcal{E}_0 := \mathcal{D}_{E^*} \ominus \operatorname{Ran}\hat{\iota}$, and the unitary operator $u$ on $\mathcal{D}_{E^*}$ can be decomposed as

$$u \cong \begin{bmatrix} \hat{\iota}M_\zeta\hat{\iota}^* & 0 \\ 0 & u|_{\mathcal{E}_0} \end{bmatrix} \quad \text{acting on } \begin{bmatrix} \operatorname{Ran}\hat{\iota} \\ \mathcal{E}_0 \end{bmatrix}. \tag{3.63}$$

However, as observed previously, $\hat{\iota}\colon \overline{\Delta_\Theta L^2(\mathcal{F})} \to \operatorname{Ran}\hat{\iota}$ is a unitary transformation implementing a unitary equivalence between $M_\zeta$ on $\overline{\Delta_\Theta L^2(\mathcal{F})}$ and $u|_{\operatorname{Ran}\hat{\iota}}$. Hence, up to unitary equivalence, we may simply drop $\hat{\iota}$, and view $\operatorname{Ran}\hat{\iota}$ as equal to $\overline{\Delta_\Theta L^2(\mathcal{F})}$ and $u|_{\operatorname{Ran}\hat{\iota}}$ as equal to $M_\zeta|_{\overline{\Delta_\Theta L^2(\mathcal{F})}}$. If we now set $u_{\mathcal{E}_0} = u|_{\mathcal{E}_0}$, we see that (3.63) becomes

$$u \cong \begin{bmatrix} M_\zeta & 0 \\ 0 & u_{\mathcal{E}_0} \end{bmatrix} \quad \text{on} \quad \begin{bmatrix} \overline{\Delta_\Theta L^2(\mathcal{F})} \\ \mathcal{E}_0 \end{bmatrix}$$

and we see that $(V_1'', V_2'')$ as in (3.56) is unitarily equivalent to $(V_1''', V_2''')$ as in (3.62).

Before formalizing the final result, it is convenient to introduce the following notion of the *Bercovici–Douglas–Foias tuple* (or BDF-tuple for short) analogous to the notion of the BCL-tuple introduced in Section 3.1 in connection with the BCL-model theory.

**Definition 3.2.1** We say that any collection of objects $\Xi$ of the form $\Xi = (\mathcal{F}, \mathcal{E}_0, \Theta, u_{\mathcal{E}_0}, Y_1, Y_2)$ where

(i) $\mathcal{F}$ and $\mathcal{E}_0$ are coefficient Hilbert spaces,
(ii) $(\mathcal{F}, \Theta)$ is a contractive analytic function,
(iii) $u_{\mathcal{E}_0}$ is a unitary operator on $\mathcal{E}_0$,
(iv) $(Y_1, Y_2)$ is a commuting unitary operator pair on another Hilbert space $\mathcal{K}_u$

is a *BDF-tuple*.

Given a BDF-tuple $\Xi := (\mathcal{F}, \mathcal{E}_0, \Theta, u_{\mathcal{E}_0}, Y_1, Y_2)$, if we define an operator pair given by

$$(V_{\Xi,1}, V_{\Xi,2}) =$$
$$\left( \begin{bmatrix} M_z^{\mathcal{F}} & 0 & 0 & 0 \\ 0 & I_{H^2} \otimes M_\zeta & 0 & 0 \\ 0 & 0 & I_{H^2} \otimes u_{\mathcal{E}_0} & 0 \\ 0 & 0 & 0 & Y_1 \end{bmatrix}, \begin{bmatrix} M_\Theta & 0 & 0 & 0 \\ \Delta_\Theta|_{H^2(\mathcal{F})} & M_z^{\overline{\Delta_\Theta L^2(\mathcal{F})}} & 0 & 0 \\ 0 & 0 & M_z^{\mathcal{E}_0} & 0 \\ 0 & 0 & 0 & Y_2 \end{bmatrix} \right) \tag{3.64}$$

acting on the space $\mathcal{K} = \begin{bmatrix} H^2(\mathcal{F}) \\ H^2(\overline{\Delta_\Theta L^2(\mathcal{F})}) \\ H^2(\mathcal{E}_0) \\ \mathcal{K}_u \end{bmatrix}$, where $\Delta_\Theta$ is the operator of multiplication by the function $\Delta_\Theta(\zeta) = (I_{\mathcal{F}} - \Theta(\zeta)^*\Theta(\zeta))^{1/2}$ with values acting on $L^2(\mathcal{F})$, then we say that $(V_{\Xi,1}, V_{\Xi,2})$ (3.64) is the *BDF1-model* operator pair associated with the BDF-tuple $\Xi$.

Alternatively, suppose that $\Xi' = \left(\mathcal{F}', \mathcal{E}_0', \Theta', u'_{\mathcal{E}_0} Y_1', Y_2'\right)$ is a possibly different BDF-tuple and we associate the operator pair $\left(V'_{\Xi',1}, V'_{\Xi',2}\right)$ given by

$$\left(V'_{\Xi',1}, V'_{\Xi',2}\right) =$$
$$\left( \begin{bmatrix} M_{\Theta'} & 0 & 0 & 0 \\ \Delta_{\Theta'}|_{H^2(\mathcal{F}')} & M_z^{\overline{\Delta_{\Theta'} L^2(\mathcal{F}')}} & 0 & 0 \\ 0 & 0 & M_z^{\mathcal{E}_0'} & 0 \\ 0 & 0 & 0 & Y_1' \end{bmatrix}, \begin{bmatrix} M_z^{\mathcal{F}'} & 0 & 0 & 0 \\ 0 & I_{H^2} \otimes M_\zeta & 0 & 0 \\ 0 & 0 & I_{H^2} \otimes u_{\mathcal{E}_0'} & 0 \\ 0 & 0 & 0 & Y_2' \end{bmatrix} \right) \tag{3.65}$$

acting on the space $\mathcal{K}' = \begin{bmatrix} H^2(\mathcal{F}') \\ H^2(\overline{\Delta_{\Theta'} L^2(\mathcal{F}')}) \\ H^2(\mathcal{E}_0') \\ \mathcal{K}_u' \end{bmatrix}$. Then we say that $\left(V'_{\Xi',1}, V'_{\Xi',2}\right)$ (3.65) is the *BDF2-model* operator pair associated with the BDF-tuple $\Xi' = \left(\mathcal{F}', \mathcal{E}', \Theta', u', Y_1', Y_2'\right)$.

Note that when $\Xi = \Xi'$ we have

$$\left(V'_{\Xi,1}, V'_{\Xi,2}\right) = (V_{\Xi,2}, V_{\Xi,1}). \tag{3.66}$$

The next result gives a compact summary of the previous discussion.

**Theorem 3.2.2** *Given any commuting isometric operator pair $(V_1, V_2)$, we have the following:*

(i) *There is a BDF-tuple $\Xi = (\mathcal{F}, \mathcal{E}_0, \Theta, u_{\mathcal{E}_0}, Y_1, Y_2)$ so that $(V_1, V_2)$ is unitarily equivalent to the BDF1-functional-model commuting isometric pair $(V_{\Xi,1}, V_{\Xi,2})$ as in* (3.64).

(ii) *There is a BDF-tuple $\Xi' = \left(\mathcal{F}', \mathcal{E}', \Theta', u'_{\mathcal{E}}, Y'_1, Y'_2\right)$ so that $(V_1, V_2)$ is unitarily equivalent to the BDF2-functional-model commuting isometric pair $(V_{\Xi,'1}, V_{\Xi',2})$ as in* (3.65).

*Conversely, given any BDF-tuple, the associated BDF1-model as well as the associated BDF2-model are commuting isometric pairs.*

### 3.2.2 Decompositions of the Model

Note that the BDF1-model for a commuting isometric operator pair (3.64) breaks up as the orthogonal direct sum of three pieces, namely

$$\begin{aligned}
\left(V_1^{(1)}, V_2^{(1)}\right)_1 &= \left(\begin{bmatrix} M_z^{\mathcal{F}} & 0 \\ 0 & I_{H^2} \otimes M_\zeta \end{bmatrix}, \begin{bmatrix} M_\Theta & 0 \\ \Delta_\Theta|_{H^2(\mathcal{F})} & M_z^{\overline{\Delta_\Theta L^2(\mathcal{F})}} \end{bmatrix}\right) \\
&\quad \text{on } \mathcal{K}^{(1)} = \begin{bmatrix} H^2(\mathcal{F}) \\ H^2(\overline{\Delta_\Theta L^2(\mathcal{F})}) \end{bmatrix}, \\
\left(V_1^{(1)}, V_2^{(1)}\right)_2 &= \left(I_{H^2} \otimes u_{\mathcal{E}_0}, M_z^{\mathcal{E}_0}\right) \quad \text{on } \mathcal{K}^{(2)} := H^2(\mathcal{E}_0), \\
\left(V_1^{(1)}, V_2^{(1)}\right)_3 &= (Y_1, Y_2) \quad \text{on } \mathcal{K}^{(3)} := \mathcal{K}_u.
\end{aligned} \tag{3.67}$$

Note that the second component $\left(V_1^{(1)}, V_2^{(1)}\right)_2$ is characterized as being a unitary-shift commuting pair while the first component $\left(V_1^{(1)}, V_1^{(1)}\right)_1$ is characterized as being *1-(completely nonunitary)*, or 1-c.n.u. for short, by which we mean the following (also see [30, Section 5]).

**Definition 3.2.3** A commuting isometric pair $(V_1, V_2)$ is said to be *1-(completely nonunitary)*, or 1-c.n.u. for short, if there is no nonzero $(V_1, V_2)$-jointly reducing subspace $\mathcal{M}$ such that $V_1|_{\mathcal{M}}$ is unitary. Similarly, a commuting isometric pair $(V_1, V_2)$ is said to be *2-(completely nonunitary)*, or 2-c.n.u. for short, if there is no nonzero $(V_1, V_2)$-jointly reducing subspace $\mathcal{M}$ such that $V_2|_{\mathcal{M}}$ is unitary.

Similarly, the BDF2-model for a commuting isometric operator pair (3.65) breaks up as the orthogonal direct sum of three pieces, namely

$$\left(V_1^{(2)}, V_2^{(2)}\right)_1 = \left( \begin{bmatrix} M_{\Theta'} & 0 \\ \Delta_{\Theta'}|_{H^2(\mathcal{F}')} & M_z^{\overline{\Delta_{\Theta'} L^2(\mathcal{F}')}} \end{bmatrix}, \begin{bmatrix} M_z^{\mathcal{F}'} & 0 \\ 0 & I_{H^2} \otimes M_\zeta \end{bmatrix} \right)$$

$$\text{on } \mathcal{K}^{(1)\prime} = \begin{bmatrix} H^2(\mathcal{F}') \\ H^2(\overline{\Delta_{\Theta'} L^2(\mathcal{F}')}) \end{bmatrix},$$

$$\left(V_1^{(2)}, V_2^{(2)}\right)_2 = \left(M_z^{\mathcal{E}_0'}, I_{H^2} \otimes u_{\mathcal{E}_0'}\right) \quad \text{on } \mathcal{K}^{(2)\prime} := H^2(\mathcal{E}_0'),$$

$$\left(V_1^{(2)}, V_2^{(2)}\right)_3 = (Y_1', Y_2') \quad \text{on } \mathcal{K}^{(3)\prime} := \mathcal{K}_u'. \tag{3.68}$$

Note that $\left(V_1^{(2)} V_2^{(2)}\right)_1$ is 2-c.n.u., while $\left(V_1^{(2)}, V_2^{(2)}\right)_2$ is shift/unitary and $\left(V_1^{(2)}, V_2^{(2)}\right)_3$ is unitary/unitary.

We therefore arrive at Corollary 3.2.4.

**Corollary 3.2.4** *Let $(V_1, V_2)$ be a commuting isometric operator pair on a Hilbert space $\mathcal{K}$.*

(i) *Then there is an orthogonal decomposition $\mathcal{K} = \mathcal{K}_{1\text{cnu}} \oplus \mathcal{K}_{\text{us}} \oplus \mathcal{K}_{\text{uu}}$ with each of $\mathcal{K}_{1\text{cnu}}$, $\mathcal{K}_{\text{us}}$, $\mathcal{K}_{\text{uu}}$ jointly reducing for $(V_1, V_2)$ such that*
   - (a) $(V_1, V_2)|_{\mathcal{K}_{1\text{cnu}}}$ *is 1-c.n.u.,*
   - (b) $(V_1, V_2)|_{\mathcal{K}_{\text{us}}}$ *is unitary/shift,*
   - (c) $(V_1, V_2)|_{\mathcal{K}_{\text{uu}}}$ *is unitary/unitary.*

(ii) *Then there is an orthogonal decomposition $\mathcal{K} = \mathcal{K}'_{2\text{cnu}} \oplus \mathcal{K}'_{\text{su}} \oplus \mathcal{K}'_{\text{uu}}$ with each of $\mathcal{K}'_{2\text{cnu}}$, $\mathcal{K}'_{\text{su}}$, $\mathcal{K}'_{\text{uu}}$ jointly reducing for $(V_1, V_2)$ such that*
   - (a) $(V_1, V_2)|_{\mathcal{K}'_{2\text{cnu}}}$ *is 2-c.n.u.,*
   - (b) $(V_1, V_2)|_{\mathcal{K}'_{\text{su}}}$ *is shift/unitary,*
   - (c) $(V_1, V_2)|_{\mathcal{K}'_{\text{uu}}}$ *is unitary/unitary.*

**Remark 3.2.5** In general, if a commuting isometric pair $(V_1, V_2)$ is 1-c.n.u. or unitary/shift, essentially by definition it cannot have a nontrivial jointly reducing subspace on which its restriction is unitary/unitary, and similarly for a 2-c.n.u. or shift/unitary commuting isometric pair. Furthermore, as a consequence of Step 1 in the construction of the BCL-models (3.2) or (3.3), we see that the maximal reducing subspace on which $(V_1, V_2)$ is unitary/unitary is uniquely determined as the ambient space for the unitary part in the Wold decomposition of the product isometry $V = V_1 V_2 = V_2 V_1$. Hence in Corollary 3.2.4 we must have $\mathcal{K}_{\text{uu}} = \mathcal{K}'_{\text{uu}}$. Hence also, the restriction of the product isometry $V_1 V_2 = V_2 V_1$ to any of the subspaces $\mathcal{K}_{1\text{cnu}}$ or $\mathcal{K}_{\text{us}}$ in item (i) of Corollary 3.2.4, as well as the restriction of the product isometry $V_1 V_2 = V_2 V_1$ to any of the subspaces $\mathcal{K}'_{2\text{cnu}}$ or $\mathcal{K}'_{\text{su}}$ in item (2) of Corollary 3.2.4, must necessarily be a shift.

It is possible to get a more satisfactory synthesis of the preceding two results as follows. Let $(V_1, V_2)$ be a commuting isometric pair acting on the Hilbert space $\mathcal{K}$. In the following discussion, we use the notation $\mathcal{K}_x$ for a subspace of $\mathcal{K}$ and for $x$ equal to any of the symbols 1cnu, 2cnu, 1&2cnu, us, su, uu to indicate that $(V_1, V_2)|_{\mathcal{K}_x}$ has property $x$ (i.e., 1-c.n.u., 2-c.n.u., 1&2-c.n.u., unitary/shift, shift/unitary, unitary/unitary). As a consequence of Corollary 3.2.4 we get two orthogonal decompositions of $\mathcal{K}$:

$$\begin{aligned} \mathcal{K} &= \mathcal{K}_{1\text{cnu}} \oplus \mathcal{K}_{\text{us}} \oplus \mathcal{K}_{\text{uu}} \\ &= \mathcal{K}'_{2\text{cnu}} \oplus \mathcal{K}'_{\text{su}} \oplus \mathcal{K}'_{\text{uu}}. \end{aligned} \tag{3.69}$$

As a consequence of Remark 3.2.5 we know that $\mathcal{K}_{\text{uu}} = \mathcal{K}'_{\text{uu}}$, so we may write

$$\begin{aligned} \mathcal{K} \ominus \mathcal{K}_{\text{uu}} &= \mathcal{K}_{1\text{cnu}} \oplus \mathcal{K}_{\text{us}} \\ &= \mathcal{K}'_{2\text{cnu}} \oplus \mathcal{K}'_{\text{su}}. \end{aligned}$$

Let us next apply the 1-c.n.u. analysis to the space $\mathcal{K}'_{2\text{cnu}}$ rather than to all of $\mathcal{K}$ as we did at the start of this analysis. By part (i) of Corollary (3.2.4) applied to $(V_1, V_2)|_{\mathcal{K}'_{2\text{cnu}}}$ we see that

$$\mathcal{K}'_{2\text{cnu}} = \mathcal{K}''_{1\text{cnu}} \oplus \mathcal{K}''_{\text{us}} \oplus \mathcal{K}''_{\text{uu}}.$$

However, we have also seen that $\mathcal{K}''_{\text{uu}} \subset \mathcal{K}_{\text{uu}}$ and that $\mathcal{K}'_{2\text{cnu}} \perp \mathcal{K}_{\text{uu}}$, so this last decomposition can be refined to

$$\mathcal{K}'_{2\text{cnu}} = \mathcal{K}''_{1\text{cnu}} \oplus \mathcal{K}''_{\text{us}}. \tag{3.70}$$

This exhibits the 1-c.n.u. subspace $\mathcal{K}''_{1\text{cnu}}$ as a subspace of the 2-c.n.u. subspace $\mathcal{K}'_{2\text{cnu}}$, so we see that $(V_1, V_2)|_{\mathcal{K}''_{1\text{cnu}}}$ is both 1-c.n.u. and 2-c.n.u., enabling us to write $\mathcal{K}_{1\&2\text{cnu}}$ in place of $\mathcal{K}''_{1\text{cnu}}$; plugging this substitution into (3.70) gives us

$$\mathcal{K}'_{2\text{cnu}} = \mathcal{K}_{1\&2\text{cnu}} \oplus \mathcal{K}''_{\text{us}}.$$

Finally, plugging this substitution into (3.69) gives us

$$\mathcal{K} = \mathcal{K}_{1\&2\text{cnu}} \oplus \mathcal{K}''_{\text{us}} \oplus \mathcal{K}'_{\text{su}} \oplus \mathcal{K}_{\text{uu}}.$$

In this way we arrive at the following result.

**Corollary 3.2.6** *Let $(V_1, V_2)$ be a commuting isometric operator pair on the Hilbert space $\mathcal{K}$. Then there is an orthogonal decomposition*

$$\mathcal{K} = \mathcal{K}_{1\&2\text{cnu}} \oplus \mathcal{K}_{\text{us}} \oplus \mathcal{K}_{\text{su}} \oplus \mathcal{K}_{\text{uu}} \tag{3.71}$$

*with each of $\mathcal{K}_{1\&2\text{cnu}}$, $\mathcal{K}_{\text{us}}$, $\mathcal{K}_{\text{su}}$, $\mathcal{K}_{\text{uu}}$ jointly reducing for $(V_1, V_2)$ such that*

(i) $(V_1, V_2)|_{\mathcal{K}_{1\&2cnu}}$ *is both 1-c.n.u. and 2-c.n.u.,*

(ii) $(V_1, V_2)|_{\mathcal{K}_{su}}$ *is shift/unitary,*

(iii) $(V_1, V_2)|_{\mathcal{K}_{us}}$ *is unitary/shift,*

(iv) $(V_1, V_2)|_{\mathcal{K}_{uu}}$ *is unitary/unitary.*

**Remark 3.2.7** We note that the class of commuting isometric pairs that are both 1-c.n.u. and 2-c.n.u. (1&2-c.n.u. for short) is the same as what is called a *weak bishift* in [23]. In any case this decomposition recovers the Wold decomposition for a general commuting isometric pair originally due to Gaşper–Gaşper [63] as well as in work of Popovici [76], Bercovici–Douglas–Foias [30, Corollary 4.3] and the authors [23, Theorem 3.13]. Our work in [23] derived the result as an application of the BCL-model theory while that of [30] and our work here derive the result as an application of the BDF-model.

As a corollary of the construction of the BDF1-model for a 1-c.n.u. commuting isometric pair, we obtain the following special case of the Ando Lifting Theorem.

**Corollary 3.2.8** *Suppose that* $\mathbf{T} = (T_1, T_2)$ *is a commuting shift/contractive operator pair. Then* $\mathbf{T}$ *has an Ando lift* $\mathbf{V} = (V_1, V_2)$*, which is a 1-c.n.u. commuting isometric pair. If we model* $(T_1, T_2)$ *as*

$$(T_1, T_2) = \left(M_z^{\mathcal{F}}, M_\Theta\right) \quad \text{on } H^2(\mathcal{F}),$$

*where* $(\mathcal{F}, \Theta)$ *is a contractive analytic function, then the 1-c.n.u. Ando lift* $\mathbf{V} = (V_1, V_2)$ *is given by*

$$(V_1, V_2) = \left(\begin{bmatrix} M_z^{\mathcal{F}} & 0 \\ 0 & I_{H^2} \otimes M_\zeta \end{bmatrix}, \begin{bmatrix} M_\Theta & 0 \\ \Delta_\Theta|_{H^2(\mathcal{F})} & M_z^{\overline{\Delta_\Theta L^2(\mathcal{F})}} \end{bmatrix}\right) \quad \text{on } \begin{bmatrix} H^2(\mathcal{F}) \\ \overline{\Delta_\Theta L^2(\mathcal{F})} \end{bmatrix}.$$

At this stage we have explicit functional models for the following classes of commuting isometric operator pairs: 1-c.n.u., 2-c.n.u., shift/unitary, unitary/shift, unitary/unitary. What appears to be missing is an explicit functional model for a commuting isometric pair that is 1&2-c.n.u. As a first step, we obtain the following characterization of the $(M_z^{\mathcal{F}}, M_\Theta)$-joint reducing subspaces $\mathcal{M}$ of $H^2(\mathcal{F})$ such that $M_\Theta|_{\mathcal{M}}$ is unitary.

**Theorem 3.2.9** *Let* $\mathcal{F}$ *be a coefficient Hilbert space.*

(i) *Then reducing subspaces* $\mathcal{M}$ *for the shift operator* $M_z^{\mathcal{F}}$ *on* $H^2(\mathcal{F})$ *are of the form* $\mathcal{M} = H^2(\mathcal{F}_0)$ *for any subspace* $\mathcal{F}_0$*. Here* $H^2(\mathcal{F}_0)$ *is identified as a subspace of* $H^2(\mathcal{F})$ *in the canonical way:*

$$H^2(\mathcal{F}_0) = \left\{ \sum_{n=0}^{\infty} f_n^{(0)} z^n \colon f_n^{(0)} \in \mathcal{F}_0 \text{ and } \sum_{n=0}^{\infty} \left\| f_n^{(0)} \right\|^2 < \infty \right\}$$
$$\subset H^2(\mathcal{F}) = \left\{ \sum_{n=0}^{\infty} f_n z^n \colon f_n \in \mathcal{F} \text{ and } \sum_{n=0}^{\infty} \|f_n\|^2 < \infty \right\}.$$

(ii) *Let $(\mathcal{F}, \Theta)$ be a contractive analytic function. Then the jointly $(M_z^{\mathcal{F}}, M_\Theta)$-reducing subspaces $\mathcal{M}$ of $H^2(\mathcal{F}_0)$ such that $M_\Theta|_{\mathcal{M}}$ is unitary correspond to those subspaces $\mathcal{F}_0 \subset \mathcal{F}$ such that $\Theta(z)|_{\mathcal{F}_0} = \Theta(0)|_{\mathcal{F}_0}$ for all $z \in \mathbb{D}$, where $\Theta(0)|_{\mathcal{F}_0}$ is a unitary operator on $\mathcal{F}_0$.*

(iii) *The maximum joint $(M_z, M_\Theta)$-reducing subspace $\mathcal{M} \subset H^2(\mathcal{F})$ such that $M_\Theta|_{\mathcal{M}}$ is unitary is equal to $H^2(\mathcal{F}_{\rm u})$, where $\mathcal{F}_{\rm u} \subset H^2(\mathcal{F})$ is the subspace of $\mathcal{F}$ that comes up in the canonical decomposition of $\Theta(0)$ as an operator on $\mathcal{F}$:*

$$\Theta(0) \underset{\rm u}{\cong} \begin{bmatrix} \Theta(0)_{\rm u} & 0 \\ 0 & \Theta(0)_{\rm cnu} \end{bmatrix},$$

*where $\Theta(0)_{\rm u}$ is unitary on $\mathcal{F}_{\rm u}$ and $\Theta(0)_{\rm cnu}$ is c.n.u. on $\mathcal{F}_{\rm cnu}$, and $\mathcal{F} \underset{\rm u}{\cong} \begin{bmatrix} \mathcal{F}_{\rm u} \\ \mathcal{F}_{\rm cnu} \end{bmatrix}$.*

*Proof of (i)* A proof of this statement can be found in the first paragraph of the proof of Theorem 3.3 in [23] but we repeat the proof here for completeness. Suppose that $\mathcal{M} \subset H^2(\mathcal{F})$ is reducing for $M_z^{\mathcal{F}}$. Then, in particular, $\mathcal{M}$ and $\mathcal{M}^\perp$ are invariant for $M_z^{\mathcal{F}}$ and hence have Beurling–Lax representations $\mathcal{M} = \Psi_1 H^2(\mathcal{E}_1)$ and $\mathcal{M}^\perp = \Psi_2 H^2(\mathcal{E}_2)$ for inner functions $(\mathcal{E}_1, \mathcal{F}, \Psi_1)$ and $(\mathcal{E}_2, \mathcal{F}, \Psi_2)$. As $\mathcal{M}$ and $\mathcal{M}^\perp$ are mutually orthogonal in $H^2(\mathcal{F})$ it follows that

$$\left( \begin{bmatrix} \mathcal{E}_1 \\ \mathcal{E}_2 \end{bmatrix}, \mathcal{F}, \begin{bmatrix} \Psi_1 & \Psi_2 \end{bmatrix} \right)$$

is inner as well. As $\mathcal{M} \oplus \mathcal{M}^\perp = H^2(\mathcal{F})$, it follows that this last inner function is a Beurling–Lax representer for the whole space $H^2(\mathcal{F})$, which trivially has Beurling–Lax representer $(\mathcal{F}, I_{\mathcal{F}})$. By the uniqueness assertion in the Beurling–Lax theorem it follows that there is a constant unitary $\begin{bmatrix} \Psi_1' & \Psi_2' \end{bmatrix}$ so that $\begin{bmatrix} \Psi_1 & \Psi_2 \end{bmatrix} = I_{\mathcal{F}} \begin{bmatrix} \Psi_1' & \Psi_2' \end{bmatrix}$, which is to say that $\begin{bmatrix} \Psi_1 & \Psi_2 \end{bmatrix}$ is a constant unitary matrix. Now it remains only to set $\mathcal{F}_0 = \operatorname{Ran} \Psi_1$ and $\mathcal{F}_0^\perp = \operatorname{Ran} \Psi_2$ to see that $\mathcal{M} = H^2(\mathcal{F}_0)$ and $\mathcal{M}^\perp = H^2\left(\mathcal{F}_0^\perp\right)$ as claimed.

*Proof of (ii)* Let $\mathcal{M}$ be a jointly $\left(M_z^{\mathcal{F}}, M_\Theta\right)$-reducing subspace of $H^2(\mathcal{F})$. Since, in particular, by assumption $\mathcal{M}$ is reducing for $M_z^{\mathcal{F}}$, by part (i) already proved we know that $\mathcal{M}$ has the form $\mathcal{M} = H^2(\mathcal{F}_0)$ for some subspace $\mathcal{F}_0 \subset \mathcal{F}$. It remains to understand when a subspace of the form $H^2(\mathcal{F}_0)$ is not only reducing for $M_\Theta$

but also has the property that the restriction operator $M_\Theta|_{H^2(\mathcal{F}_0)}$ is unitary as an operator on $H^2(\mathcal{F}_0)$.

We therefore suppose that $\mathcal{F}_0$ is such a subspace. In particular, $H^2(\mathcal{F}_0)$ is invariant for the multiplication operator $M_\Theta$. If we let $M_\Theta$ act on a constant function $f_0 \in \mathcal{F}_0$, the result $\Theta(z)f_0$ is an element of $H^2(\mathcal{F}_0)$, so $\Theta|_{\mathcal{F}_0}$ can be viewed as a $\mathcal{B}(\mathcal{F}_0)$-valued function. As the action of $M_\Theta$ on $H^2(\mathcal{F}_0)$ is unitary by assumption, in particular it is isometric, so the contractive analytic function $(\mathcal{F}_0, \Theta|_{\mathcal{F}_0})$ is inner. Hence $\Theta|_{\mathcal{F}_0}$ can be considered as a Beurling–Lax representer for the subspace $\Theta \cdot H^2(\mathcal{F}_0)$. By assumption, $M_\Theta|_{H^2(\mathcal{F}_0)}$ is actually unitary, so we have $\Theta \cdot H^2(\mathcal{F}_0) = H^2(\mathcal{F}_0)$. Again by the uniqueness assertion in the Beurling–Lax theorem, this forces $\Theta(z)|_{\mathcal{F}_0} = \Theta(0)|_{\mathcal{F}_0}$ to be a unitary constant on $\mathcal{F}_0$ as claimed.

*Proof of (iii)* By the result of part (ii) we know that any subspace $\mathcal{F}_0 \subset \mathcal{F}$ such that the space $H^2(\mathcal{F}_0)$ reduces $M_\Theta$ while maintaining the additional property that the restriction $M_\Theta|_{\mathcal{M}}$ is unitary is such that $\mathcal{M} = H^2(\mathcal{F}_0)$, where $\mathcal{F}_0 \subset \mathcal{F}$ is such that $\Theta(z)|_{\mathcal{F}_0} = \Theta(0)|_{\mathcal{F}_0}$ with $\Theta(0)|_{\mathcal{F}_0}$ equal to a unitary operator on $\mathcal{F}_0$. To get the largest subspace $\mathcal{F}_0$ for which these properties hold amounts to the unitary part in the Sz.-Nagy–Foias canonical decomposition for the operator $\Theta(0)$ acting on $\mathcal{F}$. □

The following is a quick corollary in view of the Sz.-Nagy–Foias canonical decomposition applied to the contraction $\Theta(0)$.

**Corollary 3.2.10** *If $(\mathcal{F}, \Theta)$ is a c.n.u. contractive analytic function, then there is no nonzero joint $(M_z^{\mathcal{F}}, M_\Theta)$-reducing subspace $\mathcal{M} \subset H^2(\mathcal{F})$ such that $M_\Theta|_{\mathcal{M}}$ is unitary. In particular, when $(\mathcal{F}, \Theta)$ is a c.n.u. inner function, then the commuting isometric pair $(M_z, M_\Theta)$ is 1&2-c.n.u.*

**Remark 3.2.11** Let us consider the intermediate special case of Theorem 3.2.9, where $(\mathcal{F}, \Theta)$ is inner but not necessarily c.n.u. Then the Sz.-Nagy–Foias canonical decomposition of $\Theta(0)$ ($\Theta(0) \underset{\mathrm{u}}{\cong} \begin{bmatrix} \Theta(0)_{\mathrm{u}} & 0 \\ 0 & \Theta(0)_{\mathrm{cnu}} \end{bmatrix}$, where $\Theta(0)_{\mathrm{u}}$ is unitary on $\mathcal{F}_{\mathrm{u}}$ while $\Theta(0)_{\mathrm{cnu}}$ is c.n.u. on $\mathcal{F}_{\mathrm{cnu}}$ and $\mathcal{F} \underset{\mathrm{u}}{\cong} \begin{bmatrix} \mathcal{F}_{\mathrm{u}} \\ \mathcal{F}_{\mathrm{cnu}} \end{bmatrix}$, leads to the identification of $H^2(\mathcal{F}_{\mathrm{u}})$ as the maximum joint $(M_z, M_\Theta)$-reducing subspace $\mathcal{M}_{\mathrm{u}} \subset H^2(\mathcal{F})$ such that $M_\Theta|_{\mathcal{M}_{\mathrm{u}}}$ is unitary. However, in general, this does not say much about the structure of $M_\Theta|_{\mathcal{M}_{\mathrm{u}}^\perp}$, apart from the observation that $M_\Theta|_{\mathcal{M}_{\mathrm{u}}^\perp}$ is an isometry. In general, $M_\Theta|_{\mathcal{M}_{\mathrm{u}}^\perp}$ can have its own nontrivial Wold decomposition; examples are explored in [23, Example 4.3] as well as in [49, Example 4.3]. However, if we assume that $\dim \mathcal{F} < \infty$, it turns out that $M_\Theta|_{\mathcal{M}_{\mathrm{u}}^\perp}$ is a shift, and $\mathcal{M}_{\mathrm{u}}^\perp \oplus \mathcal{M}_{\mathrm{u}}$ gives the Wold decomposition for the isometry $M_\Theta$. One can see this result as an application of Theorem 3.3 in [49]. As a corollary one can see that 1&2-c.n.u. is the same as shift/shift when $\dim \mathcal{F} < \infty$.

Now we are ready to identify what additional condition $\Theta$ must satisfy for the 1-c.n.u. model associated with $\Theta$ to be 2-c.n.u. as well, or the other way around.

**Theorem 3.2.12** *Let $(\mathcal{F}, \Theta)$ be a contractive analytic function.*

(i) *Consider the functional-model 1-c.n.u. commuting isometric pair corresponding to $\Theta$:*

$$\left( \begin{bmatrix} M_z^{\mathcal{F}} & 0 \\ 0 & I_{H^2} \otimes M_\zeta \end{bmatrix}, \begin{bmatrix} M_\Theta & 0 \\ \Delta_\Theta|_{H^2(\mathcal{F})} & M_z^{\overline{\Delta_\Theta L^2(\mathcal{F})}} \end{bmatrix} \right) \quad on \quad \begin{bmatrix} H^2(\mathcal{F}) \\ H^2(\overline{\Delta_\Theta L^2(\mathcal{F})}) \end{bmatrix}. \tag{3.72}$$

*If the contractive analytic function $(\mathcal{F}, \Theta)$ is c.n.u. (see Definition 2.2.8), then the isometric pair in* (3.72) *is 1&2-c.n.u. Conversely, any 1&2-c.n.u. commuting isometric pair $(V_1, V_2)$ is unitarily equivalent to a pair of the form* (3.72) *with the characteristic function $\Theta$ being c.n.u.*

(ii) *Similarly, consider the 2-c.n.u. commuting isometric pair corresponding to the contractive analytic function $\Theta$:*

$$\left( \begin{bmatrix} M_\Theta & 0 \\ \Delta_\Theta|_{H^2(\mathcal{F})} & M_z^{\overline{\Delta_\Theta L^2(\mathcal{F})}} \end{bmatrix}, \begin{bmatrix} M_z^{\mathcal{F}} & 0 \\ 0 & I_{H^2} \otimes M_\zeta \end{bmatrix} \right) \quad on \quad \begin{bmatrix} H^2(\mathcal{F}) \\ H^2(\overline{\Delta_\Theta L^2(\mathcal{F})}) \end{bmatrix}. \tag{3.73}$$

*If the contractive analytic function $(\mathcal{F}, \Theta)$ is c.n.u., then $(V_1, V_2)$ is 1&2-c.n.u. Conversely, any 1&2-c.n.u. commuting isometric pair $(V_1, V_2)$ is unitarily equivalent to a pair of the form* (3.73) *with the characteristic function $\Theta$ being c.n.u.*

*Proof* It is enough to prove one of the items; we choose to prove item (i). We prove the converse direction first. So suppose that $(V_1, V_2)$ is 1&2-c.n.u. Since $(V_1, V_2)$ is a 1-c.n.u. isometric pair, by Theorem 3.2.2 there is a contractive analytic function $(\mathcal{F}, \Theta)$ such that $(V_1, V_2)$ is of the form (3.72). We note that $\Theta(0)$ must be c.n.u., because if it is not, then there exists a nontrivial decomposition $\mathcal{F} = \mathcal{F}_0 \oplus \mathcal{F}_1$ that is reducing for $\Theta(0)$, $\Theta(0)|_{\mathcal{F}_1} : \mathcal{F}_1 \to \mathcal{F}_1$ is c.n.u., and $\Theta(0)|_{\mathcal{F}_0} : \mathcal{F}_0 \to \mathcal{F}_0$ is unitary. As a consequence of the fact that $z \mapsto \|\Theta(z)u_0\|$ is subharmonic and hence satisfies the maximum principle, it follows that for $u_0 \in \mathcal{F}_0$ the function $\Theta(z)|_{\mathcal{F}_0}$ must be the constant function $\Theta(z)_{\mathcal{F}_0} = \Theta(0)|_{\mathcal{F}_0}$ for all $z \in \mathbb{D}$. Therefore, $H^2(\mathcal{F}_0)$ is a joint reducing subspace for $(V_1, V_2)$, where $V_2$ is unitary. This contradicts the fact that $(V_1, V_2)$ is 2-c.n.u.

For the forward direction, let us suppose that the contractive analytic function $(\mathcal{F}, \Theta)$ is such that $\Theta(0)$ is c.n.u. We shall show that the 1-c.n.u. isometric pair in (3.72) is 2-c.n.u. as well. Let us note that by Corollary 3.2.6, any commuting isometric pair $(V_1, V_2)$ decomposes into at most four pieces:

$$(V_1, V_2) = 1\&2\text{-c.n.u.} \oplus \text{shift/unitary} \oplus \text{unitary/shift} \oplus \text{unitary/unitary}.$$

For a 1-c.n.u. commuting isometric pair $(V_1, V_2)$, the existence of a nontrivial, joint invariant subspace on which $(V_1, V_2)$ is unitary/shift or unitary/unitary is precluded. Hence, for the 1-c.n.u. case, we are left with

$$(V_1, V_2) = 1\&2\text{-c.n.u.} \oplus \text{shift/unitary}.$$

Thus, if the 1-c.n.u. commuting isometric pair (3.72) is not 2-c.n.u. as well, then by the previous discussion, there must be a joint reducing subspace $\mathcal{M}$ on which the pair is of the form shift/unitary. By inspecting the formula for $V_1$ in formula (3.72), we see from the fact that $V_1|_{\mathcal{M}}$ is prescribed to be a shift that it must be the case that

$$\mathcal{M} \subset \begin{bmatrix} H^2(\mathcal{F}) \\ \{0\} \end{bmatrix}. \tag{3.74}$$

Furthermore, as $V_1|_{\left[\begin{smallmatrix} H^2(\mathcal{F}) \\ 0 \end{smallmatrix}\right]} = \left[\begin{smallmatrix} I_{H^2(\mathcal{F})} \\ 0 \end{smallmatrix}\right] M_z^{\mathcal{F}}$, from the prescription that $\mathcal{M}$ must be reducing for $V_1$ we see by item (i) in Theorem 3.2.9 that $\mathcal{M}$ must have the form

$$\mathcal{M} = \begin{bmatrix} H^2(\mathcal{F}_0) \\ 0 \end{bmatrix} \quad \text{for some subspace } \mathcal{F}_0 \subset \mathcal{F}.$$

Recall that we are also imposing the requirement that $\mathcal{M}$ be reducing for $V_2$. The weaker requirement that $\mathcal{M}$ be invariant for $V_2$ imposes the restriction that

$$\Delta_\Theta(\zeta) f_0 = 0 \quad \text{for a.e. } \zeta \in \mathbb{T} \quad \text{for each } f_0 \in \mathcal{F}_0. \tag{3.75}$$

Note that the requirement that $M_\Theta|_{H^2(\mathcal{F}_0)}$ be unitary on $H^2(\mathcal{F}_0)$ also leads to condition (3.75). We conclude that the problem of characterizing when $\left[\begin{smallmatrix} H^2(\mathcal{F}_0) \\ 0 \end{smallmatrix}\right]$ is reducing for $V_2$ with restriction to $\left[\begin{smallmatrix} H^2(\mathcal{F}_0) \\ 0 \end{smallmatrix}\right]$ equal to a unitary operator is equivalent to characterizing when $H^2(\mathcal{F}_0)$ is a reducing subspace for $M_\Theta$ such that the restriction $M_\Theta|_{H^2(\mathcal{F}_0)}$ is unitary. But this latter problem is handled in item (ii) of Theorem 3.2.9: It suffices to demand that $\Theta(z)|_{\mathcal{F}_0} = \Theta(0)|_{\mathcal{F}_0}$, where $\Theta(0)|_{\mathcal{F}_0}$ is unitary on $\mathcal{F}_0$. From the definitions we see that $\Theta$ being c.n.u. (meaning that $\Theta(0)$ is c.n.u. on $\mathcal{F}$) is the condition that forces any such subspace $\mathcal{F}_0$ to be the zero subspace, thereby proving that $(V_1, V_2)$ is 2-c.n.u. (as well as 1-c.n.u.). This completes the proof. □

We thus have identified two canonical functional models for the class of 1&2-c.n.u. commuting isometric pairs $(V_1, V_2)$, which we label as follows.

**Definition 3.2.13** We shall call the functional model appearing in item (i) of Theorem 3.2.12 the *BDF1-functional model for 1&2-c.n.u. commuting isometric pairs.*

Analogously, we shall call the functional model appearing in item (ii) of Theorem 3.2.12, the *BDF2-functional model for 1&2-c.n.u. commuting isometric pairs.*

Theorem 3.2.12 gives a finer BDF-model for a commuting isometric pair as follows.

**Definition 3.2.14** We say that any collection of objects $\Xi$ of the form $\Xi = \left(\mathcal{F}_0, \mathcal{F}_1, \mathcal{E}_0, u_{\mathcal{F}_0}, \Theta_1, u_{\mathcal{E}_0}, Y_1, Y_2\right)$ is a *refined BDF-tuple* where

(i) $\mathcal{F}_0, \mathcal{F}_1$, and $\mathcal{E}_0$ are coefficient Hilbert spaces,

(ii) $u_{\mathcal{F}_0}$ is a unitary operator on $\mathcal{F}_0$,

(iii) $(\mathcal{F}_1, \Theta_1)$ is a c.n.u. contractive analytic function (see Definition 2.2.8),

(iv) $u_{\mathcal{E}_0}$ is a unitary operator $\mathcal{E}_0$,

(v) $(Y_1, Y_2)$ is a commuting unitary operator pair on another Hilbert space $\mathcal{K}_u$.

Given a refined BDF-tuple $\Xi = \left(\mathcal{F}_0, \mathcal{F}_1, \mathcal{E}_0, u_{\mathcal{F}_0}, \Theta_1, u_{\mathcal{E}_0}, Y_1, Y_2\right)$, define an operator pair $(V_{\Xi,1}, V_{\Xi,2})$ given by

$$V_{\Xi,1} = \begin{bmatrix} M_z^{\mathcal{F}_0} & 0 & 0 & 0 & 0 \\ 0 & M_z^{\mathcal{F}_1} & 0 & 0 & 0 \\ 0 & 0 & I_{H^2} \otimes M_\zeta & 0 & 0 \\ 0 & 0 & 0 & I_{H^2} \otimes u_{\mathcal{E}_0} & 0 \\ 0 & 0 & 0 & 0 & Y_1 \end{bmatrix},$$

$$V_{\Xi_2} = \begin{bmatrix} I_{H^2} \otimes u_{\mathcal{F}_0} & 0 & 0 & 0 & 0 \\ 0 & M_{\Theta_1} & 0 & 0 & 0 \\ 0 & \Delta_{\Theta_1}|_{H^2(\mathcal{F}_1)} & M_z^{\overline{\Delta_{\Theta_1} L^2(\mathcal{F}_1)}} & 0 & 0 \\ 0 & 0 & 0 & M_z^{\mathcal{E}_0} & 0 \\ 0 & 0 & 0 & 0 & Y_2 \end{bmatrix} \quad \text{on } \mathcal{K} = \begin{bmatrix} H^2(\mathcal{F}_0) \\ H^2(\mathcal{F}_1) \\ H^2\left(\overline{\Delta_{\Theta_1} L^2(\mathcal{F}_1)}\right) \\ H^2(\mathcal{E}_0) \\ \mathcal{K}_u \end{bmatrix}, \tag{3.76}$$

where $\Delta_{\Theta_1}$ is the operator of multiplication by the function $\Delta_{\Theta_1}(\zeta) = (I_{\mathcal{F}_1} - \Theta_1(\zeta)^*\Theta_1(\zeta))^{1/2}$ with values acting on $L^2(\mathcal{F}_1)$. We say that $(V_{\Xi,1}, V_{\Xi,2})$ (3.64) is the *refined BDF1-model* operator pair associated with the refined BDF-tuple $\Xi$. Alternatively, given a refined BDF-tuple $\Xi'$

$$\Xi' = \left(\mathcal{F}_0', \mathcal{F}_1', \mathcal{E}_0', u'_{\mathcal{F}_0}, \Theta_0', u'_{\mathcal{E}_0}, Y_1', Y_2'\right),$$

we associate the operator pair $\left(V'_{\Xi',1}, V'_{\Xi',2}\right)$ given by

$$V'_{\Sigma',1} = \begin{bmatrix} I_{H^2} \otimes u_{\mathcal{F}'_0} & 0 & 0 & 0 & 0 \\ 0 & M_{\Theta'_0} & 0 & 0 & 0 \\ 0 & \Delta_{\Theta'_0}|_{H^2(\mathcal{F}'_1)} & M_z^{\overline{\Delta_{\Theta'_0} L^2(\mathcal{F}'_1)}} & 0 & 0 \\ 0 & 0 & 0 & M_z^{\mathcal{E}'_0} & 0 \\ 0 & 0 & 0 & 0 & Y'_2 \end{bmatrix},$$

$$V'_{\Sigma',2} = \begin{bmatrix} M_z^{\mathcal{F}'_0} & 0 & 0 & 0 & 0 \\ 0 & M_z^{\mathcal{F}'_1} & 0 & 0 & 0 \\ 0 & 0 & I_{H^2} \otimes M_\zeta & 0 & 0 \\ 0 & 0 & 0 & I_{H^2} \otimes u_{\mathcal{E}'_0} & 0 \\ 0 & 0 & 0 & 0 & Y'_1 \end{bmatrix} \quad \text{on } \mathcal{K}' = \begin{bmatrix} H^2(\mathcal{F}'_0) \\ H^2(\mathcal{F}'_1) \\ H^2\left(\overline{\Delta_{\Theta'_0} L^2(\mathcal{F}'_1)}\right) \\ H^2(\mathcal{E}'_0) \\ \mathcal{K}'_u \end{bmatrix}. \tag{3.77}$$

Then we say that $(V'_{\Xi',1}, V'_{\Xi',2})$ is the *refined BDF2-model* operator pair associated with the refined BDF-tuple $\Xi'$.

If we wish to emphasize that the BDF-tuple $\Xi := (\mathcal{F}, \mathcal{E}, \Theta, u, Y_1, Y_2)$ (in the sense of Definition 3.2.1) does not carry the additional structure displayed in a refined BDF-tuple as discussed here, we shall say that $\Xi$ is a *coarse BDF-tuple.*

Note that, as in the case of coarse BDF-models, if we have $\Xi = \Xi'$ in the previous definition, then we have

$$(V'_{\Xi,1}, V'_{\Xi,2}) = (V_{\Xi,2}, V_{\Xi,1}). \tag{3.78}$$

**Remark 3.2.15** It is worthwhile to note the point of distinction between a refined BDF-tuple and a coarse BDF-tuple. A refined BDF-tuple $(\mathcal{F}_0, \mathcal{F}_1, \mathcal{E}_0, u_{\mathcal{F}_0}, \Theta, u_{\mathcal{E}_0}, Y_1, Y_2)$ consists of a BDF-tuple $(\mathcal{F}, \Theta, u_{\mathcal{E}}, Y_1, Y_2)$ with $\Theta$ required to be c.n.u. together with an additional unitary operator $u_{\mathcal{F}_0}$ acting on $\mathcal{F}_0$. Note that it is possible that the additional component $(\mathcal{F}_0, u_{\mathcal{F}_0})$ is absent in a refined BDF-tuple. This is precisely the case of a coarse BDF-tuple $(\mathcal{F}, \mathcal{E}, \Theta, u_{\mathcal{E}}, Y_1, Y_2)$ having the additional property that $\Theta(0)$ is c.n.u. Associated to every coarse BDF-tuple $(\mathcal{F}, \mathcal{E}_0, \Theta, u_{\mathcal{E}_0}, Y_1, Y_2)$ is a refined BDF-tuple $(\mathcal{F}_0, \mathcal{F}_1, \mathcal{E}_0, u_{\mathcal{F}}, \Theta|_{\mathcal{F}_1}, u_{\mathcal{E}}, Y_1, Y_2)$, obtained simply by applying the canonical decomposition to the contraction $\Theta(0)$: Decompose $\mathcal{F}$ as $\mathcal{F} = \mathcal{F}_0 \oplus \mathcal{F}_1$, where $\Theta(0)|_{\mathcal{F}_0}$ is unitary and $\Theta(0)|_{\mathcal{F}_1}$ is c.n.u., and then set $u_{\mathcal{F}_0} = \Theta(0)|_{\mathcal{F}_0}$ and $\Theta_0(z) = \Theta(z)|_{\mathcal{F}_1}$.

As for the coarse BDF functional models, let us note that, given a refined BDF-tuple $(\mathcal{F}_0, \mathcal{F}_1, \mathcal{E}_0, u_{\mathcal{F}_0}, \Theta_0, Y_1, Y_2)$, if we set

$$\mathcal{F} = \begin{bmatrix} \mathcal{F}_0 \\ \mathcal{F}_1 \end{bmatrix}$$

and consider the contractive analytic function $(\mathcal{F}, \Theta)$ given by

$$\Theta(z) = \begin{bmatrix} u_{\mathcal{F}_0} & 0 \\ 0 & \Theta_0(z) \end{bmatrix} : \begin{bmatrix} \mathcal{F}_0 \\ \mathcal{F}_1 \end{bmatrix} \to \begin{bmatrix} \mathcal{F}_0 \\ \mathcal{F}_1 \end{bmatrix},$$

then the refined BDF1-model as in (3.76) corresponding to the refined BDF-tuple $(\mathcal{F}_0, \mathcal{F}_1, \mathcal{E}_0, u_{\mathcal{F}_0}, \Theta_0, u_{\mathcal{E}_0}, Y_1, Y_2)$ is the same as the BDF1-model (3.64) corresponding to the coarse BDF-tuple $(\mathcal{F}, \mathcal{E}_0, \Theta, u_{\mathcal{E}_0}, Y_1, Y_2)$. The same assertion holds when coarse BDF-tuple and coarse BDF1-model are replaced by refined BDF-tuple and refined BDF2-model, respectively.

Remark 3.2.15 and Theorem 3.2.2 imply the following as a consequence.

**Corollary 3.2.16** *Given any commuting isometric operator pair $(V_1, V_2)$, we have the following.*

(i) *There is a refined BDF-tuple $\Xi = (\mathcal{F}_0, \mathcal{F}_1, \mathcal{E}_0, u_{\mathcal{F}_0}, \Theta_0, u_{\mathcal{E}_0}, Y_1, Y_2)$ so that $(V_1, V_2)$ is unitarily equivalent to the refined BDF1-model commuting isometric pair $(V_{\Xi,1}, V_{\Xi,2})$ as in* (3.76).
(ii) *There is a refined BDF-tuple $\Xi' = \left(\mathcal{F}_0', \mathcal{F}_1', \mathcal{E}_0', u_{\mathcal{F}_0'}, \Theta_0', u_{\mathcal{E}_0'}, Y_1', Y_2'\right)$ so that $(V_1, V_2)$ is unitarily equivalent to the refined BDF2-model commuting isometric pair $(V_{\Xi,'1}, V_{\Xi',2})$ as in* (3.77).

*Conversely, given any refined BDF-tuple, the associated refined BDF1-model as well as the associated refined BDF2-model are commuting isometric pairs.*

Let us note that Corollary 3.2.6 can also be seen as a consequence of Corollary 3.2.16 and Theorem 3.2.12.

### 3.2.3 BDF-Tuple as a Complete Unitary Invariant

It is also possible to characterize unitary equivalence of two commuting isometric operator pairs in terms of BDF-tuples as follows. This result is the analogue for the BDF setting of Theorem 3.1.6 obtained for the BCL-setting and can also be considered as a more complete formulation of Corollary 5.1 in [30].

**Theorem 3.2.17** *Let $(V_1, V_2)$ on $\mathcal{K}$ and $\left(\widetilde{V}_1, \widetilde{V}_2\right)$ on $\widetilde{\mathcal{K}}$ be two commuting isometric pairs. Suppose $(V_1, V_2)$ is unitarily equivalent to the BDF1-model* (3.64) *operator pair associated with the BDF-tuple $\Xi = (\mathcal{F}, \mathcal{E}_0, \Theta, u_{\mathcal{E}_0}, Y_1, Y_2)$ while $\left(\widetilde{V}_1, \widetilde{V}_2\right)$ is unitarily equivalent to BDF1-model* (3.64) *associated with the BDF-tuple $\widetilde{\Xi} = \left(\widetilde{\mathcal{F}}_0, \widetilde{\mathcal{E}}_0, \widetilde{\Theta}, u_{\widetilde{\mathcal{E}}_0}, \widetilde{Y}_1, \widetilde{Y}_2\right)$. Then the following are equivalent.*

(i) $(V_1, V_2)$ *and* $\left(\widetilde{V}_1, \widetilde{V}_2\right)$ *are jointly unitarily equivalent.*

(ii) *The associated BDF-tuples* $\Xi$ *and* $\widetilde{\Xi}$ *are unitarily equivalent as BDF-tuples in the following sense.*

*Given two BDF-tuples* $\Xi = (\mathcal{F}, \mathcal{E}_0, \Theta, u_{\mathcal{E}_0}, Y_1, Y_2)$ *and* $\widetilde{\Xi} = (\widetilde{\mathcal{F}}, \widetilde{\mathcal{E}}_0, \widetilde{\Theta}, u_{\widetilde{\mathcal{E}}_0}, \widetilde{Y}_1, \widetilde{Y}_2)$, *we say that* $\Xi$ *is unitarily equivalent to* $\widetilde{\Xi}$ *if there exist unitary operators*

$$\tau_{\mathcal{F}} \colon \mathcal{F} \to \widetilde{\mathcal{F}}, \quad \tau_{\mathcal{E}_0} \colon \mathcal{E}_0 \to \widetilde{\mathcal{E}}_0, \quad \tau_u \colon \mathcal{K} \to \widetilde{\mathcal{K}},$$

*so that*

$$\begin{aligned} &\tau_{\mathcal{F}}\Theta(z) = \widetilde{\Theta}(z)\tau_{\mathcal{F}} \quad \text{for all } z \in \mathbb{D}, \\ &\tau_{\mathcal{E}_0} u_{\mathcal{E}_0} = u_{\widetilde{\mathcal{E}}_0}\tau_{\mathcal{E}_0}, \\ &\tau_u(Y_1, Y_2) = \left(\widetilde{Y}_1, \widetilde{Y}_2\right)\tau_u. \end{aligned} \tag{3.79}$$

*Consequently there is a bijective correspondence between unitary equivalence classes of commuting isometric pairs* $(V_1, V_2)$ *and unitary equivalence classes of BDF-tuples (in the sense of* (3.79)*).*

*An analogous statement holds with the BDF1-model replaced by the BDF2-model* (3.65).

*Proof* We discuss only the BDF1-model as the BDF2-model is handled in a completely parallel way.

The direction (ii)⟹(i) is the easier. Since $\tau_{\mathcal{F}}\Theta(z) = \widetilde{\Theta}(z)\tau_{\mathcal{F}}$, it follows that $\tau_{\mathcal{F}}\Theta(\zeta)^*\Theta(\zeta) = \widetilde{\Theta}(\zeta)^*\widetilde{\Theta}(\zeta)\tau_{\mathcal{F}}$ for almost every unimodular $\zeta$ (precisely where the radial limit of both $\Theta(\zeta)$ and $\widetilde{\Theta}(\zeta)$ exists). The unitary $\tau_{\mathcal{F}}$, therefore, gives rise to another unitary operator $\widehat{\tau_{\mathcal{F}}} \colon \overline{\Delta_\Theta L^2(\mathcal{F})} \to \overline{\Delta_{\widetilde{\Theta}} L^2\left(\widetilde{\mathcal{F}}\right)}$ defined on the dense set $\left\{\Delta_\Theta \zeta^{-n} f \colon n \ge 1, f \in H^2(\mathcal{F})\right\}$ as

$$\widehat{\tau_{\mathcal{F}}}(\Delta_\Theta \zeta^{-n} f) = \zeta^{-n}\Delta_{\widetilde{\Theta}}\tau_{\mathcal{F}} f. \tag{3.80}$$

Note that $\widehat{\tau_{\mathcal{F}}}$, so defined, has the intertwining property

$$\widehat{\tau_{\mathcal{F}}} M_\zeta|_{\overline{\Delta_\Theta L^2(\mathcal{F})}} = M_\zeta|_{\overline{\Delta_{\widetilde{\Theta}} L^2\left(\widetilde{\mathcal{F}}\right)}}\widehat{\tau_{\mathcal{F}}}.$$

It is now a matter of simple matrix computation to check using (3.80) that the diagonal unitary operator

$$\begin{bmatrix} I_{H^2} \otimes \tau_{\mathcal{F}} & & & \\ & I_{H^2} \otimes \widehat{\tau_{\mathcal{F}}} & & \\ & & I_{H^2} \otimes \tau_{\mathcal{E}_0} & \\ & & & \tau_u \end{bmatrix} \colon \begin{bmatrix} H^2(\mathcal{F}) \\ H^2(\overline{\Delta_\Theta L^2(\mathcal{F})}) \\ H^2(\mathcal{E}_0) \\ \mathcal{K}_u \end{bmatrix} \to \begin{bmatrix} H^2\left(\widetilde{\mathcal{F}}\right) \\ H^2\left(\overline{\Delta_{\widetilde{\Theta}} L^2\left(\widetilde{\mathcal{F}}\right)}\right) \\ H^2\left(\widetilde{\mathcal{E}}_0\right) \\ \widetilde{\mathcal{K}}_u \end{bmatrix} \tag{3.81}$$

identifies the BDF1-models (3.64) of $(V_1, V_2)$ and $\left(\widetilde{V}_1, \widetilde{V}_2\right)$.

For (i)⇒(ii), assume that

$$\tau\colon \begin{bmatrix} H^2(\mathcal{F}) \\ H^2(\overline{\Delta_\Theta L^2(\mathcal{F})}) \\ H^2(\mathcal{E}_0) \\ \mathcal{K}_u \end{bmatrix} =: \mathcal{K} \to \begin{bmatrix} H^2\left(\widetilde{\mathcal{F}}\right) \\ H^2\left(\overline{\Delta_{\widetilde{\Theta}} L^2\left(\widetilde{\mathcal{F}}\right)}\right) \\ H^2\left(\widetilde{\mathcal{E}}_0\right) \\ \widetilde{\mathcal{K}}_u \end{bmatrix} =: \widetilde{\mathcal{K}}$$

is a unitary and $\tau(V_1, V_2) = \left(\widetilde{V}_1, \widetilde{V}_2\right)\tau$, where $(V_1, V_2)$ and $\left(\widetilde{V}_1, \widetilde{V}_2\right)$ are in the BDF1-model form as in (3.64), corresponding to BDF-tuples $\Xi = (\mathcal{F}, \mathcal{E}_0, \Theta, u_{\mathcal{E}_0}, Y_1, Y_2)$ and $\widetilde{\Xi} = \left(\widetilde{\mathcal{F}}, \widetilde{\mathcal{E}}_0, \widetilde{\Theta}, u_{\widetilde{\mathcal{E}}_0}, \widetilde{Y}_1, \widetilde{Y}_2\right)$, respectively. We shall show that $\tau$ must be of the form (3.81). Lemma 2.3.1 will be used.

Let us note that $V_1$ and $\widetilde{V}_1$ are of the forms

$$\begin{bmatrix} M_z^{\mathcal{F}} & 0 \\ 0 & W \end{bmatrix} \quad \text{and} \quad \begin{bmatrix} M_z^{\widetilde{\mathcal{F}}} & 0 \\ 0 & \widetilde{W} \end{bmatrix},$$

where $W, \widetilde{W}$ are unitary operators. Since $\tau$ intertwines $V_1$ and $\widetilde{V}_1$, by Lemma 2.3.1, $\tau$ must be of the form

$$\tau = \begin{bmatrix} \tau_{\mathcal{F}} & 0 \\ 0 & \sigma \end{bmatrix} \tag{3.82}$$

for some unitary operators $\tau_{\mathcal{F}}\colon \mathcal{F} \to \widetilde{\mathcal{F}}$ and

$$\sigma\colon \begin{bmatrix} H^2(\overline{\Delta_\Theta L^2(\mathcal{F})}) \\ H^2(\mathcal{E}_0) \\ \mathcal{K}_u \end{bmatrix} \to \begin{bmatrix} H^2\left(\overline{\Delta_{\widetilde{\Theta}} L^2\left(\widetilde{\mathcal{F}}\right)}\right) \\ H^2\left(\widetilde{\mathcal{E}}_0\right) \\ \widetilde{\mathcal{K}}_u \end{bmatrix}$$

such that $\sigma$ intertwines

$$W = \begin{bmatrix} I_{H^2} \otimes M_\zeta|_{\overline{\Delta_\Theta L^2(\mathcal{F})}} & 0 & 0 \\ 0 & I_{H^2} \otimes u_{\mathcal{E}_0} & 0 \\ 0 & 0 & Y_1 \end{bmatrix},$$

and

$$\widetilde{W} = \begin{bmatrix} I_{H^2} \otimes M_\zeta|_{\overline{\Delta_{\widetilde{\Theta}} L^2(\widetilde{\mathcal{F}})}} & 0 & 0 \\ 0 & I_{H^2} \otimes u_{\widetilde{\mathcal{E}}_0} & 0 \\ 0 & 0 & \widetilde{Y}_1 \end{bmatrix}.$$

Since $\tau$ also intertwines $V_2$ and $\widetilde{V}_2$, an easy matrix computation shows that we must have

$$\tau_{\mathcal{F}}\Theta(z) = \widetilde{\Theta}(z)\tau_{\mathcal{F}}, \quad \widehat{\tau_{\mathcal{F}}}\Delta_\Theta|_{H^2(\mathcal{F})} = \left(\Delta_{\widetilde{\Theta}}|_{H^2(\widetilde{\mathcal{F}})}\right)\widehat{\tau_{\mathcal{F}}}, \quad \sigma\begin{bmatrix}\Delta_\Theta\\0\\0\end{bmatrix} = \begin{bmatrix}\Delta_{\widetilde{\Theta}}\\0\\0\end{bmatrix}\tau_{\mathcal{F}} \tag{3.83}$$

and

$$\sigma\begin{bmatrix}M_z^{\mathcal{G}} & 0\\ 0 & Y_2\end{bmatrix} = \begin{bmatrix}M_z^{\widetilde{\mathcal{G}}} & 0\\ 0 & \widetilde{Y}_2\end{bmatrix}\sigma,$$

where $\widehat{\tau_{\mathcal{F}}}$ is as defined in (3.80),

$$\mathcal{G} = \begin{bmatrix}\overline{\Delta_\Theta L^2(\mathcal{F})}\\ \mathcal{E}_0\end{bmatrix} \quad \text{and} \quad \widetilde{\mathcal{G}} = \begin{bmatrix}\overline{\Delta_{\widetilde{\Theta}} L^2\left(\widetilde{\mathcal{F}}\right)}\\ \widetilde{\mathcal{E}}_0\end{bmatrix}.$$

Apply Lemma 2.3.1 again to conclude that $\sigma$ must be of the form

$$\sigma = \begin{bmatrix}\sigma_{\mathcal{G}} & 0\\ 0 & \tau_u\end{bmatrix} \tag{3.84}$$

for some unitary operators $\sigma_{\mathcal{G}}\colon \mathcal{G} \to \widetilde{\mathcal{G}}$ and $\tau_u\colon \mathcal{K}_u \to \widetilde{\mathcal{K}}_u$ satisfying

$$\sigma_{\mathcal{G}} M_z^{\mathcal{G}} = M_z^{\widetilde{\mathcal{G}}}\sigma_{\mathcal{G}} \quad \text{and} \quad \tau_u Y_2 = \widetilde{Y}_2\tau_u. \tag{3.85}$$

Let us write $\sigma_{\mathcal{G}}\colon \mathcal{G} \to \widetilde{\mathcal{G}}$ as the $2 \times 2$-operator matrix

$$\sigma_{\mathcal{G}} = \begin{bmatrix}\sigma_{\mathcal{G},11} & \sigma_{\mathcal{G},12}\\ \sigma_{\mathcal{G},21} & \tau_{\mathcal{E}}\end{bmatrix} : \begin{bmatrix}\overline{\Delta_\Theta L^2(\mathcal{F})}\\ \mathcal{E}\end{bmatrix} \to \begin{bmatrix}\overline{\Delta_{\widetilde{\Theta}} L^2\left(\widetilde{\mathcal{F}}\right)}\\ \widetilde{\mathcal{E}}\end{bmatrix}.$$

Since $\sigma$ satisfies (3.83), we must have

$$\sigma_{\mathcal{G},11}\Delta_\Theta|_{H^2(\mathcal{F})} = \left(\Delta_{\widetilde{\Theta}}|_{H^2\left(\widetilde{\mathcal{F}}\right)}\right)\tau_{\mathcal{F}}. \tag{3.86}$$

As noted previously, the unitary operator $\sigma$ also intertwines $W$ and $\widetilde{W}$. This intertwining gives

$$\sigma_{\mathcal{G},11} M_\zeta|_{\overline{\Delta_\Theta L^2(\mathcal{F})}} = \left(M_\zeta|_{\overline{\Delta_{\widetilde{\Theta}} L^2\left(\widetilde{\mathcal{F}}\right)}}\right)\sigma_{\mathcal{G},11}, \quad \tau_{\mathcal{E}_0} u_{\mathcal{E}_0} = u_{\widetilde{\mathcal{E}}_0}\tau_{\mathcal{E}_0} \quad \text{and} \quad \tau_u Y_1 = \widetilde{Y}_1\tau_u. \tag{3.87}$$

Let us use the second identity in (3.83) and the identity in (3.86) to get

$$\sigma_{\mathcal{G},11}\Delta_\Theta|_{H^2(\mathcal{F})} = \tau_{\mathcal{F}}\Delta_\Theta|_{H^2(\mathcal{F})}.$$

This, together with the first intertwining in (3.87), immediately gives

$$\begin{aligned}\sigma_{\mathcal{G},11}\Delta_\Theta(\zeta)\zeta^{-n} f(\zeta) &= \zeta^{-n}\sigma_{\mathcal{G},11}\Delta_\Theta(\zeta) f(\zeta)\\ &= \zeta^{-n}\tau_{\mathcal{F}}\Delta_\Theta(\zeta) f(\zeta) = \tau_{\mathcal{F}}\Delta_\Theta(\zeta)\zeta^{-n} f(\zeta)\end{aligned}$$

for every $n \geq 0$ and $f \in H^2(\mathcal{F})$. Since the set $\{\zeta^{-n} f(\zeta) : n \geq 0,\ f \in H^2(\mathcal{F})\}$ is dense in $L^2(\mathcal{F})$, it follows that

$$\sigma_{\mathcal{G},11} = \widehat{\tau_{\mathcal{F}}},$$

where $\widehat{\tau_{\mathcal{F}}} \colon \overline{\Delta_\Theta L^2(\mathcal{F})} \to \overline{\Delta_{\widetilde{\Theta}} L^2\left(\widetilde{\mathcal{F}}\right)}$ is as in (3.80). Since $\sigma_{\mathcal{G}}$ is a unitary and its (11)-entry $\sigma_{\mathcal{G},11}$ coincides with a unitary, namely, $\widehat{\tau_{\mathcal{F}}}$, it follows that $\sigma_{\mathcal{G}}$ must be of the form

$$\sigma_{\mathcal{G}} = \begin{bmatrix} \widehat{\tau_{\mathcal{F}}} & 0 \\ 0 & \tau_{\mathcal{E}} \end{bmatrix} : \begin{bmatrix} \overline{\Delta_\Theta L^2(\mathcal{F})} \\ \mathcal{E}_0 \end{bmatrix} \to \begin{bmatrix} \overline{\Delta_{\widetilde{\Theta}} L^2\left(\widetilde{\mathcal{F}}\right)} \\ \widetilde{\mathcal{E}}_0 \end{bmatrix}$$

and, consequently, $\tau$ is as desired, that is, as in (3.81). Finally, we see from (3.83), (3.85), and (3.87) that the unitary operators $\tau_{\mathcal{F}}$, $\tau_u$, and $\tau_{\mathcal{E}}$ satisfy all the conditions that they are supposed to, that is, the conditions in (3.79). This completes the proof. □

A particular consequence of Theorem 3.2.17 is that a commuting isometric pair $(V_1, V_2)$ uniquely determines a contractive analytic function $(\mathcal{F}, \Theta)$ inside the BDF-tuple $\Xi_{V_1,V_2} = \left(\mathcal{F}, \mathcal{E}_0, \Theta, u_{\mathcal{E}_0}, Y_1, Y_2\right)$ up to unitary equivalence (of contractive analytic functions). In detail, given a BDF1-model as in (3.64), we note that the contractive analytic function $\Theta$ influences only the 1-c.n.u. component of the model, namely

$$\left(V_1^{(1)}, V_2^{(1)}\right)_1 = \left( \begin{bmatrix} M_z^{\mathcal{F}} & 0 \\ 0 & I_{H^2} \otimes M_\zeta \end{bmatrix}, \begin{bmatrix} M_\Theta & 0 \\ \Delta_\Theta|_{H^2(\mathcal{F})} & M_z^{\overline{\Delta_\Theta L^2(\mathcal{F})}} \end{bmatrix} \right)$$

$$\text{on } \mathcal{K}^{(1)} = \begin{bmatrix} H^2(\mathcal{F}) \\ H^2(\overline{\Delta_\Theta L^2(\mathcal{F})}) \end{bmatrix}.$$

Just like a coarse BDF-tuple, a refined BDF-tuple is a complete unitary invariant for commuting isometries.

**Theorem 3.2.18** *Let $(V_1, V_2)$ on $\mathcal{K}$ and $\left(\widetilde{V}_1, \widetilde{V}_2\right)$ on $\widetilde{\mathcal{K}}$ be two commuting isometric pairs. Suppose $(V_1, V_2)$ is unitarily equivalent to the refined BDF1-model* (3.76) *generated by the refined BDF-tuple $\Xi = (\mathcal{F}_0, \mathcal{F}_1, \mathcal{E}_0, u_{\mathcal{F}_0}, \Theta_0, u_{\mathcal{E}_0}, Y_1, Y_2)$ while $\left(\widetilde{V}_1, \widetilde{V}_2\right)$ is unitarily equivalent to refined BDF1-model* (3.76) *generated by the refined BDF-tuple $\widetilde{\Xi} = \left(\widetilde{\mathcal{F}}_0, \widetilde{\mathcal{F}}_1, \widetilde{\mathcal{E}}_0, u_{\widetilde{\mathcal{F}}_0} \widetilde{\Theta}_0, u_{\widetilde{\mathcal{E}}_0}, \widetilde{Y}_1, \widetilde{Y}_2\right)$. Then the following are equivalent.*

(i) *$(V_1, V_2)$ and $\left(\widetilde{V}_1, \widetilde{V}_2\right)$ are jointly unitarily equivalent.*
(ii) *The associated refined BDF-tuples $\Xi$ and $\widetilde{\Xi}$ are unitarily equivalent in the following sense: Two refined BDF-tuples*

$$\Xi = (\mathcal{F}_0, \mathcal{F}_1, \mathcal{E}_0, u_{\mathcal{F}_0}, \Theta_0, u_{\mathcal{E}_0}, Y_1, Y_2),$$
$$\widetilde{\Xi} = \left(\widetilde{\mathcal{F}}_0, \widetilde{\mathcal{F}}_1, \widetilde{\mathcal{E}}_0, u_{\widetilde{\mathcal{F}}_0}, \widetilde{\Theta}_0, u_{\widetilde{\mathcal{E}}_0}, \widetilde{Y}_1, \widetilde{Y}_2\right) \tag{3.88}$$

*are said to be unitarily equivalent if there exist unitary operators*

$$\tau_{\mathcal{F}_0} : \mathcal{F}_0 \to \widetilde{\mathcal{F}}_0, \quad \tau_{\mathcal{F}_1} : \mathcal{F}_1 \to \widetilde{\mathcal{F}}_1, \quad \tau_{\mathcal{E}_0} : \mathcal{E}_0 \to \widetilde{\mathcal{E}}_0, \quad \tau_u : \mathcal{K}_u \to \widetilde{\mathcal{K}}_u,$$

*so that*

$$\tau_{\mathcal{F}_0} u_{\mathcal{F}_0} = u_{\widetilde{\mathcal{F}}_0} \tau_{\mathcal{F}_0},$$
$$\tau_{\mathcal{F}_1} \Theta_0(z) = \widetilde{\Theta}_0(z) \tau_{\mathcal{F}_1} \quad \textit{for all } z \in \mathbb{D},$$
$$\tau_{\mathcal{E}_0} u_{\mathcal{E}_0} = u_{\widetilde{\mathcal{E}}_0} \tau_{\mathcal{E}_0},$$
$$\tau_u (Y_1, Y_2) = \left(\widetilde{Y}_1, \widetilde{Y}_2\right) \tau_u. \tag{3.89}$$

*Consequently there is a bijective correspondence between unitary equivalence classes of commuting isometric pairs* $(V_1, V_2)$ *and unitary equivalence classes of refined BDF-tuples (in the sense of* (3.89)*).*

*An analogous statement holds with the refined BDF1-model replaced by the refined BDF2-model as in* (3.77).

*Proof* The proof can be seen to follow from Theorem 3.2.17 and part (ii) of Lemma 2.3.4 as follows. Set

$$\mathcal{F} = \begin{bmatrix} \mathcal{F}_0 \\ \mathcal{F}_1 \end{bmatrix}, \quad \widetilde{\mathcal{F}} = \begin{bmatrix} \widetilde{\mathcal{F}}_0 \\ \widetilde{\mathcal{F}}_1 \end{bmatrix}$$

and define the contractive analytic functions $(\mathcal{F}, \Theta)$ and $\left(\widetilde{\mathcal{F}}, \widetilde{\Theta}\right)$ by

$$\Theta(z) = \begin{bmatrix} u_{\mathcal{F}_0} & 0 \\ 0 & \Theta_0(z) \end{bmatrix} \text{ on } \mathcal{F}, \quad \widetilde{\Theta}(z) = \begin{bmatrix} u_{\widetilde{\mathcal{F}}_0} & 0 \\ 0 & \widetilde{\Theta}_0(z) \end{bmatrix} \text{ on } \widetilde{\mathcal{F}}.$$

Then, as observed in Remark 3.2.15, the refined BDF1-models as in (3.76) for the refined BDF-tuples in (3.88) are the same as the BDF1-models for the coarse BDF-tuples

$$\Pi = (\mathcal{F}_1, \mathcal{E}_0, \Theta, u_{\mathcal{E}_0}, Y_1, Y_2), \quad \widetilde{\Pi} = \left(\widetilde{\mathcal{F}}_1, \widetilde{\mathcal{E}}_0, \widetilde{\Theta}, u_{\widetilde{\mathcal{E}}_0}, \widetilde{Y}_1, \widetilde{Y}_2\right).$$

With this identification, Theorem 3.2.18 is the same as Theorem 3.2.17. The conclusion of Theorem 3.2.17 gives unitary operators $\tau_{\mathcal{F}} : \mathcal{F} \to \widetilde{\mathcal{F}}$, $\tau_{\mathcal{E}} : \mathcal{E} \to \widetilde{\mathcal{E}}$, and $\tau_u : \mathcal{K}_u \to \widetilde{\mathcal{K}}_u$ so that the last two equations in (3.89) hold and such that

$$\tau_{\mathcal{F}} \Theta(z) = \widetilde{\Theta}(z) \tau_{\mathcal{F}} \quad \text{for all } z \in \mathbb{D}.$$

In particular, for $z = 0$, we have

$$\tau_{\mathcal{F}} \begin{bmatrix} u_{\mathcal{F}_0} & 0 \\ 0 & \Theta_0(0) \end{bmatrix} = \begin{bmatrix} u_{\widetilde{\mathcal{F}}_0} & 0 \\ 0 & \widetilde{\Theta}_0(0) \end{bmatrix} \tau_{\mathcal{F}}.$$

Since $\Xi$ and $\widetilde{\Xi}$ as in the statement of Theorem 3.2.18 are refined BDF-tuples, $\Theta(0)$ and $\widetilde{\Theta}(0)$ are c.n.u. contractions. Therefore, by part (ii) of Lemma 2.3.4, $\tau_{\mathcal{F}}$ must decompose as

$$\tau_{\mathcal{F}} = \begin{bmatrix} \tau_{\mathcal{F}_0} & 0 \\ 0 & \tau_{\mathcal{F}_1} \end{bmatrix} : \begin{bmatrix} \mathcal{F}_0 \\ \mathcal{F}_1 \end{bmatrix} \to \begin{bmatrix} \widetilde{\mathcal{F}}_0 \\ \widetilde{\mathcal{F}}_1 \end{bmatrix}$$

for some unitary operators $\tau_{\mathcal{F}_0} : \mathcal{F}_0 \to \widetilde{\mathcal{F}}_0$ and $\tau_{\mathcal{F}_1} : \mathcal{F}_1 \to \widetilde{\mathcal{F}}_1$. This decomposed form of $\tau_{\mathcal{F}}$ together with the intertwining $\tau_{\mathcal{F}}\Theta(z) = \widetilde{\Theta}(z)\tau_{\mathcal{F}}$ for all $z \in \mathbb{D}$ gives the remaining two equations in (3.89). This completes the proof. □

## 3.3 Synthesis of BCL- and BDF-Models

At this stage we know that any commuting isometric pair $\mathbf{V} = (V_1, V_2)$ has a BCL-model $(V_1, V_2)_{\mathrm{BCL}}$ as well as a BDF-model $(V_1, V_2)_{\mathrm{BDF}}$, so $(V_1, V_2)$ is unitarily equivalent to both $(V_1, V_2)_{\mathrm{BCL}}$ and $(V_1, V_2)_{\mathrm{BDF}}$. More concretely, if $(V_1, V_2)_{\mathrm{BCL}}$ is a BCL-functional model, then there must be a BDF-functional model $(V_1, V_2)_{\mathrm{BDF}}$ so that $(V_1, V_2)_{\mathrm{BCL}}$ is unitarily equivalent to $(V_1, V_2)_{\mathrm{BDF}}$. On the other hand, we also know that BCL-models $(V_1, V_2)_{\mathrm{BCL}}$ are in one-to-one correspondence (up to unitary equivalence) with BCL-tuples $\Xi_{\mathrm{BCL}} = (\mathcal{F}, P, U)$, and similarly BDF-models $(V_1, V_2)_{\mathrm{BDF}}$ are in one-to-one correspondence (up to unitary equivalence) with BDF-tuples $\Xi_{\mathrm{BDF}} = (\mathcal{F}_0, \mathcal{F}_1, \mathcal{E}_0, u_{\mathcal{F}_0}, \Theta(z), u_{\mathcal{E}_0})$. In this section we give a recipe for converting a BCL-tuple $\Xi_{\mathrm{BCL}}$ to a companion BDF-tuple that we label as $\Xi_{\mathrm{BCL}}^{\times}$ so that the BCL-model operator pair generated by $\Xi_{\mathrm{BCL}}$ is unitarily equivalent to the BDF-model operator pair generated by $\Xi_{\mathrm{BCL}}^{\times}$, and vice versa: One can start with a BDF-tuple $\Xi_{\mathrm{BDF}}$ and identify a companion BCL-tuple labeled as $\Xi_{\mathrm{BDF}}^{\times}$ so that the BDF-model isometric pair generated by $\Xi_{\mathrm{BDF}}$ is unitarily equivalent to the BCL-model isometric pair generated by $\Xi_{\mathrm{BDF}}^{\times}$. The details involve understanding how to pull an appropriate unitary $2 \times 2$-block matrix $\left[\begin{smallmatrix} A & B \\ C & D \end{smallmatrix}\right]$ out of a contractive analytic function $(\mathcal{F}, \Theta(z))$ so that $\Theta(z)$ can be represented as $\Theta(z) = A + zB(I - zD)^{-1}C$, a version of the realization problem from mathematical systems theory.

### 3.3.1 BDF-Characteristic Function and Unitary System-Node Realizations of Contractive Analytic Functions

Let $(V_1, V_2)$ be a commuting isometric pair. To begin the construction of the BDF-model for $(V_1, V_2)$ (see Section 3.2.1), the first step is the observation that the projection of $V_2$ on the space $\mathcal{K}_{1s}$ is a contraction operator commuting with the shift $V_1|_{\mathcal{K}_{1s}}$. The observability operator $\mathcal{O}_{\mathcal{D}_{V_1^*}, V_1^*} : \mathcal{K} \to H^2\left(\mathcal{D}_{V_1^*}\right)$ defined by

$$\mathcal{O}_{\mathcal{D}_{V_1^*}, V_1^*} : k \mapsto D_{V_1^*}\left(I - zV_1^*\right)^{-1} k$$

is a partial isometry from $\mathcal{K}$ onto $H^2\left(\mathcal{D}_{V_1^*}\right)$ with initial space equal to $\mathcal{K}_{1s}$ that implements the intertwining

$$\mathcal{O}_{\mathcal{D}_{V_1^*}, V_1^*} V_1 = M_z^{\mathcal{D}_{V_1^*}} \mathcal{O}_{\mathcal{D}_{V_1^*}, V_1^*}$$

(see, e.g., Theorem 2.1.4). The condition that $M_\Theta$ should be the representative of $P_{\mathcal{K}_{1s}} V_2$ on $H^2\left(\mathcal{D}_{V_1^*}\right)$ after this change of coordinates then implies that we must have

$$\mathcal{O}_{\mathcal{D}_{V_1^*}, V_1^*} V_2 = M_\Theta \mathcal{O}_{\mathcal{D}_{V_1^*}, V_1^*}$$

or, equivalently,

$$D_{V_1^*}\left(I - zV_1^*\right)^{-1} V_2 = \Theta(z) D_{V_1^*}\left(I - zV_1^*\right)^{-1}.$$

If we restrict this formula to the subspace $\mathcal{D}_{V_1^*} \subset \mathcal{K}_{1s}$, we wind up with the formula

$$\Theta(z) = D_{V_1^*}\left(I - zV_1^*\right)^{-1} V_2|_{\mathcal{D}_{V_1^*}} \tag{3.90}$$

for the values of the multiplier $\left(\mathcal{D}_{V_1^*}, \Theta\right)$ at the point $z \in \mathbb{D}$. Plugging into this formula the identity

$$\left(I - zV_1^*\right)^{-1} = I + zV_1^*\left(I - zV_1^*\right)^{-1}$$

then leads us to the alternate formula

$$\Theta(z) = D_{V_1^*} V_2|_{\mathcal{D}_{V_1^*}} + zD_{V_1^*}\left(I - zV_1^*\right)^{-1}\left(V_1^* V_2\right)|_{\mathcal{D}_{V_1^*}}. \tag{3.91}$$

The calculations leading to the formula (3.91) have implications in several directions, which we now explore. If we set

$$\begin{bmatrix} A_{\mathbf{V}} & B_{\mathbf{V}} \\ C_{\mathbf{V}} & D_{\mathbf{V}} \end{bmatrix} = \begin{bmatrix} D_{V_1^*} V_2|_{\mathcal{D}_{V_1^*}} & D_{V_1^*}|_{\mathcal{D}_{V_2^*}} \\ V_1^* V_2|_{\mathcal{D}_{V_1^*}} & V_1^*|_{\mathcal{D}_{V_2^*}} \end{bmatrix} \tag{3.92}$$

then the formula (3.91) takes on the form

$$\Theta(z) = A_{\mathbf{V}} + zB_{\mathbf{V}}(I - zD_{\mathbf{V}})^{-1} C_{\mathbf{V}}, \tag{3.93}$$

where the matrix $U = \begin{bmatrix} A_{\mathbf{V}} & B_{\mathbf{V}} \\ C_{\mathbf{V}} & D_{\mathbf{V}} \end{bmatrix}$ is also the unitary operator in the canonical BCL-tuple for the commuting isometric pair $\mathbf{V}$ and hence in particular is unitary.

In general, in mathematical systems theory, any representation of a matrix or operator-valued function $\Theta$ of the form $\Theta(z) = A + zB(I - zD)^{-1}C$ is said to be the *realization* of $\Theta(z)$ as the *transfer function* for the discrete-time linear input/state/output linear system

$$\begin{cases} y(n) &= Au(n) + Bx(n), \\ x(n+1) &= Cu(n) + Dx(n), \end{cases} \tag{3.94}$$

generated by what we shall call the *system node* $U = \left[\begin{smallmatrix} A & B \\ C & D \end{smallmatrix}\right]$. The terminology comes from the fact that if $\{u(n), x(n), y(n)\}_{n\geq 0}$ is a system trajectory of the system (3.94) with state initial condition $x(0) = 0$ imposed, then the $Z$-transform $\widehat{u}(z) = \sum_{n=0}^{\infty} u(n)z^n$ of the input signal determines the $Z$-transform $\widehat{y}(z) = \sum_{n=0}^{\infty} u(n)z^n$ of the output signal according to the formula for input/output transfer

$$\widehat{y}(z) = T_{\left[\begin{smallmatrix} A & B \\ C & D \end{smallmatrix}\right]}(z) \cdot \widehat{u}(z),$$

where $T_{\left[\begin{smallmatrix} A & B \\ C & D \end{smallmatrix}\right]}(z) = A + zB(I - zD)^{-1}C$ is the *transfer function* of the system (3.94). It is well known by other methods (see item number 1 in the Notes at the end of the chapter (Section 3.5)) that any contractive analytic function $(\mathcal{F}, \Theta)$ can be realized as the transfer function of a unitary system node $U = \left[\begin{smallmatrix} A & B \\ C & D \end{smallmatrix}\right]$, but now we have a new proof based on the BDF-model theory for commuting isometric pairs as follows.

**Theorem 3.3.1** *Any contractive analytic function $(\mathcal{F}, \Theta)$ can be realized as the transfer function of a unitary system node $U = \left[\begin{smallmatrix} A & B \\ C & D \end{smallmatrix}\right]$:*

$$\Theta(z) = A + zB(I - zD)^{-1}C, \quad \textit{where} \begin{bmatrix} A & B \\ C & D \end{bmatrix} \textit{ on } \begin{bmatrix} \mathcal{F} \\ \mathcal{E} \end{bmatrix} \textit{ is unitary.}$$

*Proof* We start with the contractive analytic function $(\mathcal{F}, \Theta)$. Then we pair up $\Theta$ with the shift operator $M_z^{\mathcal{F}}$ on $H^2(\mathcal{F})$ to get the commuting shift/contractive pair $\left(M_z^{\mathcal{F}}, M_\Theta\right)$ on $H^2(\mathcal{F})$. By Corollary 3.2.8 we see that $\left(M_z^{\mathcal{F}}, M_\Theta\right)$ can be lifted to the commuting isometric pair

$$\mathbf{V} = (V_1, V_2) = \left( \begin{bmatrix} M_z^{\mathcal{F}} & 0 \\ 0 & I_{H^2} \otimes M_\zeta \end{bmatrix}, \begin{bmatrix} M_\Theta & 0 \\ \Delta_\Theta|_{H^2(\mathcal{F})} & M_z^{\overline{\Delta_\Theta L^2(\mathcal{F})}} \end{bmatrix} \right). \tag{3.95}$$

It is a consequence of the computations leading to the formula (3.91) that we recover $\Theta(z)$ via the formula $\Theta(z) = T_U(z)$ with system node $U = \begin{bmatrix} A_{\mathbf{V}} & B_{\mathbf{V}} \\ C_{\mathbf{V}} & D_{\mathbf{V}} \end{bmatrix}$ unitary. □

**Remark 3.3.2** It is worth pointing out the following fact concerning transfer functions of unitary system nodes: *If* $U = \left[\begin{smallmatrix} A & B \\ C & D \end{smallmatrix}\right]$ *is unitary and* $D$ *has a nontrivial canonical decomposition*

$$D \underset{\mathrm{u}}{\cong} \begin{bmatrix} D_1 & 0 \\ 0 & D_0 \end{bmatrix} \quad \text{on} \quad \begin{bmatrix} \mathcal{E}_1 \\ \mathcal{E}_0 \end{bmatrix}$$

*with* $D_1$ *c.n.u. on* $\mathcal{E}_1$ *and* $D_0$ *unitary on* $\mathcal{E}_0$ *with* $\mathcal{E}_0 \neq \{0\}$, *then the transfer function* $T_{\left[\begin{smallmatrix} A & B \\ C & D \end{smallmatrix}\right]}(z)$ *is independent of the value of the unitary part* $D_0$. Indeed, in this situation the splitting of $\mathcal{E}$ as $\mathcal{E} \underset{\mathrm{u}}{\cong} \left[\begin{smallmatrix} \mathcal{E}_1 \\ \mathcal{E}_0 \end{smallmatrix}\right]$ induces a partitioning of $\left[\begin{smallmatrix} A & B \\ C & D \end{smallmatrix}\right]$ as

$$\begin{bmatrix} A & B \\ C & D \end{bmatrix} \underset{\mathrm{u}}{\cong} \begin{bmatrix} A & B_1 & 0 \\ C_1 & D_1 & 0 \\ 0 & 0 & D_0 \end{bmatrix},$$

implying that the transfer function $T_{\left[\begin{smallmatrix} A & B \\ C & D \end{smallmatrix}\right]}(z)$ assumes the form

$$\begin{aligned} T_{\left[\begin{smallmatrix} A & B \\ C & D \end{smallmatrix}\right]}(z) &= A + z \begin{bmatrix} B_1 & 0 \end{bmatrix} \begin{bmatrix} (I - zD_1)^{-1} & 0 \\ 0 & (I - zD_0)^{-1} \end{bmatrix} \begin{bmatrix} C_1 \\ 0 \end{bmatrix} \\ &= A + zB_1(I - zD_1)^{-1}C_1 =: T_{\left[\begin{smallmatrix} A & B_1 \\ C_1 & D_1 \end{smallmatrix}\right]}(z), \qquad \text{where } D_1 \text{ is c.n.u.} \end{aligned}$$

Thus, as far as calculation of the transfer function of a unitary system node $\left[\begin{smallmatrix} A & B \\ C & D \end{smallmatrix}\right]$ is concerned, we may assume without loss of generality that $D$ is c.n.u. For this reason we shall say that the unitary realization is *minimal* if the operator $D$ is c.n.u. When talking about realizations of a given contractive analytic function, it makes sense to restrict ourselves to such minimal realizations.

We next show how it is possible to express the canonical realization (3.92) directly in terms of the function $(\mathcal{F}, \Theta)$ rather than in terms of an Ando lift $\mathbf{V} = (V_1, V_2)$ (3.95) of $(M_z^{\mathcal{F}}, M_\Theta)$.

**Theorem 3.3.3** *Let* $(\mathcal{F}, \Theta)$ *be a contractive analytic function. Set* $(V_1, V_2)$ *equal to the 1-c.n.u. commuting isometric pair generated by the degenerate BDF-tuple* $(\mathcal{F}, \Theta)$ *generating the 1-c.n.u. BDF1-model* $(V_1, V_2)_{\mathrm{BDF1}}$ *given by*

$$\left( \begin{bmatrix} M_z^{\mathcal{F}} & 0 \\ 0 & I_{H^2} \otimes M_\zeta \end{bmatrix}, \begin{bmatrix} M_\Theta & 0 \\ \Delta_\Theta|_{H^2(\mathcal{F})} & M_z^{\overline{\Delta_\Theta L^2(\mathcal{F})}} \end{bmatrix} \right) \text{ on } \begin{bmatrix} H^2(\mathcal{F}) \\ H^2(\overline{\Delta_\Theta L^2(\mathcal{F})}) \end{bmatrix}. \tag{3.96}$$

*Recall the Sz.-Nagy–Foias functional model space associated with the contractive function* $(\mathcal{F}, \Theta)$ *(see* (2.61)*)*

$$\mathcal{H}_\Theta = \begin{bmatrix} H^2(\mathcal{F}) \\ \overline{\Delta_\Theta L^2(\mathcal{F})} \end{bmatrix} \ominus \begin{bmatrix} \Theta \\ \Delta_\Theta \end{bmatrix} H^2(\mathcal{F}).$$

*Then the canonical unitary system-node realizing* $\Theta$ *via the auxiliary commuting isometric pair* $(V_1, V_2)$ *is given by*

$$\begin{bmatrix} A_\Theta & B_\Theta \\ C_\Theta & D_\Theta \end{bmatrix} = \begin{bmatrix} \Theta(0) & \mathbf{ev}^{(1)}_{0,\mathcal{F}}|_{\mathcal{H}_\Theta} \\ \left[ \begin{bmatrix} \frac{\Theta(z)-\Theta(0)}{z} \\ \overline{\zeta}\Delta_\Theta(\zeta) \end{bmatrix} \right]\Big|_{\mathcal{F}} & \begin{bmatrix} (M_z^{\mathcal{F}})^* & 0 \\ 0 & M_{\overline{\zeta}} \end{bmatrix} \end{bmatrix} \quad \text{acting on} \quad \begin{bmatrix} \mathcal{F} \\ \mathcal{H}_\Theta \end{bmatrix} = \begin{bmatrix} \mathcal{D}_{V_1^*} \\ \mathcal{D}_{V_2^*} \end{bmatrix}. \tag{3.97}$$

*Here the notation* $\mathbf{ev}^{(1)}_{0,\mathcal{F}}$ *is the map of evaluation at zero in the first component acting on an element* $\left[\begin{smallmatrix} h \\ k \end{smallmatrix}\right] \in \mathcal{H}_\Theta := \left[\begin{smallmatrix} H^2(\mathcal{F}) \\ \Delta_\Theta L^2(\mathcal{F}) \end{smallmatrix}\right] \ominus \left[\begin{smallmatrix} \Theta \\ \Delta_\Theta \end{smallmatrix}\right] H^2(\mathcal{F})$.

*Proof* As a first step we compute the defect spaces $\mathcal{D}_{V_1^*}$ and $\mathcal{D}_{V_2^*}$ as follows. First note that

$$\mathcal{D}_{V_1^*} = \begin{bmatrix} \mathbf{ev}^*_{0,\mathcal{F}}\mathcal{F} \\ 0 \end{bmatrix} \underset{\text{u}}{\cong} \begin{bmatrix} \mathcal{F} \\ 0 \end{bmatrix} \underset{\text{u}}{\cong} \mathcal{F}.$$

Then note that

$$\operatorname{Ran} V_2 = \begin{bmatrix} M_\Theta \\ \Delta_\Theta \end{bmatrix} H^2(\mathcal{F}) \bigoplus \begin{bmatrix} 0 \\ \overline{z \cdot H^2(\Delta_\Theta L^2(\mathcal{F}))} \end{bmatrix}$$

and hence the defect space $\mathcal{D}_{V_2^*} = (\operatorname{Ran} V_2)^\perp$ is given by

$$\begin{aligned} \mathcal{D}_{V_2^*} &= \left( \begin{bmatrix} M_\Theta \\ \Delta_\Theta \end{bmatrix} H^2(\mathcal{F}) \right)^\perp \bigcap \left( \begin{bmatrix} 0 \\ \overline{z \cdot H^2(\Delta_\Theta L^2(\mathcal{F}))} \end{bmatrix} \right)^\perp \\ &= \left( \begin{bmatrix} M_\Theta \\ \Delta_\Theta \end{bmatrix} H^2(\mathcal{F}) \right)^\perp \bigcap \begin{bmatrix} H^2(\mathcal{F}) \\ \overline{\Delta_\Theta L^2(\mathcal{F})} \end{bmatrix} \\ &= \begin{bmatrix} H^2(\mathcal{F}) \\ \overline{\Delta_\Theta L^2(\mathcal{F})} \end{bmatrix} \ominus \begin{bmatrix} \Theta \\ \Delta_\Theta \end{bmatrix} \cdot H^2(\mathcal{F}) \\ &=: \mathcal{H}_\Theta. \end{aligned}$$

Then we compute

$$\begin{aligned} A_\Theta &= D_{V_1^*} V_2|_{\mathcal{D}_{V_1^*}} : f \underset{\text{u}}{\cong} \begin{bmatrix} f \\ 0 \end{bmatrix} \mapsto \begin{bmatrix} \Theta(0)f \\ 0 \end{bmatrix} \underset{\text{u}}{\cong} \Theta(0)f \quad \text{for } f \in \mathcal{F}, \\ B_\Theta &= D_{V_1^*}|_{\mathcal{D}_{V_2^*}} : \begin{bmatrix} h \\ k \end{bmatrix} \mapsto h(0), \quad \text{where } \begin{bmatrix} h \\ k \end{bmatrix} \in \mathcal{H}_\Theta, \\ C_\Theta &= V_1^* V_2|_{\mathcal{D}_{V_1^*}} : \begin{bmatrix} f \\ 0 \end{bmatrix} \mapsto \begin{bmatrix} \frac{\Theta(z)-\Theta(0)}{z} f \\ \overline{\zeta}\Delta_\Theta(\zeta) f \end{bmatrix} \quad \text{for } f \in \mathcal{F}, \\ D_\Theta &= V_1^*|_{\mathcal{D}_{V_2^*}} : \begin{bmatrix} h(z) \\ k(\zeta) \end{bmatrix} \mapsto \begin{bmatrix} \frac{h(z)-h(0)}{z} \\ \overline{\zeta} k(\zeta) \end{bmatrix} \quad \text{for } \begin{bmatrix} h(z) \\ k(\zeta) \end{bmatrix} \in \mathcal{H}_\Theta. \end{aligned} \tag{3.98}$$

Now it is a matter of working with the definitions to arrive at the formula (3.97) for an equivalent form of the canonical unitary system-node expressed directly in terms of $\Theta$ rather than in terms of $\mathbf{V} = (V_1, V_2)$ (3.95) □

What's missing in the statement of Theorem 3.3.3 is a more concrete version of the formula (3.24) for the map

$$\tau_\Theta \colon \begin{bmatrix} H^2(\mathcal{F}) \\ H^2(\overline{\Delta_\Theta L^2(\mathcal{F})}) \end{bmatrix} \to H^2\left(\begin{bmatrix} \mathcal{F} \\ \mathcal{H}_\Theta \end{bmatrix}\right)$$

implementing the unitary equivalence of the BDF1-model pair $(V_1, V_2)_{\rm BDF1}$ (3.96) with the BCL1-model

$$(V_1, V_2)_{\rm BCL} = \left( \begin{bmatrix} M_z \otimes A_\Theta^* & I_{H^2} \otimes C_\Theta^* \\ M_z \otimes B_\Theta^* & I_{H^2} \otimes D_\Theta^* \end{bmatrix}, \begin{bmatrix} I_{H^2} \otimes A_\Theta & I_{H^2} \otimes B_\Theta \\ M_z \otimes C_\Theta & M_z \otimes D_\Theta \end{bmatrix} \right) \quad \text{on} \quad \begin{bmatrix} H^2(\mathcal{F}) \\ H^2(\mathcal{H}_\Theta) \end{bmatrix}. \tag{3.99}$$

To get a clean explicit formula for $\tau_\Theta$ for the general case depends on having a concrete formula (tailored to this example) for the Wold decomposition for the product isometry

$$V = V_1 V_2 = \begin{bmatrix} M_{z \cdot \Theta} & 0 \\ (I_{H^2} \otimes M_\zeta)\Delta_\Theta|_{H^2(\mathcal{F})} & M_z \otimes M_\zeta|_{H^2(\overline{\Delta_\Theta L^2(\mathcal{F})})} \end{bmatrix} \quad \text{on} \quad \begin{bmatrix} H^2(\mathcal{F}) \\ H^2(\overline{\Delta_\Theta L^2(\mathcal{F})}) \end{bmatrix}.$$

If we assume that $(\mathcal{F}, \Theta)$ is inner (i.e., $\Delta_\Theta(\zeta) \equiv 0$), then all the formulas simplify and become more tractable. We then have that the auxiliary commuting isometric pair is $(V_1, V_2) = (M_z^{\mathcal{F}}, M_\Theta)$ acting on $\mathcal{H}^2(\mathcal{F})$. Then $\mathcal{D}_{V_1^*} = \operatorname{Ran} \mathbf{ev}_{0,\mathcal{F}}^* \underset{\rm u}{\cong} \mathcal{F}$, while $\mathcal{D}_{V_2^*} = H^2(\mathcal{F}) \ominus \Theta H^2(\mathcal{F})$, which we denote as $\mathfrak{H}(\Theta)$ equal to the de Branges–Rovnyak functional-model space associated with the inner function $\Theta$. Then the formulas (3.98) reduce to

$$\begin{aligned}
A_\Theta &= D_{V_1^*} V_2|_{\mathcal{D}_{V_1^*}} \colon f \mapsto \Theta(0) f \quad \text{for } f \in \mathcal{F} \underset{u}{=} \mathcal{D}_{V_1^*}, \\
B_\Theta &= D_{V_1^*}|_{\mathcal{D}_{V_2^*}} \colon h \mapsto h(0) \quad \text{for } h \in \mathfrak{H}(\Theta) = \mathcal{D}_{V_2^*}, \\
C_\Theta &= V_1^* V_2|_{\mathcal{D}_{V_2^*}} \colon f \mapsto \frac{\Theta(z) - \Theta(0)}{z} f \quad \text{for } f \in \mathcal{F}, \\
D_\Theta &= V_1^*|_{\mathcal{D}_{V_2^*}} \colon h(z) \mapsto \frac{h(z) - h(0)}{z} \quad \text{for } h \in \mathfrak{H}(\Theta).
\end{aligned} \tag{3.100}$$

Note that the wandering subspace $\mathcal{D}_{V^*}$ for the product isometry $V = M_{z \cdot \Theta}$ is given by

$$\mathcal{D}_{V^*} = H^2(\mathcal{F}) \ominus z \cdot \Theta(z) H^2(\mathcal{F}),$$

which has the internal direct-sum decomposition

$$\mathcal{D}_{V^*} = V_2 \mathcal{D}_{V_1^*} \oplus \mathcal{D}_{V_2^*} = \Theta \cdot \mathcal{F} \oplus \mathfrak{H}(\Theta).$$

Then the Wold decomposition for $V = M_{z\Theta(z)}$ on $H^2(\mathcal{Y})$ has the concrete form

$$f(z) = \sum_{n=0}^{\infty} z^n \Theta(z)^n (\Theta(z) y_n + h_n(z))$$

for any $f \in H^2(\mathcal{Y})$, where $y_n \in \mathcal{F}$ and $h_n \in \mathfrak{H}(\Theta)$ are determined by

$$\Theta(z) y_n + h_n(z) = P_{\mathcal{D}_{V^*}} V^{*n} h.$$

We then define an identification map $\tau_\Theta \colon H^2(\mathcal{Y}) \to H^2\left(\begin{bmatrix} \mathcal{Y} \\ \mathfrak{H}(\Theta) \end{bmatrix}\right) \cong \begin{bmatrix} H^2(\mathcal{Y}) \\ H^2(\mathfrak{H}(\Theta)) \end{bmatrix}$ by

$$\tau_\Theta \colon \sum_{n=0}^{\infty} z^n \Theta(z)^n (\Theta(z) y_n + h_n(z)) \mapsto \sum_{n=0}^{\infty} z^n \begin{bmatrix} y_n \\ h_n(w) \end{bmatrix} =: \begin{bmatrix} y(z) \\ h(z,w) \end{bmatrix}, \quad (3.101)$$

where we set

$$y(z) = \sum_{n=0}^{\infty} y_n z^n \in H^2(\mathcal{F}), \quad h(z,w) = \sum_{n=0}^{\infty} z^n h_n(w) \in H^2(\mathfrak{H}(\Theta)).$$

Then it is straightforward to see that $\tau_\Theta$ implements the intertwining identity

$$\tau_\Theta M_{z\Theta(z)} = M_z \tau_\Theta.$$

It is less obvious to identify by inspection the inner operator pencils $\Psi_1(z)$ and $\Psi_2(z)$ so that

$$\tau_\Theta M_z = M_{\Psi_1(z)} \tau_\Theta, \quad \tau_\Theta M_\Theta = M_{\Psi_2(z)} \tau_\Theta. \quad (3.102)$$

However, from the general formulas obtained in Theorem 3.1.7 and Corollary 3.1.8 we know that

$$(M_{\Psi_1}, M_{\Psi_2}) \text{ with } \Psi_1(z) = U^* P^\perp + z U^* P, \quad \Psi_2(z) = PU + z P^\perp U, \quad (3.103)$$

where we choose

$$P = \begin{bmatrix} I_\mathcal{F} & 0 \\ 0 & 0 \end{bmatrix}, \quad U = \begin{bmatrix} \mathbf{ev}_0 M_\Theta|_\mathcal{Y} & \mathbf{ev}_0|_{\mathfrak{H}(\Theta)} \\ M_w^* M_\Theta|_\mathcal{Y} & M_w^*|_{\mathfrak{H}(\Theta)} \end{bmatrix} =: \begin{bmatrix} A_\Theta & B_\Theta \\ C_\Theta & D_\Theta \end{bmatrix} \quad \text{on} \quad \begin{bmatrix} \mathcal{F} \\ \mathcal{H}(\Theta) \end{bmatrix} \quad (3.104)$$

acting on $\begin{bmatrix} \mathcal{F} \\ \mathcal{H}(\Theta) \end{bmatrix} = \begin{bmatrix} \mathcal{D}_{V_1^*} \\ \mathcal{D}_{V_2^*} \end{bmatrix}$ does the job (where we now use $w$ as the independent variable for functions in $\mathfrak{H}(\Theta)$), as is seen by specializing the formulas in (3.23) to the case where

$$\mathcal{D}_{V_1^*} = \mathcal{Y}, \quad \mathcal{D}_{V_2^*} = \mathfrak{H}(\Theta), \quad V_1 = M_z, \quad V_2 = M_{\Theta(z)}.$$

Note that writing out $\Psi_1(z)$ and $\Psi_2(z)$ as $2 \times 2$-block matrices compatible with the $2 \times 2$-block matrix format of $P$ and $U$ then gives us the alternative version of (3.103):

$$(\Psi_1(z), \Psi_2(z)) = \left( \begin{bmatrix} zA_\Theta^* & C_\Theta^* \\ zB_\Theta^* & D_\Theta^* \end{bmatrix}, \begin{bmatrix} A_\Theta & B_\Theta \\ zC_\Theta & zD_\Theta \end{bmatrix} \right). \tag{3.105}$$

In summary, the pair of formulas (3.103), (3.104) or equivalently (3.105), gives the explicit conversion of the BDF-model $(M_z, M_\Theta)$ to a BCL1-model $(M_{\Psi_1}, M_{\Psi_2})$. We summarize all these observations for the case where $\Theta$ is inner as follows.

**Corollary 3.3.4** *Suppose that the contractive analytic function* $(\mathcal{F}, \Theta)$ *is inner and set* $U = \left[\begin{smallmatrix} A & B \\ C & D \end{smallmatrix}\right]$, *where* $A, B, C, D$ *are given by* (3.100), *or*

$$U = \begin{bmatrix} \Theta(0) & \mathbf{ev}_{0,\mathcal{F}}|_{\mathfrak{H}(\Theta)} \\ \frac{\Theta(z)-\Theta(0)}{z}\Big|_{\mathcal{F}} & \left(M_z^{\mathcal{F}}\right)^*\Big|_{\mathfrak{H}(\Theta)} \end{bmatrix}. \tag{3.106}$$

*Then* $U$ *is a unitary system-node realization of* $\Theta$, *that is, we recover* $\Theta(z)$ *as*

$$\Theta(z) = T_U(z) = A + zB(I - zD)^{-1}C$$

*with* $U$ *as in* (3.106).

*Moreover, in this case, the unitary similarity* $\tau_\Theta \colon H^2(\mathcal{F}) \to H^2\left(\left[\begin{smallmatrix} \mathcal{F} \\ \mathfrak{H}(\Theta) \end{smallmatrix}\right]\right)$ *implementing the unitary equivalence of the BDF1-model* $\left(M_z^{\mathcal{F}}, M_\Theta\right)$ *with the BCL1-model* $M_{\Psi_1}, M_{\Psi_2})$ (3.105) *on* $H^2\left(\left[\begin{smallmatrix} \mathcal{F} \\ \mathfrak{H}(\Theta) \end{smallmatrix}\right]\right)$ *is given by the formula* (3.101).

**Remark 3.3.5** The de Branges–Rovnyak realization for the inner case appears in the work of de Branges–Rovnyak [51, 52] with generalization to coisometric realizations for model spaces contained only contractively in $H^2(\mathcal{F})$.

Similar results hold for the Sz.-Nagy–Foias model theory. We note that the formula for the Sz.-Nagy–Foias characteristic function of the Hilbert space contraction operator $T$ acting on the Hilbert space $\mathcal{H}$ (see (2.50)),

$$\Theta_T(z) = \left(-T + zD_{T^*}(I - zT^*)^{-1}D_T\right)|_{\mathcal{D}_T} \colon \mathcal{D}_T \to \mathcal{D}_{T^*},$$

exhibits the Sz.-Nagy–Foias characteristic function as the transfer function of the (nonsquare) unitary system-node (sometimes called the *Halmos unitary dilation* of $T^*$):

$$U_{T^*} = \begin{bmatrix} -T & D_{T^*} \\ D_T & T^* \end{bmatrix} \colon \begin{bmatrix} \mathcal{D}_T \\ \mathcal{H} \end{bmatrix} \to \begin{bmatrix} \mathcal{D}_{T^*} \\ \mathcal{H} \end{bmatrix}.$$

If $(\mathcal{F}, \mathcal{F}_*, \Theta)$ is a (nonsquare) contractive analytic function that is *purely contractive* in the Sz.-Nagy–Foias sense ($\|\Theta(0)d\| < \|d\|$ for all $0 \neq d \in \mathcal{F}$), then it is shown in [92, Section VI.3] that $\Theta$ coincides with the characteristic function $\Theta_{T(\Theta)}$ of the model contraction operator defined on the associated model space $\mathcal{H}(\Theta)$. This is finally enough to show that any such contractive analytic function $(\mathcal{F}, \mathcal{F}_*, \Theta)$ can be realized as the transfer function of a (nonsquare) unitary system node. Because of the "nonsquareness" of these other model theories, direct comparisons with the BDF theory are not possible. This connection between the realization of contractive analytic functions and Sz.-Nagy–Foias model theory was first observed by J. W. Helton in [65].

We have seen in Remark 3.1.16 that the form for a normalized BCL-tuple generating a 1&2-c.n.u. BCL1-model is

$$\Xi^{\mathrm{BCL}}_{1\&2\mathrm{cnu}} = \left( \begin{bmatrix} \mathcal{F}_1 \\ \mathcal{E}_1 \end{bmatrix}, \begin{bmatrix} I_{\mathcal{F}_1} & 0 \\ 0 & 0 \end{bmatrix}, \begin{bmatrix} A_1 & B_1 \\ C_1 & D_1 \end{bmatrix} \right), \tag{3.107}$$

where we require that $A_1$ is c.n.u. on $\mathcal{F}_1$ and that $D_1$ is c.n.u. on $\mathcal{E}_1$, generating the 1&2-c.n.u. BCL1-model

$$(V_1, V_2)_{1\&2\mathrm{cnu}} = \left( \begin{bmatrix} M_z \otimes A_1^* & I_{H^2} \otimes C_1^* \\ M_z \otimes B_1^* & I_{H^2} \otimes D_1^* \end{bmatrix}, \begin{bmatrix} I_{H^2} \otimes A_1 & I_{H^2} \otimes B_1 \\ M_z \otimes C_1 & M_z \otimes D_1 \end{bmatrix} \right) \tag{3.108}$$

acting on $H^2\left(\begin{bmatrix} \mathcal{F}_1 \\ \mathcal{E}_1 \end{bmatrix}\right)$. On the other hand we know that the BDF-tuple generating a 1&2-c.n.u. commuting isometric pair has the degenerate form

$$\Xi^{\mathrm{BDF}}_{1\&2\mathrm{cnu}} = (\mathcal{F}_1, \Theta_1), \tag{3.109}$$

where in addition we require that $\Theta_1(0)$ is c.n.u. on $\mathcal{F}_1$, generating the 1-c.n.u. BDF1-model pair

$$(V_1, V_2)_{\mathrm{BDF1},1\&2\mathrm{cnu}} = \left( \begin{bmatrix} M_z^{\mathcal{F}_1} & 0 \\ 0 & I_{H^2} \otimes M_\zeta \end{bmatrix}, \begin{bmatrix} M_{\Theta_1} & 0 \\ \Delta_{\Theta_1}|_{H^2(\mathcal{F}_1)} & M_z^{\overline{\Delta_{\Theta_1} H^2(\mathcal{F}_1)}} \end{bmatrix} \right) \tag{3.110}$$

acting on $\begin{bmatrix} H^2(\mathcal{F}_1) \\ \overline{H^2(\Delta_{\Theta_1} L^2(\mathcal{F}_1))} \end{bmatrix}$. We would like to understand better how these two models for 1&2-c.n.u. commuting isometric pairs are related to each other. Toward this end the following terminology is useful.

**Definition 3.3.6** Given the 1&2-c.n.u. normalized BCL-tuple $\Xi^{\mathrm{BCL}}_{1\&2\mathrm{cnu}}$ (3.107), we assign a companion 1&2-c.n.u. BDF-tuple $\Xi^{\mathrm{BCL},\times}_{1\&2\mathrm{cnu}}$ according to the rule

$$\Xi^{\mathrm{BCL},\times}_{1\&2\mathrm{cnu}} = (\mathcal{F}_1, \Theta_1(z)) \quad \text{where } \Theta_1(z) = A_1 + zB_1(I - zD_1)^{-1}C_1, \tag{3.111}$$

where $A_1, B_1, C_1, D_1$ are read from $\Xi^{\mathrm{BCL}}_{1\&2\mathrm{cnu}}$.

Similarly, given a 1&2-c.n.u. BDF-tuple $\Xi^{\rm BDF}_{1\&2{\rm cnu}}$ (3.109), we assign a companion 1&2-c.n.u. BCL-tuple according to the rule

$$\Xi^{\rm BDF,\times}_{1\&2{\rm cnu}} = \left( \begin{bmatrix} \mathcal{F}_1 \\ \mathcal{H}_{\Theta_1} \end{bmatrix}, \begin{bmatrix} P_{\mathcal{F}_1} & 0 \\ 0 & 0 \end{bmatrix}, \begin{bmatrix} A_{\Theta_1} & B_{\Theta_1} \\ C_{\Theta_1} & D_{\Theta_1} \end{bmatrix} \right), \tag{3.112}$$

where $A_{\Theta_1}, B_{\Theta_1}, C_{\Theta_1}, D_{\Theta_1}$ are given by (3.97) (with $\Theta_1$ in place of $\Theta$).

With this terminology in place, we have the following result.

**Theorem 3.3.7** (i) *Assume that the 1&2-c.n.u. BDF-tuple $\Xi^{\rm BCL,\times}_{1\&2{\rm cnu}}$ is a BDF-companion of the 1&2-c.n.u. BCL-tuple $\Xi^{\rm BCL}_{1\&2{\rm cnu}}$. Then the 1&2-c.n.u. BDF1-model generated by $\Xi^{\rm BCL,\times}_{1\&2{\rm cnu}}$ is unitarily equivalent to the BCL1-model generated by $\Xi^{\rm BCL}_{1\&2{\rm cnu}}$.*

(ii) *Assume that the 1&2-c.n.u. BCL-tuple $\Xi^{\rm BDF,\times}_{1\&2{\rm cnu}}$ is a BCL-companion of the 1&2-c.n.u. BDF-tuple $\Xi^{\rm BDF}_{1\&2{\rm cnu}}$. Then the BCL1-model generated by $\Xi^{\rm BDF,\times}_{1\&2{\rm cnu}}$ is unitarily equivalent to the BDF1-model generated by $\Xi^{\rm BDF}_{1\&2{\rm cnu}}$.*

*Proof of (i)* Let $\mathbf{V} = (V_1, V_2)$ be the BCL1-model pair generated by $\Xi^{\rm BCL}_{1\&2{\rm cnu}}$, that is, $(V_1, V_2)$ is given by (3.108). The content of the discussion leading up formula (3.93) is that we can get a 1&2-c.n.u. BDF1-model by taking $\Theta(z)$ equal to the BDF-characteristic function for $\mathbf{V}$, namely $\Theta_{\mathbf{V}} = A_{\mathbf{V}} + zB_{\mathbf{V}}(I - z\mathbf{A})^{-1}C_{\mathbf{V}}$ with values in $\mathcal{D}_{V_1^*}$. However, as (3.107) is a BCL-tuple generating the BCL1-model $\mathbf{V} = (V_1, V_2)$ it must be that the BCL-tuple (3.107) is unitarily equivalent to canonical BCL-tuple $\left( \begin{bmatrix} \mathcal{F}_{\mathbf{V}} \\ \mathcal{E}_{\mathbf{V}} \end{bmatrix}, \begin{bmatrix} I_{\mathcal{F}_{\mathbf{V}}} & 0 \\ 0 & 0 \end{bmatrix}, \begin{bmatrix} A_{\mathbf{V}} & B_{\mathbf{V}} \\ C_{\mathbf{V}} & D_{\mathbf{V}} \end{bmatrix} \right)$. In particular, this means that there exist unitary transformations

$$\tau_{\mathcal{F}_1} : \mathcal{F}_1 \to \mathcal{D}_{V_1^*}, \quad \tau_{\mathcal{E}_1} : \mathcal{E}_1 \to \mathcal{D}_{V_2^*}$$

so that

$$\begin{bmatrix} \tau_{\mathcal{F}_1} & 0 \\ 0 & \tau_{\mathcal{E}_1} \end{bmatrix} \begin{bmatrix} A_1 & B_1 \\ C_1 & D_1 \end{bmatrix} = \begin{bmatrix} A_{\mathbf{V}} & B_{\mathbf{V}} \\ C_{\mathbf{V}} & D_{\mathbf{V}} \end{bmatrix} \begin{bmatrix} \tau_{\mathcal{F}_1} & 0 \\ 0 & \tau_{\mathcal{E}_1} \end{bmatrix},$$

giving us the relations

$$\tau_{\mathcal{F}_1} A_1 = A_{\mathbf{V}} \tau_{\mathcal{F}_1}, \quad \tau_{\mathcal{F}_1} B_1 = B_{\mathbf{V}} \tau_{\mathcal{E}_1}, \quad \tau_{\mathcal{E}_1} C_1 = C_{\mathbf{V}} \tau_{\mathcal{F}_1}, \quad \tau_{\mathcal{E}_1} D_1 = D_{\mathbf{V}} \tau_{\mathcal{E}_1}.$$

If we set $\Theta_1(z) = A_1 + zB_1(I - zD_1)^{-1}C_1$, we therefore have

$$\begin{aligned} \tau_{\mathcal{F}_1} \Theta_1(z) &= \tau_{\mathcal{F}_1} \left( A_1 + zB_1(I - zD_1)^{-1} C_1 \right) \\ &= A_{\mathbf{V}} \tau_{\mathcal{F}_1} + zB_{\mathbf{V}} \tau_{\mathcal{E}_1} \left( I - z\tau_{\mathcal{E}_1}^* D_{\mathbf{V}} \tau_{\mathcal{E}_1} \right)^{-1} \tau_{\mathcal{E}_1}^* C_{\mathbf{V}} \tau_{\mathcal{F}_1} \\ &= \left( A_{\mathbf{V}} + zB_{\mathbf{V}}(I - zD_{\mathbf{V}})^{-1} C_{\mathbf{V}} \right) \tau_{\mathcal{F}_1} \\ &= \Theta_{\mathbf{V}}(z) \tau_{\mathcal{F}_1}, \end{aligned}$$

that is, $(\mathcal{F}_1, \Theta_1)$ and $\left(\mathcal{D}_{V_1^*}, \Theta_{\mathbf{V}}\right)$ are equivalent as contractive analytic functions and hence generate unitarily equivalent BDF1-models. Hence it follows that the model generated by the 1&2-c.n.u. BDF-tuple $(\mathcal{F}_1, \Theta_1)$ is unitarily equivalent to the 1&2-c.n.u. BDF-model generated by the 1&2-c.n.u. BDF-tuple $\left(\mathcal{D}_{V_1^*}, \Theta_{\mathbf{V}}\right)$. We have already seen that this latter object is unitarily equivalent to the original BCL-model generated by the 1&2-c.n.u. BCL-tuple (3.107). Hence by transitivity of unitary equivalence, we conclude that $\mathbf{V}$ (the BCL-model pair generated by $\Xi^{\rm BCL}_{\rm 1\&2cnu}$) is unitarily equivalent to the BDF1-model generated by $(\mathcal{F}_1, \Theta_1)$ as claimed.

*Proof of (ii)* We let $\mathbf{V} = (V_1, V_2)$ be the 1&2-c.n.u. BDF1-model operator pair generated by the 1&2-c.n.u. BDF-tuple $(\mathcal{F}_1, \Theta_1)$. By construction (see Theorem 3.3.3), the companion BCL-tuple (3.112) is unitarily equivalent to the canonical BCL-tuple for $\mathbf{V}$. But then this implies that $\mathbf{V}$ is unitarily equivalent to the BCL1-model operator pair generated by $\Xi^{\rm BDF, \times}_{\rm 1\&2cnu}$ as desired. □

### 3.3.2 Refined BCL/BDF Companion Pairs and Transcriptions between BCL- and BDF-Models

It remains to extend Theorem 3.3.7 to the general setting of pure commuting isometric pairs (i.e., commuting isometric pairs with trivial unitary/unitary component).

Suppose that $\Xi^{\rm BCL}_{\rm ref} = (\mathcal{F}, P, U)$ given by

$$\mathcal{F} = \begin{bmatrix} \mathcal{F}_0 \\ \mathcal{F}_1 \\ \mathcal{E}_1 \\ \mathcal{E}_0 \end{bmatrix}, \quad P = \begin{bmatrix} I_{\mathcal{F}_0} & 0 & 0 & 0 \\ 0 & I_{\mathcal{F}_1} & 0 & 0 \\ 0 & 0 & 0 & 0 \\ 0 & 0 & 0 & 0 \end{bmatrix}, \quad U = \begin{bmatrix} A_0 & 0 & 0 & 0 \\ 0 & A_1 & B_1 & 0 \\ 0 & C_1 & D_1 & 0 \\ 0 & 0 & 0 & D_0 \end{bmatrix} \tag{3.113}$$

is a refined BCL-tuple (see Definition 3.1.15), that is, where $A_0$ and $D_0$ are unitary, $A_1$ and $D_1$ are c.n.u., and $\left[\begin{smallmatrix} A_1 & B_1 \\ C_1 & D_1 \end{smallmatrix}\right]$ is unitary. Given such a refined BCL-tuple, we define its refined BDF-companion, denoted as $\Xi^{\rm BCL, \times}_{\rm ref}$, by

$$\Xi^{\rm BCL, \times}_{\rm ref} = \left(\mathcal{F}_0, \mathcal{F}_1, \mathcal{E}_0, A_0, \Theta_1(z), D_0^*\right), \tag{3.114}$$

where $A_0$ on $\mathcal{F}_0$ and $D_0$ on $\mathcal{E}_0$ are read from $\Xi^{\rm BCL}_{\rm ref}$ and where we make use of the unitary matrix $\left[\begin{smallmatrix} A_1 & B_1 \\ C_1 & D_1 \end{smallmatrix}\right]$ embedded in the unitary $U$ in $\Xi^{\rm BCL}_{\rm ref}$ to define the contractive analytic function $(\mathcal{F}_1, \Theta_1)$ via

$$\Theta_1(z) = A_1 + zB_1(I - zD_1)^{-1}C_1, \tag{3.115}$$

satisfying the requirement $\Theta_1(0)$ is c.n.u. for a refined BDF-tuple.

Conversely, given a refined BDF-tuple

$$\Xi^{\mathrm{BDF}}_{\mathrm{ref}} = (\mathcal{F}_0, \mathcal{F}_1, \mathcal{E}_0, u_{\mathcal{F}_0}, \Theta_1(z), u_{\mathcal{E}_0}), \tag{3.116}$$

we define its BCL-tuple companion $\Xi^{\mathrm{BDF},\times}_{\mathrm{ref}}$ to be

$$\Xi^{\mathrm{BDF},\times}_{\mathrm{ref}} = \left( \begin{bmatrix} \mathcal{F}_0 \\ \mathcal{F}_1 \\ \mathcal{H}_{\Theta_1} \\ \mathcal{E}_0 \end{bmatrix}, \begin{bmatrix} I_{\mathcal{F}_0} & 0 & 0 & 0 \\ 0 & I_{\mathcal{F}_1} & 0 & 0 \\ 0 & 0 & 0 & 0 \\ 0 & 0 & 0 & 0 \end{bmatrix}, \begin{bmatrix} u_{\mathcal{F}_0} & 0 & 0 & 0 \\ 0 & A_1 & B_1 & 0 \\ 0 & C_1 & D_1 & 0 \\ 0 & 0 & 0 & u^*_{\mathcal{E}_0} \end{bmatrix} \right), \tag{3.117}$$

where $u_{\mathcal{F}_0}$ and $u_{\mathcal{E}_0}$ are seen from $\Xi^{\mathrm{BDF}}_{\mathrm{ref}}$, $\mathcal{H}_{\Theta_1}$ is the Sz.-Nagy–Foias functional model space (see (2.61)) associated with the contractive analytic function $(\mathcal{F}_1, \Theta_1)$, and the unitary operator matrix

$$\begin{bmatrix} A_{\Theta_1} & B_{\Theta_1} \\ C_{\Theta_1} & D_{\Theta_1} \end{bmatrix} \quad \text{on} \quad \begin{bmatrix} \mathcal{F}_1 \\ \mathcal{H}_{\Theta_1} \end{bmatrix}$$

is given by the formulas (3.97) (with $\Theta_1$ in place of $\Theta$).

With these explanations out of the way, we arrive at the following result.

**Theorem 3.3.8** (i) *Assume that the refined BDF-tuple $\Xi^{\mathrm{BCL},\times}_{\mathrm{ref}}$ is a BDF-companion of the refined BCL-tuple $\Xi^{\mathrm{BCL}}_{\mathrm{ref}}$. Then the BDF1-model generated by $\Xi^{\mathrm{BCL},\times}_{\mathrm{ref}}$ is unitarily equivalent to the BCL1-model generated by $\Xi^{\mathrm{BCL}}_{\mathrm{ref}}$.*

(ii) *Assume that the refined BCL-tuple $\Xi^{\mathrm{BDF},\times}_{\mathrm{ref}}$ is a BCL-companion of the refined BDF-tuple $\Xi^{\mathrm{BDF}}_{\mathrm{ref}}$. Then the BCL1-model generated by $\Xi^{\mathrm{BDF},\times}_{\mathrm{ref}}$ is unitarily equivalent to the BDF1-model generated by $\Xi^{\mathrm{BDF}}_{\mathrm{ref}}$.*

*Proof of (i)* Let us first note that the refined BCL-tuple (3.113) splits as the orthogonal direct sum of three components:

$$\begin{aligned} \Xi^{\mathrm{BCL}}_1 &= (\mathcal{F}_0, I_{\mathcal{F}_0}, A_0), \\ \Xi^{\mathrm{BCL}}_2 &= \left( \begin{bmatrix} \mathcal{F}_1 \\ \mathcal{E}_1 \end{bmatrix}, \begin{bmatrix} I_{\mathcal{F}_1} & 0 \\ 0 & 0 \end{bmatrix}, \begin{bmatrix} A_1 & B_1 \\ C_1 & D_1 \end{bmatrix} \right), \\ \Xi^{\mathrm{BCL}}_3 &= (\mathcal{E}_0, 0, D_0), \end{aligned}$$

with associated BCL1-models (refer to (3.2))

$$\begin{aligned} \left(V_1^{(1)}, V_2^{(1)}\right)_{\mathrm{BCL1}} &= (M_z \otimes A_0^*, I_{H^2} \otimes A_0), \\ \left(V_1^{(2)}, V_2^{(2)}\right)_{\mathrm{BCL1}} &= \left( \begin{bmatrix} M_z \otimes A_1^* & I_{H^2} \otimes C_1^* \\ M_z \otimes B_1^* & I_{H^2} \otimes D_1^* \end{bmatrix}, \begin{bmatrix} I_{H^2} \otimes A_1 & I_{H^2} \otimes B_1 \\ M_z \otimes C_1 & M_z \otimes D_1 \end{bmatrix} \right), \\ \left(V_1^{(3)}, V_2^{(3)}\right)_{\mathrm{BCL1}} &= (I_{H^2} \otimes D_0^*, M_z \otimes D_0), \end{aligned}$$

while the companion BDF-tuple $\Xi_{\rm ref}^{\rm BCL,\times}$ as in (3.114) splits as the orthogonal direct sum of three component BDF-tuples

$$\Xi_1^{\rm BCL,\times} = (\mathcal{F}_0, A_0),$$
$$\Xi_2^{\rm BCL,\times} = (\mathcal{F}_1, \Theta_1(z)),$$
$$\Xi_3^{\rm BCL,\times} = (\mathcal{E}_0, D_0^*),$$

with corresponding BDF1-model commuting isometric pairs (refer to (3.76))

$$\left(V_1^{(1)}, V_2^{(1)}\right)_{\rm BCL1,\times} = \left(M_z^{\mathcal{F}_0}, I_{H^2} \otimes A_0\right),$$
$$\left(V_1^{(2)}, V_2^{(2)}\right)_{\rm BCL1,\times} = \left(\begin{bmatrix} M_z^{\mathcal{F}_1} & 0 \\ 0 & I_{H^2} \otimes M_\zeta \end{bmatrix}, \begin{bmatrix} M_{\Theta_1} & 0 \\ \Delta_{\Theta_1}|_{H^2(\mathcal{F}_1)} & M_z^{\overline{\Delta_{\Theta_1} L^2(\mathcal{F}_1)}} \end{bmatrix}\right),$$
$$\left(V_1^{(3)}, V_2^{(3)}\right)_{\rm BCL1,\times} = \left(I_{H^2} \otimes D_0^*, M_z^{\mathcal{E}_0}\right).$$

To verify item (i) it suffices to show that

$$\left(V_1^{(j)}, V_2^{(j)}\right)_{\rm BCL1} \underset{\rm u}{\cong} \left(V_1^{(j)}, V_2^{(j)}\right)_{\rm BCL1,\times}$$

for $j = 1, 2, 3$. We deal with each case in turn.

**Case 1** $j = 1$ We must show that the BCL1-model operator pair $(M_z \otimes A_0^*, I_{H^2} \otimes A_0)$ is unitarily equivalent to the BDF1-model operator pair $\left(M_z^{\mathcal{F}_0}, I_{H^2} \otimes A_0\right)$. While these two operator pairs appear to be different, they are in fact unitarily equivalent. One can define a unitary operator $\omega$ on $H^2(\mathcal{F}_0)$ by

$$\omega\colon \sum_{n=0}^\infty z^n f_n \mapsto \sum_{n=0}^\infty z^n A_0^n f_n, \quad \omega^*\colon \sum_{n=0}^\infty z^n f_n \mapsto \sum_{n=0}^\infty z^n A_0^{*n} f_n,$$

and then check that

$$\omega(M_z \otimes A_0^*)\omega^* = M_z^{\mathcal{F}_0}, \quad \omega(I_{H^2} \otimes A_0)\omega^* = I_{H^2} \otimes A_0,$$

thereby demonstrating the unitary equivalence.

**Case 2** $j = 2$ This case is exactly what is handled by Theorem 3.3.7.

**Case 3** $j = 3$ The task is to show that the operator pair $\left(I_{H^2} \otimes D_0^*, M_z \otimes D_0\right)$ is unitarily equivalent to the operator pair $\left(I_{H^2} \otimes D_0^*, M_z^{\mathcal{E}_0}\right)$. This is the same as Case 1 with the order of the operators in each pair reversed and with $D_0^*$ in place of $A_0$. This completes the proof of item (i).

*Proof of (ii)* Note that the refined BDF-tuple $\Xi_{\rm ref}^{\rm BDF}$ as in (3.116) splits into the orthogonal direct sum of three more basic BDF-tuples

$$\Xi_1^{\rm BDF} = (\mathcal{F}_0, u_{\mathcal{F}_0}),$$
$$\Xi_2^{\rm BDF} = (\mathcal{F}_1, \Theta_1(z)),$$
$$\Xi_2^{\rm BDF} = (\mathcal{E}_0, u_{\mathcal{E}_0}),$$

with associated BDF1-model operator pairs (refer to (3.76))

$$\left(V_1^{(1)}, V_2^{(1)}\right)_{\rm BDF1} = \left(M_z^{\mathcal{F}_0}, I_{H^2} \otimes u_{\mathcal{F}_0}\right),$$

$$\left(V_1^{(2)}, V_2^{(2)}\right)_{\rm BDF1} = \left(\begin{bmatrix} M_z^{\mathcal{F}_1} & 0 \\ 0 & I_{H^2} \otimes M_\zeta \end{bmatrix}, \begin{bmatrix} M_{\Theta_1} & 0 \\ \Delta_{\Theta_1}|_{H^2(\mathcal{F}_1)} & M_z^{\overline{\Delta_{\Theta_1} L^2(\mathcal{F}_1)}} \end{bmatrix}\right),$$

$$\left(V_1^{(3)}, V_2^{(3)}\right)_{\rm BDF1} = \left(I_{H^2} \otimes u_{\mathcal{E}_0}, M_z^{\mathcal{E}_0}\right),$$

while the BCL-tuple $\Xi_{\rm ref}^{\rm BDF,\times}$ as in (3.117) breaks up into the three orthogonal direct summands

$$\Xi_1^{\rm BDF,\times} = (\mathcal{F}_0, I_{\mathcal{F}_0}, u_{\mathcal{F}_0}),$$

$$\Xi_2^{\rm BDF,\times} = \left(\begin{bmatrix} \mathcal{F}_1 \\ \mathcal{H}_{\Theta_1} \end{bmatrix}, \begin{bmatrix} I_{\mathcal{F}_1} & 0 \\ 0 & 0 \end{bmatrix}, \begin{bmatrix} A_1 & B_1 \\ C_1 & D_1 \end{bmatrix}\right)$$

$$= \left(\begin{bmatrix} \mathcal{F}_{\mathbf{V}} \\ \mathcal{E}_{\mathbf{V}} \end{bmatrix}, \begin{bmatrix} I_{\mathcal{F}} & 0 \\ 0 & 0 \end{bmatrix}, \begin{bmatrix} A_{\mathbf{V}} & B_{\mathbf{V}} \\ C_{\mathbf{V}} & D_{\mathbf{V}} \end{bmatrix}\right) \text{ given by (3.23) with } \mathbf{V} = \left(V_1^{(2)}, V_2^{(2)}\right)_{\rm BDF1},$$

$$\Xi_3^{\rm BDF,\times} = \left(\mathcal{E}_0, 0, u_{\mathcal{E}_0}^*\right),$$

with associated BCL1-model operator pairs (refer to (3.2))

$$\left(V_1^{(1)}, V_2^{(1)}\right)_{\rm BDF1,\times} = \left(M_z \otimes u_{\mathcal{F}_0}^*, , I_{H^2} \otimes u_{\mathcal{F}_0}\right),$$

$$\left(V_1^{(2)}, V_2^{(2)}\right)_{\rm BDF1,\times} = \left(\begin{bmatrix} M_z \otimes A_1^* & I_{H^2} \otimes C_1^* \\ M_z \otimes B_1^* & I_{H^2} \otimes D_1^* \end{bmatrix}, \begin{bmatrix} I_{H^2} \otimes A_1 & I_{H^2} \otimes B_1 \\ M_z \otimes C_1 & M_z \otimes D_1 \end{bmatrix}\right),$$

$$\left(V_1^{(3)}, V_2^{(3)}\right)_{\rm BDF1,\times} = \left(I_{H^2} \otimes u_{\mathcal{E}_0}, M_z \otimes u_{\mathcal{E}_0}^*\right).$$

It remains to show that

$$\left(V_1^{(j)}, V_2^{(j)}\right)_{\rm BDF1} \underset{\rm u}{\cong} \left(V_1^{(j)}, V_2^{(j)}\right)_{\rm BDF1,\times}$$

for $j = 1, 2, 3$. For $j = 1, 3$, the verification is much like the corresponding point in the proof of item (i) above. The case $j = 2$ is again handled by Theorem 3.3.7. □

## 3.4 BDF-Model for $(V_1, V_2)$ versus BDF-Model for $(V_2, V_1)$

The next result is the BDF-analogue of the corresponding result for the BCL-models given by Theorem 3.1.13. As the proof of the result amounts to pulling together various particular results already done in detail earlier in this chapter, we leave the proof as an exercise for the interested reader. For convenience, we will do the 1&2-c.n.u. special case, as tacking on commuting shift/unitary, unitary/shift, and unitary/unitary direct summands is routine.

**Theorem 3.4.1** *Suppose that $(V_1, V_2)$ is a 1&2-c.n.u. commuting operator tuple unitarily equivalent to the BDF1-model* $\mathbf{V} = (V_1, V_2)$ *given by*

$$\left( \begin{bmatrix} M_z^{\mathcal{F}_1} & 0 \\ 0 & I_{H^2} \otimes M_\zeta \end{bmatrix}, \begin{bmatrix} M_{\Theta_1} & 0 \\ \Delta_{\Theta_1}|_{H^2(\mathcal{F}_1)} & M_z^{\overline{\Delta_\Theta L^2(\mathcal{F}_1)}} \end{bmatrix} \right) \quad on \quad \begin{bmatrix} H^2(\mathcal{F}_1) \\ \overline{H^2(\Delta_{\Theta_1} L^2(\mathcal{F}_1))} \end{bmatrix},$$

*where $(\mathcal{F}_1, \Theta_1)$ is a contractive analytic function with $\Theta_1(0)$ c.n.u. Then $\Theta_1$ has a canonical unitary system-node realization*

$$\Theta_1(z) = A_{\Theta_1} + zB_{\Theta_1}(I - zD_{\Theta_1})^{-1}C_{\Theta_1},$$

*where $A_{\Theta_1}, B_{\Theta_1}, C_{\Theta_1}, D_{\Theta_1}$ are given by* (3.97) *(with $\Theta_1$ in place of $\Theta$). Let $(\mathcal{H}_{\Theta_1}, \Theta_1')$ be the $\mathcal{H}_{\Theta_1}$-valued contractive analytic function given by*

$$\Theta_1'(z) = D_{\Theta_1}^* + zB_{\Theta_1}^* \left( I - zA_{\Theta_1}^* \right)^{-1} C_{\Theta_1}^*.$$

*Then* $\mathbf{V}' = (V_2, V_1)$ *is unitarily equivalent to the BDF1-model*

$$\left( \begin{bmatrix} M_z^{\mathcal{H}_{\Theta_1}} & 0 \\ 0 & I_{H^2} \otimes M_\zeta \end{bmatrix}, \begin{bmatrix} M_{\Theta_1'} & 0 \\ \Delta_{\Theta_1'}|_{H^2(\mathcal{H}_{\Theta_1})} & M_z^{\overline{\Delta_{\Theta_1'} H^2(\mathcal{H}_{\Theta_1})}} \end{bmatrix} \right).$$

A more digestible special case is the case where $\Theta$ is inner. Then one has the de Branges–Rovnyak canonical functional-model realization for inner $\Theta$ rather than the Sz.-Nagy–Foias canonical functional-model realization for a general contractive analytic function $\Theta$.

## 3.5 Notes to Chapter 3

**1.** A general method for finding unitary realization for contractive analytic functions is the *lurking isometry method*; this method has proved to be quite adaptable to several-variable settings (see, e.g., [4, 16, 39]).

**2.** A foreshadowing of the general principle developed in Section 3.3.2 can be found in the paper of Bhattacharyya–Kumar–Sau [35]. The main emphasis in [35] is on connections with distinguished varieties and associated dilation theory; here we strip out the distinguished varieties and encode the structure that remains.

# 4

# Ando's Dilation and Commutant Lifting Theorems

In this chapter we give a general discussion surrounding Ando's dilation theorem. We present two proofs of Ando's theorem, namely: (i) Ando's original proof and (ii) a proof via the Sz.-Nagy–Foias commutant lifting theorem. This chapter also includes some new related results.

## 4.1 Preliminaries

We begin by formally defining the main theme of this monograph.

**Definition 4.1.1** Let $(T_1, T_2)$ be a commuting pair of operators on $\mathcal{H}$. A triple $(\Pi, V_1, V_2)$ is said to be a (joint) *lift* of $(T_1, T_2)$ if (i) there is a Hilbert space $\mathcal{K}$ such that $\Pi\colon \mathcal{H} \to \mathcal{K}$ is an isometric embedding of $\mathcal{H}$ into $\mathcal{K}$ and (ii) $(V_1, V_2)$ is a commuting operator pair on $\mathcal{K}$ such that

$$V_j^*\Pi = \Pi T_j^* \quad \text{for } j = 1, 2.$$

A lift $(\Pi, V_1, V_2)$ of $(T_1, T_2)$ is said to be an *Ando isometric lift* or, simply an *Ando lift*, if $V_1, V_2$ are isometries.

Let $T$ be an operator acting on $\mathcal{H}$. An operator $V$ acting on $\mathcal{K} \supset \mathcal{H}$ is said to be an extension (respectively, a coextension) of $T$ if $V|_{\mathcal{H}} = T$ (respectively, if $V^*|_{\mathcal{H}} = T^*$). It is a trivial exercise to see that any (joint-) coextension (but not extension) of a lift is also a lift. To escape redundancy, we shall be particularly interested in the case where the Ando lift is *minimal* in the following precise sense.

**Definition 4.1.2** An Ando isometric lift is said to be *minimal* if

$$\mathcal{K} = \bigvee_{n_1, n_2 \geq 0} V_1^{n_1} V_2^{n_2} \operatorname{Ran} \Pi. \tag{4.1}$$

We say that two Ando lifts $(\Pi, V_1, V_2)$ acting on $\mathcal{K}$ and $(\Pi', V_1', V_2')$ acting on $\mathcal{K}'$ of a commuting contractive pair $(T_1, T_2)$ are *unitarily equivalent* if there is a unitary operator $\tau\colon \mathcal{K} \to \mathcal{K}'$ such that

$$\tau\Pi = \Pi', \quad \tau V_1 = V_1'\tau, \quad \tau V_2 = V_2'\tau. \tag{4.2}$$

The following lemma is of independent interest. The result remains true for more (or fewer) than two variables but we confine ourselves here to just the two-variable scenario.

**Lemma 4.1.3** *Let $(T_1, T_2)$ and $(S_1, S_2)$ be two commuting contractive operator pairs acting on $\mathcal{H}_T$ and $\mathcal{H}_S$, respectively. Suppose that the pairs are jointly unitarily equivalent via a unitary similarity $\tau\colon \mathcal{H}_T \to \mathcal{H}_S$. If $(\Pi, V_1, V_2)$ is an Ando isometric lift of $(T_1, T_2)$, then $(\Pi\tau^*, V_1, V_2)$ is an Ando isometric lift of $(S_1, S_2)$. Moreover, if $(\Pi, V_1, V_2)$ is minimal, then so is $(\Pi\tau^*, V_1, V_2)$.*

*Proof* For the lift part, we need to check $V_j^*\Pi\tau^* = S_j^*\Pi\tau^*$ for $j = 1, 2$. Let us use the fact that $(\Pi, V_1, V_2)$ is a lift of $(T_1, T_2)$ and $\tau(T_1, T_2) = (S_1, S_2)\tau$ to see that, for each $j = 1, 2$,

$$V_j^*\Pi\tau^* = \Pi T_j^*\tau^* = \Pi\tau^* S_j^*.$$

For the minimality, let us note that

$$\mathcal{K} = \bigvee_{n_1, n_2 \geq 0} V_1^{n_1} V_2^{n_2} \Pi\mathcal{H}_T = \bigvee_{n_1, n_2 \geq 0} V_1^{n_1} V_2^{n_2} \Pi\tau^*\tau\mathcal{H}_T = \bigvee_{n_1, n_2 \geq 0} V_1^{n_1} V_2^{n_2} \Pi\tau^*\mathcal{H}_S.$$

This completes the proof. □

In the single-variable case we have seen that any two minimal isometric lifts of a contractive operator are unitarily equivalent (see Theorem 2.1.3). This result fails in the bivariate setting of Ando lifts for a commuting contractive pair. We next provide a concrete example exhibiting this nonuniqueness phenomenon. Additional examples can be found in Examples 5.5.2 and 5.5.3 to come.

**Example 4.1.4** Take $(T_1, T_2) = (0, 0)$ on $\mathcal{H} = \mathbb{C}$. Then both $(V_1, V_2) = (M_z, M_z)$ on $H^2$ and $(V_1, V_2) = (M_{z_1}, M_{z_2})$ on $H^2_{\mathbb{D}^2}$ are Ando lifts of $(0, 0)$, where in both the instances the isometric embedding of $\mathbb{C}$ is given by the map $\Pi\colon z_0 \mapsto f_c(\cdot) \equiv c$. Note that these two lifts cannot be unitarily equivalent as the second lift is doubly commuting ($V_2^*V_1 = V_1V_2^*$ and $V_1V_2 = V_2V_1$) while the first one is only commuting ($V_1V_2 = V_2V_1$).

It is therefore natural to wonder what it is that makes two minimal Ando lifts nonunitarily equivalent! It appears that there are additional invariants that must be equivalent in an appropriate sense before two minimal Ando lifts can be unitarily equivalent. We defer the discussion regarding this uniqueness problem until we have described the models of Ando lifts in Chapters 5 and 6. What is true is summed up in the following lemma.

**Lemma 4.1.5** *Given an Ando lift $(\Pi, V_1, V_2)$ acting on $\mathcal{K}$ of $(T_1, T_2)$ acting on $\mathcal{H}$, there is always a unitarily equivalent Ando lift $(\Pi', V_1, V_2)$ acting on $\mathcal{K}'$ so that $\mathcal{H}$ is equal to a subspace of $\mathcal{K}'$ and $\Pi' : \mathcal{H} \to \mathcal{K}'$ is just the inclusion map.*

*Proof* To see this, simply set $\mathcal{K}' = \left[\begin{smallmatrix} \mathcal{H} \\ (I-\Pi\Pi^*)\mathcal{K} \end{smallmatrix}\right]$ and observe that the map

$$\tau = \begin{bmatrix} \Pi^* \\ I - \Pi\Pi^* \end{bmatrix} : \mathcal{K} \to \mathcal{K}'$$

is unitary. If we then set

$$\Pi' = \left[\begin{smallmatrix} I_{\mathcal{H}} \\ 0 \end{smallmatrix}\right] : \mathcal{H} \to \mathcal{K}', \quad V_i' = \tau V_i \tau^* \text{ for } i = 1, 2,$$

we see that all of the conditions (4.2) are satisfied. Furthermore, identifying $\mathcal{H}$ with $\left[\begin{smallmatrix} \mathcal{H} \\ 0 \end{smallmatrix}\right] \subset \mathcal{K}'$ makes $\mathcal{H}$ a subspace of $\mathcal{K}'$ and then $\Pi'$ is just the inclusion map. □

When $\Pi : \mathcal{H} \to \mathcal{K}$ is an inclusion map we write simply $(V_1, V_2)$ rather than $(\iota_{\mathcal{H}\to\mathcal{K}}, V_1, V_2)$ for the Ando isometric lift.

## 4.2 The Original Proof of Ando

The original proof of Ando takes the embedding operator $\Pi$ to be the inclusion map. We include a proof of Ando's dilation theorem here to make the book self-contained.

**Theorem 4.2.1** *Let $(T_1, T_2)$ be a commuting contractive operator pair acting on $\mathcal{H}$. There exists a Hilbert space $\mathcal{K}$ containing $\mathcal{H}$ as a subspace and a commuting isometric pair $(V_1, V_2)$ such that*

$$(V_1, V_2) = \left( \begin{bmatrix} T_1 & 0 \\ * & * \end{bmatrix}, \begin{bmatrix} T_2 & 0 \\ * & * \end{bmatrix} \right) : \begin{bmatrix} \mathcal{H} \\ \mathcal{K} \ominus \mathcal{H} \end{bmatrix} \to \begin{bmatrix} \mathcal{H} \\ \mathcal{K} \ominus \mathcal{H} \end{bmatrix}.$$

*Proof* Let $S_j$ be the isometric lift of $T_j$ constructed by Schäffer as in (2.30), for $j = 1, 2$. For our purposes here it is instructive to view each $S_j$ as an operator on the much larger space $\mathcal{H} \oplus H^2(\mathcal{H})$ rather than the much smaller space $\mathcal{H} \oplus H^2\left(\mathcal{D}_{T_j}\right)$ and to write out each operator $S_j$ in the expanded form

$$S_j = \begin{bmatrix} T_j & & & \\ D_{T_j} & & & \\ & I_{\mathcal{H}} & & \\ & & I_{\mathcal{H}} & \\ & & & \ddots \end{bmatrix}.$$

Here each unspecified entry is equal to the zero operator. A routine matrix computation shows that $S_1$, $S_2$ do not commute. Of course, if they did, then every tuple of commuting contractions would have a commuting isometric lift, contrary to the result of Parrott [74] that there is a triple of commuting contractions that fails to have a commuting isometric lift! To amend the lack of commutativity of $S_1, S_2$, Ando employs the following ingenious idea: Keep the first two rows of $S_j$ as they are, insert a zero third row, and then shift the original third row one place down. With this modification implemented we arrive at

$$\widetilde{S}_j = \begin{bmatrix} T_j & & & \\ D_{T_j} & & & \\ 0 & 0 & \cdots & \\ & I_{\mathcal{H}} & & \\ & & I_{\mathcal{H}} & \\ & & & \ddots \end{bmatrix}. \tag{4.3}$$

This modified isometric pair remains a lift of $(T_1, T_2)$ but still does not commute; yet the insertion of this additional zero row enigmatically helps serve the purpose as we see next.

Let $U$ be a unitary operator on

$$\mathcal{E} = \mathcal{H} \oplus \mathcal{H} \oplus \mathcal{H} \oplus \mathcal{H}$$

and define the unitary operator

$$\widetilde{U} = \begin{bmatrix} I_{\mathcal{H}} & 0 \\ 0 & I_{H^2} \otimes U \end{bmatrix} : \begin{bmatrix} \mathcal{H} \\ H^2(\mathcal{E}) \end{bmatrix} \to \begin{bmatrix} \mathcal{H} \\ H^2(\mathcal{E}) \end{bmatrix}.$$

We now view $\widetilde{S}_j$ as operators on $\mathcal{H} \oplus H^2(\mathcal{E})$ and define

$$(V_1, V_2) = \left(\widetilde{U}\widetilde{S}_1, \widetilde{S}_2\widetilde{U}^*\right). \tag{4.4}$$

It is obvious that $V_1, V_2$ are isometries because each is a product of isometries. To explore the commutativity of $V_1, V_2$, we compute

$$
\begin{aligned}
&V_1V_2\,(h,(h_1,h_2,h_3,h_4),\dots)\\
&= \widetilde{U}\widetilde{S}_1\widetilde{S}_2\widetilde{U}^*\,(h,(h_1,h_2,h_3,h_4),\dots)\\
&= \widetilde{U}\widetilde{S}_1\widetilde{S}_2\,(h,U^*(h_1,h_2,h_3,h_4),\dots)\\
&= \widetilde{U}\widetilde{S}_1\,\big(T_2h,D_{T_2}h,0,U^*(h_1,h_2,h_3,h_4),\dots\big)\\
&= \widetilde{U}\,\big(T_1T_2h,D_{T_1}T_2h,0,D_{T_2}h,0,U^*(h_1,h_2,h_3,h_4),\dots\big)\\
&= \big(T_1T_2h,U(D_{T_1}T_2h,0,D_{T_2}h,0),(h_1,h_2,h_3,h_4),\dots\big)
\end{aligned}
$$

while

$$
\begin{aligned}
&V_2V_1\,(h,(h_1,h_2,h_3,h_4),\dots)\\
&= \widetilde{S}_2\widetilde{S}_1\,(h,(h_1,h_2,h_3,h_4),\dots)\\
&= \widetilde{S}_2\,(T_1h,D_{T_1}h,0,h_1,h_2,\dots)\\
&= \big(T_2T_1h,D_{T_2}T_1h,0,D_{T_1}h,0,h_1,h_2,\dots\big)\,.
\end{aligned}
$$

Thus $V_1V_2 = V_2V_1$ if and only if we have a unitary $U$ on $\mathcal{E} = \mathcal{H}\oplus\mathcal{H}\oplus\mathcal{H}\oplus\mathcal{H}$ such that

$$U(D_{T_1}T_2h,0,D_{T_2}h,0) = (D_{T_2}T_1h,0,D_{T_1}h,0).$$

To analyze this condition, let us introduce linear manifolds

$$\mathcal{M} = \big\{(D_{T_1}T_2h,0,D_{T_2}h,0)\colon h\in\mathcal{H}\big\},\quad \mathcal{N} = \big\{(D_{T_2}T_1h,0,D_{T_1}h,0)\colon h\in\mathcal{H}\big\}$$

and define a linear operator $U_0\colon \mathcal{M}\to\mathcal{N}$ by

$$U_0\colon (D_{T_1}T_2h,0,D_{T_2}h,0)\mapsto (D_{T_2}T_1h,0,D_{T_1}h,0). \tag{4.5}$$

At this point we use the commutativity of $T_1, T_2$ to see that[1]

$$\|D_{T_1}T_2h\|^2 + \|D_{T_2}h\|^2 = \|D_{T_2}T_1h\|^2 + \|D_{T_1}h\|^2 \quad \text{for all } h\in\mathcal{H}, \tag{4.6}$$

that is, the formula (4.5) defines an isometry from $\mathcal{M}$ onto $\mathcal{N}$. By taking limits we can uniquely extend the formula (4.5) to an isometry from the closure $\overline{\mathcal{M}}$ of $\mathcal{M}$ onto the closure $\overline{\mathcal{N}}$ of $\mathcal{N}$:

$$U_0\colon \overline{\mathcal{M}} \underset{\text{onto isom}}{\to} \overline{\mathcal{N}}.$$

The preceding analysis shows that part of the problem of finding an Ando lift $(V_1,V_2)$ of $(T_1,T_2)$ is to find a unitary operator $U\colon \mathcal{E}\to\mathcal{E}$ so that

$$U|_{\overline{\mathcal{M}}} = U_0\colon \overline{\mathcal{M}} \underset{\text{onto isom}}{\to} \overline{\mathcal{N}}.$$

[1] Note that the isometric version of this identity has already appeared as identity (3.20) in connection with the proof of Theorem 3.1.7.

By elementary Hilbert space theory, the necessary and sufficient condition for this to be possible is that

$$\dim\left(\mathcal{E}\ominus\overline{\mathcal{M}}\right)=\dim\left(\mathcal{E}\ominus\overline{\mathcal{N}}\right),$$

as then we can define $U|_{\mathcal{E}\ominus\overline{\mathcal{M}}}$ to be any isometry $U_1$ from $\mathcal{E}\ominus\overline{\mathcal{M}}$ onto $\mathcal{E}\ominus\overline{\mathcal{N}}$, then define $U|_{\overline{\mathcal{M}}}=U_0$, $U|_{\mathcal{E}\ominus\overline{\mathcal{M}}}=U_1$, and then extend by linearity. As $\dim\overline{\mathcal{M}}=\dim\overline{\mathcal{N}}$, since $U_0$ is an onto isometry between these two spaces, when $\dim\mathcal{E}<\infty$ (or equivalently $\dim\mathcal{H}=\frac{1}{4}\dim\mathcal{E}<\infty$), we can conclude that

$$\dim\left(\mathcal{E}\ominus\overline{\mathcal{M}}\right)=\dim\mathcal{E}-\dim\overline{\mathcal{M}}=\dim\mathcal{E}-\dim\overline{\mathcal{N}}=\dim\left(\mathcal{E}\ominus\overline{\mathcal{N}}\right),$$

as needed. However, if $\dim\mathcal{H}$ is not finite, then $\dim\mathcal{E}=4\cdot\dim\mathcal{H}=\dim\mathcal{H}$, so we are unable to conclude via the preceding argument that $\dim\left(\mathcal{E}\ominus\overline{\mathcal{M}}\right)=\dim\left(\mathcal{E}\ominus\overline{\mathcal{N}}\right)$ as desired, and we need some other argument.

It is at this point that the clever insertion of the zero row comes in handy. Indeed, note that both $\mathcal{E}\ominus\mathcal{M}$ and $\mathcal{E}\ominus\mathcal{N}$ contain a copy of the original Hilbert space $\mathcal{H}$, and therefore

$$\dim\mathcal{H}\geq\dim\left(\mathcal{E}\ominus\overline{\mathcal{M}}\right)\geq\dim\mathcal{H},\quad\dim\mathcal{H}\geq\dim\left(\mathcal{E}\ominus\overline{\mathcal{N}}\right)\geq\dim\mathcal{H},$$

assuring us that indeed $\dim\left(\mathcal{E}\ominus\overline{\mathcal{M}}\right)=\dim\left(\mathcal{E}\ominus\overline{\mathcal{N}}\right)=\dim\mathcal{H}$.

Finally, the fact that the commuting isometric pair $(V_1,V_2)$ is a lift of $(T_1,T_2)$ follows from the matrix representations of $(\tilde{S}_1,\tilde{S}_2)$ and of $\widetilde{U}$; we leave the routine details to the reader. □

## 4.3 The Unitary Part of an Ando Lift

Let us recall the definition of unitary part of an isometric pair from Definition 3.1.5: *Given a commuting isometric pair $(V_1,V_2)$ on $\mathcal{K}$, the* unitary part *$(W_1,W_2)$ is the commuting unitary operator pair given by $(W_1,W_2)=(V_1|_{\mathcal{K}_u},V_2|_{\mathcal{K}_u})$, where $V_1V_2|_{\mathcal{K}_u}$ gives the unitary part in the Wold decomposition of the product isometry $V_1V_2$, that is,*

$$V_1V_2=(V_1V_2)|_{\mathcal{K}_u^{\perp}}\oplus(V_1V_2)|_{\mathcal{K}_u},$$

*where $(V_1V_2)|_{\mathcal{K}_u^{\perp}}$ is a shift and $(V_1V_2)|_{\mathcal{K}_u}$ is unitary.*

We have seen in Example 4.1.4 that two minimal Ando isometric lifts of a given commutative contractive pair need not be unitarily equivalent. In this section we shall prove that the unitary parts of any two minimal isometric lifts, however, are always unitarily equivalent.

**Theorem 4.3.1** *The unitary part of any minimal isometric lift of a commuting contractive pair is unique up to unitary equivalence.*

*Proof* Let $(V_1, V_2)$ on $\mathcal{K}$ be a minimal isometric lift of the commuting contractive pair $(T_1, T_2)$ on $\mathcal{H}$ via the embedding $\Pi\colon \mathcal{H} \to \mathcal{K}$. We can assume without loss of generality that the pair $(V_1, V_2)$ is in the Berger–Coburn–Lebow decomposition (see Theorem 3.1.4):

$$(V_1, V_2) = \left( \begin{bmatrix} S_1 & 0 \\ 0 & W_1 \end{bmatrix}, \begin{bmatrix} S_2 & 0 \\ 0 & W_2 \end{bmatrix} \right) \quad \text{on} \begin{bmatrix} H^2(\mathcal{F}) \\ \mathcal{K}_u \end{bmatrix},$$

where $(S_1, S_2)$ on $H^2(\mathcal{F})$ is the pure part and $(W_1, W_2)$ on $\mathcal{K}_u$ is the unitary part of $(V_1, V_2)$. The explicit form of $(S_1, S_2)$ as in the Berger–Coburn–Lebow theorem will not be used here, hence the brevity of the notation. Using the identities $\left(V_1^*, V_2^*\right) \Pi = \Pi \left(T_1^*, T_2^*\right)$ we get

$$f\left(W_1^*, W_2^*\right) P_{\mathcal{K}_u} \Pi = P_{\mathcal{K}_u} \Pi f\left(T_1^*, T_2^*\right) \quad \text{for every } f \in \mathbb{C}[z_1, z_2]. \tag{4.7}$$

The minimality of the lift means

$$\begin{bmatrix} H^2(\mathcal{F}) \\ \mathcal{K}_u \end{bmatrix} = \bigvee_{n_1, n_2 \geq 0} \begin{bmatrix} S_1^{n_1} S_2^{n_2} & 0 \\ 0 & W_1^{n_1} W_2^{n_2} \end{bmatrix} \Pi \mathcal{H},$$

which in particular implies

$$\mathcal{K}_u = \bigvee_{n_1, n_2 \geq 0;\, h \in \mathcal{H}} W_1^{n_1} W_2^{n_2} P_{\mathcal{K}_u} \Pi h.$$

For notational convenience, we write

$$W_1^{n_1} W_2^{n_2} = \boldsymbol{W}^{\boldsymbol{n}}, \text{ where } \boldsymbol{n} = (n_1, n_2).$$

For uniqueness up to unitary equivalence, it is thus enough to show that the norm of vectors of the form

$$\sum_{j=1}^{N} \boldsymbol{W}^{\boldsymbol{n}_j} P_{\mathcal{K}_u} \Pi h_{\boldsymbol{n}_j} \tag{4.8}$$

does not depend on the particular lift $(V_1, V_2)$ in any way. Toward this end we use (4.7) for monomials to note that, for any choice of nonnegative integers $m, n, p, q$, and any choice of $h, k$ in $\mathcal{H}$, the inner product

$$\left\langle \boldsymbol{W}^{(m,n)} P_{\mathcal{K}_u} \Pi h, \boldsymbol{W}^{(p,q)} P_{\mathcal{K}_u} \Pi k \right\rangle$$

is given by the following case-by-case formula:

$$\begin{cases} \left\langle P_{\mathcal{K}_u}\Pi h, P_{\mathcal{K}_u}\Pi T_1^{*(m-p)}T_2^{*(n-q)}k\right\rangle & \text{if } m \geq p, n \geq q, \\ \left\langle P_{\mathcal{K}_u}\Pi T_1^{*(p-m)}h, P_{\mathcal{K}_u}\Pi T_2^{*(n-q)}k\right\rangle & \text{if } p \geq m, n \geq q, \\ \left\langle P_{\mathcal{K}_u}\Pi T_2^{*(q-n)}h, P_{\mathcal{K}_u}\Pi T_1^{*(m-p)}k\right\rangle & \text{if } m \geq p, q \geq n, \\ \left\langle P_{\mathcal{K}_u}\Pi T_1^{*(p-m)}T_2^{*(q-n)}h, P_{\mathcal{K}_u}\Pi k\right\rangle & \text{if } p \geq m, q \geq n. \end{cases} \tag{4.9}$$

In this computation, the double commutativity of $(W_1, W_2)$ is crucially used. For the product contraction $T = T_1T_2$, let $Q_{T^*}$ be as in (2.21), that is, $Q_{T^*}$ is the limit of the operator sequence $T^nT^{*n}$ in the strong operator topology. We show that

$$\left\langle P_{\mathcal{K}_u}\Pi h, P_{\mathcal{K}_u}\Pi k\right\rangle = \langle Q_{T^*}h, Q_{T^*}k\rangle \quad \text{for every } h, k \in \mathcal{H}.$$

As usual, let us denote $V = V_1V_2$, $S = S_1S_2$, and $W = W_1W_2$ and compute

$$\begin{aligned} \langle Q_{T^*}h, Q_{T^*}k\rangle &= \lim_n \langle T^{*n}h, T^{*n}k\rangle = \lim_n \langle \Pi T^{*n}h, \Pi T^{*n}k\rangle = \lim_n \langle V^{*n}\Pi h, V^{*n}\Pi k\rangle \\ &= \lim_n \left(\left\langle S^{*n}P_{H^2(\mathcal{F})}\Pi h, S^{*n}P_{H^2(\mathcal{F})}\Pi k\right\rangle + \left\langle W^{*n}P_{\mathcal{K}_u}\Pi h, W^{*n}P_{\mathcal{K}_u}\Pi k\right\rangle\right) \\ &= \left\langle P_{\mathcal{K}_u}\Pi h, P_{\mathcal{K}_u}\Pi k\right\rangle. \end{aligned}$$

This shows that all the inner products in (4.9) depend only on the pair $(T_1, T_2)$. This proves the stated uniqueness. □

Let us extend Definition 3.1.5 made for commuting isometric pairs to commuting operator pairs.

**Definition 4.3.2** We say that a commuting operator pair $(T_1, T_2)$ is a *pure pair*, if the product operator $T = T_1T_2$ is pure, that is, if $T^{*n} \to 0$ in the strong operator topology. More strongly, if the pair $(T_1, T_2)$ is such that both $T_1$ and $T_2$ are pure, then the pair is said to be *bipure*.

**Example 4.3.3** Let us note that while a bipure pair is obviously pure, a pure pair need not be bipure. Take, for example, $(M_z, I)$ on $H^2$ or, more interestingly perhaps, the commuting isometric pair $(V_1, V_2) = (M_{P^\perp U + zPU}, M_{U^*P + zU^*P^\perp})$ on $H^2(\mathbb{C}^2)$, where

$$(P, U) = \left(\begin{bmatrix} 1 & 0 \\ 0 & 0 \end{bmatrix}, \begin{bmatrix} 1 & 0 \\ 0 & 1 \end{bmatrix}\right).$$

Here

$$P^\perp U = \begin{bmatrix} 0 & 0 \\ 0 & 1 \end{bmatrix} \quad \text{and} \quad U^*P = \begin{bmatrix} 1 & 0 \\ 0 & 0 \end{bmatrix}.$$

Therefore, for $(c_1, c_2) \in \mathbb{C}^2$, we note that

$$V_1^{*n}\left(1 \otimes \begin{bmatrix} c_1 \\ c_2 \end{bmatrix}\right) = I_{H^2} \otimes (U^*P^{\perp})^{*n}\left(1 \otimes \begin{bmatrix} c_1 \\ c_2 \end{bmatrix}\right) = 1 \otimes \begin{bmatrix} 0 \\ c_2 \end{bmatrix} \quad \text{and}$$

$$V_2^{*n}\left(1 \otimes \begin{bmatrix} c_1 \\ c_2 \end{bmatrix}\right) = I_{H^2} \otimes (PU)^{*n}\left(1 \otimes \begin{bmatrix} c_1 \\ c_2 \end{bmatrix}\right) = 1 \otimes \begin{bmatrix} c_1 \\ 0 \end{bmatrix}.$$

This shows that none of $V_1, V_2$ is pure but the product isometry $V = V_1V_2 = M_z$ is a shift operator.

**Remark 4.3.4** We used an interesting fact in the proof of Theorem 4.3.1, namely, *if* $(\Pi, V)$ *is an isometric lift acting on* $\mathcal{K}$ *of a contraction* $T$ *acting on* $\mathcal{H}$, *then*

$$\left\langle P_{\mathcal{K}_u}\Pi h, P_{\mathcal{K}_u}\Pi k\right\rangle = \langle Q_{T^*}h, Q_{T^*}k\rangle \quad \text{for all } h, k \in \mathcal{H}. \tag{4.10}$$

Let us note that this was observed in item (vi) of Proposition 2.1.1 where we analyzed the coordinate-free form of an isometric lift of a contraction operator. As a consequence of (4.10), it therefore follows that *if* $(T_1, T_2)$ *is a pure commuting contractive pair, then* $Q_{T^*} = 0$, and hence in this case there is no unitary part of *any* minimal Ando isometric lift of $(T_1, T_2)$. Consequently, *every minimal Ando isometric lift of a pure commuting contractive pair is pure.* It is instructive to have an independent proof of this fact. We next provide such a proof.

**Proposition 4.3.5** *Let* $(T_1, T_2)$ *be a pure commuting contractive pair acting on* $\mathcal{H}$, *that is, the product contraction* $T = T_1T_2$ *is pure. Then every minimal isometric lift of* $(T_1, T_2)$ *is pure, that is, no minimal isometric lift of* $(T_1, T_2)$ *should have a unitary part.*

*Proof* Let $(\Pi, V_1, V_2)$ acting on $\mathcal{K}$ be a minimal isometric lift of $(T_1, T_2)$. By Theorem 3.1.4, $(V_1, V_2)$ is unitarily equivalent to

$$(V_1', V_2') = \left(\begin{bmatrix} S_1 & 0 \\ 0 & W_1 \end{bmatrix}, \begin{bmatrix} S_2 & 0 \\ 0 & W_2 \end{bmatrix}\right) \quad \text{acting on} \begin{bmatrix} H^2(\mathcal{D}_{V^*}) \\ \mathcal{Q}_{V^*} \end{bmatrix}$$

via the unitary transformation

$$\tau = \begin{bmatrix} D_{V^*}(I - zV^*)^{-1} \\ Q_{V^*} \end{bmatrix} : \mathcal{H} \to \begin{bmatrix} H^2(\mathcal{D}_{V^*}) \\ \mathcal{Q}_{V^*} \end{bmatrix}.$$

Then it is trivial to see that $\left(\tau\Pi, V_1', V_2'\right)$ is also a minimal isometric lift of $(T_1, T_2)$ unitarily equivalent to the original lift $(\Pi, V_1, V_2)$. With $V' = V_1'V_2'$ and $\Pi' := \tau\Pi$, let us use the relation $\Pi'T^* = V'^*\Pi'$ successively to note that $V'^{*n}\Pi' = \Pi'T^{*n}$ and so with $W = W_1W_2$ we must have

$$W'^{*n}\tau\Pi = \tau\Pi T^{*n}.$$

The right-hand side converges to the zero operator strongly, whereas the operators on the left-hand side are isometric operators. This is not possible unless the space $\mathcal{Q}_{V^*} = \{0\}$. This completes the proof. □

Given the uniqueness of the unitary part of an Ando isometric lift, it is desirable to identify a concrete form of it. In Section 4.4, we construct a canonical pair of commuting unitaries associated to a commuting contractive pair that appears as the unitary part of any minimal Ando isometric lift. Indeed, we shall arrive at the same uniqueness conclusion as in Theorem 4.3.1 in Chapter 5, where we shall show that every minimal Ando isometric lift is unitarily equivalent to a model Ando lift whose unitary part is exactly the canonical pair of commuting unitaries, which we present next.

## 4.4 Canonical Commuting Unitary Pair Induced by Commuting Contractive Pair

Let us recall (see part (vii) of Proposition 2.1.1) that, given a contraction operator $T$ on a Hilbert space $\mathcal{H}$, we may define a positive-semidefinite operator $Q_{T^*}^2$ on $\mathcal{H}$ as the strong limit

$$Q_{T^*}^2 := \text{SOT-}\lim_{n\to\infty} T^n T^{*n}. \tag{4.11}$$

We set $Q_{T^*}$ equal to the positive-semidefinite square root of $Q_{T^*}^2$. As explained in Section 2.2.2 (see (2.35)), the identity $TQ_{T^*}^2 T^* = Q_{T^*}^2$ implies that the formula

$$X^* Q_{T^*} h = Q_{T^*} T^* h \tag{4.12}$$

extends by continuity to a well-defined isometry $X^*$ on $\overline{\text{Ran}}\, Q_{T^*}$ that has a minimal unitary extension, denoted as $W_D^*$, on the Hilbert space $\mathcal{Q}_{T^*}$ equal to the closure of $\cup_{n=0}^{\infty} W_D^n \,\text{Ran}\, Q_T$ defined densely as in formula (2.37):

$$W_D^* W_D^n Q_{T^*} h = W_D^{n-1} Q_{T^*} h \quad \text{for } n \geq 1, \quad W_D^* Q_{T^*} h = X^* Q_{T^*} h = Q_{T^*} T^* h.$$

Then, as explained in Section 2.2.2, $(\Pi_D, V_D)$ is the Douglas-minimal isometric lift for the single contraction operator $T$, where the isometry $V_D$ on $\begin{bmatrix} H^2(\mathcal{D}_{T^*}) \\ \mathcal{Q}_{T^*} \end{bmatrix}$ is defined as $V_D = \begin{bmatrix} M_z & 0 \\ 0 & W_D \end{bmatrix}$ and $\Pi_D \colon \mathcal{H} \to \begin{bmatrix} H^2(\mathcal{D}_{T^*}) \\ \mathcal{Q}_{T^*} \end{bmatrix}$ is the isometric embedding of $\mathcal{H}$ defined as $\Pi_D = \begin{bmatrix} D_{T^*}(I - zT^*)^{-1} \\ Q \end{bmatrix}$.

We shall need the following simple but telling result of Douglas. Here we use the standard notation: If $X$ and $Y$ are operators on a Hilbert space $\mathcal{H}$, we write $X \preceq Y$ if $Y - X$ is *positive semidefinite*, that is, $\langle (Y - X)h, h\rangle_{\mathcal{H}} \geq 0$ for all $h \in \mathcal{H}$.

**Lemma 4.4.1** (Douglas lemma [53]) *Let $A$ and $B$ be two bounded operators on a Hilbert space $\mathcal{H}$. Then there exists a contraction $C$ such that $A = BC$ if and only if*

$$AA^* \preceq BB^*.$$

The Douglas model minimal isometric lift for the single contraction operator $T$ as summarized in the introductory part of Section 2.2.2, but now applied to the case where $T = T_1T_2$, is the product operator coming from the commuting contractive operator pair $(T_1, T_2)$, leads to the following result, which will play a significant role in what follows.

**Theorem 4.4.2** *Given a commuting contractive operator pair $(T_1, T_2)$ on $\mathcal{H}$, set $T = T_1T_2$ and let $Q_{T^*}$ be the positive contraction operator whose square is given by* (4.11). *Then there exist a pair of unitary operators $(W_{\flat 1}, W_{\flat 2})$ on $\mathcal{Q}_{T^*}$ so that*

$$W_{\flat 1}W_{\flat 2} = W_D \quad \text{and} \quad W_{\flat j}^*Q_{T^*} = Q_{T^*}T_j^*, \quad \text{for each } j = 1, 2. \tag{4.13}$$

*Proof* First note that when $Q_{T^*}$ is as in (2.21) and $T = T_1T_2 = T_2T_1$, then

$$\left\langle T_iQ_{T^*}^2T_i^*h, h\right\rangle = \lim_{n\to\infty} \left\langle T^n(T_iT_i^*)T^{*n}h, h\right\rangle \leq \lim_{n\to\infty} \langle T^nT^{*n}h, h\rangle = \left\langle Q_{T^*}^2h, h\right\rangle$$

for all $h \in \mathcal{H}$, from which we conclude that

$$T_1Q_{T^*}^2T_1^* \preceq Q_{T^*}^2 \quad \text{and} \quad T_2Q_{T^*}^2T_2^* \preceq Q_{T^*}^2. \tag{4.14}$$

By Lemma 4.4.1, the inequalities in (4.14) imply that there exist two contraction operators $X_1^*$ and $X_2^*$ on $\overline{\text{Ran}}\ Q_{T^*}$ such that

$$X_1^*Q_{T^*} = Q_{T^*}T_1^*, \quad X_2^*Q_{T^*} = Q_{T^*}T_2^*. \tag{4.15}$$

From the equalities in (4.15) and (4.12) it is clear that $X_1$ and $X_2$ commute and that

$$X^* = X_1^*X_2^*. \tag{4.16}$$

Since $X^*$ is an isometry, both $X_1^*$ and $X_2^*$ are isometries, as a consequence of the general fact that, whenever $T$ is an isometry with factorization $T = T_1T_2$ for some commuting contractions $T_1$ and $T_2$, then in fact $T_1$ and $T_2$ are also isometries; one way to see this is to recall identity (4.6), which comes up in Ando's original proof of his dilation theorem: For any commuting contractive pair $(T_1, T_2)$ we have

$$\|D_{T_1}T_2h\|^2 + \|D_{T_2}h\|^2 = \|D_Th\|^2 = \|D_{T_1}h\|^2 + \|D_{T_2}T_1h\|^2 \quad \text{for all } h \in \mathcal{H}.$$

By Lemma 3.1.19 we get a commuting unitary extension $\left(W_{\flat 1}^*, W_{\flat 2}^*\right)$ of the commuting isometric pair $(X_1^*, X_2^*)$ on $\mathcal{Q}_{T^*} = \overline{\text{span}}\left\{W_D^n x : x \in \text{Ran}\, Q_{T^*} \text{ and } n \geq 0\right\}$. Thus the product $W_D^* = W_{\flat 1}^* W_{\flat 2}^*$ is the minimal unitary extension of the product $X^* = X_1^* X_2^*$. □

**Definition 4.4.3** The pair $(W_{\flat 1}, W_{\flat 2})$ as constructed in the proof of Theorem 4.4.2 will be referred to as the *canonical pair of commuting unitaries* associated with the commuting contractive pair $(T_1, T_2)$.

The canonicity of a canonical pair of unitaries $(W_{\flat 1}, W_{\flat 2})$ for a given commuting contractive pair $(T_1, T_2)$ is illustrated in the following result.

**Lemma 4.4.4** *Let $(T_1, T_2)$ on $\mathcal{H}$ and $(T_1', T_2')$ on $\mathcal{H}'$ be two pairs of commuting contractions and let $(W_{\flat 1}, W_{\flat 2})$ on $\mathcal{Q}_{T^*}$ and $\left(W_{\flat 1}', W_{\flat 2}'\right)$ on $\mathcal{Q}_{T'^*}$ be the respective canonical pairs of commuting unitaries. If $(T_1, T_2)$ is unitarily equivalent to $(T_1', T_2')$ via the unitary similarity $\phi : \mathcal{H} \to \mathcal{H}'$, then $(W_{\flat 1}, W_{\flat 2})$ and $\left(W_{\flat 1}', W_{\flat 2}'\right)$ are unitarily equivalent via the induced unitary transformation $\tau_\phi : \mathcal{Q}_{T^*} \to \mathcal{Q}_{T'^*}$ determined by $\tau_\phi : W_D^n Q_{T^*} h \to W_D'^n Q_{T'^*} \phi h$. In particular, if $(T_1, T_2) = (T_1', T_2')$, then $(W_{\flat 1}, W_{\flat 2}) = \left(W_{\flat 1}', W_{\flat 2}'\right)$.*

*Proof* Let $\phi : \mathcal{H} \to \mathcal{H}'$ be a unitary that intertwines $(T_1, T_2)$ and $(T_1', T_2')$. Let us denote $T = T_1 T_2$ and $T' = T_1' T_2'$. Let $Q_{T^*}$ and $Q_{T'^*}$ be the limits of $T^n T^{*n}$ and $T'^n T'^{*n}$, respectively, in the strong operator topology. Clearly, $\phi$ intertwines $Q_{T^*}$ and $Q_{T'^*}$. Therefore $\phi$ takes $\overline{\text{Ran}}\, Q_{T^*}$ onto $\overline{\text{Ran}}\, Q_{T'^*}$. We denote the restriction of $\phi$ to $\overline{\text{Ran}}\, Q_T$ by $\phi$ itself. Let $(X_1, X_2)$ on $\overline{\text{Ran}}\, Q_{T^*}$ and $(X_1', X_2')$ on $\overline{\text{Ran}}\, Q_{T'^*}$ be the pairs of commuting coisometries corresponding to the pairs $(T_1, T_2)$ and $(T_1', T_2')$ as in (4.15), respectively. It is easy to see from the definition that

$$\phi(X_1, X_2) = (X_1', X_2')\phi.$$

Let $(W_{\flat 1}, W_{\flat 2})$ on $\mathcal{Q}_{T^*}$ and $\left(W_{\flat 1}', W_{\flat 2}'\right)$ on $\mathcal{Q}_{T'^*}$ be the pairs of commuting unitaries corresponding to $(T_1, T_2)$ and $(T_1', T_2')$, respectively. Remembering the formula (2.36) for the spaces $\mathcal{Q}_{T^*}$ and $\mathcal{Q}_{T'^*}$, we can densely define $\tau_\phi : \mathcal{Q}_{T^*} \to \mathcal{Q}_{T'^*}$ by

$$\tau_\phi : W_D^n x \mapsto W_D'^n \phi x, \text{ for every } x \in \overline{\text{Ran}}\, Q_{T^*} \text{ and } n \geq 0 \tag{4.17}$$

and extend by taking limits. Trivially, $\tau_\phi$ is unitary and intertwines $W_D$ with $W_D'$. For a nonnegative integer $n$ and $x$ in $\mathcal{Q}_{T^*}$, we have (using (4.13))

$$\begin{aligned}\tau_\phi W_{\flat 1}\left(W_D^n x\right) &= \tau_\phi W_D^{n+1}\left(W_{\flat 2}^* x\right) = W_D'^{n+1}\phi\left(X_2^* x\right)\\ &= W_D'^{n+1} W_{\flat 2}'^* \phi x = W_{\flat 1}' W_D'^n \phi x = W_{\flat 1}' \tau_\phi \left(W_D^n x\right).\end{aligned}$$

A similar computation shows that $\tau_\phi$ intertwines $W_{\flat 2}$ with $W_{\flat 2}'$ as well.

Note that when $(T_1, T_2) = (T_1', T_2')$, then

$$Q_{T^*} = Q_{T'^*}, \quad \phi = I_{\mathcal{H}}, \quad \text{and} \quad \tau_\phi = I_{Q_{T^*}}.$$

This completes the proof of the lemma. □

The following is a straightforward consequence of the above lemma.

**Corollary 4.4.5** *Let $(T_1, T_2)$ be a commuting contractive operator pair and let $Q_{T^*}$ be as in (4.11), where $T = T_1T_2$. Let $(W_1, W_2)$ be some pair of commuting unitaries on $Q_{T^*}$ such that*

$$W_D = W_1 W_2 \quad \textit{and} \quad W_j^* Q_{T^*} = Q_{T^*} T_j^* \quad \textit{for } j = 1, 2. \tag{4.18}$$

*Then $(W_1, W_2) = (W_{\flat 1}, W_{\flat 2})$.*

## 4.5 Commutative Contractive Pair with Unique Minimal Ando Lift

Given an orthogonal projection $P$ and a unitary $U$ acting on a Hilbert space $\mathcal{F}$, we show in Theorem 4.5.1 that any two minimal isometric lifts for the commuting contractive pair $(U^*P^\perp, PU)$ are unitarily equivalent as lifts. Let us note that pairs of this form were given an algebraic characterization in Lemma 3.1.1. Also note that the pair $(U^*P^\perp, PU)$ is jointly unitarily equivalent to $(P^\perp U^*, UP)$ via the unitary $U$; hence, by Lemma 4.1.3, the two pairs have the same set of minimal Ando isometric lifts. Therefore the conclusion of the next result is true for both pairs once it is true for one of them. Therefore it suffices to work with just one of the pairs.

**Theorem 4.5.1** *Let $P$ be any orthogonal projection and let $U$ be any unitary acting on a Hilbert space $\mathcal{F}$. Then any two minimal isometric lifts of the commuting contractive pair $(U^*P^\perp, PU)$ are the same up to unitary equivalence of lifts, that is, for this case, there is only one minimal Ando lift up to unitary equivalence.*

*Proof* Let us first note that the commuting isometric pair

$$(M_{U^*P^\perp + zU^*P}, M_{PU + zP^\perp U}) \quad \text{acting on } H^2(\mathcal{F})$$

is an isometric lift of $(U^*P^\perp, PU)$ acting on $\mathcal{F}$ via the isometric embedding of the coefficient space $\mathcal{F}$ into the subspace of constant functions in $H^2(\mathcal{F})$. Since the product of $P^\perp U$ and $U^*P$ is zero, by Remark 4.3.4 or Proposition 4.3.5, *no* minimal isometric lift of $(P^\perp U, U^*P)$ can have a nontrivial unitary part. Let

$$\left(M_{U_*^* P_*^\perp + zU_*^* P_*}, M_{P_* U_* + zP_*^\perp U_*}\right) \quad \text{acting on } H^2(\mathcal{F}_*) \tag{4.19}$$

be an Ando isometric lift of $(P^\perp U, U^*P)$ via an isometric embedding $\Pi : \mathcal{F} \to H^2(\mathcal{F}_*)$. Then, by definition of lift, we have

$$M^*_{U^*_* P^\perp_* + zU^*_* P_*}\Pi = \Pi U^* P^\perp \quad \text{and} \quad M^*_{P_* U_* + zP^\perp_* U_*}\Pi = \Pi PU. \tag{4.20}$$

Since $M_{U^*_* P^\perp_* + zU^*_* P_*} M_{P_* U_* + zP^\perp_* U_*} = M_z^{\mathcal{F}_*}$, by multiplying the two operator equations in (4.20), we get

$$\begin{aligned} M_z^*\Pi &= M^*_{U^*_* P^\perp_* + zU^*_* P_*} M^*_{P_* U_* + zP^\perp_* U_*}\Pi \\ &= M^*_{U_* P^\perp_* + zU_* P_*}\Pi PU = \Pi U^* P^\perp PU = 0. \end{aligned}$$

This shows that $\operatorname{Ran}\Pi$ consists of the constant functions in $H^2(\mathcal{F}_*)$ and therefore the isometric embedding $\Pi$ may be viewed as of the form

$$\Pi = \Lambda_* : \mathcal{F} \to \mathcal{F}_*$$

for some isometry $\Lambda_*$. In view of this, the lift equations (4.20) now take the form

$$P^\perp_* U_* \Lambda_* = \Lambda_* P^\perp U \quad \text{and} \quad U^*_* P_* \Lambda_* = \Lambda_* U^* P. \tag{4.21}$$

Multiplication by $\Lambda^*_*$ on the left of both the equations in (4.21) gives us

$$P^\perp U = \Lambda^*_* P^\perp_* U_* \Lambda_* \quad \text{and} \quad U^* P = \Lambda^*_* U^*_* P_* \Lambda_*.$$

Adding the adjoint of the first to the second in this last pair of equations then gives us

$$U = \Lambda^*_* U_* \Lambda_*. \tag{4.22}$$

We argue next that $\Lambda_* \mathcal{F}$ is invariant under both $U_*$ and $U^*_*$ as follows. First note that, for $f \in \mathcal{F}$, we have

$$\|f\| = \|Uf\| = \|\Lambda^*_* U_* \Lambda_* f\| \le \|U_* \Lambda_* f\| = \|f\|,$$

forcing equality throughout and hence $\|\Lambda^*_* U_* \Lambda_* f\| = \|U_* \Lambda_* f\|$. This in turn forces $U_* \Lambda_* f \in \Lambda_* \mathcal{F}$, that is, $\Lambda_* \mathcal{F}$ is invariant under $U_*$.

A similar analysis verifies the invariance of $\Lambda_* \mathcal{F}$ under $U^*_*$ as follows. Since $U_*$ is a unitary and $\operatorname{Ran}\Lambda_*$ is invariant under $U_*$, we must have $U_*(\operatorname{Ran}\Lambda_*) = \operatorname{Ran}\Lambda_*$. Then by equations (4.21) we conclude that $\operatorname{Ran}\Lambda_*$ is invariant under both $P_*$ and $P^\perp_*$. This together with the minimality of the lift (4.19) implies that

$$\begin{aligned} H^2(\mathcal{F}_*) &= \bigvee_{m,n \ge 0} \left(M_{U^*_* P^\perp_* + zU^*_* P_*}\right)^m \left(M_{P_* U_* + zP^\perp_* U_*}\right)^n \Lambda_* \mathcal{F} \\ &\subseteq H^2(\operatorname{Ran}\Lambda_*) \subseteq H^2(\mathcal{F}_*). \end{aligned}$$

This happens only if $\operatorname{Ran}\Lambda_* = \mathcal{F}_*$, that is, when $\Lambda_*\colon \mathcal{F} \to \mathcal{F}_*$ is a unitary. Therefore we have (4.22) in the stronger form

$$U_*\Lambda_* = \Lambda_* U. \tag{4.23}$$

This together with the second equation in (4.21) gives

$$P_*\Lambda_* = \Lambda_* P. \tag{4.24}$$

Equations (4.23) and (4.24) together establish the desired unitary equivalence of the two lifts via the unitary $I_{H^2} \otimes \Lambda_*\colon H^2(\mathcal{F}) \to H^2(\mathcal{F}_*)$. □

## 4.6 The Commutant Lifting Theorem

In this section, we present another proof of Ando's dilation theorem 4.2.1. This time the tool that is used – the so-called commutant lifting theorem – actually turns out to be equivalent to Ando's dilation theorem, that is, not only does the commutant lifting theorem give a proof of Ando's dilation theorem, but the converse is also true. We exhibit both directions in this section. We begin by formally stating the commutant lifting theorem. Readers are referred to the classic book of Foias–Frazho [62] for a bird's-eye view of this theme along with a number of proofs thereof.

**Theorem 4.6.1** *Let $T$ be a contractive operator and let $A$ be a bounded operator acting on a Hilbert space $\mathcal{H}$ such that $AT = TA$. If $V$ acting on $\mathcal{K}$ is an isometric lift (minimal or not) of $T$ via the isometric embedding $\iota\colon \mathcal{H} \to \mathcal{K}$ (the inclusion), then there exists an operator $B$ in $\mathcal{B}(\mathcal{K})$ such that*

(i) *$B$ is in the commutant of $V$, that is, $BV = VB$,*
(ii) *$B$ is a lifting of $A$, that is, $AP_{\mathcal{H}} = P_{\mathcal{H}}B$ or equivalently $B^*|_{\mathcal{H}} = A^*$, and*
(iii) *$B$ is a norm-preserving lifting of $A$, that is, $\|B\| = \|A\|$.*

We shall use Ando's dilation theorem to prove the following variant of Theorem 4.6.1.

**Theorem 4.6.1′** *Let $T$ and $A'$ be contractive operators acting on the Hilbert space $\mathcal{H}$ such that $A'T = TA'$. Let $V'$ acting on $\mathcal{K}'$ be a minimal isometric lift of $T$ via the isometric embedding $\iota\colon \mathcal{H} \to \mathcal{K}$, the inclusion map. Then there exists a contractive operator $B'$ in $\mathcal{B}(\mathcal{K}')$ such that*

(i) *$B'$ is in the commutant of $V'$, that is, $B'V' = V'B'$, and*
(ii) *$B'$ is a lifting of $A'$, that is, $A'P_{\mathcal{H}} = P_{\mathcal{H}}B'$ or equivalently $B'^*|_{\mathcal{H}} = A'^*$.*

Once Theorem 4.6.1′ is proved, our original goal of proving Theorem 4.6.1 follows from verifying the next claim.

**Claim** To prove Theorem 4.6.1, it is enough to prove Theorem 4.6.1′.

As a first step we shall verify the claim.

*Proof of the claim* To establish the claim, we assume the validity of Theorem 4.6.1′ and deduce Theorem 4.6.1 as a consequence. Let $(T, A, V)$ be as in the hypotheses of Theorem 4.6.1 (so $A$ is not necessarily a contraction). Set $A' = A/\|A\|$. Then $A'$ is a contraction and it commutes with $V$. Let us consider the space

$$\mathcal{K}_0 = \bigvee_{n\geq 0} V^n\mathcal{H}.$$

By Lemma 2.3.7, $\mathcal{K}_0$ is reducing for $V$ and $V_0 := V|_{\mathcal{K}_0}$ is a minimal isometric lift of $T$ via the inclusion map $\iota\colon \mathcal{H} \to \mathcal{K}_0$. Suppose $V$ decomposes as

$$V = \begin{bmatrix} V_0 & 0 \\ 0 & V_0' \end{bmatrix} : \begin{bmatrix} \mathcal{K}_0 \\ \mathcal{K}\ominus\mathcal{K}_0 \end{bmatrix} \to \begin{bmatrix} \mathcal{K}_0 \\ \mathcal{K}\ominus\mathcal{K}_0 \end{bmatrix} \tag{4.25}$$

for some isometry $V_0'$ acting on $\mathcal{K}\ominus\mathcal{K}_0$. Apply Theorem 4.6.1′ to the data $(T, A', V_0)$ to obtain a contractive operator $B'$ acting on $\mathcal{K}_0$ such that conclusions (i) and (ii) of Theorem 4.6.1′ hold, that is,

$$B'V_0 = V_0B', \quad B'^*|_{\mathcal{H}} = A'^*.$$

Rescaling these identities with $\|A\|$ then gives

$$(\|A\|B')V_0 = V_0(\|A\|B'), \quad (\|A\|B'^*)|_{\mathcal{H}} = \|A\|A'^* = A^*. \tag{4.26}$$

Here we use the fact that $A' = A/\|A\|$ to obtain the last equality in (4.26). Consider

$$B = \begin{bmatrix} \|A\|B' & 0 \\ 0 & \|A\|I_{\mathcal{K}\ominus\mathcal{K}_0} \end{bmatrix} : \begin{bmatrix} \mathcal{K}_0 \\ \mathcal{K}\ominus\mathcal{K}_0 \end{bmatrix} \to \begin{bmatrix} \mathcal{K}_0 \\ \mathcal{K}\ominus\mathcal{K}_0 \end{bmatrix}.$$

It is then easy to see from the first identity in (4.26) that the operator $B$ commutes with $V$ as in (4.25), that is, conclusion (i) in Theorem 4.6.1 holds. Since $\mathcal{H} \subset \mathcal{K}_0$, it also follows from the second identity in (4.26) that $B^*|_{\mathcal{H}} = A^*$, that is, conclusion (ii) in Theorem 4.6.1 holds. Finally, from this last conclusion and the definition of $B$, we see that

$$\|A^*\| \leq \|B^*\| = \|B\| = \max\{\|A\|, \|A\|\|B'\|\} \leq \|A\|.$$

The last inequality is drawn from the fact that $\|B'\| \leq 1$. Thus $\|B\| = \|A\|$, and conclusion (iii) of Theorem 4.6.1 holds. This completes the proof of the claim. □

Let us now record how Theorem 4.6.1′ is a consequence of Ando's theorem.

*Proof of Theorem 4.6.1′* Suppose that $(A', T, V')$ is as in the hypotheses of Theorem 4.6.1′. Then $(A', T)$ is a commuting contractive pair and thus has an Ando commuting isometric lift $(V_1, V_2)$ acting on a larger Hilbert space $\mathcal{K} \supset \mathcal{H}$. Then, in particular, $V_1$ is an isometric lift of $A'$ and $V_2$ is an isometric lift (possibly not minimal) of $T$. Apply Lemma 2.3.7 to the isometric lift $V_2$ of $T$ to get it decomposed as

$$V_2 = \begin{bmatrix} V_{20} & 0 \\ 0 & V'_{20} \end{bmatrix} : \begin{bmatrix} \mathcal{K}_{20} \\ \mathcal{K} \ominus \mathcal{K}_{20} \end{bmatrix} \to \begin{bmatrix} \mathcal{K}_{20} \\ \mathcal{K} \ominus \mathcal{K}_{20} \end{bmatrix},$$

where

$$\mathcal{K}_{20} = \bigvee_{n \geq 0} V_2^n \mathcal{H} \quad \text{and} \quad V_{20} = V_2|_{\mathcal{K}_{20}}. \tag{4.27}$$

With respect to this decomposition, let us write $V_1$ as

$$V_1 = \begin{bmatrix} B & C \\ D & E \end{bmatrix} : \begin{bmatrix} \mathcal{K}_{20} \\ \mathcal{K} \ominus \mathcal{K}_{20} \end{bmatrix} \to \begin{bmatrix} \mathcal{K}_{20} \\ \mathcal{K} \ominus \mathcal{K}_{20} \end{bmatrix}, \tag{4.28}$$

where $\widehat{V}' := V_{20}$ is a *minimal* isometric lift of $T$ with embedding operator equal to the inclusion map $\iota \colon \mathcal{H} \to \mathcal{K}_0$.

Let us assume for the moment the following.

**Clairvoyance assumption** *We were able to anticipate the construction of* $V_{20}$ *as in* (4.27) *and we assume that we can take the minimal isometric lift* $V'$ *of* $T$ *prescribed in the hypotheses of Theorem 4.6.1′ to be* $\widehat{V}' := V_{20}$ *as in* (4.27).

Thus the data set from the hypotheses is $\left(T, A', \widehat{V}' = V_{20}\right)$ and, to complete the proof of Theorem 4.6.1′, we must find $\widehat{B}' \in \mathcal{B}(\mathcal{K}_{20})$ so that

$$\widehat{B}' V_{20} = V_{20} \widehat{B}', \quad A' P_{\mathcal{H}} = P_{\mathcal{H}} \widehat{B}'. \tag{4.29}$$

Let us set $\widehat{B}' = B$, where $B$ is as in (4.27). Then the first equation in (4.29) is an immediate consequence of the fact that $V_1 V_2 = V_2 V_1$ while the second is an immediate consequence of the fact that $V_1^*|_{\mathcal{H}} = A'^*$ and $\mathcal{H} \subset \mathcal{K}_{20}$.

We can remove the clairvoyance assumption as follows. Suppose that $V' \in \mathcal{B}(\mathcal{K}')$ is any minimal isometric lift of $T$. But then $\widehat{V}' := V_{20} \in \mathcal{B}(\mathcal{K}_{20})$ is another minimal isometric lift of $T$, where both lifts use an inclusion isometric embedding operator. By Theorem 2.1.3, there is a unitary $\tau \colon \mathcal{K}_{20} \to \mathcal{K}'$ such that

$$\tau V_{20} = \widehat{V}' \tau \quad \text{and} \quad \tau|_{\mathcal{H}} = I_{\mathcal{H}}. \tag{4.30}$$

Consider

$$B' = \tau B \tau^* : \mathcal{K}' \to \mathcal{K}'.$$

Then $\|B'\| = \|B\| \leq 1$. Moreover, in view of (4.30), we have $B'^*|_{\mathcal{H}} = B^*|_{\mathcal{H}} = A'^*$ and

$$B'V' = \tau B \tau^* \tau V_{20} \tau^* = \tau B V_{20} \tau^* = \tau V_{20} B \tau^* = \tau V_{20} \tau^* \tau B \tau^* = V'B'.$$

This completes the proof of Theorem 4.6.1′. □

The choice of isometric embedding for the isometric lift $V$ in Theorem 4.6.1 is not a restriction. If one takes a general isometric embedding $\Pi : \mathcal{H} \to \mathcal{K}$ then item (ii) in Theorem 4.6.1 changes to $A\Pi^* = \Pi^* B$. We have taken $\Pi$ to be the inclusion map because it makes the proof slightly less technical.

There is an "intertwining" version of Theorem 4.6.1. This is what one might call an intertwining lifting theorem.

**Theorem 4.6.2** *Suppose $T$ and $T'$ are contractive operators acting on $\mathcal{H}$ and $\mathcal{H}'$, respectively. Let $X : \mathcal{H} \to \mathcal{H}'$ be a bounded operator (not necessarily contractive) such that $XT = T'X$. Suppose $W$ on $\mathcal{K}$ and $W'$ on $\mathcal{K}'$ are isometric lifts of $T$ and $T'$, respectively, both via the inclusion embedding of the underlying Hilbert spaces. Then there exists a bounded operator $Y : \mathcal{K} \to \mathcal{K}'$ such that*

(i) *$Y$ intertwines $W$ and $W'$, that is, $YW = W'Y$;*
(ii) *$Y$ is a lifting of $X$, that is, $XP_{\mathcal{H}} = P_{\mathcal{H}}Y$ or equivalently $Y^*|_{\mathcal{H}'} = X^*$;*
(iii) *$Y$ is a norm-preserving lifting of $X$, that is, $\|Y\| = \|X\|$.*

*Proof* This can be seen to follow from Theorem 4.6.1 as follows. Take

$$T = \begin{bmatrix} T & 0 \\ 0 & T' \end{bmatrix} \quad \text{and} \quad A = \begin{bmatrix} 0 & 0 \\ X & 0 \end{bmatrix} \quad \text{acting on} \begin{bmatrix} \mathcal{H} \\ \mathcal{H}' \end{bmatrix} = \mathcal{G}.$$

Then we have $A$, a bounded operator, commuting with the contractive operator $T$. Let us note that

$$V = \begin{bmatrix} W & 0 \\ 0 & W' \end{bmatrix} : \begin{bmatrix} \mathcal{K} \\ \mathcal{K}' \end{bmatrix} \to \begin{bmatrix} \mathcal{K} \\ \mathcal{K}' \end{bmatrix}$$

is an isometric lift of $T$ via the natural inclusion embedding of $\mathcal{G}$ into $\mathcal{L} = \left[ \begin{smallmatrix} \mathcal{K} \\ \mathcal{K}' \end{smallmatrix} \right]$. By Theorem 4.6.1, there exists a bounded operator

$$B = \begin{bmatrix} B_{11} & B_{12} \\ Y & B_{22} \end{bmatrix} : \begin{bmatrix} \mathcal{K} \\ \mathcal{K}' \end{bmatrix} \to \begin{bmatrix} \mathcal{K} \\ \mathcal{K}' \end{bmatrix}$$

such that

$$BV = VB, \quad B^*|_{\mathcal{G}} = A^* \quad \text{and} \quad \|B\| = \|X\|.$$

From the commutativity of $Y$ and $V$ and their matrix representations, it follows that the bounded operator $Y: \mathcal{K} \to \mathcal{K}'$ satisfies (i). From $B^*|_{\mathcal{G}} = A^*$ and again from the matrix representations of $B$ and $A$, it follows that $Y^*|_{\mathcal{H}'} = X^*$. This proves (ii). To see (iii), let us note using $Y^*|_{\mathcal{H}'} = X^*$ and the matrix representation of $B$ that

$$\|X\| \le \|Y\| \le \|B\| = \|X\|.$$

This completes the proof. □

We next tackle the reverse direction: *the commutant lifting theorem implies Ando's theorem.* The proof can be seen to have two parts. Lemma 4.6.3, borrowed from the Foias–Frazho book [62, Section 6, Chapter VII], is one half of the story: *If the operator A in Theorem 4.6.1 is an isometry, then B can be chosen to be an isometry as well.*

**Lemma 4.6.3** *Let $T$ be a contraction and let $X$ be an isometry acting on $\mathcal{H}$ such that $XT = TX$. Let $V$ acting on $\mathcal{K}$ be a minimal isometric lift of $T$ via the inclusion map $\iota: \mathcal{H} \to \mathcal{K}$. Then there is an isometry $Y$ acting on $\mathcal{K}$ such that $YV = VY$ and $Y$ is a lift of $X$, that is, $XP_{\mathcal{H}} = P_{\mathcal{H}}Y$ or equivalently, $Y^*|_{\mathcal{H}} = X^*$.*

**Remark 4.6.4** One can see Theorem 3.3.3 as a special case of Lemma 4.6.3, as follows. In the setup of Theorem 3.3.3 we are given a contractive analytic function $(\mathcal{F}, \Theta)$. Let us set $T = M_\Theta$ and $X = M_z^{\mathcal{F}}$ on $\mathcal{H} = H^2(\mathcal{F})$ (so $X$ is a shift rather than a general isometry),

$$V = \begin{bmatrix} M_\Theta & 0 \\ \Delta_\Theta|_{H^2(\mathcal{F})} & M_z^{\overline{\Delta_\Theta L^2(\mathcal{F})}} \end{bmatrix} \quad \text{on } \mathcal{K} = \begin{bmatrix} H^2(\mathcal{F}) \\ H^2(\overline{\Delta_\Theta L^2(\mathcal{F})}) \end{bmatrix}$$

with the obvious inclusion map

$$\iota = \begin{bmatrix} I_{H^2(\mathcal{F})} \\ 0 \end{bmatrix} : H^2(\mathcal{F}) \to \begin{bmatrix} H^2(\mathcal{F}) \\ H^2(\overline{\Delta_\Theta L^2(\mathcal{F})}) \end{bmatrix}.$$

Note that $V$ is a minimal isometric lift of $M_\Theta$ in the Schäffer form. Then for this setting

$$Y = \begin{bmatrix} M_z^{\mathcal{F}} & 0 \\ 0 & M_z^{\overline{\Delta_\Theta L^2(\mathcal{F})}} \end{bmatrix}$$

serves as the lift of $X$ that commutes with $V$ guaranteed by Lemma 4.6.3.

*Proof* Since every minimal isometric lift of a single contraction is unitarily equivalent (see Theorem 2.1.3), it is enough to prove it for the case when the minimal isometric lift $V$ of $T$ is in the Schäffer form:

$$V = \begin{bmatrix} T & 0 \\ \mathbf{ev}_0^* D_T & M_z^{\mathcal{D}_T} \end{bmatrix} : \begin{bmatrix} \mathcal{H} \\ H^2(\mathcal{D}_T) \end{bmatrix} \to \begin{bmatrix} \mathcal{H} \\ H^2(\mathcal{D}_T) \end{bmatrix}.$$

Since $XT = TX$ and $X$ is an isometry, we have

$$D_T^2 = I - T^*T = X^*(I - T^*T)X,$$

which implies that $\omega \colon \mathcal{D}_T \to \mathcal{D}_T$, defined by

$$\omega D_T h = D_T X h \quad \text{for all } h \in \mathcal{H}, \tag{4.31}$$

is an isometry. Set

$$Y = \begin{bmatrix} X & 0 \\ 0 & I_{H^2} \otimes \omega \end{bmatrix} : \begin{bmatrix} \mathcal{H} \\ H^2(\mathcal{D}_T) \end{bmatrix} \to \begin{bmatrix} \mathcal{H} \\ H^2(\mathcal{D}_T) \end{bmatrix}.$$

Both $Y$ and $V$ are lower triangular and $I_{H^2} \otimes \omega$ commutes with $M_z^{\mathcal{D}_T}$. Therefore, we only need to check the (21)-entries of $YV$ and $VY$ for the commutation relation $YV = VY$. We compute

$$[YV]_{21} = (I_{H^2} \otimes \omega)\mathbf{ev}_0^* D_T = \mathbf{ev}_0^* \omega D_T = \mathbf{ev}_0^* D_T X = [VY]_{21}.$$

This completes the proof. □

We are now ready to prove Ando's dilation theorem using the commutant lifting theorem.

*Proof of Ando's dilation theorem 4.2.1 via the commutant lifting theorem* Suppose that $(T_1, T_2)$ is a commuting contractive operator pair. By the Sz.-Nagy lifting theorem, there is an isometric operator $B_1$ acting on $\mathcal{K}' \supset \mathcal{H}$ that is a lift of $T_1$ via the inclusion map $\iota \colon \mathcal{H} \to \mathcal{K}'$, that is,

$$B_1^*|_{\mathcal{H}} = T_1^*. \tag{4.32}$$

By the commutant lifting theorem the contractive operator $T_2$ in the commutant of $T_1$ has a contractive lifting $B_2$ to the commutant of $B_1$, that is,

$$\text{there exists } B_2 \in \mathcal{B}(\mathcal{K}'), \quad B_1 B_2 = B_2 B_1, \quad B_2^*|_{\mathcal{H}} = T_2^*. \tag{4.33}$$

Again by the Sz.-Nagy lifting theorem the contraction operator $B_2$ on $\mathcal{K}'$ can be lifted to an isometry $V_2$ acting on $\mathcal{K} \supset \mathcal{K}'$ (with the isometric embedding operator taken to be the inclusion map):

$$V_2^*|_{\mathcal{K}'} = B_2^*. \tag{4.34}$$

We now apply Lemma 4.6.3 for the choice $(X, T) = (B_1, B_2)$ to get an isometry $V_1$ such that $V_1 V_2 = V_2 V_1$ and

$$V_1^*|_{\mathcal{K}'} = B_1^*. \tag{4.35}$$

It remains to show that the commuting isometric pair $(V_1, V_2)$ is a lift of $(T_1, T_2)$. For this, let us note that $\mathcal{H} \subset \mathcal{K}' \subset \mathcal{K}$ and, therefore, using (4.32) and (4.35) we get

$$V_1^*|_{\mathcal{H}} = B_1^*|_{\mathcal{H}} = T_1^*,$$

while using (4.33) and (4.34) we get

$$V_2^*|_{\mathcal{H}} = B_2^*|_{\mathcal{H}} = T_2^*.$$

This completes the proof. □

## 4.7 Notes to Chapter 4

**1.** Theorems 4.3.1 and 4.5.1 here have played an important role in the recent paper [50], which is concerned with the problem of when a commuting contractive pair $(T_1, T_2)$ has an affine variety of the bidisk as a complete spectral set. While the problem still remains open in general, it is shown that the answer is affirmative under the assumption that the commuting contractive operator pair has finite-dimensional defect spaces, more precisely, when $\dim \mathcal{D}_{T_j^*} < \infty$ for $j = 1, 2$.

**2.** The liftings $B$ in Theorem 4.6.1 or $Y$ in Theorem 4.6.2 are not unique. The reader is referred to [62] for classifications of all the liftings, and to [55] for a step-by-step construction of the lifting.

**3.** The equivalence of the 1963 Ando dilation theorem [10] with the 1968 Sz.-Nagy–Foias commutant lifting theorem [89] was realized much later in 1978 by Parrott [75].

**4.** While there are many proofs of the Sz.-Nagy dilation theorem (see number 1 in the Notes to Chapter 2), there are not so many proofs of Ando's dilation theorem. Besides Ando's original proof (Section 4.2) and the aforementioned proof via commutant lifting, let us mention the independent proof of Ball–Sadosky–Vinnikov from looking at the geometry of two-evolution Lax–Phillips scattering systems (see Corollary 5.3 in [19]). See number 1 in the Notes to Chapter 5 for references to earlier papers where the authors use the identity (4.6) to obtain new proofs of the Ando dilation theorem for various special cases.

# 5

# Douglas-Type Model for Ando Isometric Lifts

The Ando dilation theorem ([10]), coming 10 years after the Sz.-Nagy dilation theorem, provides a two-variable analogue of the Sz.-Nagy result: *Given a commuting pair of contractions $(T_1, T_2)$ on a Hilbert space $\mathcal{H}$, there is a commuting pair of isometric operators $(V_1, V_2)$ on a larger Hilbert space $\mathcal{K} \supset \mathcal{H}$ so that $V_j$ is a lift of $T_j$ for each $j = 1, 2$.* The proof there, as reviewed in Section 4.2, fails to shed much light on the geometry of the dilation space; it is to be expected that such a geometry must be much richer than in the single-operator context due to the lack of uniqueness up to a notion of unitary equivalence in the bivariate context. The bidisk setting has been plowed by stalwarts for its operator theory and function theory, but there has been essentially no follow-up to Ando's result in the direction of a Sz.-Nagy–Foias-type model theory for a commuting contractive operator pair, although there have now been some preliminary results in this direction (see [5, 24, 48]).

In this chapter we find functional models for minimal Ando lifts of a commuting contractive operator pair $(T_1, T_2)$ by exploring how a copy of a Douglas model for the minimal Sz.-Nagy lift of the product contraction operator $T_1T_2$ must embed into the Berger–Coburn–Lebow (BCL) model for the minimal Ando lift of $(T_1, T_2)$.

## 5.1 Ando Tuples: The Key Ingredient

We now define the additional invariants alluded to in the discussion preceding Lemma 4.1.5.

**Definition 5.1.1** A collection of objects of the form $(\mathcal{F}, \Lambda, P, U)$ is said to be an *Ando tuple* for a commuting contractive pair $(T_1, T_2)$ if $\mathcal{F}$ is a Hilbert space, $\Lambda\colon \mathcal{D}_{T_1T_2} \to \mathcal{F}$ is an isometry, $P$ is a projection operator on $\mathcal{F}$, and $U$ is a unitary operator on $\mathcal{F}$.

We shall see in the following chapters that not only do Ando tuples describe the entire class of Ando isometric lifts of a commuting contractive pair but they govern when two minimal Ando isometric lifts are unitarily equivalent under a natural notion of coincidence, defined as follows.

**Definition 5.1.2** Two Ando tuples $(\mathcal{F}, \Lambda, P, U)$ and $(\mathcal{F}', \Lambda', P', U')$ for $(T_1, T_2)$ are said to *coincide* if there is a unitary operator $\tau\colon \mathcal{F} \to \mathcal{F}'$ such that

$$\tau\Lambda = \Lambda', \quad \tau P\tau^* = P', \quad \tau U\tau^* = U'.$$

Thus an Ando tuple amounts to a BCL-tuple (see Definition 3.1.2), but with the added ingredient of the isometry $\Lambda\colon \mathcal{D}_{T_1T_2} \to \mathcal{F}$, while coincidence of Ando tuples is a natural extension of the notion of unitary equivalence of BCL-tuples in the sense of Theorem 3.1.6 (with the possible commuting pairs of unitary operators $(W_1, W_2)$ and $(W_1', W_2')$ ignored).

We shall require mainly two distinct types of Ando tuples – the Type I Ando tuples (see Definition 5.1.5) and the Type II Ando tuples (see Definition 6.1.2). But there is a special class of Ando tuples, called *special Ando tuples*, each of which is simultaneously of both Type I and Type II. We defer the construction of such special Ando tuples to Section 5.3, and the verification that any special Ando tuple is of Type I or of Type II to Sections 5.4 and 6.2, respectively.

In the sequel we shall need various notions of structural properties for Ando tuples depending on the context. Let us now introduce for the record two such closely related but distinct properties: *Douglas minimality* and *irreducibility*.

**Definition 5.1.3** **1.** An Ando tuple $(\mathcal{F}, \Lambda, P, U)$ is said to be *Douglas minimal* if the smallest joint invariant subspace for the BCL1-model operator pair $\big(M_{U^*(P^\perp + zP)}, M_{(P+zP^\perp)U}\big)$ containing the space $H^2(\operatorname{Ran}\Lambda)$ is the whole space $H^2(\mathcal{F})$.

**2.** An Ando tuple $(\mathcal{F}, \Lambda, P, U)$ is said to be *irreducible* if the smallest subspace of $\mathcal{F}$ invariant under $U$, $U^*$, $P$ and containing $\operatorname{Ran}\Lambda$ is the whole space $\mathcal{F}$.

**Remark 5.1.4** In view of Theorem 3.1.18, the Ando tuple $(\mathcal{F}, \Lambda, P, U)$ is irreducible if and only if the only reduced sub-BCL-tuple $(\mathcal{F}_0, P_0, U_0)$ of the BCL-tuple $(\mathcal{F}, P, U)$ such that $\mathcal{F}_0 \supset \operatorname{Ran}\Lambda$ is the whole Ando tuple $(\mathcal{F}, P, U)$. Another equivalent statement is: *$(\mathcal{F}, \Lambda, P, U)$ is an irreducible Ando tuple if and only if the smallest joint reducing subspace for $\big(M_{U^*(P^\perp + zP)}, M_{(P+zP^\perp)U}\big)$ containing $H^2(\operatorname{Ran}\Lambda)$ is the whole space $H^2(\mathcal{F})$*. Since any reducing subspace is also invariant, it is at the level of a tautology to see that the minimality of an Ando tuple implies its irreducibility. In all the examples that we have checked, the converse also holds but, to this point, we have not been able to determine if the converse holds in general.

We shall see that minimal Ando lifts of a given commuting contractive operator pair $(T_1, T_2)$ are in one-to-one correspondence with Douglas-minimal Ando tuples for $(T_1^*, T_2^*)$ that satisfy some additional conditions as follows.

**Definition 5.1.5** Given a commuting contractive operator pair $(T_1, T_2)$ and an Ando tuple $(\mathcal{F}_*, \Lambda_*, P_*, U_*)$ for $(T_1^*, T_2^*)$, we say that $(\mathcal{F}_*, \Lambda_*, P_*, U_*)$ is a *Type I Ando tuple* for $(T_1^*, T_2^*)$ if the following additional conditions are satisfied:

$$P_*^\perp U_* \Lambda_* D_{T^*} + P_* U_* \Lambda_* D_{T^*} T^* = \Lambda_* D_{T^*} T_1^*, \tag{5.1}$$

$$U_*^* P_* \Lambda_* D_{T^*} + U_*^* P_*^\perp \Lambda_* D_{T^*} T^* = \Lambda_* D_{T^*} T_2^*. \tag{5.2}$$

If a Type I Ando tuple $(\mathcal{F}_*, \Lambda_*, P_*, U_*)$ for $(T_1^*, T_2^*)$ is Douglas minimal (see Definition 5.1.3), we say that it is a *minimal Type I Ando tuple* for $(T_1^*, T_2^*)$.

We postpone the question of whether Type I Ando tuples exist (i.e., whether one can find Ando tuples $(\mathcal{F}_*, \Lambda_*, P_*, U_*)$ for a given contractive commutative operator pair $(T_1, T_2)$ for which the set of equations (5.1) and (5.2) is satisfied) until the discussion concerning regular factorizations and special Ando tuples in Section 5.3 (see specifically Corollary 5.3.4 combined with Theorem 5.4.1).

Let us note that the notion of a Type I Ando tuple for a given commuting contractive pair $(T_1, T_2)$ is *coordinate-free* in the sense that if $(\mathcal{F}_*, \Lambda_*, P_*, U_*)$ is a Type I Ando tuple for $(T_1, T_2)$ and if $(\mathcal{F}_*', \Lambda_*', P_*', U_*')$ is an Ando tuple for $(T_1, T_2)$ that coincides with $(\mathcal{F}_*, \Lambda_*, P_*, U_*)$ (in the sense of Definition 5.1.2), then $(\mathcal{F}_*', \Lambda_*', P_*', U_*')$ is also a Type I Ando tuple for $(T_1, T_2)$.

## 5.2 Douglas-Type Structure of a General Ando Lift

The next result provides a functional model for minimal Ando lifts $(V_1, V_2)$ of a given commuting contractive operator pair $(T_1, T_2)$ in terms of minimal Type I Ando tuples for $(T_1^*, T_2^*)$. This result when combined with the results of Corollary 5.3.4 and Theorem 5.4.1 to come then provides an independent proof of Ando's dilation theorem.

**Theorem 5.2.1** *Let $(T_1, T_2)$ be a commuting contractive operator pair and let $(\mathcal{F}_*, \Lambda_*, P_*, U_*)$ be a minimal Type I Ando tuple for $(T_1^*, T_2^*)$. Set $T = T_1 T_2$, let the unitary operator $W_D$ on the Hilbert space $Q_{T^*}$ be as in the Douglas model for the minimal isometric lift of $T$ (see Section 2.2.2), and let $W_{\flat 1}$, $W_{\flat 2}$ be the* *canonical pair of commuting unitaries associated with* $(T_1, T_2)$ *as in Theorem 4.4.2. Define operators and spaces by*

$$\mathcal{K}_D = \begin{bmatrix} H^2(\mathcal{F}_*) \\ Q_{T^*} \end{bmatrix},$$

$$(\mathbf{V}_{D,1}, \mathbf{V}_{D,2}) = \left( \begin{bmatrix} M_{U_*^*(P_*^\perp + zP_*)} & 0 \\ 0 & W_{\flat 1} \end{bmatrix}, \begin{bmatrix} M_{(P_* + zP_*^\perp)U_*} & 0 \\ 0 & W_{\flat 2} \end{bmatrix} \right) \quad \textit{acting on } \mathcal{K}_D,$$

$$\mathbf{\Pi}_D = \begin{bmatrix} (I_{H^2} \otimes \Lambda_*) \, \mathcal{O}_{D_{T^*}, T^*} \\ Q_{T^*} \end{bmatrix} : \mathcal{H} \to \mathcal{K}_D. \tag{5.3}$$

*Then* $(\mathbf{\Pi}_D, \mathbf{V}_{D,1}, \mathbf{V}_{D,2})$ *is a minimal Ando lift for* $(T_1, T_2)$.

*Conversely, given any minimal Ando lift* $(\mathbf{\Pi}, \mathbf{V}_1, \mathbf{V}_2)$ *for* $(T_1, T_2)$, *there is a minimal Type I Ando tuple* $(\mathcal{F}_*, \Lambda_*, P_*, U_*)$ *for* $(T_1^*, T_2^*)$ *so that the lift* $(\mathbf{\Pi}, \mathbf{V}_1, \mathbf{V}_2)$ *is unitarily equivalent to the Douglas model Ando lift (5.3) associated with* $(\mathcal{F}_*, \Lambda_*, P_*, U_*)$ *as in (5.3).*

*Proof* Suppose that $(\mathcal{F}_*, \Lambda_*, P_*, U_*)$ is a minimal Type I Ando tuple for the commuting pair of operators $(T_1^*, T_2^*)$ acting on $\mathcal{H}$ and let $(\mathbf{\Pi}_D, \mathbf{V}_{D,1}, \mathbf{V}_{D,2})$ be as in (5.3). Note that $(\mathbf{V}_{D,1}, \mathbf{V}_{D,2})$ on $\mathcal{K}_D$ is a pair of commuting isometries since this pair has the form of the BCL-model for commuting isometric operator pairs. Note that $\mathbf{\Pi}_D$ can be factored

$$\mathbf{\Pi}_D = \begin{bmatrix} I_{H^2} \otimes \Lambda_* & 0 \\ 0 & I_{Q_{T^*}} \end{bmatrix} \begin{bmatrix} \mathcal{O}_{D_{T^*}, T^*} \\ Q_{T^*} \end{bmatrix}$$

as the product of isometries, and hence is itself an isometric embedding $\mathbf{\Pi}_D : \mathcal{H} \to \mathcal{K}_D$. To show that $(\mathbf{\Pi}, \mathbf{V}_{D,1}, \mathbf{V}_{D,2})$ is an Ando lift of $(T_1, T_2)$ it remains only to verify the intertwining relations $\mathbf{V}_{D,j}^* \Pi = \mathbf{\Pi} T_j^*$ for $j = 1, 2$, that is,

$$\begin{bmatrix} (M_{U_*^* P_*^\perp + z U_*^* P_*})^* & 0 \\ 0 & W_{\flat 1}^* \end{bmatrix} \begin{bmatrix} (I_{H^2} \otimes \Lambda_*) \, \mathcal{O}_{D_{T^*}, T^*} \\ Q_{T^*} \end{bmatrix} = \begin{bmatrix} (I_{H^2} \otimes \Lambda_*) \, \mathcal{O}_{D_{T^*}, T^*} \\ Q_{T^*} \end{bmatrix} T_1^*,$$

$$\begin{bmatrix} (M_{P_* U_* + z P_*^\perp U_*})^* & 0 \\ 0 & W_{\flat 2}^* \end{bmatrix} \begin{bmatrix} (I_{H^2} \otimes \Lambda_*) \, \mathcal{O}_{D_{T^*}, T^*} \\ Q_{T^*} \end{bmatrix} = \begin{bmatrix} (I_{H^2} \otimes \Lambda_*) \, \mathcal{O}_{D_{T^*}, T^*} \\ Q_{T^*} \end{bmatrix} T_2^*,$$

or equivalently, the system of equations

$$\left(I_{H^2} \otimes P_*^\perp U_* + M_z^* \otimes P_* U_*\right) (I_{H^2} \otimes \Lambda_*) \, \mathcal{O}_{D_{T^*}, T^*} = (I_{H^2} \otimes \Lambda_*) \, \mathcal{O}_{D_{T^*}, T^*} T_1^*, \tag{5.4}$$

$$W_{\flat 1}^* Q_{T^*} = Q_{T^*} T_1^*, \tag{5.5}$$

$$\left(I_{H^2} \otimes U_*^* P_* + M_z^* \otimes U_*^* P_*^\perp\right) (I_{H^2} \otimes \Lambda_*) \, \mathcal{O}_{D_{T^*}, T^*} = (I_{H^2} \otimes \Lambda_*) \, \mathcal{O}_{D_{T^*}, T^*} T_2^*, \tag{5.6}$$

$$W_{\flat 2}^* Q_{T^*} = Q_{T^*} T_2^*. \tag{5.7}$$

Note that equations (5.5) and (5.7) follow immediately from Theorem 4.4.2 (see the second identity in (4.13)). Applying (5.4) to a generic vector $h \in \mathcal{H}$ gives

$$\sum_{n=0}^{\infty} \left( P_*^{\perp} U_* \Lambda_* D_{T^*} T^{*n} h + P_* U_* \Lambda_* D_{T^*} T^{*(n+1)} h \right) z^n = \sum_{n=0}^{\infty} \left( \Lambda_* D_{T^*} T^{*n} T_1^* h \right) z^n .$$

This identity between power series is the same as the matching of power series coefficients of the two sides holding for all $h \in \mathcal{H}$:

$$P_*^{\perp} U_* \Lambda_* D_{T^*} T^{*n} + P_* U_* \Lambda_* D_{T^*} T^{*(n+1)} = \Lambda_* D_{T^*} T^{*n} T_1^* \quad \text{for all } n = 0, 1, 2, \ldots . \tag{5.8}$$

Since $T_1^*$ and $T^*$ commute, we see that the identity for the case of a general $n$ follows from the special case of $n = 0$ by multiplying on the right by $T^{*n}$; hence the system of equations (5.8) is equivalent to the single equation

$$P_*^{\perp} U_* \Lambda_* D_{T^*} + P_* U_* \Lambda_* D_{T^*} T^* = \Lambda_* D_{T^*} T_1^*,$$

which amounts to the operator equation (5.1). In a similar way, condition (5.6) reduces to (5.2).

With notation as in (5.3) let us set

$$\mathcal{K}_{D,0} = \bigvee_{n_1, n_2 \geq 0} \mathbf{V}_{D,1}^{n_1} \mathbf{V}_{D,2}^{n_2} \operatorname{Ran} \mathbf{\Pi}_D .$$

To show that $(\mathbf{\Pi}_D, \mathbf{V}_{D,1}, \mathbf{V}_{D,2})$ is a minimal Ando lift for $(T_1, T_2)$, it remains only to show that $\mathcal{K}_D = \mathcal{K}_{D,0}$.

As an intermediate step, we shall first identify the smaller space

$$\mathcal{K}_{D,00} := \bigvee_{n \geq 0} \mathbf{V}_{D,1}^{n} \mathbf{V}_{D,2}^{n} \operatorname{Ran} \mathbf{\Pi}_D = \bigvee_{n \geq 0} \mathbf{V}_D^n \operatorname{Ran} \mathbf{\Pi}_D \subset \mathcal{K}_{D,0}, \tag{5.9}$$

where $\mathbf{V}_D = \mathbf{V}_{D,1} \mathbf{V}_{D,2} = \begin{bmatrix} M_z^{\mathcal{F}_*} & 0 \\ 0 & W_D \end{bmatrix}$ on $\begin{bmatrix} H^2(\mathcal{F}_*) \\ Q_{T^*} \end{bmatrix}$. Once $\mathcal{K}_{D,00}$ is identified, we can find $\mathcal{K}_{D,0}$ via the formula

$$\mathcal{K}_{D,0} = \bigvee_{n_1, n_2 = 0,1,2,\ldots} \mathbf{V}_{D,1}^{n_1} \mathbf{V}_{D,2}^{n_2} \mathcal{K}_{D,00}. \tag{5.10}$$

Note that $\mathbf{\Pi}_D$ given by (5.3) factors as

$$\mathbf{\Pi}_D = \begin{bmatrix} I_{H^2} \otimes \Lambda_* & 0 \\ 0 & I_{Q_{T^*}} \end{bmatrix} \Pi_D,$$

where

$$\Pi_D = \begin{bmatrix} \mathcal{O}_{D_{T^*}, T^*} \\ Q_{T^*} \end{bmatrix} : \mathcal{H} \to \mathcal{K}_D = \begin{bmatrix} H^2(\mathcal{D}_{T^*}) \\ Q_{T^*} \end{bmatrix}$$

is the embedding operator for the Douglas model isometric lift of $T$ as in (2.40). Let us also note the intertwining

$$\mathbf{V}_D^n \begin{bmatrix} I_{H^2} \otimes \Lambda_* & 0 \\ 0 & I_{Q_{T^*}} \end{bmatrix} = \begin{bmatrix} I_{H^2} \otimes \Lambda_* & 0 \\ 0 & I_{\mathcal{D}_{T^*}} \end{bmatrix} V_D^n,$$

where $V_D = \begin{bmatrix} M_z^{\mathcal{D}_{T^*}} & 0 \\ 0 & W_D \end{bmatrix}$ is the isometric lift of $T$ in the Douglas model as in (2.39). Hence we see that, for $h \in \mathcal{H}$,

$$\mathbf{V}_D^n \mathbf{\Pi}_D h = \mathbf{V}_D^n \begin{bmatrix} I_{H^2} \otimes \Lambda_* & 0 \\ 0 & I_{Q_{T^*}} \end{bmatrix} \Pi_D h = \begin{bmatrix} I_{H^2} \otimes \Lambda_* & 0 \\ 0 & I_{Q_{T^*}} \end{bmatrix} V_D^n \Pi_D h,$$

where

$$\bigvee_{n=0,1,2,\ldots,\, h \in \mathcal{H}} V_D^n \Pi_D h = \mathcal{K}_D := \begin{bmatrix} H^2(\mathcal{D}_{T^*}) \\ Q_{T^*} \end{bmatrix}$$

since the Douglas model lift of $T$ is minimal, as seen in Proposition 2.2.3. We have now identified the space $\mathcal{K}_{D,00}$ as being exactly equal to

$$\mathcal{K}_{D,00} = \begin{bmatrix} I_{H^2} \otimes \Lambda_* & 0 \\ 0 & I_{Q_{T^*}} \end{bmatrix} \begin{bmatrix} H^2(\mathcal{D}_{T^*}) \\ Q_{T^*} \end{bmatrix} \tag{5.11}$$

$$= \begin{bmatrix} H^2(\operatorname{Ran} \Lambda_*) \\ Q_{T^*} \end{bmatrix}. \tag{5.12}$$

We now use the formula (5.10) to see that

$$\mathcal{K}_{D,0} = \bigvee_{n_1,\, n_2 \geq 0} \mathbf{V}_{D,1}^{n_1} \mathbf{V}_{D,2}^{n_2} \begin{bmatrix} H^2(\operatorname{Ran} \Lambda_*) \\ Q_{T^*} \end{bmatrix} = \begin{bmatrix} \bigvee_{n_1,\, n_2 \geq 0} M_1^{n_1} M_2^{n_2} H^2(\operatorname{Ran} \Lambda_*) \\ Q_{T^*} \end{bmatrix},$$

where we use the short-hand notation

$$M_1 := M_{U_*^*(P_*^\perp + zP_*)}, \quad M_2 := M_{(P_* + zP_*^\perp)U_*}. \tag{5.13}$$

This space is equal to the whole space $\mathcal{K}_D = \begin{bmatrix} H^2(\mathcal{F}_*) \\ Q_{T^*} \end{bmatrix}$ exactly when

$$\bigvee_{n_1,\, n_2 \geq 0} M_1^{n_1} M_2^{n_2} H^2(\operatorname{Ran} \Lambda_*) = H^2(\mathcal{F}_*),$$

that is, when the Type I Ando tuple $(\mathcal{F}_*, P_*, U_*, \Lambda_*)$ is minimal.

Conversely, suppose that $(\mathbf{\Pi}, \mathbf{V}_1, \mathbf{V}_2)$ is any Ando lift for the commuting contractive operator pair $(T_1, T_2)$ on $\mathcal{H}$. Thus $(\mathbf{V}_1, \mathbf{V}_2)$ is a commuting isometric operator pair on some Hilbert space $\mathcal{K}$ and $\mathbf{\Pi}$ is an isometric embedding of $\mathcal{H}$ into $\mathcal{K}$ so that

$$\mathbf{V}_j^* \mathbf{\Pi} = \mathbf{\Pi} T_j^* \quad \text{for } j = 1, 2.$$

By Theorem 3.1.4 we know that the commuting pair of isometries $(\mathbf{V}_1, \mathbf{V}_2)$ is unitarily equivalent to its BCL1-model as in (3.9) acting on the space $\mathcal{K} := \left[\begin{smallmatrix} H^2(\mathcal{D}_{\mathbf{V}^*}) \\ \mathcal{Q}_{\mathbf{V}^*} \end{smallmatrix}\right]$ via the unitary identification map $\tau_{\mathrm{BCL}} \colon \mathbf{\mathcal{K}} \to \mathcal{K}$ given by

$$\tau_{\mathrm{BCL}} = \begin{bmatrix} \mathcal{O}_{D_{\mathbf{V}^*}, \mathbf{V}^*} \\ Q_{\mathbf{V}^*} \end{bmatrix} : \mathbf{k} \mapsto \begin{bmatrix} \mathcal{O}_{D_{\mathbf{V}^*}, \mathbf{V}^*} \\ Q_{\mathbf{V}^*} \end{bmatrix} \mathbf{k}.$$

Here $\mathbf{V} := \mathbf{V}_1 \mathbf{V}_2$ satisfies the intertwining relation $\mathcal{O}_{D_{\mathbf{V}^*}, \mathbf{V}^*} \mathbf{V} = M_z^{\mathcal{D}_{\mathbf{V}^*}} \mathcal{O}_{D_{\mathbf{V}^*}, \mathbf{V}^*}$. Moreover, we have the intertwining relations involving the operators $(\mathbf{V}_1, \mathbf{V}_2)$ in the factorization of $\mathbf{V}$:

$$\mathcal{O}_{D_{\mathbf{V}^*}, \mathbf{V}^*} \mathbf{V}_1 = M_1 \mathcal{O}_{D_{\mathbf{V}^*}, \mathbf{V}^*}, \quad \mathcal{O}_{D_{\mathbf{V}^*}, \mathbf{V}^*} \mathbf{V}_2 = M_2 \mathcal{O}_{D_{\mathbf{V}^*}, \mathbf{V}^*},$$

where $M_1, M_2$ are as in (5.13). Let us also introduce operators $W_1$, $W_2$, $W$ on $\mathcal{Q}_{\mathbf{V}^*} = \mathcal{K}_u$ according to

$$W_1 = \mathbf{V}_1|_{\mathcal{Q}_{\mathbf{V}^*}}, \quad W_2 = \mathbf{V}_2|_{\mathcal{Q}_{\mathbf{V}^*}}, \quad W = \mathbf{V}|_{\mathcal{Q}_{\mathbf{V}^*}},$$

and define an isometric embedding map $\Pi \colon \mathcal{H} \to \mathcal{K} = \left[\begin{smallmatrix} H^2(\mathcal{D}_{\mathbf{V}^*}) \\ \mathcal{Q}_{\mathbf{V}^*} \end{smallmatrix}\right]$ according to

$$\Pi = \tau_{\mathrm{BCL}} \mathbf{\Pi} = \begin{bmatrix} \mathcal{O}_{D_{\mathbf{V}^*}, \mathbf{V}^*} \\ Q_{\mathbf{V}^*} \end{bmatrix} \mathbf{\Pi}.$$

Then it is easily checked that

$$\left( \Pi, \begin{bmatrix} M_1 & 0 \\ 0 & W_1 \end{bmatrix}, \begin{bmatrix} M_2 & 0 \\ 0 & W_2 \end{bmatrix} \right) \tag{5.14}$$

is also an Ando lift of $(T_1, T_2)$ that is unitarily equivalent (as an Ando lift) to our original more abstract Ando lift $(\mathbf{\Pi}, \mathbf{V}_1, \mathbf{V}_2)$.

Apply Theorem 2.3.5 to the isometric lift $(\Pi, \mathbf{V} = \mathbf{V}_1 \mathbf{V}_2)$ of $T = T_1 T_2$ to obtain an isometry $\Lambda_* \colon \mathcal{D}_{T^*} \to \mathcal{D}_{\mathbf{V}^*}$ so that

$$\Lambda_* \colon D_{T^*} h \mapsto D_{\mathbf{V}^*} \mathbf{\Pi} h, \tag{5.15}$$

and another isometry $\omega \colon \mathcal{Q}_{T^*} \to \mathcal{Q}_{\mathbf{V}^*}$ so that $\operatorname{Ran} \omega$ is reducing for $W = \mathbf{V}|_{\mathcal{Q}_{\mathbf{V}^*}}$, and

$$\Pi = \begin{bmatrix} \mathcal{O}_{D_{\mathbf{V}^*}, \mathbf{V}^*} \\ Q_{\mathbf{V}^*} \end{bmatrix} \mathbf{\Pi} = \begin{bmatrix} (I_{H^2} \otimes \Lambda_*) \mathcal{O}_{D_{T^*}, T^*} \\ \omega Q_{T^*} \end{bmatrix}. \tag{5.16}$$

Moreover, as noted in (2.72), the ambient space

$$\mathcal{K}_{00} := \bigvee_{n \geq 0} \begin{bmatrix} M_z^{\mathcal{D}_{\mathbf{V}^*}} & 0 \\ 0 & W \end{bmatrix}^n \Pi \mathcal{H}$$

for the minimal isometric lift $\left[\begin{smallmatrix} M_z^{\mathcal{D}_{\mathbf{V}^*}} & 0 \\ 0 & W \end{smallmatrix}\right]\Big|_{\mathcal{K}_{00}}$ of $T = T_1T_2$ is given by

$$\mathcal{K}_{00} = \widehat{\Gamma}\begin{bmatrix} H^2(\mathcal{D}_{T^*}) \\ \mathcal{Q}_{T^*} \end{bmatrix} = \begin{bmatrix} (I_{H^2} \otimes \Lambda_*)\, H^2(\mathcal{D}_{T^*}) \\ \omega \mathcal{Q}_{T^*} \end{bmatrix}, \tag{5.17}$$

that is, we see that the subspace $\mathcal{K}_{00}$ splits with respect to the orthogonal block-decomposition appearing in the ambient space $\mathcal{K} = \left[\begin{smallmatrix} H^2(\mathcal{D}_{\mathbf{V}^*}) \\ \mathcal{Q}_{\mathbf{V}^*} \end{smallmatrix}\right]$. The lifting property for the collection (5.14) can be rewritten as

$$\begin{bmatrix} M_j^* & 0 \\ 0 & W_j^* \end{bmatrix}\begin{bmatrix} (I_{H^2} \otimes \Lambda_*)\, \mathcal{O}_{D_{T^*},T^*} \\ \omega \mathcal{Q}_{T^*} \end{bmatrix} = \begin{bmatrix} (I_{H^2} \otimes \Lambda_*)\, \mathcal{O}_{D_{T^*},T^*} \\ \omega \mathcal{Q}_{T^*} \end{bmatrix} T_j^* \quad \text{for } j = 1, 2,$$

$$\begin{bmatrix} \left(M_z^{\mathcal{D}_{\mathbf{V}^*}}\right)^* & 0 \\ 0 & W^* \end{bmatrix}\begin{bmatrix} (I_{H^2} \otimes \Lambda_*)\, \mathcal{O}_{D_{T^*},T^*} \\ \omega \mathcal{Q}_{T^*} \end{bmatrix} = \begin{bmatrix} (I_{H^2} \otimes \Lambda_*)\, \mathcal{O}_{D_{T^*},T^*} \\ \omega \mathcal{Q}_{T^*} \end{bmatrix} T^*. \tag{5.18}$$

Finally, let us note that the fact that we have now identified $\Lambda_* \colon \mathcal{D}_{T^*} \to \mathcal{D}_{\mathbf{V}^*}$ as an isometry means that the tuple $(\mathcal{D}_{\mathbf{V}^*}, P_*, U_*, \Lambda_*)$ is an Ando tuple for $\left(T_1^*, T_2^*\right)$. As in the forward direction (the analysis following equations (5.4)–(5.7)), a comparison of the coefficients of

$$M_j^*(I_{H^2} \otimes \Lambda_*)\mathcal{O}_{D_{T^*},T^*} = (I_{H^2} \otimes \Lambda_*)\mathcal{O}_{D_{T^*},T^*}T_j^* \quad \text{for } j = 1, 2$$

yields that $(\mathcal{D}_{\mathbf{V}^*}, \Lambda_*, P_*, U_*)$ is actually a Type I Ando tuple for $\left(T_1^*, T_2^*\right)$.

Next, we would like to analyze the copy of the minimal Ando isometric lift for the commuting, contractive pair $(T_1, T_2)$ obtained by restricting $\left[\begin{smallmatrix} M_1 & 0 \\ 0 & W_1 \end{smallmatrix}\right], \left[\begin{smallmatrix} M_2 & 0 \\ 0 & W_2 \end{smallmatrix}\right]$ ($(M_1, M_2)$ as in (5.13)) to the subspace $\mathcal{K}_0$ given by

$$\begin{aligned} \mathcal{K}_0 &:= \bigvee_{n_1,n_2 \geq 0} \left\{ \begin{bmatrix} M_1 & 0 \\ 0 & W_1 \end{bmatrix}^{n_1} \begin{bmatrix} M_2 & 0 \\ 0 & W_2 \end{bmatrix}^{n_2} \begin{bmatrix} \mathcal{O}_{D_{\mathbf{V}^*},\mathbf{V}^*} \\ \mathcal{Q}_{\mathbf{V}^*} \end{bmatrix} \operatorname{Ran} \mathbf{\Pi} \right\} \\ &= \bigvee_{n_1,n_2 \geq 0} \left\{ \begin{bmatrix} M_1 & 0 \\ 0 & W_1 \end{bmatrix}^{n_1} \begin{bmatrix} M_2 & 0 \\ 0 & W_2 \end{bmatrix}^{n_2} \begin{bmatrix} (I_{H^2} \otimes \Lambda_*)\, \mathcal{O}_{D_{T^*},T^*} \\ \omega \mathcal{Q}_{T^*} \end{bmatrix} \mathcal{H} \right\}. \end{aligned}$$

As $\mathcal{K}_0$ a priori is given by the same formula but with the nonnegative integers $n_1, n_2$ constrained to satisfy $n_1 = n_2$, it is clear that we have the nesting of subspaces

$$\mathcal{K}_{00} \subset \mathcal{K}_0 \subset \mathcal{K}.$$

Owing to the splitting appearing in the formula (5.17) for $\mathcal{K}_{00}$, we see that $\mathcal{K}_0$ can be rewritten as

$$\mathcal{K}_0 = \bigvee_{n_1,n_2\geq 0} \left\{ \begin{bmatrix} M_1 & 0 \\ 0 & W_1 \end{bmatrix}^{n_1} \begin{bmatrix} M_2 & 0 \\ 0 & W_2 \end{bmatrix}^{n_2} \begin{bmatrix} (I_{H^2} \otimes \Lambda_*)\, H^2(\mathcal{D}_{T^*}) \\ \omega \mathcal{Q}_{T^*} \end{bmatrix} \right\}$$
$$= \begin{bmatrix} \bigvee_{n_1,n_2\geq 0} M_1^{n_1} M_2^{n_2} \Gamma\, H^2(\mathcal{D}_{T^*}) \\ \bigvee_{n_1,n_2\geq 0} W_1^{n_1} W_2^{n_2} \omega\, \mathcal{Q}_{T^*} \end{bmatrix}. \tag{5.19}$$

We first analyze the bottom component

$$\bigvee_{n_1,n_2\geq 0} W_1^{n_1} W_2^{n_2} \omega\, \mathcal{Q}_{T^*}.$$

Let us recall that a conclusion of Theorem 2.3.5 is that $\operatorname{Ran}\omega$ is reducing for $W = W_1 W_2$. We argue that it is reducing for $W_j$ ($j = 1, 2$) as well and that the unitary operator $\omega\colon \mathcal{Q}_{T^*} \to \operatorname{Ran}\omega$ implements a unitary equivalence between the operator $W_{\flat j}$ on $\mathcal{Q}_{T^*}$ (given by Theorem 4.4.2) and $W_j|_{\operatorname{Ran}\omega}$ on $\operatorname{Ran}\omega$. Indeed, from (5.18) together with the intertwining given in Theorem 4.4.2, we see that, for $j = 1, 2$,

$$W_j^* \omega \mathcal{Q}_{T^*} = \omega \mathcal{Q}_{T^*} T_j^* = \omega W_{\flat j}^* \mathcal{Q}_{T^*}$$

and, more generally, for $k = 0, 1, 2, \ldots$,

$$W_j^* \omega \left( W_D^k \mathcal{Q}_{T^*} \right) = W^k W_j^* \omega \mathcal{Q}_{T^*} = W^k \omega W_{\flat j}^* \mathcal{Q}_{T^*} = \omega W_D^k W_{\flat j}^* \mathcal{Q}_{T^*}$$
$$= \omega W_{\flat j}^* \left( W_D^k \mathcal{Q}_{T^*} \right). \tag{5.20}$$

As $\cup_{k=0}^{\infty} \operatorname{Ran} W_D^k \mathcal{Q}_{T^*}$ is dense in $\mathcal{Q}_{T^*}$, these computations show not only that $\operatorname{Ran}\omega$ is invariant under $W_j^*$ but also the intertwining: $W_j^* \omega = \omega W_{\flat j}^*$ for $j = 1, 2$. Also, by taking the adjoint of these intertwining relations and using the fact that the operators $W_{\flat j}$ and $W_j$ (for $j = 1, 2$) all are unitary operators, we see that $\omega W_{\flat j} = W_j \omega$ for $j = 1, 2$. This implies that $\operatorname{Ran}\omega$ is invariant under $W_1$ and $W_2$ as well.

Let us now recall the expression (5.19) for $\mathcal{K}_0$. From the preceding analysis, we see that

$$\mathcal{K}_0 = \begin{bmatrix} \bigvee_{n_1,n_2\geq 0} M_1^{n_1} M_2^{n_2} H^2(\mathcal{D}_{T^*}) \\ \operatorname{Ran}\omega \end{bmatrix}.$$

Note that minimality of the Ando lift (5.14) just means that

$$\mathcal{K}_0 = \mathcal{K} = \begin{bmatrix} H^2(\mathcal{D}_{\mathbf{V}^*}) \\ \mathcal{D}_{\mathbf{V}^*} \end{bmatrix},$$

which is equivalent to the following two conditions holding simultaneously:

$$H^2(\mathcal{D}_{\mathbf{V}^*}) = \bigvee_{n_1,n_2\geq 0} M_1^{n_1} M_2^{n_2} (I_{H^2} \otimes \Lambda_*)\, H^2(\mathcal{D}_{T^*}), \tag{5.21}$$

$$\mathcal{Q}_{\mathbf{V}^*} = \omega \mathcal{Q}_{T^*}. \tag{5.22}$$

By definition, condition (5.21) holds exactly when the Ando tuple

$$(\mathcal{D}_{\mathbf{V}^*}, \Lambda_*, P_*, U_*)$$

is *minimal*, while the second condition (5.22) holds exactly when $\omega$ is *unitary* (i.e., a surjective isometry). When this is the case, the preceding analysis shows that we can use $\omega$ to identify $\mathcal{Q}_{T^*}$ with $\mathcal{Q}_{\mathbf{V}^*}$, and that under this unitary identification the operators $W$, $W_1$, $W_2$ on $\mathcal{Q}_{\mathbf{V}^*}$ become the operators $W_D$, $W_{\flat,1}$, $W_{\flat,2}$ on $\mathcal{Q}_{T^*}$ and the collection (5.14) has the form of the functional-model Ando lift (5.3) based on the Type I Ando tuple $(\mathcal{D}_{\mathbf{V}^*}, \Lambda_*, P_*, U_*)$ for $(T_1^*, T_2^*)$.

Finally, let us recall that the Ando lift (5.14) is constructed so as to be unitarily equivalent to the original abstract Ando lift $(\mathbf{\Pi}, \mathbf{V}_1, \mathbf{V}_2)$. Hence minimality of $(\mathbf{\Pi}, \mathbf{V}_1, \mathbf{V}_2)$ is equivalent to minimality of the collection (5.14) as an Ando lift of $(T_1, T_2)$. Thus, if we start with a minimal abstract Ando lift, we have shown that it is unitarily equivalent to a functional model Ando lift (5.3) constructed from a minimal Type I Ando tuple for $(T_1^*, T_2^*)$, in this case $(\mathcal{D}_{\mathbf{V}^*}, \Lambda_*, P_*, U_*)$. □

## 5.3 Regular Factorizations and Special Ando Tuples

In this section, starting with a commuting contractive operator pair $(T_1, T_2)$, we construct a class of Ando tuples for $(T_1, T_2)$, called *special Ando tuples*, for $(T_1, T_2)$, which we show by direct calculation are of Type I (i.e., the operator equations (5.1)–(5.2) are satisfied). This, combined with Theorem 5.2.1, provides yet another proof of Ando's theorem with the bonus that it also provides a functional model for the lift.

The first step is the constructive definition of *special Ando tuples*. We think it instructive to push the formalism as far as possible without a commutativity assumption, and then see the further consequences arising once the commutativity assumption is invoked.

Suppose that $T\colon \mathcal{H}_2 \to \mathcal{H}_0$ is a (possibly nonsquare) Hilbert space contraction operator having a factorization

$$T = T'T'',$$

where $T'\colon \mathcal{H}_1 \to \mathcal{H}_0$ and $T''\colon \mathcal{H}_2 \to \mathcal{H}_1$ are also contraction operators. Then the identity

$$D_T^2 = I - T^*T = I - T''^*T'^*T'T'' = T''^*(I - T'^*T')T'' + I - T''^*T''$$
$$= T''^* D_{T'}^2 T'' + D_{T''}^2 \tag{5.23}$$

implies that the map $Z\colon \mathcal{D}_T \to \mathcal{D}_{T'} \oplus \mathcal{D}_{T''}$ defined densely by

$$Z\colon D_T h \mapsto D_{T'}T''h \oplus D_{T''}h \quad \text{for } h \in \mathcal{H} \tag{5.24}$$

is an isometry. Such isometries come up in the characterization of invariant subspaces for a contraction operator in terms of its Sz.-Nagy–Foias functional model (see [92, Chapter VII] and Section 9.5 later). Following Sz.-Nagy–Foias, we make the following definition.

**Definition 5.3.1** Given a contraction operator $T\colon \mathcal{H}_0 \to \mathcal{H}$ with factorization $T = T' \cdot T''$ for contraction operators $T'\colon \mathcal{H}_1 \to \mathcal{H}$ and $T''\colon \mathcal{H}_0 \to \mathcal{H}_2$, we say that the factorization $T = T' \cdot T''$ is *regular* if it is the case that the operator $Z\colon \mathcal{D}_T \to \mathcal{D}_{T'} \oplus \mathcal{D}_{T''}$ in (5.24) is surjective (so $Z\colon \mathcal{D}_T \to \mathcal{D}_{T'} \oplus \mathcal{D}_{T''}$ is unitary).

Now let us suppose that $(T_1, T_2)$ is a commuting contractive pair on $\mathcal{H}$ and that we set $T = T_1 \cdot T_2 = T_2 \cdot T_1$. Then we have two versions of (5.23) corresponding to setting $T' = T_1, T'' = T_2$ or the reverse $T' = T_2, T'' = T_1$:

$$D_T^2 = T_2^* D_{T_1}^2 T_2 + D_{T_2}^2, \quad D_T^2 = D_{T_1}^2 + T_1^* D_{T_2}^2 T_1 \tag{5.25}$$

from which we conclude, in particular, that

$$T_2^* D_{T_1}^2 T_2 + D_{T_1}^2 = D_{T_1}^2 + T_1^* D_{T_2}^2 T_1. \tag{5.26}$$

As a consequence of the first identity in (5.25), we see that the operator $\Lambda_\dagger\colon \mathcal{D}_T \to \mathcal{D}_{T_1} \oplus \mathcal{D}_{T_2}$ defined densely by

$$\Lambda_\dagger\colon D_T h \mapsto D_{T_1}T_2 h \oplus D_{T_2} h \quad \text{for all } h \in \mathcal{H} \tag{5.27}$$

is an isometry from $\mathcal{D}_T$ into $\mathcal{D}_{T_1} \oplus \mathcal{D}_{T_2}$. Let us define an operator $U_0$ with respective domain $\mathrm{Dom}(U_0)$ and range $\mathrm{Ran}(U_0)$ given by

$$\begin{aligned} \mathrm{Dom}(U_0) &:= \mathrm{clos.}\{D_{T_1}T_2 h \oplus D_{T_2} h\colon h \in \mathcal{H}\} \subset \mathcal{D}_{T_1} \oplus \mathcal{D}_{T_2}, \\ \mathrm{Ran}(U_0) &:= \mathrm{clos.}\{D_{T_1} h \oplus D_{T_2}T_1 h\colon h \in \mathcal{H}\} \subset \mathcal{D}_{T_1} \oplus \mathcal{D}_{T_2}. \end{aligned} \tag{5.28}$$

As a consequence of (5.26), we see that the operator $U_0\colon \mathrm{Dom}(U_0) \to \mathrm{Ran}(U_0)$ defined densely by

$$U_0\colon D_{T_1}T_2 h \oplus D_{T_2} h \mapsto D_{T_1} h \oplus D_{T_2}T_1 h \quad \text{for all } h \in \mathcal{H} \tag{5.29}$$

is unitary. If the orthogonal complements $\mathrm{Dom}(U_0)^\perp$ and $\mathrm{Ran}(U_0)^\perp$ of the spaces $\mathrm{Dom}(U_0)$ and $\mathrm{Ran}(U_0)$ in $\mathcal{D}_{T_1} \oplus \mathcal{D}_{T_2}$, respectively, have the same dimension, we can find a unitary identification map $U_{00}\colon \operatorname{Dom}(U_0)^\perp \to \operatorname{Ran}(U_0)^\perp$ and then define a unitary operator

$$U_\dagger\colon \mathcal{D}_{T_1} \oplus \mathcal{D}_{T_2} \to \mathcal{D}_{T_1} \oplus \mathcal{D}_{T_2}$$

by setting

$$U_\dagger|_{\operatorname{Dom}(U_0)} = U_0, \quad U_\dagger|_{\operatorname{Dom}(U_0)^\perp} = U_{00} \tag{5.30}$$

and then extending by linearity. If it is the case that the orthogonal complements $\operatorname{Dom}(U_0)^\perp$ and $\operatorname{Ran}(U_0)^\perp$ in $\mathcal{D}_{T_1} \oplus \mathcal{D}_{T_2}$ have different dimensions, we may change the ambient space $\mathcal{D}_{T_1} \oplus \mathcal{D}_{T_2}$ to $\mathcal{D}_{T_1} \oplus \mathcal{D}_{T_2} \oplus \mathcal{H}$, where we recall that $\mathcal{H}$ is the Hilbert space where $T_1, T_2$ act and therefore the defect spaces $\mathcal{D}_{T_j}$ are subspaces of $\mathcal{H}$. With this change of the ambient space, each of the orthogonal complements $\operatorname{Dom}(U_0)^\perp$ and $\operatorname{Ran}(U_0)^\perp$ will contain the subspace $\{0\} \oplus \{0\} \oplus \mathcal{H}$, which has the same dimension as $\mathcal{H}$. Since the ambient space also has the same dimension as that of $\mathcal{H}$, it follows that $\operatorname{Dom}(U_0)^\perp$ and $\operatorname{Ran}(U_0)^\perp$ will have the same dimension. So we have a unitary identification map $U_{00}\colon \operatorname{Dom}(U_0)^\perp \to \operatorname{Ran}(U_0)^\perp$, and thereby obtain a unitary operator $U_\dagger$ on the larger space $\mathcal{F}_\dagger := \mathcal{D}_{T_1} \oplus \mathcal{D}_{T_2} \oplus \mathcal{H}$ by setting

$$U_\dagger|_{\operatorname{Dom}(U_0)} = U_0, \quad U_\dagger|_{\operatorname{Dom}(U_0)^\perp} = U_{00}$$

and extending by linearity.

In addition we let $P_\dagger$ be any projection operator on $\mathcal{F}_\dagger$ that is an extension of the projection operator

$$P_0\colon d \oplus r \mapsto d \oplus 0$$

defined on $\mathcal{D}_{T_1} \oplus \mathcal{D}_{T_2}$, that is, $P_\dagger$ is any projection operator on $\mathcal{F}_\dagger$ of the form

$$P_\dagger\colon d \oplus r \oplus f_0 \mapsto d \oplus 0 \oplus 0 \quad \text{for } d \oplus r \oplus f_0 \in \mathcal{D}_{T_1} \oplus \mathcal{D}_{T_2} \oplus \mathcal{H} = \mathcal{F}_\dagger. \tag{5.31}$$

For future reference we now introduce the formal definition of a *special Ando tuple*.

**Definition 5.3.2** Given a commuting contractive operator pair $(T_1, T_2)$ on a Hilbert space $\mathcal{H}$, any collection of spaces and operators $(\mathcal{F}_\dagger, \Lambda_\dagger, P_\dagger, U_\dagger)$ constructed as in (5.27), (5.30), and (5.31) will be called a *canonical special Ando tuple* for the pair $(T_1, T_2)$. We shall say that the Ando tuple $\left(\mathcal{F}'_\dagger, \Lambda'_\dagger, P'_\dagger, U'_\dagger\right)$ coinciding (in the sense of Definition 5.1.2) with a canonical special Ando tuple $(\mathcal{F}_\dagger, \Lambda_\dagger, P_\dagger, U_\dagger)$ for $(T_1, T_2)$ is simply a *special Ando tuple* for $(T_1, T_2)$ (not necessarily in canonical form). Canonical special Ando tuples for the adjoint pair $(T_1^*, T_2^*)$ will often be denoted by $(\mathcal{F}_{\dagger *}, \Lambda_{\dagger *}, P_{\dagger *}, U_{\dagger *})$.

We shall say that $\Xi := (\mathcal{F}_\dagger, \Lambda_\dagger, P_\dagger, U_\dagger)$ is an *irreducible special Ando tuple* if it is a special Ando tuple that is also irreducible as an Ando tuple (see Definition 5.1.3 (1)), that is, if $\Xi$ is a special Ando tuple for which it is the case that the smallest subspace of $\mathcal{F}_\dagger$ containing $\operatorname{Ran}\Lambda_\dagger$ that is invariant for $U_\dagger$, $U_\dagger^*$, and $P_\dagger$ is the whole space $\mathcal{F}_\dagger$. If $(\mathcal{F}_\dagger, \Lambda_\dagger, P_\dagger, U_\dagger)$ is a canonical special Ando tuple, irreducibility means that the smallest reducing subspace for $U_\dagger$, $U_\dagger^*$, $P_\dagger$ containing $\mathcal{D}_{T_1} \oplus \mathcal{D}_{T_2}$ is the whole space $\mathcal{F}_\dagger$.

Note that the point of the distinction between a *canonical special Ando tuple* and a *special Ando tuple* for $(T_1, T_2)$ is that the notion of a *canonical special Ando tuple* is not coordinate-free. The enlarged class of *special Ando tuples* as in Definition 5.3.2 then is the coincidence envelope of the *canonical special Ando tuples*. In other words, the class of special Ando tuples is the smallest collection of Ando tuples containing the canonical special Ando tuples that is invariant under the relation of coincidence.

**Remark 5.3.3** We note that canonical special Ando tuples are easily constructed. There are various scenarios possible, which we discuss in turn.

**Scenario 1 $T = T_1 \cdot T_2$ and $T = T_2 \cdot T_1$ are both regular factorizations** We have seen that the first identity in (5.25) implies that $\Lambda_\dagger$ given by (5.27) is an isometry. Let us now also note that the second identity in (5.25) implies that the map $\Lambda'_\dagger \colon \mathcal{D}_T \to \mathcal{D}_{T_1} \oplus \mathcal{D}_{T_2}$ given by

$$\Lambda'_\dagger \colon D_T h \mapsto D_{T_1} h \oplus D_{T_2} T_1 h \quad \text{for } h \in \mathcal{H}$$

is an isometry. Note also from the definitions that

$$\operatorname{Ran}\Lambda_\dagger = \operatorname{Dom}(U_0), \quad \operatorname{Ran}\Lambda'_\dagger = \operatorname{Ran}(U_0).$$

By definition, $T = T_1 \cdot T_2$ is a regular factorization exactly when $\operatorname{Ran}\Lambda_\dagger$ is the whole space $\mathcal{D}_{T_1} \oplus \mathcal{D}_{T_2}$, that is,

$$\operatorname{Dom}(U_0) = \mathcal{D}_{T_1} \oplus \mathcal{D}_{T_2}.$$

Similarly, $T = T_2 \cdot T_1$ is a regular factorization exactly when $\operatorname{Ran}\Lambda'_\dagger = \mathcal{D}_{T_1} \oplus \mathcal{D}_{T_2}$, that is, when

$$\operatorname{Ran}(U_0) = \mathcal{D}_{T_1} \oplus \mathcal{D}_{T_2}.$$

Thus, when it is the case that both $T = T_1 \cdot T_2$ and $T = T_2 \cdot T_1$ are regular factorizations, $U_0$ given by (5.29) actually already defines a unitary operator on $\mathcal{F}_\dagger = \mathcal{D}_{T_1} \oplus \mathcal{D}_{T_2}$ and conversely: If $\operatorname{Dom}(U_0) = \operatorname{Ran}(U_0) = \mathcal{D}_{T_1} \oplus \mathcal{D}_{T_2}$ so $U_0$ already defines a unitary operator on $\mathcal{D}_{T_1} \oplus \mathcal{D}_{T_2}$, then both $T = T_1 \cdot T_2$ and $T = T_2 \cdot T_1$ are regular factorizations. If this is the case, then any other unitary extension $\widetilde{U}$ must agree with $U_0$ on $\operatorname{Dom}(U_0) = \mathcal{D}_{T_1} \oplus \mathcal{D}_{T_2}$ and hence

$\mathcal{D}_{T_1} \oplus \mathcal{D}_{T_2}$ is already reducing for $\widetilde{U}$ forcing us to the conclusion that $\widetilde{U} = U_0$. We conclude: *When both $T_1 \cdot T_2$ and $T_2 \cdot T_1$ are regular factorizations, then the commuting contractive pair $(T_1, T_2)$ has a unique irreducible special Ando tuple in canonical form.*

**Scenario 2** $\dim \mathrm{Dom}(U_0)^\perp = \dim \mathrm{Ran}(U_0)^\perp > 0$ **in the ambient space** $\mathcal{D}_{T_1} \oplus \mathcal{D}_{T_2}$ In this case, we can extend $U_0$ to a unitary operator $U$ on all of $\mathcal{F}_\dagger := \mathcal{D}_{T_1} \oplus \mathcal{D}_{T_2}$ by choosing any unitary identification map $U_{00}\colon \mathrm{Dom}(U_0)^\perp \to \mathrm{Ran}(U_0)^\perp$, then defining

$$U|_{\mathrm{Dom}(U_0)} = U_0\colon \mathrm{Dom}(U_0) \to \mathrm{Ran}(U_0), \quad \text{and}$$
$$U|_{\mathrm{Dom}(U_0)^\perp} = U_{00}\colon \mathrm{Dom}(U_0)^\perp \to \mathrm{Ran}(U_0)^\perp$$

and then extending linearly to a unitary operator $U$ on all of

$$\mathcal{D}_{T_1} \oplus \mathcal{D}_{T_2} = \mathrm{Dom}(U_0) \oplus \mathrm{Dom}(U_0)^\perp = \mathrm{Ran}(U_0) \oplus \mathrm{Ran}(U_0)^\perp.$$

Then any such $U$ induces an irreducible Ando tuple $(\mathcal{F}_\dagger = \mathcal{D}_{T_1} \oplus \mathcal{D}_{T_2}, P, U)$ since the subspace $\mathcal{D}_{T_1} \oplus \mathcal{D}_{T_2}$ is already reducing for $U$. However, we see that there is a freedom in the choice of the operator $U_{00}\colon \mathrm{Dom}(U_0)^\perp \to \mathrm{Ran}(U_0)^\perp$. We conclude that in this case *irreducible special Ando tuples exist but do not have a unique representative in canonical form.* In fact, as we shall see in examples to come, there are canonical minimal special Ando tuples with the ambient Hilbert space $\mathcal{F}_\dagger$ properly containing $\mathcal{D}_{T_1} \oplus \mathcal{D}_{T_2}$.

**Scenario 3** $\dim \mathrm{Dom}(U_0)^\perp \neq \dim \mathrm{Ran}(U_0)^\perp$ **in the ambient space** $\mathcal{D}_{T_1} \oplus \mathcal{D}_{T_2}$ In this case, as discussed in the discussion preceding Definition 5.3.2, we may enlarge the ambient space $\mathcal{D}_{T_1} \oplus \mathcal{D}_{T_2}$ to the space $\mathcal{F}_\dagger := \mathcal{D}_{T_1} \oplus \mathcal{D}_{T_2} \oplus \mathcal{H}$ so as to get

$$\dim \mathrm{Dom}(U_0)^\perp = \dim \mathrm{Ran}(U_0)^\perp = \dim \mathcal{H}.$$

Here, the orthogonal complement is now taken in the ambient space $\mathcal{F}_\dagger = \mathcal{D}_{T_1} \oplus \mathcal{D}_{T_2} \oplus \mathcal{H}$ rather than just $\mathcal{D}_{T_1} \oplus \mathcal{D}_{T_2}$. We may then proceed as in Scenario 2 to construct a unitary extension of $U_0$. We can always arrange for the resulting canonical-form special Ando tuple $(\mathcal{F}_\dagger, \Lambda_\dagger, U_\dagger, P_\dagger)$ to be irreducible by cutting $\mathcal{F}_\dagger$ down to the smallest reducing subspace for $U_\dagger$ containing $\mathcal{D}_{T_1} \oplus \mathcal{D}_{T_2}$. Owing to the freedom in the choice of unitary extension of the partially defined $U_0$, it is clear that irreducible special Ando tuples for $(T_1, T_2)$ are not unique.

As a corollary of the extended discussion in Remark 5.3.3 we get the following.

**Corollary 5.3.4** *(1) A commuting contractive pair $(T_1, T_2)$ always has a special Ando tuple.*

*(2) $(T_1, T_2)$ has a unique (up to coincidence) irreducible special Ando tuple if and only if both $T = T_1 \cdot T_2$ and $T = T_2 \cdot T_1$ are regular factorizations.*

The question arises as to whether it suffices to assume that only one of the factorizations $T_1 \cdot T_2$ and $T_2 \cdot T_1$ is regular. This issue is resolved by the following result.

**Theorem 5.3.5** *Let $(T_1, T_2)$ be a commuting contractive pair of contraction operators on a Hilbert space $\mathcal{H}$.*

*(1) Assume that all defect spaces $\mathcal{D}_T$, $\mathcal{D}_{T_1}$, $\mathcal{D}_{T_2}$ are finite-dimensional. Then $T_1 \cdot T_2$ is a regular factorization if and only if $T_2 \cdot T_1$ is a regular factorization.*

*(2) In the infinite-dimensional setting, it is possible for one of the factorizations $T_1 \cdot T_2$ to be regular while the other $T_2 \cdot T_1$ is not regular.*

*(3) Suppose that $(T_1, T_2)$ is a commuting pair of isometries. Then both $T_1^* \cdot T_2^*$ and $T_2^* \cdot T_1^*$ are regular factorizations.*

*Proof of (1)* Let us write $\Lambda_r$ for the counterpart of $\Lambda$ when the roles of $T_1$ and $T_2$ are reversed:

$$\Lambda_r : D_T h \mapsto D_{T_1} T_2 h \oplus D_{T_2} h.$$

Then, as a consequence of the second identity in (5.25), we see that $\Lambda_r$ is an isometry from $\mathcal{D}_T$ onto $\mathcal{R}(U_0)$. Moreover, if $T_1 \cdot T_2$ is a regular factorization, then

$$\begin{aligned}
\dim \mathcal{D}_T &= \dim \mathcal{D}(U_0) \text{ (since } \Lambda \text{ is an isometry from } \mathcal{D}_T \text{ onto } \mathrm{Dom}(U_0)) \\
&= \dim(\mathcal{D}_{T_1} \oplus \mathcal{D}_{T_2}) \text{ (by regularity of } T_1 \cdot T_2) \\
&\geq \dim \mathrm{Ran}(U_0) \text{ (since } \mathrm{Ran}(U_0) \subset \mathcal{D}_{T_1} \oplus \mathcal{D}_{T_2}) \\
&= \dim \mathcal{D}_T \text{ (since } \Lambda_r \text{ is an isometry from } \mathcal{D}_T \text{ onto } \mathrm{Ran}(U_0)).
\end{aligned}$$

Thus we see that necessarily the inequality in the third line must be equality. If we are in the finite-dimensional setting ($\mathcal{D}_T$, $\mathcal{D}_{T_1}$, $\mathcal{D}_{T_2}$ all finite-dimensional), we necessarily then have $\mathrm{Ran}(U_0) = \mathcal{D}_{T_1} \oplus \mathcal{D}_{T_2}$, which is the statement that the factorization $T_2 \cdot T_1$ is also regular.

*Proof of (2)* Let $(T_1, T_2)$ be the following pair of contractions on $\mathcal{H} := H^2 \oplus H^2$:

$$(T_1, T_2) = \left( \begin{bmatrix} 0 & 0 \\ I & 0 \end{bmatrix}, \begin{bmatrix} T_z & 0 \\ 0 & T_z \end{bmatrix} \right),$$

where $T_z$ is the Toeplitz operator $f(z) \mapsto zf(z)$ on $H^2$. Note that

$$T_1 T_2 = \begin{bmatrix} 0 & 0 \\ T_z & 0 \end{bmatrix} = T_2 T_1 := T.$$

Then

$$D_T^2 = \begin{bmatrix} I_{H^2} & 0 \\ 0 & I_{H^2} \end{bmatrix} - \begin{bmatrix} 0 & T_z^* \\ 0 & 0 \end{bmatrix} \begin{bmatrix} 0 & 0 \\ T_z & 0 \end{bmatrix} = \begin{bmatrix} 0 & 0 \\ 0 & I_{H^2} \end{bmatrix} = D_T, \quad \mathcal{D}_T = \begin{bmatrix} 0 \\ H^2 \end{bmatrix},$$

$$D_{T_1}^2 = \begin{bmatrix} I & 0 \\ 0 & I \end{bmatrix} - \begin{bmatrix} 0 & I \\ 0 & 0 \end{bmatrix} \begin{bmatrix} 0 & 0 \\ I & 0 \end{bmatrix} = \begin{bmatrix} 0 & 0 \\ 0 & I \end{bmatrix} = D_{T_1}, \quad \mathcal{D}_{T_1} = \begin{bmatrix} 0 \\ H^2 \end{bmatrix},$$

$$D_{T_2}^2 = \begin{bmatrix} I & 0 \\ 0 & I \end{bmatrix} - \begin{bmatrix} T_z^* & 0 \\ 0 & T_z^* \end{bmatrix} \begin{bmatrix} T_z & 0 \\ 0 & T_z \end{bmatrix} = \begin{bmatrix} 0 & 0 \\ 0 & 0 \end{bmatrix}, \quad \mathcal{D}_{T_2} = \{0\},$$

and $\mathcal{D}_{T_1} \oplus \mathcal{D}_{T_2} = \left[\begin{smallmatrix} 0 \\ H^2 \end{smallmatrix}\right]$.

We let $\Lambda$ be the map associated with the factorization $T_1 \cdot T_2$ given by (5.27), while $\Lambda_r$ is the same map associated with the factorization $T_2 \cdot T_1$ (i.e., (5.27) but with the indices and then the components interchanged). For $h = \left[\begin{smallmatrix} h_1 \\ h_2 \end{smallmatrix}\right] \in \mathcal{H} = \left[\begin{smallmatrix} H^2 \\ H^2 \end{smallmatrix}\right]$ we compute

$$\Lambda \colon D_T h = \begin{bmatrix} 0 \\ h_2 \end{bmatrix} \mapsto D_{T_1} T_2 h \oplus D_{T_2} h = \begin{bmatrix} 0 \\ T_z h \end{bmatrix}$$

and we conclude that

$$\operatorname{Ran} \Lambda = \begin{bmatrix} 0 \\ zH^2 \end{bmatrix} \underset{\neq}{\subset} \begin{bmatrix} 0 \\ H^2 \end{bmatrix} = \mathcal{D}_{T_1} \oplus \mathcal{D}_{T_2},$$

implying that $T_1 \cdot T_2$ is not a regular factorization. On the other hand,

$$\Lambda_r \colon D_T h = \begin{bmatrix} 0 \\ h_2 \end{bmatrix} \mapsto D_{T_1} h \oplus D_{T_2} T_1 h = \begin{bmatrix} 0 \\ h_2 \end{bmatrix} \oplus 0,$$

from which we see that

$$\operatorname{Ran} \Lambda_r = \begin{bmatrix} 0 \\ H^2 \end{bmatrix} = \mathcal{D}_{T_1} \oplus \mathcal{D}_{T_2},$$

implying that the factorization $T_2 \cdot T_1$ is regular.

*Proof of (3)* This is an immediate corollary of Proposition VII.3.2(b) from [92], which asserts: For a contractive pair $(T_1, T_2)$ on a Hilbert space $\mathcal{H}$, *the factorization* $T = T_1 \cdot T_2$ *is regular whenever* $T_1$ *or* $T_2^*$ *is isometric.* However, this fact in turn can be seen as an immediate corollary of the following alternative characterization of regular factorization discovered somewhat later (see the Proposition in [91]): $T_1 \cdot T_2$ *is a regular factorization if and only if* $\mathcal{D}_{T_1} \cap \mathcal{D}_{T_2^*} = \{0\}$. Let us note that this criterion also comes up in the characterization of triviality of overlapping spaces in the de Branges–Rovnyak model theory (see [13]). For a direct proof of item (3) in Theorem 5.3.5 , see [83, Lemma 27]. □

**Remark 5.3.6** The general issue arising here is the following: *Given subspaces* $\mathrm{Dom}(U_0)$ *and* $\mathrm{Ran}(U_0)$ *of some Hilbert space* $\mathcal{H}_0$ *and a unitary map* $U_0\colon \mathrm{Dom}(U_0) \to \mathrm{Ran}(U_0)$, *find a unitary extension* $U\colon \mathcal{H} \to \mathcal{H}$ *on a possibly larger Hilbert space* $\mathcal{H} \supset \mathcal{H}_0$ *so that* $U|_{\mathrm{Dom}(U_0)} = U_0\colon \mathrm{Dom}(U_0) \to \mathrm{Ran}(U_0)$ *and U is* minimal *in the sense that the smallest reducing subspace for U containing* $\mathcal{H}_0$ *is all of* $\mathcal{H}$. This problem is the core of the *lurking isometry technique* in interpolation and realization theory (see, e.g., Theorems 2.1 and 2.2 in [14]). We refer to the paper of Arov–Grossman [11] for a thorough treatment of the general problem.

## 5.4 Ando's Dilation Theorem via Type I Ando Tuples

We note that the definitions of Type I Ando tuples is existential: It is not clear whether or not one can solve equations (5.1) and(5.2) for $\Lambda_*$, $P_*$, $U_*$. On the other hand, the preceding discussion shows, in particular, that it is always possible to construct special Ando tuples (for $(T_1, T_2)$ or for $(T_1^*, T_2^*)$) and, in general, there are many such choices: The possibilities are parametrized by a choice of unitary extension $U$ for the partially defined isometry $U_0$. We shall see in Theorem 5.4.1 that any special Ando tuple for $(T_1^*, T_2^*)$ is also a Type I Ando tuple for $(T_1^*, T_2^*)$, thereby confirming the existence of Type I Ando tuples. Conversely, for each computable simple example that we have been able to work out the converse holds: Any Type I Ando tuple for $(T_1^*, T_2^*)$ turns out to be special. However, the validity of the converse statement in full generality remains an open problem.

**Theorem 5.4.1** *Let* $(T_1, T_2)$ *be any commuting contractive operator pair on* $\mathcal{H}$. *Then any special Ando tuple for* $(T_1^*, T_2^*)$ *is also a Type I Ando tuple for* $(T_1^*, T_2^*)$. *In particular, the latter class is not empty and hence Ando lifts of* $(T_1, T_2)$ *exist by Theorem 5.2.1*

*Proof* As the notion of Type I Ando tuple for $(T_1^*, T_2^*)$ is coordinate-free, we may without loss of generality suppose that the special Ando tuple $(\mathcal{F}_{\dagger *}, \Lambda_{\dagger *}, P_{\dagger *}, U_{\dagger *})$ for $(T_1^*, T_2^*)$ is given in canonical form. Clearly this is an Ando tuple for $(T_1^*, T_2^*)$. To show that this collection is a Type I Ando tuple, we need only verify conditions (5.1) and (5.2). We deal in detail only with (5.1), as verification of (5.2) is completely analogous. Making use of the defining properties of a canonical special Ando tuple gives, for each $h \in \mathcal{H}$,

$$
\begin{aligned}
&P_{\dagger*}^{\perp} U_{\dagger*} \Lambda_{\dagger*} D_{T^*} h + P_{\dagger*} U_{\dagger*} \Lambda_{\dagger*} D_{T^*} T^* h \\
&= P_{\dagger*}^{\perp} U_{\dagger*} \left( D_{T_1^*} T_2^* h \oplus D_{T_2^*} h \right) + P_{\dagger*} U_{\dagger*} \left( D_{T_1^*} T_2^* T^* h \oplus D_{T_2^*} T^* h \right) \\
&= P_{\dagger*}^{\perp} \left( D_{T_1^*} h \oplus D_{T_2^*} T_1^* h \right) + P_{\dagger*} \left( D_{T_1^*} T^* h \oplus D_{T_2^*} T_1^* T^* h \right) \\
&= \left( 0 \oplus D_{T_2^*} T_1^* h \right) + \left( D_{T_1^*} T_2^* T_1^* h \oplus 0 \right) = \Lambda_{\dagger*} D_{T^*} T_1^* h.
\end{aligned}
$$

As $h \in \mathcal{H}$ is arbitrary, this verifies (5.1), as desired.

As we have seen in the extended Remark 5.3.3, there always exist special Ando tuples for any commuting contractive pair $(T_1^*, T_2^*)$. As the class of special Ando tuples for $(T_1^*, T_2^*)$ forms a subclass of the class of Type I Ando tuples by the first part of the theorem, it follows that these latter classes are all nonempty. Then the construction in Theorem 5.2.1 based on a special Ando tuple as a starting point leads to an explicit Ando lift for $(T_1, T_2)$. □

## 5.5 Examples: Douglas-Type Model of Ando Lifts

A particularly transparent illustration of Theorem 5.2.1 is the case where the commuting contractive pair $(T_1, T_2)$ is, in fact, an isometric pair and hence $(T_1, T_2)$ can be considered as an Ando lift of itself: $(T_1, T_2) = (V_1, V_2)$ on $\mathcal{K} = \mathcal{H}$. In this case Type I Ando tuples for $(T_1^*, T_2^*)$ are closely associated with BCL-tuples for $(V_1, V_2) = (T_1, T_2)$ giving rise to the BCL1-model, as explored in Theorem 5.5.1. Recall the notation $\mathbf{ev}_{0,\mathcal{F}}$ appearing as number 3 in Notation 2.0.1.

**Theorem 5.5.1** *Let $(\mathcal{F}, P, U)$ be a BCL-tuple (i.e., $P$ is a projection and $U$ is a unitary operator on the Hilbert space $\mathcal{F}$). Then there is a choice of a commuting isometric pair $(T_1, T_2)$ and a unitary operator $\Lambda \colon \mathcal{D}_{T^*} \to \mathcal{F}$ ($T = T_1 T_2$) so that $(\mathcal{F}, \Lambda, P, U)$ is a special (and hence also Type I by Theorem 5.4.1) Ando tuple for $(T_1^*, T_2^*)$. Indeed one can take $(T_1, T_2)$ to be the BCL1-model associated with the BCL-tuple $(\mathcal{F}, P, U)$.*

*Conversely, if $(T_1, T_2)$ is a commuting isometric pair such that $T = T_1 T_2$ is a shift, then any irreducible Type I Ando tuple $(\mathcal{F}, \Lambda, P, U)$ for $(T_1^*, T_2^*)$ has the property that $\Lambda \colon \mathcal{D}_{T^*} \to \mathcal{F}$ is unitary and $(\mathcal{F}, P, U)$ is a BCL-tuple generating a BCL1-model for the commuting isometric pair $(T_1, T_2)$.*

*Proof* Define $(T_1, T_2)$ as the BCL1-model commuting isometric pair generated by the BCL-tuple $(\mathcal{F}, P, U)$:

$$(T_1, T_2) = \left( M_{U^*(P^{\perp} + zP)}, M_{(P + zP^{\perp})U} \right) \text{ on } H^2(\mathcal{F}),$$

or, equivalently, in tensor-product form more convenient for computations,

$$T_1 = (I_{H^2} \otimes U^* P^\perp) + (M_z \otimes U^* P), \quad T_2 = (I_{H^2} \otimes PU) + (M_z \otimes P^\perp U). \tag{5.32}$$

Then $T = M_z^{\mathcal{F}}$ and $D_{T^*} = \mathbf{ev}_{0,\mathcal{F}}^* \mathbf{ev}_{0,\mathcal{F}}$ is the projection onto the constant functions in $H^2(\mathcal{F})$. Hence there is a unitary identification map $\Lambda$, namely

$$\Lambda = \mathbf{ev}_{0,\mathcal{F}}|_{\mathcal{D}_{T^*}} : \mathcal{D}_{T^*} \to \mathcal{F},$$

between the defect space $\mathcal{D}_{T^*}$ for $T^*$ and the coefficient Hilbert space $\mathcal{F}$. Thus the tuple $(\mathcal{F}, \Lambda, P, U)$ is an Ando tuple for $(T_1^*, T_2^*)$ having the additional property that the embedding operator $\Lambda \colon \mathcal{D}_{T^*} \to \mathcal{F}$ is unitary. It is possible to show directly that $(\mathcal{F}, \Lambda, P, U)$ is a Type I Ando tuple for $(T_1^*, T_2^*)$ (see Remark 5.5.6); however, by Theorem 5.4.1 this is not necessary once we verify that $(\mathcal{F}, \Lambda, P, U)$ is a special Ando tuple for $(T_1^*, T_2^*)$, as claimed.

As a first step towards this goal, observe that by part (3) of Theorem 5.3.5, we know that both $T_1^* \cdot T_2^*$ and $T_2^* \cdot T_1^*$ are regular factorizations of $T^*$. Hence by Scenario 1 in Remark 5.3.3 we see that $(T_1^*, T_2^*)$ has a unique irreducible canonical special Ando tuple

$$\left(\mathcal{F}_\dagger, \Lambda_\dagger, P_\dagger, U_\dagger\right) \tag{5.33}$$

with

$$\mathcal{F}_\dagger = \mathcal{D}_{T_1^*} \oplus \mathcal{D}_{T_2^*}, \quad \Lambda_\dagger \colon D_{T^*} h \mapsto D_{T_1^*} T_2^* h \oplus D_{T_2^*} h,$$
$$P_\dagger \colon D_{T_1^*} h_1 \oplus D_{T_2^*} h_2 \mapsto D_{T_1^*} h_1 \oplus 0, \; U_\dagger \colon D_{T_1^*} T_2^* h \oplus D_{T_2^*} h \mapsto D_{T_1^*} h \oplus D_{T_2^*} T_1^* h.$$

We wish to show that the canonical special Ando tuple (5.33) coincides with the Ando tuple $(\mathcal{F}, \Lambda, P, U)$, from which it then follows that $(\mathcal{F}, \Lambda, P, U)$ is a special Ando tuple for $(T_1^*, T_2^*)$.

An easy calculation based on the formulas (5.32) for $(T_1, T_2)$ gives us

$$I_{H^2(\mathcal{F})} - T_1 T_1^* = \mathbf{ev}_0^* \mathbf{ev}_0 \otimes U^* P U, \quad I_{H^2(\mathcal{F})} - T_2 T_2^* = \mathbf{ev}_0^* \mathbf{ev}_0 \otimes P^\perp,$$
$$I_{H^2(\mathcal{F})} - T T^* = \mathbf{ev}_0^* \mathbf{ev}_0 \otimes I_{\mathcal{F}}.$$

Since each of these operators is a projection, we can alternatively write these last identities as

$$D_{T_1^*} = \mathbf{ev}_0^* \mathbf{ev}_0 \otimes U^* P U, \quad D_{T_2^*} = \mathbf{ev}_0^* \mathbf{ev}_0 \otimes P^\perp, \quad D_{T^*} = \mathbf{ev}_0^* \mathbf{ev}_0 \otimes I_{\mathcal{F}},$$

which in turn can be rewritten without tensor products as

$$D_{T_1^*} = \mathbf{ev}_{0,\mathcal{F}}^* U^* P U \mathbf{ev}_{0,\mathcal{F}}, \quad D_{T_2^*} = \mathbf{ev}_{0,\mathcal{F}}^* P^\perp \mathbf{ev}_{0,\mathcal{F}}, \quad D_{T^*} = \mathbf{ev}_{0,\mathcal{F}}^* \mathbf{ev}_{0,\mathcal{F}}. \tag{5.34}$$

Then the evaluation-at-zero map $\mathbf{ev}_{0,\mathcal{F}}$ provides a unitary identification of the defect spaces $\mathcal{D}_{T_1^*}$, $\mathcal{D}_{T_2^*}$, $\mathcal{D}_{T^*}$ with corresponding subspaces of $\mathcal{F}$:

$$\begin{aligned}\mathbf{ev}_{0,\mathcal{F}}\colon \mathcal{D}_{T_1^*} &\to \operatorname{Ran} U^*PU \subset \mathcal{F},\\ \mathbf{ev}_{0,\mathcal{F}}\colon \mathcal{D}_{T_2^*} &\to \operatorname{Ran} P^\perp \subset \mathcal{F},\\ \mathbf{ev}_{0,\mathcal{F}}\colon \mathcal{D}_{T^*} &\to \mathcal{F},\end{aligned}$$

and we have corresponding formulas

$$\mathbf{ev}_{0,\mathcal{F}} D_{T_1^*} h = U^*PUh(0), \quad \mathbf{ev}_{0,\mathcal{F}} D_{T_2^*} h = P^\perp h(0), \quad \mathbf{ev}_{0,\mathcal{F}} D_{T^*} h = h(0).$$

Furthermore, one can use the formulas (5.32) to check that

$$\mathbf{ev}_{0,\mathcal{F}} D_{T_1^*} T_2^* h = U^*Ph(0), \quad \mathbf{ev}_{0,\mathcal{F}} D_{T_2^*} T_1^* h = P^\perp Uh(0).$$

All this suggests that we define a unitary map $\tau$ from $\mathcal{D}_{T_2^*} \oplus \mathcal{D}_{T_1^*}$ onto $\mathcal{F}$ according to the formula

$$\tau\colon D_{T_1^*} h_1 \oplus D_{T_2^*} h_2 \mapsto PUh_1(0) + P^\perp h_2(0)$$

for $h_1, h_2 \in H^2(\mathcal{F})$. Note that $\tau$ is unitary since by previous observations all of the maps

$$\begin{aligned}D_{T_1^*} h \in \mathcal{D}_{T_1^*} &\mapsto PUh(0) \in \operatorname{Ran} PU,\\ D_{T_2^*} h \in \mathcal{D}_{T_2^*} &\mapsto P^\perp h(0) \in \operatorname{Ran} P^\perp\end{aligned}$$

are unitary and, since $U$ is unitary, we also have that

$$\operatorname{Ran} PU = \operatorname{Ran} P$$

is the orthogonal complement of $\operatorname{Ran} P^\perp$ in $\mathcal{F}$ and hence $\operatorname{Ran} PU$ and $\operatorname{Ran} P^\perp$ form an orthogonal decomposition of the space $\mathcal{F}$.

We now use $\tau$ as the unitary operator required to transform the canonical special Ando tuple $\left(\mathcal{D}_{T_2^*} \oplus \mathcal{D}_{T_1^*}, \Lambda_\dagger, P_\dagger, U_\dagger\right)$ to a new Ando tuple

$$(\mathcal{F}, \tau\Lambda_\dagger, \tau P_\dagger \tau^*, \tau U_\dagger \tau^*),$$

which coincides with the canonical special Ando tuple (and hence, according to our terminology, is itself special). In fact, we shall show that this new Ando tuple is the same as our original Ando tuple $(\mathcal{F}, \Lambda, P, U)$, that is, we shall verify the following identities:

$$\tau\Lambda_\dagger = \Lambda, \tag{5.35}$$

$$\tau P_\dagger = P\tau, \tag{5.36}$$

$$\tau U_\dagger = U\tau. \tag{5.37}$$

Once this is done, the proof of the first part of the theorem will be complete.
**Verification of** (5.35) We compute, for $h \in H^2(\mathcal{F})$,

$$\begin{aligned}\tau\Lambda_{\dagger}D_{T^*}h &= \tau\left(D_{T_1^*}T_2^*h \oplus D_{T_2^*}h\right) = Ph(0) + P^{\perp}h(0)\\ &= h(0) = \Lambda D_{T^*}h,\end{aligned}$$

and (5.35) follows.

**Verification of** (5.36) We compute $\tau P_{\dagger}$ acting on elements of the form $D_{T_1^*}T_2^*h \oplus D_{T_2^*}h$ for $h \in H^2(\mathcal{F})$, which form a dense subset of $\mathcal{D}_{T_1^*} \oplus \mathcal{D}_{T_2^*}$ owing to the regularity of the factorizations $T_1^* \cdot T_2^*$ and $T_2^* \cdot T_1^*$ as follows:

$$\tau P_{\dagger}\left(D_{T_1^*}T_2^*h \oplus D_{T_2^*}h\right) = \tau\left(D_{T_1^*}T_2^*h \oplus 0\right) = Ph(0)$$

while

$$P\tau\left(D_{T_1^*}T_2^*h \oplus D_{T_2^*}h\right) = P\tau\Lambda_{\dagger}D_{T^*}h = Ph(0),$$

verifying (5.36).

**Verification of** (5.37) We compute, again on the dense subset $\{D_{T_1^*}T_2^*h\oplus D_{T_2^*}h\colon h \in H^2(\mathcal{F})\}$ of $\mathcal{D}_{T_1^*} \oplus \mathcal{D}_{T_2^*}$,

$$\begin{aligned}\tau U_{\dagger}\left(D_{T_1^*}T_2^*h \oplus D_{T_2^*}h\right) &= \tau\left(D_{T_1^*}h \oplus D_{T_2^*}T_1^*h\right)\\ &= PUh(0)) + P^{\perp}Uh(0)\\ &= Uh(0)\end{aligned}$$

while, on the other hand,

$$\begin{aligned}U\tau\left(D_{T_1^*}T_2^*h \oplus D_{T_2^*}h\right) &= U\tau\Lambda_{\dagger}D_{T^*}h\\ &= Uh(0),\end{aligned}$$

from which (5.37) follows. We have now completed the proof of the first part of Theorem 5.5.1.

Conversely, suppose that $(\mathcal{F}', \Lambda', P', U')$ is an irreducible Type I Ando tuple for $(T_1^*, T_2^*)$. By Definition 5.1.5, this means that

$$P'^{\perp}U'\Lambda'D_{T^*} + P'U'\Lambda'D_{T^*}T^* = \Lambda'D_{T^*}T_1^*, \tag{5.38}$$

$$U'^*P'\Lambda'D_{T^*} + U'^*P'^{\perp}\Lambda'D_{T^*}T^* = \Lambda'D_{T^*}T_2^*. \tag{5.39}$$

Suppose that the BCL-tuple $(\mathcal{F}, P, U)$ generates a BCL1-model for the commuting isometric pair $(T_1, T_2)$. By the forward direction, we know that there is a choice of a unitary $\Lambda\colon \mathcal{D}_{T^*} \to \mathcal{F}$ so that $(\mathcal{F}, \Lambda, P, U)$ is a special Ando tuple for $(T_1^*, T_2^*)$ and hence is of Type I. By Definition 5.1.5, we therefore have

$$P^{\perp}U\Lambda D_{T^*} + PU\Lambda D_{T^*}T^* = \Lambda D_{T^*}T_1^*, \tag{5.40}$$

$$U^*P\Lambda D_{T^*} + U^*P^{\perp}\Lambda D_{T^*}T^* = \Lambda D_{T^*}T_2^*. \tag{5.41}$$

The operation $\Lambda'^* \times (5.38) - \Lambda^* \times (5.40)$, together with the fact that $\Lambda$ and $\Lambda'$ are isometric, gives

$$(\Lambda'^* P'^{\perp} U' \Lambda' - \Lambda^* P^{\perp} U \Lambda) D_{T^*} + (\Lambda'^* P' U' \Lambda' - \Lambda^* P U \Lambda) D_{T^*} T^* = 0, \quad (5.42)$$

while the operation $\Lambda'^* \times (5.39) - \Lambda^* \times (5.41)$, combined again with the fact that $\Lambda'$ and $\Lambda$ are isometric, gives

$$(\Lambda'^* U'^* P' \Lambda' - \Lambda^* U^* P \Lambda) D_{T^*} + (\Lambda'^* U'^* P'^{\perp} \Lambda' - \Lambda^* U^* P^{\perp} \Lambda) D_{T^*} T^* = 0. \quad (5.43)$$

If we set two operators $X_1, X_2 \in \mathcal{B}(\mathcal{D}_{T^*})$ as

$$X_1 := \Lambda'^* P'^{\perp} U' \Lambda' - \Lambda^* P^{\perp} U \Lambda \quad \text{and} \quad X_2 := \Lambda'^* U'^* P' \Lambda' - \Lambda^* U^* P \Lambda,$$

then the two equations above are the same as

$$X_1 D_{T^*} + X_2^* D_{T^*} T^* = 0 \quad \text{and} \quad X_2 D_{T^*} + X_1^* D_{T^*} T^* = 0.$$

Applying these two equations repeatedly, we get

$$D_{T^*} X_1 D_{T^*} = -D_{T^*} X_2^* D_{T^*} T^* = T D_{T^*} X_1 D_{T^*} T^* = \cdots = T^n D_{T^*} X_1 D_{T^*} T^{*n}.$$

Since $T$ is a shift operator (i.e., $T$ is an isometry and $T^{*n} \to 0$ in the strong operator topology), we must have $D_{T^*} X_1 D_{T^*} = 0$. Similar computation gives the same conclusion with $X_2$ in place of $X_1$. Since $X_1, X_2$ are operators acting on $\mathcal{D}_{T^*}$, we have $X_1 = 0 = X_2$, that is,

$$\Lambda'^* P'^{\perp} U' \Lambda' = \Lambda^* P^{\perp} U \Lambda \quad \text{and} \quad \Lambda'^* U'^* P' \Lambda' = \Lambda^* U^* P \Lambda. \quad (5.44)$$

Adding the first identity to the adjoint of the second gives

$$\Lambda^* U \Lambda = \Lambda'^* U' \Lambda'. \quad (5.45)$$

We have $\Lambda \colon \mathcal{D}_{T^*} \to \mathcal{F}$ and $U \colon \mathcal{F} \to \mathcal{F}$ both unitary and hence $\Lambda'^* U' \Lambda'$ must be a unitary on $\mathcal{D}_{T^*}$. Therefore, for every $d \in \mathcal{D}_{T^*}$,

$$\|U' \Lambda' d\| = \|d\| = \|\Lambda'^* U' \Lambda' d\| = \|\Lambda' \Lambda'^* U' \Lambda' d\|.$$

This implies that $\operatorname{Ran} \Lambda'$ must be invariant for $U'$. A similar computation with the adjoint of (5.45) gives that $\operatorname{Ran} \Lambda'$ is invariant for $U'^*$. Let us use this information to rewrite the first identity in (5.44) as

$$\Lambda'^* P'^{\perp} \Lambda' \Lambda'^* U' \Lambda' = \Lambda^* P^{\perp} \Lambda \Lambda^* U \Lambda.$$

By (5.45), $\Lambda'^* U' \Lambda'$ and $\Lambda^* U \Lambda$ are equal to the same unitary operator; hence we may cancel them off in the preceding equation to arrive at

$$\Lambda'^* P' \Lambda' = \Lambda^* P \Lambda. \quad (5.46)$$

Since $\Lambda$ is a unitary, $\Lambda^* P \Lambda$ is an orthogonal projection and therefore so is $\Lambda'^* P' \Lambda'$. With this in mind, we compute for $d \in \mathcal{D}_{T^*}$,

$$\|P'\Lambda' d\|^2 = \langle \Lambda'^* P' \Lambda' d, d\rangle = \langle \Lambda'^* P' \Lambda' \Lambda'^* P' \Lambda' d, d\rangle = \|\Lambda'\Lambda'^* P' \Lambda' d\|^2.$$

Since this is true for every $d \in \mathcal{D}_{T^*}$ we conclude that $\operatorname{Ran}\Lambda'$ is invariant under $P'$. Consequently, we have $\operatorname{Ran}\Lambda'$ invariant for $U'$, $U'^*$, and $P'$ as well. Since $(\mathcal{F}', \Lambda', P', U')$ is an irreducible Type I Ando tuple (see Definition 5.1.3), this is possible only when $\Lambda'$ is surjective. This, together with equations (5.45) and (5.46), proves the converse part because these identities show that the Ando tuples $(\mathcal{F}, \Lambda, P, U)$ and $(\mathcal{F}', \Lambda', P', U')$ coincide (see Definition 5.1.2) via the unitary operator $\Lambda'\Lambda^* : \mathcal{F} \to \mathcal{F}'$. □

To further illustrate the ideas we here set down some Ando lifts for a simple commuting pair of contractions $(T_1, T_2)$ and compute some associated minimal Ando tuples. In particular, the examples illustrate that a given commuting contractive pair can have many minimal Ando lifts, which are not unitarily equivalent as lifts.

**Example 5.5.2** Let $(T_1, T_2)$ be a commuting pair of contraction operators on a finite-dimensional Hilbert space $\mathcal{H}$, say of dimension $N$, such that neither $T_1$ nor $T_2$ has an eigenvalue of modulus equal to 1. For simplicity we assume that $(T_1, T_2)$ has a basis of joint eigenvectors. For convenience we work in detail with a basis of joint eigenvectors for the adjoint pair $\left(T_1^*, T_2^*\right)$. Let us denote by $\{v_1, \ldots, v_N\}$ the basis of joint eigenvectors for $\left(T_1^*, T_2^*\right)$ with joint eigenvalues

$$\left(\overline{\lambda}^{(1)}, \ldots, \overline{\lambda}^{(N)}\right) = \left(\left(\overline{\lambda}_1^{(1)}, \overline{\lambda}_2^{(1)}\right), \ldots, \left(\overline{\lambda}_1^{(N)}, \overline{\lambda}_2^{(N)}\right)\right).$$

By the assumption that neither $T_1$ nor $T_2$ has an eigenvalue of modulus 1, we know that each of these joint eigenvalues $\lambda^{(j)} = \left(\lambda_1^{(j)}, \lambda_2^{(j)}\right)$ is in the open bidisk $\mathbb{D}^2$. Thus we have for $r = 1, 2$ and $j = 1, \ldots, N$ that

$$T_r^* v_j = \overline{\lambda}_r^{(j)} v_j. \tag{5.47}$$

In particular, we have

$$\left\langle \left(I - T_1 T_1^*\right) v_j, v_i \right\rangle_{\mathcal{H}} = \left(1 - \lambda_1^{(i)} \overline{\lambda}_1^{(j)}\right) \langle v_j, v_i\rangle_{\mathcal{H}}.$$

As $T_1$ is a contraction, the matrix on the left (with rows indexed by $i$ and columns by $j$) is positive semidefinite, say of rank $d$. Hence there are vectors $y_1, \ldots, y_N$ in a $d$-dimensional Hilbert space $\mathcal{Y}$ so that

$$\left\langle \left(I - T_1 T_1^*\right) v_j, v_i \right\rangle_{\mathcal{H}} = \langle y_j, y_i\rangle_{\mathcal{Y}}. \tag{5.48}$$

Combining the last two displayed identities and using the assumption that each $\lambda_1^{(j)}$ is in the open unit disk, we can solve for $\langle v_j, v_i \rangle$ to get

$$\langle v_j, v_i \rangle_{\mathcal{H}} = \frac{\langle y_j, y_i \rangle_{\mathcal{Y}}}{1 - \lambda_1^{(i)} \overline{\lambda}_1^{(j)}}. \tag{5.49}$$

As we are assuming the $\{v_1, \ldots, v_N\}$ is a basis for $\mathcal{H}$, we know that in fact the matrix (5.49) is invertible.

Let us now introduce the vectorial Hardy space $H^2(\mathcal{Y})$ and the vectorial kernel functions $k_\lambda y$ (for $\lambda \in \mathbb{D}$ and $y \in \mathcal{Y}$) given by

$$(k_\lambda y)(z) = \frac{1}{1 - z\overline{\lambda}} y$$

having the reproducing kernel property

$$\langle f, k_\lambda y \rangle_{H^2(\mathcal{Y})} = \langle f(\lambda), y \rangle_{\mathcal{Y}}.$$

The identity (5.49) implies that the map

$$\mathbf{\Pi} \colon v_j \mapsto k_{\lambda_1^{(j)}} y_j \quad \text{for } j = 1, \ldots, N \tag{5.50}$$

extends by linearity to a unitary map from $\mathcal{H}$ onto the Hilbert space

$$\widetilde{\mathcal{H}} = \bigvee \left\{ k_{\lambda_1^{(j)}} y_j \colon j = 1, \ldots, N \right\} \subset H^2(\mathcal{Y}) \tag{5.51}$$

equal to the span of the kernel functions $k_{\lambda_1^{(j)}} y_j$ in the Hardy space $H^2(\mathcal{Y})$. Furthermore, the operators $T_1^*$ and $T_2^*$ are transformed via the unitary identification $\mathbf{\Pi} \colon \mathcal{H} \to \widetilde{\mathcal{H}}$ to the operators

$$\widetilde{T}_1^* \colon k_{\lambda_1^{(j)}} y_j \mapsto \overline{\lambda}_1^{(j)} k_{\lambda_1^{(j)}} y_j,$$
$$\widetilde{T}_2^* \colon k_{\lambda_2^{(j)}} y_j \mapsto \overline{\lambda}_2^{(j)} k_{\lambda_2^{(j)}} y_j.$$

Note next that the contractivity of the operator $\widetilde{T}_2^*$ implies that

$$\left\| \sum_{j=1}^N c_j k_{\lambda_1^{(j)}} y_j \right\|^2 - \left\| \sum_{j=1}^N c_j \overline{\lambda}_2^{(j)} k_{\lambda_2^{(j)}} y_j \right\|^2 \geq 0 \quad \text{for all } c_1, \ldots, c_N \in \mathbb{C}.$$

Spelling out this condition gives us the positive-semidefiniteness condition

$$\mathbb{P} := \left[ \frac{1 - \lambda_2^{(i)} \overline{\lambda}_2^{(j)}}{1 - \lambda_1^{(i)} \overline{\lambda}_1^{(j)}} \langle y_j, y_i \rangle_{\mathcal{Y}} \right]_{1 \leq i,j \leq N} \succeq 0. \tag{5.52}$$

**Identification of Ando Lifts of $(T_1, T_2)$ via Rational Matrix Inner Solutions of a Tangential Nevanlinna–Pick Interpolation Problem** By an extension of Theorem 6.15 of Agler–McCarthy [4] to the case of tangential matrix-valued interpolation, *there is a rational $N \times N$-matrix inner function $\Theta$ (i.e., a finite matrix Blaschke–Potapov product as in [77]) satisfying the left-tangential interpolation conditions*

$$\Theta\left(\lambda_1^{(j)}\right)^* y_j = \overline{\lambda}_2^{(j)} y_j \quad \text{for } j = 1, \ldots, N,$$

which we prefer to write as

$$y_j^* \Theta\left(\lambda_1^{(j)}\right) = y_j^* \lambda_2^{(j)} \quad \text{for } j = 1, \ldots, N, \tag{5.53}$$

if and only if the Pick matrix $\mathbb{P}$ is positive semidefinite.

Let us now view $\mathbf{\Pi}$ as an isometric embedding operator of $\mathcal{H}$ into $H^2(\mathcal{Y})$. The previous computations show that

$$\left(M_z^*, M_\Theta^*\right) \mathbf{\Pi} v_j = \mathbf{\Pi}\left(T_1^*, T_2^*\right) v_j \quad \text{for } j = 1, \ldots, N.$$

Since $\mathcal{H}$ is the span of $v_1, \ldots, v_N$, we can rewrite this last identity in operator form

$$\left(M_z^*, M_\Theta^*\right) \mathbf{\Pi} = \mathbf{\Pi}\left(T_1^*, T_2^*\right),$$

that is, $(\mathbf{\Pi}, \mathbf{V}_1, \mathbf{V}_2) := (\mathbf{\Pi}, M_z, M_\Theta)$ is an Ando lift for the commuting contractive pair $(T_1, T_2)$. We conclude: *If $\Theta$ is any inner solution of the interpolation conditions* (5.53)*, then $(\mathbf{\Pi}, M_z, M_\Theta)$ is an Ando lift of $(T_1, T_2)$.*

We argue next that any such Ando lift $(\mathbf{\Pi}, \mathbf{V}_1, \mathbf{V}_2) = (\mathbf{\Pi}, M_z, M_\Theta)$ is minimal. Indeed, we shall prove the stronger statement

$$\bigvee_{j=0}^{\infty} \mathbf{V}_1^j \operatorname{Ran} \mathbf{\Pi} = H^2(\mathcal{Y}). \tag{5.54}$$

To see this, observe that

$$\left(\left(I - \overline{\lambda}_1^{(j)} \mathbf{V}_1\right) k_{\lambda_1^{(j)}} y_j\right)(z) = \left(1 - \overline{\lambda}_1^{(j)} z\right) \cdot \frac{y_j}{1 - \overline{\lambda}_1^{(j)} z} = y_j$$

and hence

$$\bigvee_{j=0}^{\infty} \mathbf{V}_1^j \operatorname{Ran} \mathbf{\Pi} \supset \bigvee_{j=1}^{N} \{y_j \colon j = 1, \ldots, N\}.$$

Note that the Gramian matrix $\left[\langle y_i, y_j \rangle\right]_{i,j=1,\ldots,N}$ of the vectors $y_1, \ldots, y_N$ has rank equal to the rank $d$ of the defect operator $D_{T_1^*}$ as a consequence of the identity (5.48). But on the other hand, we have chosen the space $\mathcal{Y}$ to have dimension equal to $d$ so we can conclude that $\bigvee_{j=1}^{d}\{y_j : j = 1, \ldots, N\} = \mathcal{Y}$, and the last displayed identity can be rewritten as

$$\bigvee_{j=0}^{\infty} \mathbf{V}_1^j \operatorname{Ran} \mathbf{\Pi} \supset \bigvee_{j=1}^{N} \{y_j : j = 1, \ldots, N\} = \mathcal{Y}$$

(where here we identify $\mathcal{Y}$ with the subspace of constant functions in $H^2(\mathcal{Y})$). It then follows that

$$\bigvee_{j=0}^{\infty} \mathbf{V}_1^j \operatorname{Ran} \mathbf{\Pi} \supset \bigvee_{j=1}^{N} M_z^j \mathcal{Y} = H^2(\mathcal{Y})$$

and (5.54) follows, that is, $(\mathbf{\Pi}, \mathbf{V}_1, \mathbf{V}_2)$ in particular is a minimal Ando lift. Moreover, from the fact that $\dim \mathcal{Y} = d < \infty$, we see that $\Theta$ is a rational matrix inner function. If it were the case that $\Theta(\lambda_0)$ had a modulus-1 eigenvalue $\omega \in \mathbb{T}$ for some $\lambda_0$, then from the fact that $|\langle \Theta(\lambda)u, v\rangle|$ is a harmonic function of $\lambda$ on the unit disk, by the maximum principle for harmonic functions it would follow that $|\langle \Theta(\lambda)u, y\rangle|$ is constant on $\mathbb{D}$. Hence if $\Theta$ has an eigenvalue $\omega \in \mathbb{T}$ with eigenvector $v_0$, then $\langle \Theta(\lambda)v_0, v_0\rangle$ is independent of $\lambda$ and we conclude that $\Theta(\lambda)v_0 = \omega v_0$ for all $\lambda \in \mathbb{D}$, and hence in particular we have $\Theta\left(\lambda_1^{(j)}\right) v_0 = \omega v_0$ for each $j = 1, \ldots, N$. Then for each $j = 1, \ldots, N$ we have

$$\overline{\omega_j}\langle y_j, v_0\rangle = \langle y_j, \omega v_0\rangle = \left\langle y_j, \Theta\left(\lambda_1^{(j)}\right) v_0\right\rangle = \overline{\lambda_2^{(j)}}\langle y_j, v_0\rangle.$$

As $|\omega_j| = 1$ and $\left|\lambda_2^{(j)}\right| < 1$, we conclude that $\overline{\omega_j} \neq \overline{\lambda_2^{(j)}}$ forcing us to the conclusion that $v_0 \perp \bigvee\{y_1, \ldots, y_N\}$. But as we have already arranged that $\{y_1, \ldots, y_N\}$ spans the whole space $\mathcal{Y}$, we conclude that $v_0 = 0$. Thus $\Theta(\lambda)$ cannot have any eigenvalues of modulus 1 for any $\lambda \in \mathbb{D}$. By Remark 3.2.11 it follows that the isometry $M_\Theta$ generated by the inner function $\Theta$ is in fact a shift operator.

We are now at the starting point for the proof of the converse direction in Theorem 5.2.1. Let us note that the model $\left(M_z^{\mathcal{Y}}, M_\Theta\right)$ as the BDF1-model for a 1&2-c.n.u. commuting isometric pair for the special case where $(\mathcal{Y}, \Theta)$ is a matrix-valued inner function (rather than the general case of contractive operator valued). Hence Theorem 3.3.7 tells us how to convert the BDF1-model to a BCL1-model, namely: The BCL-tuple generating a BCL1-model unitarily equivalent to $(M_z^{\mathcal{Y}}, M_\Theta)$ is given by

$$\left(\mathcal{F} = \begin{bmatrix} \mathcal{Y} \\ \mathfrak{H}(\Theta) \end{bmatrix}, \begin{bmatrix} I_{\mathcal{Y}} & 0 \\ 0 & 0 \end{bmatrix}, \begin{bmatrix} A_\Theta & B_\Theta \\ C_\Theta & D_\Theta \end{bmatrix}\right),$$

where $\left[\begin{smallmatrix} A_\Theta & B_\Theta \\ C_\Theta & D_\Theta \end{smallmatrix}\right]$ acting on $\left[\begin{smallmatrix} \mathcal{Y} \\ \mathfrak{H}(\Theta) \end{smallmatrix}\right]$ is given by (3.100) (with $\mathcal{Y}$ in place of $\mathcal{F}$) with associated BCL1-model $(M_1, M_2) = (M_{\Psi_1}, M_{\Psi_2})$ on $H^2\left(\left[\begin{smallmatrix} \mathcal{Y} \\ \mathfrak{H}(\Theta) \end{smallmatrix}\right]\right)$ with the inner matrix pencils $\Psi_1$ and $\Psi_2$ given by

$$(\Psi_1(z), \Psi_2(z)) = \left(\begin{bmatrix} zA_\Theta^* & C_\Theta^* \\ zB_\Theta^* & D_\Theta^* \end{bmatrix}, \begin{bmatrix} A_\Theta & B_\Theta \\ zC_\Theta & zD_\Theta \end{bmatrix}\right)$$

and the implementing unitary identification map $\tau_\Theta$ as given in (3.101). The next step is to observe that $\Pi := \tau_\Theta \mathbf{\Pi}$ is an isometric embedding of $\mathcal{H}$ into $H^2(\mathcal{Y} \oplus \mathfrak{H}(\Theta))$ and the collection

$$(\Pi, M_1, M_2) := (\tau_\Theta \mathbf{\Pi}, M_{U^*P^\perp + zU^*P}, M_{PU + zP^\perp U})$$

is again a lift of $(T_1, T_2)$, which is unitarily equivalent (via $\tau_\Theta$) to the previously discussed lift $(\mathbf{\Pi}, M_z, M_\Theta)$ on $H^2(\mathcal{Y})$, having the additional property that the commuting isometric pair $(M_1, M_2)$ giving the Ando lift is in the BCL1-model form on $H^2(\mathcal{Y} \oplus \mathfrak{H}(\Theta))$. Note also that here we are in the somewhat simpler case where the product isometry $M_1 \cdot M_2 = M_z$ on $H^2(\mathcal{Y} \oplus \mathfrak{H}(\Theta))$ is a shift, and hence our model space involves only the top component of the block $2 \times 1$-column matrices appearing for the general case. Specializing the explanation given in the proof of Theorem 5.2.1 to the situation here, we see that there is an isometric embedding

$$\Gamma \colon H^2(\mathcal{D}_{T^*}) \to H^2\left(\left[\begin{smallmatrix} \mathcal{Y} \\ \mathfrak{H}(\Theta) \end{smallmatrix}\right]\right)$$

such that

$$\Gamma M_z^{\mathcal{D}_{T^*}} = M_z^{\mathcal{Y} \oplus \mathfrak{H}(\Theta)} \Gamma, \quad \Gamma \mathcal{O}_{D_{T^*}, T^*} = \Pi \text{ (here } T = T_1 T_2),$$
$$\operatorname{Ran} \Gamma = \bigvee_{n \geq 0} \left(M_z^{\mathcal{Y} \oplus \mathfrak{H}(\Theta)}\right)^n \operatorname{Ran} \Pi =: \mathcal{K}_{\min} \subset H^2(\mathcal{Y} \oplus \mathfrak{H}(\Theta)),$$

and such that $(\Pi, M_z|_{\operatorname{Ran} \Gamma})$ is a version of the essentially unique minimal isometric lift for the single contraction operator $T$. Furthermore, $\Gamma$ has the form of a multiplication by a constant $\Gamma = I_{H^2} \otimes \Lambda$ for an isometry from $\Lambda \colon \mathcal{D}_{T^*} \to \mathcal{Y} \oplus \mathfrak{H}(\Theta)$. The explicit formula (5.15) for $\Lambda$ here can be given the form

$$\Lambda \colon D_{T^*} v_j \mapsto \left(\tau_\Theta k_{\lambda_1^{(j)}} y_j\right)(0) \quad \text{for } j = 1, 2, \ldots, N. \tag{5.55}$$

It is then this $\Lambda$ that serves as the embedding operator for the Type I Ando tuple

$$\left(\begin{bmatrix} \mathcal{Y} \\ \mathfrak{H}(\Theta) \end{bmatrix}, \Lambda, P, U\right) \quad \text{with } U = \begin{bmatrix} A_\Theta & B_\Theta \\ C_\Theta & D_\Theta \end{bmatrix}, P = \begin{bmatrix} I_\mathcal{Y} & 0 \\ 0 & 0 \end{bmatrix} \text{ on } \begin{bmatrix} \mathcal{Y} \\ \mathfrak{H}(\Theta) \end{bmatrix}. \tag{5.56}$$

This then gives us the parameter set needed to build the Douglas model Ando lift for the original commuting contractive pair $(T_1, T_2)$ specified in terms of joint eigenvectors for $\left(T_1^*, T_2^*\right)$ (5.47). Since, as observed in the previous paragraph, the Ando lift $(\Pi, \mathbf{V}_1, \mathbf{V}_2)$ is minimal, it follows that this Type I Ando tuple for $\left(T_1^*, T_2^*\right)$ is minimal as an Ando tuple as well.

Finally we show that a minimal Ando lift of $(T_1, T_2)$ need not be unique (up to unitary equivalence of lifts). Toward this end, we impose an additional hypothesis. We have observed in the previous discussion that the matrix (5.49) is strictly positive definite but that the Pick matrix (5.52) appears to be only semidefinite. Let us impose the following additional hypothesis.

**A1 Invertibility assumption** The Pick matrix (5.52) is positive definite (positive semidefinite and invertible).

It is conceivable that one can arrange for this Assumption **A1** to hold via an arbitrarily small perturbation of the basis of eigenvectors $\{v_1, \ldots, v_N\}$ but we do not pursue this point here.

**Verification of Lack of Uniqueness of Ando Lift when Assumption A1 Holds**

**Step 1 Positive definiteness of augmented Pick matrices** Assuming that the Pick matrix (5.52) is (strictly) positive definite, we would like to analyze when we can solve the augmented interpolation problem of the form

$$\Theta\left(\lambda_1^{(j)}\right)^* y_j = \overline{\lambda}_2^{(j)} y_j \quad \text{for } j = 1, \ldots, N, N+1 \tag{5.57}$$

for a rational matrix inner function $\Theta$, where the additional data points $\lambda_1^{(N+1)}$, $\lambda_2^{(N+1)} \in \mathbb{D}$, and $y_{N+1} \in \mathbb{C}^{N+1}$ encode an additional interpolation condition beyond what is already specified in (5.53), namely

$$\Theta\left(\lambda_1^{(N+1)}\right)^* y_{N+1} = \overline{\lambda_2^{(N+1)}} y_{N+1}.$$

For the additional interpolation condition to be nontrivial, we require that $y_{N+1} \neq 0$; by a rescaling we can then assume that $\|y_{N+1}\| = 1$. By the adaptation of the result [4, Theorem 6,15] mentioned in connection with the interpolation conditions (5.53), we know that the augmented interpolation problem (5.57) has a rational matrix inner solution $\Theta$ if and only if the augmented Pick matrix is positive semidefinite:

$$\mathbb{P}_{\text{aug}} = \left[\frac{1 - \lambda_2^{(i)} \overline{\lambda}_2^{(j)}}{1 - \lambda_1^{(i)} \overline{\lambda}_1^{(j)}} \langle y_j, y_i \rangle_{\mathcal{Y}}\right]_{1 \le i, j \le N+1} \succeq 0. \tag{5.58}$$

We shall show that, in fact, given that $\mathbb{P} \succ 0$, we can find $\lambda_1^{(N+1)}, \lambda_2^{(N+1)}, y_{N+1}$ so that in fact $\mathbb{P}_{\text{aug}} \succ 0$ (strictly positive definite).

Toward this goal we do a Schur-complement analysis as follows. Write out $\mathbb{P}_{\rm aug}$ as

$$\mathbb{P}_{\rm aug} = \begin{bmatrix} \mathbb{P} & B \\ B^* & d \end{bmatrix},$$

where $B$ and $d$ are matrices of respective sizes $N \times 1$ and $1 \times 1$ given by

$$B = \begin{bmatrix} \frac{1-\lambda_2^{(i)}\overline{\lambda}_2^{(N+1)}}{1-\lambda_1^{(i)}\overline{\lambda}_1^{(N+1)}} \langle y_{N+1}, y_i \rangle \end{bmatrix}_{1 \le i \le N},$$

$$d = \frac{1-\lambda_2^{(N+1)}\overline{\lambda}_2^{(N+1)}}{1-\lambda_1^{(N+1)}\overline{\lambda}_1^{(N+1)}} \langle y_{N+1}, y_{N+1} \rangle = \frac{1-|\lambda_2^{(N+1)}|^2}{1-|\lambda_1^{(N+1)}|^2}.$$

Let us fix the unit vector $y_{N+1} \in \mathbb{C}^N$ but let $\left(\lambda_1^{(N+1)}, \lambda_2^{(N+1}\right)$ be a free parameter in the bidisk subject to additional restrictions still to be determined.

By the Schur-complement test, given that the scalar matrix $d$ is positive, the matrix $\mathbb{P}_{\rm aug}$ is invertible if and only if the Schur-complement matrix with pivot in the $(N+1, N+1)$-entry is positive definite, that is,

$$\mathbb{S} := \mathbb{P} - Bd^{-1}B^* \succ 0.$$

We have already given that $\mathbb{P}$ is positive definite, that is, there is $\epsilon_{\mathbb{P}} > 0$ so that

$$\mathbb{P} \succ \epsilon_{\mathbb{P}} I_N.$$

Thus it remains only to show that we can choose $\left(\lambda_1^{(N+1)}, \lambda_2^{(N+1)}\right) \in \mathbb{D}^2$ so that

$$Bd^{-1}B^* \preceq \epsilon_{\mathbb{P}} I_N. \tag{5.59}$$

Note next that, since $\lambda_1^{(i)}$ and $\lambda_2^{(i)}$ are fixed points in the open disk $\mathbb{D}$ for $1 \le i \le N$, elementary estimates show that there is a finite positive constant $M_B < \infty$ so that

$$\|B\| \le M_B \quad \text{for all } \left(\lambda_1^{(N+1)}, \lambda_2^{(N+1)}\right) \in \mathbb{D}^2.$$

As $Bd^{-1}B^* \preceq \|B\|^2 d^{-1} I_N \le M_B^2 d^{-1} I_N$, it suffices to find $\left(\lambda_1^{(N+1)}, \lambda_2^{(N+1)}\right) \in \mathbb{D}^2$ so that $N^2 M_B^2 d^{-1} < \epsilon_{\mathbb{P}}$, that is,

$$d := \frac{1 - \left|\lambda_2^{(N+1)}\right|^2}{1 - \left|\lambda_1^{(N+1)}\right|^2} > \frac{N^2 M_B^2}{\epsilon_{\mathbb{P}}},$$

or equivalently,

$$1 > \left|\lambda_2^{(N+1)}\right|^2 + \frac{N^2 M_B^2}{\epsilon_{\mathbb{P}}} \left(1 - \left|\lambda_1^{(N+1)}\right|^2\right). \tag{5.60}$$

Clearly this condition can be arranged by demanding that $|\lambda_1^{(N+1)}|$ is sufficiently close to 1 and then that $|\lambda_2^{(N+1)}|$ be sufficiently close to 0; let us say that any such $\left(\lambda_1^{(N+1)}, \lambda_2^{(N+1)}\right)$ is *admissible*.

**Step 2 Lack of uniqueness of Ando lift** In the terminology at the end of Step 1, let $(\mu_1, \mu_2)$ be one choice of admissible $\left(\lambda_1^{(N+1)}, \lambda_2^{(N+1)}\right)$. Again by the adapted result of Agler–McCarthy [4, Theorem 6.15], we can solve the associated interpolation problem

$$y_j^* \Theta_1 \left(\lambda_1^{(j))}\right)^* = y_j^* \overline{\lambda}_2^{(j)} \text{ for } j = 1, \ldots, N,$$
$$y_{N+1}^* \Theta_1(\mu_1)^* = y_{N+1}^* \overline{\mu}_2 \tag{5.61}$$

for a rational matrix inner function $\Theta_1$. In particular we see that $\mu_2$ is an eigenvalue for $\Theta_1(\mu_1)$. Let us consider a second interpolation problem for $\Theta_2$:

$$y_j^* \Theta_2 \left(\lambda_1^{(j)}\right)^* = y_j^* \overline{\lambda}_2^{(j)} \text{ for } j = 1, \ldots, N,$$
$$y_{N+1}^* \Theta_2(\mu_1)^* = y_{N+1}^* \overline{\mu}_2' \tag{5.62}$$

for an admissible pair $(\mu_1, \mu_2')$, where $\mu_2' \neq \mu_2$. By the robustness of the admissibility condition (5.60), we can find such a $\mu_2'$ so that $\mu_2'$ is not among the finitely many eigenvalues of $\Theta(\mu_1)$. Again by the Agler–McCarthy result [4, Theorem 6.15], we can solve for a second rational matrix inner function $\Theta_2(z)$ satisfying the interpolation conditions (5.62).

Then by the first part of the proof both $\left(\mathbf{\Pi}, M_z^{\mathcal{Y}}, M_{\Theta_1}\right)$ and $\left(\mathbf{\Pi}, M_z^{\mathcal{Y}}, M_{\Theta_2}\right)$ are minimal lifts of $(T_1, T_2)$. As both $\Theta_1$ and $\Theta_2$ are rational inner functions satisfying the interpolation conditions (5.53), it is also clear from the first part of the proof that both $(\mathbf{\Pi}, M_z, M_{\Theta_1})$ and $(\mathbf{\Pi}, M_z, M_{\Theta_2})$ are minimal Ando lifts for the commuting contractive pair $(T_1, T_2)$. Recall that unitary equivalence of lifts requires in particular that the commuting isometric pairs $\left(M_z^{\mathcal{Y}}, M_{\Theta_1}\right)$ and $\left(M_z^{\mathcal{Y}}, M_{\Theta_2}\right)$ be jointly unitarily equivalent. By Theorem 3.2.17, joint unitary equivalence of $\left(M_z^{\mathcal{Y}}, M_{\Theta_1}\right)$ with $\left(M_z^{\mathcal{Y}}, M_{\Theta_2}\right)$ happens exactly when there is a (constant) unitary operator $u$ on $\mathcal{F}$ so that

$$u\Theta_1(z) = \Theta_2(z)u,$$

for each $z \in \mathbb{D}$, that is, $\Theta_1(z)$ is unitarily equivalent to $\Theta_2(z)$ with the implementing unitary operator $u$ on $\mathcal{F}$ independent of $z$. But this fails for $z = \mu_1$ since the eigenvalues of $\Theta_1(\mu_1)$ fail to match up with the eigenvalues of $\Theta_2(\mu_1)$. We conclude that indeed the two minimal lifts $\left(\mathbf{\Pi}, M_z^{\mathcal{F}}, M_{\Theta_1}\right)$ and $\left(\mathbf{\Pi}, M_z^{\mathcal{F}}, M_{\Theta_1}\right)$ of $(T_1, T_2)$ are not unitarily equivalent as lifts.

**Example 5.5.3** We again let $(T_1, T_2)$ be a commuting contractive pair on a Hilbert space $\mathcal{H}$ of finite dimension $N$ with a basis of joint eigenvectors (5.47).

With some additional hypotheses in place we shall construct a lift $(\Pi, M_{z_1}^{\mathcal{Y}}, M_{z_2}^{\mathcal{Y}})$ with the commuting isometric pair $(M_{z_1}^{\mathcal{Y}}, M_{z_2}^{\mathcal{Y}})$ equal to the coordinate-function shift operators on the Hardy space over the bidisk $H^2_{\mathbb{D}^2}(\mathcal{Y})$ for an appropriate coefficient Hilbert space $\mathcal{Y}$. We shall then find a Type I Ando tuple $(\mathcal{F}, \Lambda, P, U)$ for $(T_1^*, T_2^*)$ that provides the set of parameters to build a Douglas model Ando lift unitarily equivalent to the bidisk Ando lift $(\Pi, M_{z_1}^{\mathcal{Y}}, M_{z_2}^{\mathcal{Y}})$.

We first introduce the required added hypotheses. It is known that there is a couple of extra conditions required for a given commuting contractive pair $(T_1, T_2)$ to have a lift $(\Pi, M_{z_1}^{\mathcal{Y}}, M_{z_2}^{\mathcal{Y}})$ to the bidisk shift tuple $(M_{z_1}^{\mathcal{Y}}, M_{z_2}^{\mathcal{Y}})$ acting on $H^2_{\mathbb{D}^2}(\mathcal{Y})$, namely, see Theorem 3.16 in [46] for $n = 2$ and with $T_j$ replaced there by $T_j^*$.

(i) The bidisk squared-defect operator

$$D^2_{T_1^*, T_2^*} := I - T_1T_1^* - T_2T_2^* + T_1T_2T_2^*T_1^* \tag{5.63}$$

should be positive semidefinite:

$$D_{T_1^*, T_2^*} \succeq 0.$$

(ii) The pair $(T_1, T_2)$ should be *bipure*, that is, recall from Definition 4.3.2 that both $T_1$ and $T_2$ should be *pure*:

$$\lim_{N\to\infty} \left\| T_j^{*n} h \right\|^2 = 0 \quad \text{for all } h \in \mathcal{H} \quad \text{for } j = 1, 2.$$

**A Quick Digression** We digress here for a couple of related remarks.

**Remark 5.5.4** Let us note that the bidisk defect operator $D_{T_1^*, T_2^*}$ can be viewed as an application of a version of the Agler hereditary functional calculus

$$f(\boldsymbol{\lambda}, \boldsymbol{\mu}) := \sum_{\mathbf{n},\mathbf{m}\in\mathbb{Z}_+^2} a_{\mathbf{n},\mathbf{m}} \boldsymbol{\lambda}^{\mathbf{n}} \overline{\boldsymbol{\mu}}^{\mathbf{m}} \mapsto \sum_{\mathbf{n},\mathbf{m}\in\mathbb{Z}_+^2} a_{\mathbf{n},\mathbf{m}} \mathbf{T}^{\mathbf{n}} \mathbf{T}^{*\mathbf{m}}$$

(here $\mathbf{n} = (n_1, n_2)$, $\mathbf{m} = (m_1, m_2)$, $\boldsymbol{\lambda} = (\lambda_1, \lambda_2)$, $\boldsymbol{\mu} = (\mu_1, \mu_2)$, $\boldsymbol{\lambda}^{\mathbf{n}} = \lambda_1^{n_1} \lambda_2^{n_2}$, $\overline{\boldsymbol{\mu}}^{\mathbf{n}} = \overline{\mu}_1^{n_1} \overline{\mu}_2^{n_2}$ with similarly conventions for operators: $\mathbf{T} = (T_1, T_2)$, $\mathbf{T}^* = (T_1^*, T_2^*)$, $\mathbf{T}^{\mathbf{n}} = T_1^{n_1} T_2^{n_2}$ and similarly for $\mathbf{T}^* = (T_1^*, T_2^*)$, where here $(T_1, T_2)$ is a commuting operator pair) applied to the function

$$f(\boldsymbol{\lambda}, \boldsymbol{\mu}) = (1 - \lambda_1\overline{\mu}_1)(1 - \lambda_2\overline{\mu}_2) = 1 - \lambda_1\overline{\mu}_1 - \lambda_2\overline{\mu}_2 + \lambda_1\lambda_2\overline{\mu}_1\overline{\mu}_2. \tag{5.64}$$

Here $T_1, T_2$ and $T_1^*, T_2^*$ commute but $T_j$ does not necessarily commute with $T_k^*$ for any pair of indices $j, k \in \{1, 2\}$. The hereditary functional calculus gives an ad hoc rule (in this case adjoint powers of $T_j$ on the right as in (5.63)) for plugging in noncommuting operator arguments into a function having commuting scalar arguments. The operator calculus of Ambrozie–Engliš–Müller [9] gets around this by defining a functional calculus on operators: Define $L_{T_j}, R_{T_j} \in \mathcal{B}(\mathcal{B}(\mathcal{H}))$ for $j = 1, 2$ by

$$L_{T_j} : X \mapsto T_j X, \quad R_{T_j^*} : X \mapsto X T_j^* \quad \text{for } X \in \mathcal{B}(\mathcal{H}).$$

Then the set of operators $L_{T_1}, L_{T_2}, R_{T_1^*}, R_{T_2^*}$ is a commuting set of operators in $\mathcal{B}(\mathcal{B}(\mathcal{H}))$ (given that $(T_1, T_2)$ is a commuting operator pair), and the function $f\left(L_{T_1}, L_{T_2}, R_{T_1}, R_{T_2}\right)$ is well-defined (here we use the substitution $\overline{\mu}_j \mapsto L_{T_j^*}$). Then the desired operator $D^2_{T_1^*, T_2^*}$ resulting from the hereditary functional calculus using the function (5.64) can be seen as applying the function (5.64) in the standard well-defined way to the commuting operator tuple $\left(L_{T_1}, L_{T_2}, R_{T_1^*}, R_{T_2^*}\right)$ and then evaluating the result on the identity operator $I_{\mathcal{H}}$:

$$f(\lambda, \mu) \mapsto f(L_{\mathbf{T}}, R_{\mathbf{T}^*})(I_{\mathcal{H}}) = D^2_{T_1^*, T_2^*} \quad \text{if } f \text{ is given by (5.64).}$$

**Remark 5.5.5** It turns out that the same condition $D_{T_1^*, T_2^*} \succeq 0$ is necessary and sufficient for the commuting contractive pair $(T_1, T_2)$ to have a regular unitary dilation, as originally discussed by Brehmer (see [92]). This connection between the existence of polydisk shift dilation and a regular unitary dilation is also discussed in Curto–Vasilescu [46] and Timotin [93].

**Continuation of Example 5.5.3** We now proceed as in Example 5.5.2 but with an adaptation to get a bidisk shift lift $\left(M_{z_1}^{\mathcal{Y}}, M_{z_2}^{\mathcal{Y}}\right)$ on $H^2_{\mathbb{D}^2}(\mathcal{Y})$ rather than a BDF-model lift $\left(M_z^{\mathcal{Y}}, M_\Theta\right)$ on $H^2(\mathcal{Y})$. We are given $(T_1, T_2)$ on a finite-dimensional Hilbert space $\mathcal{H}$ with a basis $\{v_1, \ldots, v_n\}$ of joint eigenvectors for $(T_1^*, T_2^*)$ with associated joint eigenvalues

$$\overline{\lambda}^{(j)} = \left(\overline{\lambda}_1^{(j)}, \overline{\lambda}_2^{(j)}\right) \in \mathbb{D}^2 \quad \text{for } 1 \le j \le N \text{ as in (5.47).}$$

Then we see that

$$\begin{aligned}\left\langle D^2_{T_1^*, T_2^*} v_j, v_i \right\rangle_{\mathcal{H}} &= \left\langle \left(1 - \lambda_1^{(i)} \overline{\lambda}_1^{(j)} - \lambda_2^{(i)} \overline{\lambda}_2^{(j)} + \lambda_1^{(i)} \lambda_2^{(i)} \overline{\lambda}_1^{(j)} \overline{\lambda}_2^{(j)}\right) v_j, v_i \right\rangle_{\mathcal{H}} \\ &= \left(1 - \lambda_1^{(i)} \overline{\lambda}_1^{(j)}\right)\left(1 - \lambda_2^{(j)} \overline{\lambda}_2^{(j)}\right) \langle v_j, v_i \rangle .\end{aligned}$$

Let us set $d = \operatorname{rank} D^2_{T_1^*, T_2^*}$. If we assume that $D^2_{T_1^*, T_2^*} \succeq 0$ (as we know must be the case if $(T_1, T_2)$ is to have a lift to the bidisk shift pair $\left(M_{z_1}^{\mathcal{Y}}, M_{z_2}^{\mathcal{Y}}\right)$ on $H^2_{\mathbb{D}^2}(\mathcal{Y})$ for some coefficient Hilbert space $\mathcal{Y}$), we see that the matrix on the left (rows indexed by $i$, columns indexed by $j$) is positive semidefinite. Hence there are vectors $y_1, \ldots, y_N$ in a $d$-dimensional Hilbert space $\mathcal{Y}$ so that

$$\left\langle D^2_{T_1^*, T_2^*} v_j, v_i \right\rangle_{\mathcal{H}} = \langle y_j, y_i \rangle_{\mathcal{Y}}.$$

By combining various of the preceding displayed identities and using the assumption that each $\lambda_{j,1}$ and $\lambda_{j,2}$ is in the open unit disk, we see that

$$\langle v_j, v_i \rangle_{\mathcal{H}} = \frac{\langle y_j, y_i \rangle_{\mathcal{Y}}}{\left(1 - \lambda_1^{(i)} \overline{\lambda}_1^{(j)}\right)\left(1 - \lambda_2^{(i)} \overline{\lambda}_2^{(j)}\right)}. \tag{5.65}$$

Let us now introduce the vectorial Hardy space over the bidisk $H^2_{\mathbb{D}^2}(\mathcal{Y})$ consisting of functions $f(z_1, z_2) = \sum_{n,m \geq 0} \widehat{f}_{n,m} z_1^n z_2^m$ with Fourier coefficients $\widehat{f}_{n,m} \in \mathcal{Y}$ subject to $\|f\|^2 := \sum_{n,m \geq 0} \|\widehat{f}_{n,m}\|^2_{\mathcal{Y}} < \infty$. This is a reproducing kernel Hilbert space with vectorial kernel functions $k_{\boldsymbol{\lambda}} y$ (for $\boldsymbol{\lambda} = (\lambda_1, \lambda_2) \in \mathbb{D}^2$ and $y \in \mathcal{Y}$) given by

$$(k_{\boldsymbol{\lambda}} y)(\mathbf{z}) = \frac{y}{\left(1 - z_1 \overline{\lambda}_1\right)\left(1 - z_2 \overline{\lambda}_2\right)}, \qquad \text{where we set } \mathbf{z} = (z_1, z_2)$$

having the reproducing kernel property:

$$\langle f, k_{\boldsymbol{\lambda}} y \rangle_{H^2_{\mathbb{D}^2}(\mathcal{Y})} = \langle f(\boldsymbol{\lambda}), y \rangle_{\mathcal{Y}}.$$

The identity (5.65) shows that the map

$$\Pi \colon v_j \mapsto k_{\boldsymbol{\lambda}^{(j)}} y_j \tag{5.66}$$

extends by linearity to a unitary map from $\mathcal{H}$ to the Hilbert space

$$\widetilde{\mathcal{H}} = \bigvee \{k_{\boldsymbol{\lambda}^{(j)}} y_j \colon j = 1, \ldots, N\} \subset H^2_{\mathbb{D}^2}(\mathcal{Y}). \tag{5.67}$$

If we let $M^{\mathcal{Y}}_{z_1}$ and $M^{\mathcal{Y}}_{z_2}$ be the coordinate multiplication operators

$$M^{\mathcal{Y}}_{z_1} \colon f(\mathbf{z}) \mapsto z_1 f(\mathbf{z}), \quad M^{\mathcal{Y}}_{z_2} \colon f(\mathbf{z}) \mapsto z_2 f(\mathbf{z})$$

on $H^2_{\mathbb{D}^2}$, then we see that

$$M^{\mathcal{Y}*}_{z_i} \Pi v_j = M^{\mathcal{Y}*}_{z_i}\left(k_{\boldsymbol{\lambda}^{(j)}} y_j\right) = \overline{\lambda}_i^{(j)} k_{\boldsymbol{\lambda}^{(j)}} y_j,$$

while

$$\Pi T_i^* v_j = \Pi\left(\overline{\lambda}_i^{(j)} v_j\right) = \overline{\lambda}_i^{(j)} (\Pi v_j) = \overline{\lambda}_i^{(j)} k_{\boldsymbol{\lambda}^{(j)}} y_j$$

showing that

$$\Pi T_i^* = M^{\mathcal{Y}*}_{z_j} \Pi \quad \text{on } \mathcal{H}.$$

We conclude that the operator triple

$$(\Pi, \mathbf{V}_1, \mathbf{V}_2) := \left(\Pi, M^{\mathcal{Y}}_{z_1}, M^{\mathcal{Y}}_{z_2}\right)$$

is an Ando lift for the commuting contractive pair $(T_1, T_2)$.

Furthermore we can see that $(\Pi, \mathbf{V}_1, \mathbf{V}_2)$ is a minimal lift for $(T_1, T_2)$ as follows. Note that

$$\left(I - \overline{\lambda}_1^{(j)} \mathbf{V}_1\right) \left(I - \overline{\lambda}_2^{(j)} \mathbf{V}_2\right) k_{\lambda^{(j)}} y_j = y_j \in \bigvee_{n_1, n_2 \in \mathbb{Z}_+} \mathbf{V}_1^{n_1} \mathbf{V}_2^{n_2} \operatorname{Ran} \mathbf{\Pi} =: \mathcal{K}_0$$

(where here we view each $y_j$ as a constant function in $H^2_{\mathbb{D}^2}(\mathcal{Y})$, the ambient subspace for the minimal lift contained inside $(\mathbf{\Pi}, \mathbf{V}_1, \mathbf{V}_2)$). We conclude that

$$\bigvee \{y_j : 1 \leq j \leq N\} \subset \mathcal{K}_0. \tag{5.68}$$

From (5.65) and the two displayed formulas preceding it, we see that

$$\langle y_i, y_j \rangle_{\mathcal{Y}} = \left\langle D^2_{T_1^*, T_2^*} v_i, v_j \right\rangle,$$

implying that the Gramian matrix for $y_1, \ldots, y_d$ has the same rank as $\operatorname{rank} D^2_{T_1^*, T_2^*} = d$. As we chose $\mathcal{H}$ to have $\dim \mathcal{H} = d$, we see that the rank of the Gramian matrix $\langle y_i, y_j \rangle_{\mathcal{Y}}$ is the same as the dimension of the whole space $\mathcal{Y}$, implying in turn that the span of the vectors $y_1, \ldots, y_N$ is equal to the whole space $\mathcal{Y}$. Combining with (5.68) then gives us

$$\mathcal{Y} \subset \mathcal{K}_0.$$

But then also

$$H^2_{\mathbb{D}^2}(\mathcal{Y}) = \bigvee_{n_1, n_2} \mathbf{V}_1^{n_1} \mathbf{V}_2^{n_2} \mathcal{Y} \subset \mathcal{K}_0 \subset H^2_{\mathbb{D}^2}(\mathcal{Y})$$

forcing the equality

$$\mathcal{K}_0 = H^2_{\mathbb{D}^2}(\mathcal{Y}),$$

that is, the lift $(\mathbf{\Pi}, V_1, V_2) = \left(\mathbf{\Pi}, M_{z_1}^{\mathcal{Y}}, M_{z_2}^{\mathcal{Y}}\right)$ is minimal as a lift of $(T_1, T_2)$.

We are now at the starting point of the proof of the converse direction in Theorem 5.2.1 to find the Douglas model for the commuting isometric lift $(\mathbf{\Pi}, \mathbf{V}_1, \mathbf{V}_2)$ of $(T_1, T_2)$. As we saw in Example 3.1.20, a BCL-tuple giving rise to the BCL1-model for $\left(M_{z_1}^{\mathcal{Y}}, M_{z_2}^{\mathcal{Y}}\right)$ on $H^2_{\mathbb{D}^2}(\mathcal{Y})$ can be taken to be

$$(\mathcal{F}, P, U) = \left(\ell^2_{\mathbb{Z}}(\mathcal{Y}), P_{\ell^2_{(1,\infty)}(\mathcal{Y})}, \mathbf{S}^{\mathcal{Y}}\right),$$

where $\mathbf{S}^{\mathcal{Y}}$ is the bilateral shift acting on $\ell^2_{\mathbb{Z}}(\mathcal{Y})$, with implementation operator

$$\tau_{\mathrm{bd}, \mathcal{Y}} : H^2_{\mathbb{D}^2}(\mathcal{Y}) \to H^2(\mathcal{F}) = H^2\left(\ell^2_{\mathbb{Z}}(\mathcal{Y})\right)$$

suggested by (3.46):

$$\tau_{\mathrm{bd}, \mathcal{Y}} : z_1^i z_2^j y \mapsto \begin{cases} \mathbf{e}_{j-i} y\, z^j & \text{for } i \geq j, \\ \mathbf{e}_{j-i} y\, z^i & \text{for } i \leq j, \end{cases} \tag{5.69}$$

for $y \in \mathcal{Y}$. It remains to identify the isometric embedding operator $\Lambda\colon \mathcal{D}_{T^*} \to \mathcal{F} = \ell^2_{\mathbb{Z}}(\mathcal{Y})$ so that the resulting Type I Ando tuple $(\mathcal{F}, \Lambda, P, U)$ is the parameter set generating a Douglas model lift (5.3) unitarily equivalent to our original lift $(\mathbf{\Pi}, \mathbf{V}_1, \mathbf{V}_2) = \big(\mathbf{\Pi}, M^{\mathcal{Y}}_{z_1}, M^{\mathcal{Y}}_{z_2}\big)$ with the commuting, isometric pair $M^{\mathcal{Y}}_{z_1}, M^{\mathcal{Y}}_{z_2}$ acting on $H^2_{\mathbb{D}^2}(\mathcal{Y})$. A careful interpretation of formula (5.15) gives us

$$\Lambda\colon D_{T^*} v_j \mapsto \big(\tau_{\mathrm{bd},\mathcal{Y}} k_{\lambda_j} y_j\big)(0).$$

Finally let us note that, given a commuting pair of finite-dimensional contraction operators $(T_1, T_2)$ with $D_{T_1^*, T_2^*} \succeq 0$, we have the option to ignore the latter condition and construct a lift of the form $\big(\Pi, M^{\mathcal{Y}}_z, M_\Theta\big)$ for a contractive analytic function $(\mathcal{Y}, \Theta)$ as in the previous Example 5.5.2. Thus this commuting contractive pair $(T_1, T_2)$ has two lifts: $(\Pi_1, M_z, M_\Theta)$ on $H^2(\mathcal{Y})$, as in Example 5.5.2, as well as in the discussion here: $\big(\Pi_2, M^{\mathcal{Y}}_{z_1}, M^{\mathcal{Y}}_{z_2}\big)$ on $H^2_{\mathbb{D}^2}(\mathcal{Y})$. Again these two lifts cannot be unitarily equivalent since the associated commuting isometric pairs $(V_1, V_2) = (M_z, M_\Theta)$ on $H^2(\mathcal{Y})$ and $\big(V_1', V_2'\big) = \big(M^{\mathcal{Y}}_{z_1}, M^{\mathcal{Y}}_{z_2}\big)$ on $H^2_{\mathbb{D}^2}(\mathcal{Y})$ are not unitarily equivalent as commuting isometric pairs (the second is doubly commuting while the first is not).

We end this section with the following remark.

**Remark 5.5.6** The reader may wonder why we use a BCL1-model for the Ando lift of a given contractive pair $(T_1, T_2)$ rather than a BCL2-model in our Douglas model for an Ando lift. Had we used a BCL2-model instead, we would have arrived at a notion of what we here call a Type I$'$ Ando tuple $(\mathcal{F}', \Lambda', P', Y')$ arising as follows. The form of the model (5.3) would have the adjusted form

$$\mathcal{K} = \begin{bmatrix} H^2(\mathcal{F}') \\ \mathcal{Q}_{T^*} \end{bmatrix},$$
$$(V_1, V_2) = \left( \begin{bmatrix} M_{(P'^{\perp} + zP')U'} & 0 \\ 0 & W_{\flat 1} \end{bmatrix}, \begin{bmatrix} M_{U'^*(P' + zP'^{\perp})U_*} & 0 \\ 0 & W_{\flat 2} \end{bmatrix} \right) \text{ acting on } \mathcal{K},$$
$$\Pi = \begin{bmatrix} (I_{H^2} \otimes \Lambda') \mathcal{O}_{D_{T^*}, T^*} \\ Q_{T^*} \end{bmatrix} : \mathcal{H} \to \mathcal{K}, \tag{5.70}$$

and the operator equations characterizing when such a collection of operators and spaces would actually yield an Ando lift of $(T_1, T_2)$, that is, the analogue of (5.1) and (5.2), would be

$$U'^* P' \Lambda' D_{T^*} T^* + U'^* P'^{\perp} \Lambda' D_{T^*} = \Lambda' D_{T^*} T_1^*, \tag{5.71}$$
$$P'^{\perp} U' \Lambda' D_{T^*} T^* + P' U' \Lambda' D_{T^*} = \Lambda' D_{T^*} T_2^*. \tag{5.72}$$

These equations are obtained by replacing the part of the data $(P', U')$ with its flipped version (see (3.13)),

$$\left(P'^{\mathfrak{f}}, U'^{\mathfrak{f}}\right) = (U'^* P' U', U'^*),$$

and then plugging this transformed data set into equations (5.1) and (5.2). The drawback of this approach is that then the analogue of Theorem 5.4.1 fails, that is, it need not be the case that a special Ando tuple is a Type I′ Ando tuple.

To construct an example, proceed as follows. Let $(\mathcal{F}, \Lambda, P, U)$ be the special Ando tuple for the BCL1-model isometric pair $(T_1, T_2)$

$$T_1 = M_{U^*(P^\perp + zP)}, \quad T_2 = M_{(P+zP^\perp)U}, \quad T = T_1 \cdot T_2 = M_z \quad \text{on } H^2(\mathcal{F}),$$

as used in Theorem 5.5.1. We wish to check the validity of equations (5.71) and (5.72) for this example, that is, we wish to check the validity of the system of equations

$$\begin{aligned} U^*P\Lambda D_{T^*}T^* + U^*P^\perp \Lambda D_{T^*} &= \Lambda D_{T^*}T_1^*, \\ P^\perp U\Lambda D_{T^*}T^* + PU\Lambda D_{T^*} &= \Lambda D_{T^*}T_2^*. \end{aligned}$$

Applying the first equation to a general element $h \in H^2(\mathcal{F})$ and using the computational experience developed in the proof of Theorem 5.5.1, we see that the first equation holds if and only if for all $h \in H^2(\mathcal{F})$ we have

$$U^*Ph'(0) + U^*P^\perp h(0) = (T_1^*h)(0) := P^\perp Uh(0) + PUh'(0).$$

For this identity to hold, it must be the case that coefficients of $h(0)$ and of $h'(0)$ match:

$$U^*P^\perp = P^\perp U, \quad U^*P = PU. \tag{5.73}$$

One can easily construct counterexamples, even with $\mathcal{F} = \mathbb{C}$, for example,

$$\mathcal{F} = \mathbb{C}, \quad U^2 \neq 1, \quad P = 1.$$

Thus the analogue of Theorem 5.4.1 with a Type I′ Ando tuple in place of a Type I Ando tuple fails in general. A similar analysis holds for the second equation: Applying the second equation to a general element $h \in H^2(\mathcal{F})$ leads to

$$P^\perp Uh'(0) + PUh(0) = (T_2^*h)(0) := U^*Ph(0) + U^*P^\perp h'(0),$$

which then leads to the same system of equations (5.73).

Finally, let us remark that it is a simple exercise using these computations as a model to verify directly that the Ando tuple $(\mathcal{F}, \Lambda, P, U)$ for $(T_1^*, T_2^*)$ coming from the BCL1-model for the isometric pair $(T_1, T_2)$ (with $T_1T_2$ equal to a shift) is a Type I Ando tuple, that is, that $(\mathcal{F}, \Lambda, P, U)$ together with $(T_1^*, T_2^*)$ satisfies equations (5.1) and (5.2).

## 5.6 The Sz.-Nagy–Foias Model for an Ando Isometric Lift

In this section, we convert the preceding analysis to a functional-model form to give a functional model for Ando lifts.

Let $\omega_D$ and $\omega_{\rm NF}$ be the unitaries as defined in (2.44) and (2.51), respectively. We observed in Remark 2.2.10 that the unitary

$$\omega_{\rm NF,D} := \omega_{\rm NF}\omega_{\rm D}^* \colon \mathcal{Q}_{T^*} \to \overline{\Delta_{\Theta_T} L^2(\mathcal{D}_{\Theta_T})}$$

intertwines $W_D$ with $M_\zeta^{\mathcal{D}_T}|_{\overline{\Delta_{\Theta_T} L^2(\mathcal{D}_T)}}$. Let us adopt the notation

$$\begin{aligned}&\left(W_{\sharp 1}, W_{\sharp 2}, M_\zeta^{\mathcal{D}_T}|_{\overline{\Delta_{\Theta_T} L^2(\mathcal{D}_T)}}\right)\\ &:= \left(\omega_{\rm NF,D} W_{\flat 1}\omega_{\rm NF,D}^*, \omega_{\rm NF,D} W_{\flat 2}\omega_{\rm NF,D}^*, \omega_{\rm NF,D} W_D \omega_{\rm NF,D}^*\right),\end{aligned} \tag{5.74}$$

where $(W_{\flat 1}, W_{\flat 2})$ is the canonical pair of commuting unitaries for $(T_1, T_2)$ (see Definition (4.4.3)). Let $(\mathcal{F}_*, \Lambda_*, P_*, U_*)$ be a Type I Ando tuple for $(T_1^*, T_2^*)$ and let $(\mathbf{\Pi}_D, \mathbf{V}_{D,1}, \mathbf{V}_{D,2})$ acting on $\mathcal{K}_D = \begin{bmatrix} H^2(\mathcal{F}_*) \\ \mathcal{Q}_{T^*}\end{bmatrix}$ be the Douglas model Ando lift of $(T_1, T_2)$ corresponding to the Type I Ando tuple $(\mathcal{F}_*, \Lambda_*, P_*, U_*)$ of $(T_1^*, T_2^*)$. By Theorem 5.2.1 such a lift exists and is given as in (5.3). Consider the unitary operator

$$\mathbf{U}_{\rm NF,D} := \begin{bmatrix} I_{H^2(\mathcal{F}_*)} & 0 \\ 0 & \omega_{\rm NF,D}\end{bmatrix} \colon \begin{bmatrix} H^2(\mathcal{F}_*) \\ \mathcal{Q}_{T^*}\end{bmatrix} \to \begin{bmatrix} H^2(\mathcal{F}_*) \\ \overline{\Delta_{\Theta_T} L^2(\mathcal{D}_T)}\end{bmatrix} \tag{5.75}$$

and define

$$(\mathbf{\Pi}_{\rm NF}, \mathbf{V}_{\rm NF,1}, \mathbf{V}_{\rm NF,2}) := (\mathbf{U}_{\rm NF,D}\mathbf{\Pi}_D, \mathbf{U}_{\rm NF,D}\mathbf{V}_{D,1}\mathbf{U}_{\rm NF,D}^*, \mathbf{U}_{\rm NF,D}\mathbf{V}_{D_2}\mathbf{U}_{\rm NF,D}^*). \tag{5.76}$$

Then we note that

$$\begin{aligned}&(\mathbf{V}_{\rm NF,1}, \mathbf{V}_{\rm NF,2}, \mathbf{V}_{\rm NF,1}\mathbf{V}_{\rm NF,2})\\ &= \left(\begin{bmatrix} M_{U_*^*(P_*^\perp + zP_*)} & 0 \\ 0 & W_{\sharp 1}\end{bmatrix}, \begin{bmatrix} M_{(P_* + zP_*^\perp)U_*} & 0 \\ 0 & W_{\sharp 2}\end{bmatrix}, \begin{bmatrix} M_z^{\mathcal{F}_*} & 0 \\ 0 & M_\zeta|_{\overline{\Delta_{\Theta_T} L^2(\mathcal{D}_T)}}\end{bmatrix}\right)\end{aligned} \tag{5.77}$$

and that the isometry $\mathbf{\Pi}_{\rm NF} \colon \mathcal{H} \to \mathcal{K}_{\rm NF} := \begin{bmatrix} H^2(\mathcal{F}_*) \\ \overline{\Delta_{\Theta_T} L^2(\mathcal{D}_T)}\end{bmatrix}$ is given by

$$\mathbf{\Pi}_{\rm NF} = \mathbf{U}_{\rm NF,D}\mathbf{\Pi}_D = \begin{bmatrix} \Lambda_* \mathcal{O}_{D_{T^*}, T^*} \\ \omega_{\rm NF,D}\mathcal{Q}_{T^*}\end{bmatrix} = \begin{bmatrix} I_{H^2} \otimes \Lambda_* & 0 \\ 0 & I_{\overline{\Delta_{\Theta_T} L^2(\mathcal{D}_T)}}\end{bmatrix} \Pi_{\rm NF} h, \tag{5.78}$$

where $\Pi_{\rm NF}$ is the embedding of $\mathcal{H}$ into $\begin{bmatrix} H^2(\mathcal{D}_{T^*}) \\ \overline{\Delta_{\Theta_T} L^2(\mathcal{D}_T)}\end{bmatrix}$ given by (2.58). Consequently, by Theorem 5.2.1 we have proved the following theorem, which gives a Sz.-Nagy–Foias functional model for an Ando lift of commuting contractive operator pair constructed canonically from a Type I Ando tuple $(\mathcal{F}_*, \Lambda_*, P_*, U_*)$ for $(T_1^*, T_2^*)$.

**Theorem 5.6.1** *Let* $(T_1, T_2)$ *be a commuting pair of contractions on a Hilbert space* $\mathcal{H}$ *and let* $(\mathcal{F}_*, \Lambda_*, P_*, U_*)$ *be a Douglas-minimal Type I Ando tuple for* $(T_1^*, T_2^*)$. *Define a pair* $(\mathbf{V}_{\mathrm{NF},1}, \mathbf{V}_{\mathrm{NF},2})$ *of commuting isometries on* $\mathcal{K}_{\mathrm{NF}} := \begin{bmatrix} H^2(\mathcal{F}_*) \\ \overline{\Delta_{\Theta_T} L^2(\mathcal{D}_T)} \end{bmatrix}$ *as in (5.77) and an embedding* $\mathbf{\Pi}_{\mathrm{NF}} \colon \mathcal{H} \to \mathcal{K}_{\mathrm{NF}}$ *as in* (5.78). *Then* $(\mathbf{\Pi}_{\mathrm{NF}}, \mathbf{V}_{\mathrm{NF},1}, \mathbf{V}_{\mathrm{NF},2})$ *is a minimal Ando lift of* $(T_1, T_2)$.

*Conversely, any minimal Ando lift* $(\mathbf{\Pi}, \mathbf{V}_1, \mathbf{V}_2)$ *of* $(T_1, T_2)$ *is unitarily equivalent (as an isometric lift) to an Ando lift of the form (5.77) coming from a Douglas-minimal Type I Ando tuple for* $(T_1^*, T_2^*)$.

## 5.7 Classification of Douglas/Sz.-Nagy–Foias Models for Ando Lifts

Let $(T_1, T_2)$ be a commuting contractive operator pair on a Hilbert space $\mathcal{H}$ and let $(\mathcal{F}_*, \Lambda_*, P_*, U_*)$ be an Ando tuple of Type I for $(T_1^*, T_2^*)$. Let us recall from Section 5.2 that the Douglas model of an Ando lift of a commuting contractive pair $(T_1, T_2)$ is described in (5.3), and that the model corresponds to Type I Ando tuples for $(T_1^*, T_2^*)$. This correspondence is unique in the sense stated below.

**Theorem 5.7.1** *Let* $(T_1, T_2)$ *be a commuting contractive operator pair on a Hilbert space* $\mathcal{H}$ *and let* $(\mathcal{F}_*, \Lambda_*, P_*, U_*)$, $(\mathcal{F}_*', \Lambda_*', P_*', U_*')$ *be two Type I Ando tuples for* $(T_1^*, T_2^*)$. *Let* $(\mathbf{\Pi}_D, \mathbf{V}_{D,1}, \mathbf{V}_{D,2})$ *and* $\left(\mathbf{\Pi}_D', \mathbf{V}_{D,1}', \mathbf{V}_{D,2}'\right)$ *be the Douglas model Ando lifts of* $(T_1, T_2)$ *corresponding to* $(\mathcal{F}_*, \Lambda_*, P_*, U_*)$ *and* $(\mathcal{F}_*', \Lambda_*', P_*', U_*')$ *as in Theorem 5.2.1. Then* $(\mathbf{\Pi}_D, \mathbf{V}_{D,1}, \mathbf{V}_{D,2})$ *and* $(\mathbf{\Pi}_D', \mathbf{V}_{D,1}, \mathbf{V}_{D,2})$ *are unitarily equivalent (as lifts of* $(T_1, T_2)$*) if and only if* $(\mathcal{F}_*, \Lambda_*, P_*, U_*)$ *and* $(\mathcal{F}_*', \Lambda_*', P_*', U_*')$ *coincide (as Ando tuples).*

*Proof* We first prove the “if” part. Suppose that two Type I Ando tuples $(\mathcal{F}_*, \Lambda_*, P_*, U_*)$ and $(\mathcal{F}_*', \Lambda_*', P_*', U_*')$ of $(T_1^*, T_2^*)$ coincide, that is, by Definition 5.1.2, there exists a unitary $u_* \colon \mathcal{F}_* \to \mathcal{F}_*'$ such that

$$u_* \Lambda_* = \Lambda_*' \quad \text{and} \quad u_*(P_*, U_*) = (P_*', U_*')u_*. \tag{5.79}$$

Define the unitary

$$\tilde{u}_* \colon \begin{bmatrix} (I_{H^2} \otimes u_*) & 0 \\ 0 & I_{Q_{T^*}} \end{bmatrix} \colon \begin{bmatrix} H^2(\mathcal{F}_*) \\ Q_{T^*} \end{bmatrix} \to \begin{bmatrix} H^2(\mathcal{F}_*') \\ Q_{T^*} \end{bmatrix}.$$

Then it follows from (5.79) that

$$\tilde{u}_* \mathbf{\Pi}_D = \begin{bmatrix} I_{H^2} \otimes u_* & 0 \\ 0 & I_{Q_{T^*}} \end{bmatrix} \begin{bmatrix} (I_{H^2} \otimes \Lambda_*) \, O_{D_{T^*}, T^*} \\ Q_{T^*} \end{bmatrix} = \begin{bmatrix} (I_{H^2} \otimes u_* \Lambda_*) \, O_{D_{T^*}, T^*} \\ Q_{T^*} \end{bmatrix}$$
$$= \begin{bmatrix} (I_{H^2} \otimes \Lambda'_*) \, O_{D_{T^*}, T^*} \\ Q_{T^*} \end{bmatrix} = \mathbf{\Pi}'_D$$

and

$$\tilde{u}_* \mathbf{V}_{D,1} = \begin{bmatrix} I_{H^2} \otimes u_* & 0 \\ 0 & I_{Q_{T^*}} \end{bmatrix} \begin{bmatrix} M_{U_*^* P_*^\perp + z U_*^* P_*} & 0 \\ 0 & W_{\flat 1} \end{bmatrix}$$
$$= \begin{bmatrix} M_{u_*(U_*^* P_*^\perp + z U_*^* P_*)} & 0 \\ 0 & W_{\flat 1} \end{bmatrix} = \begin{bmatrix} M_{(U'^* P_*'^\perp + z U'^* P_*') u_*} & 0 \\ 0 & W_{\flat 1} \end{bmatrix} = \mathbf{V}'_{D,1} \tilde{u}_*.$$

The intertwining $\tilde{u}_* \mathbf{V}_{D,2} = \mathbf{V}'_{D,2} \tilde{u}_*$ follows similarly. This establishes the equivalence of $(\mathbf{\Pi}_D, \mathbf{V}_{D,1}, \mathbf{V}_{D,2})$ and $\left(\mathbf{\Pi}'_D, \mathbf{V}'_{D,1}, \mathbf{V}'_{D,2}\right)$.

Conversely, suppose that the two Ando lifts $(\mathbf{\Pi}_D, \mathbf{V}_{D,1}, \mathbf{V}_{D,2})$ and $\left(\mathbf{\Pi}'_D, \mathbf{V}'_{D,1}, \mathbf{V}'_{D,2}\right)$ of $(T_1, T_2)$ are unitarily equivalent (as lifts of $(T_1, T_2)$) and are in the model form coming from two Type I Ando tuples $(\mathcal{F}_*, \Lambda_*, P_*, U_*)$ and $(\mathcal{F}'_*, \Lambda'_*, P'_*, U'_*)$ for $(T_1, T_2)$, respectively. This means that there exists a unitary $\tau_* : \mathcal{K}_D \to \mathcal{K}'_D$ such that

$$\tau_* \mathbf{\Pi}_D = \mathbf{\Pi}'_D, \quad \tau_*(\mathbf{V}_{D,1}, \mathbf{V}_{D,2}) = \left(\mathbf{V}'_{D,1}, \mathbf{V}'_{D,2}\right) \tau_*. \tag{5.80}$$

For more detailed calculations, let us introduce that $2 \times 2$-matrix representation for the unitary $\tau_*$ and the column representations for the spaces $\mathcal{K}_D$ and $\mathcal{K}'_D$:

$$\tau_* = \begin{bmatrix} \tau' & \tau_{12} \\ \tau_{21} & \tau'' \end{bmatrix} : \mathcal{K} := \begin{bmatrix} H^2(\mathcal{F}_*) \\ Q_{T^*} \end{bmatrix} \to \mathcal{K}' := \begin{bmatrix} H^2(\mathcal{F}'_*) \\ Q_{T^*} \end{bmatrix}.$$

From the second equality in (5.80) we see that $\tau_* \mathbf{V}_{D,1} \mathbf{V}_{D,2} = \mathbf{V}'_{D,1} \mathbf{V}'_{D,2} \tau_*$, which in detail becomes

$$\begin{bmatrix} \tau' & \tau_{12} \\ \tau_{21} & \tau'' \end{bmatrix} \begin{bmatrix} M_z^{\mathcal{F}_*} & 0 \\ 0 & W_D \end{bmatrix} = \begin{bmatrix} M_z^{\mathcal{F}'_*} & 0 \\ 0 & W_D \end{bmatrix} \begin{bmatrix} \tau' & \tau_{12} \\ \tau_{21} & \tau'' \end{bmatrix}.$$

As a consequence of part (ii) of Lemma 2.3.1, we see that

$$\tau_{12} = 0, \quad \tau_{21} = 0,$$

and $\tau_*$ has the diagonal form

$$\tau_* = \begin{bmatrix} \tau' & 0 \\ 0 & \tau'' \end{bmatrix},$$

with $\tau'$ and $\tau''$ also unitary and satisfying the intertwinings

$$\tau' M_z^{\mathcal{F}_*} = M_z^{\mathcal{F}'_*}\tau', \quad \tau'' W_D = W_D \tau''.$$

The first equality forces $\tau'$ to have the form

$$\tau' = I_{H^2} \otimes u_* \quad \text{for some unitary } u_* \colon \mathcal{F}_* \to \mathcal{F}'_*.$$

Then

$$\tau_* \Pi = \begin{bmatrix} I_{H^2} \otimes u_* & 0 \\ 0 & \tau'' \end{bmatrix} \begin{bmatrix} (I_{H^2} \otimes \Lambda_*)\, \mathcal{O}_{D_{T^*}, T^*} \\ Q_{T^*} \end{bmatrix} = \begin{bmatrix} (I_{H^2} \otimes u_* \Lambda_*)\, \mathcal{O}_{D_{T^*}, T^*} \\ \tau'' Q_{T^*} \end{bmatrix}$$

while, on the other hand,

$$\mathbf{\Pi}'_D = \begin{bmatrix} (I_{H^2} \otimes \Lambda'_*)\, \mathcal{O}_{D_{T^*}, T^*} \\ Q_{T^*} \end{bmatrix}.$$

Thus the first equality in (5.80) implies that

$$(I_{H^2} \otimes u_*)\, \Lambda_* = \Lambda'_*, \quad \tau'' Q_{T^*} = Q_{T^*}. \tag{5.81}$$

The second equality in (5.80) together with the intertwining $\tau'' W_D = W_D \tau''$ implies that $\tau''$ is equal to the identity on vectors of the form $W_D^n Q_{T^*} h$ with $h \in \mathcal{H}$. As $\cup_{n=0}^{\infty} W_D^n Q_{T^*} \mathcal{H}$ is dense in $\mathcal{Q}_{T^*}$, we conclude that $\tau'' = I_{\mathcal{Q}_{T^*}}$. Since $\tau_* = \left[\begin{smallmatrix} I_{H^2} \otimes u_* & 0 \\ 0 & I_{\mathcal{Q}_{T^*}} \end{smallmatrix}\right]$ intertwines $\mathbf{V}_{D,1}$ with $\mathbf{V}'_{D,1}$, where

$$\mathbf{V}_{D,1} = \begin{bmatrix} M_{U_*^* P_*^{\perp} + z U_*^* P_*} & 0 \\ 0 & W_{\flat 1} \end{bmatrix}, \quad \mathbf{V}'_{D,1} = \begin{bmatrix} M_{U_*'^* P_*'^{\perp} + z U_*'^* P'_*} & 0 \\ 0 & W_{\flat 1} \end{bmatrix},$$

we see that

$$u_*(U_*^* P_*^{\perp} + z U_*^* P_*) = (U_*'^* P_*'^{\perp} + z U_*'^* P'_*) u_*,$$

or equivalently,

$$u_* U_*^* P_*^{\perp} = U_*'^* P_*'^{\perp} u_*, \quad u_* U_*^* P_* = U_*'^* P'_* u_*. \tag{5.82}$$

These two equations together imply

$$u_* U_*^* = u_* U_*^* (P_*^{\perp} + P_*) = U_*'^* (P_*'^{\perp} + P'_*) u_* = U_*'^* u_*. \tag{5.83}$$

This and the last equality in (5.82) together establish the intertwining

$$u_* P_* = P'_* u_*.$$

The coincidence of the two Ando tuples $(\mathcal{F}_*, \Lambda_*, P_*, U_*)$ and $(\mathcal{F}'_*, \Lambda'_*, P'_*, U'_*)$ now follows. □

We have seen in Section 5.6 that the connection between the Douglas model Ando lift $(\mathbf{\Pi}_D, \mathbf{V}_{D,1}, \mathbf{V}_{D,2})$ and the Sz.-Nagy–Foias model Ando lift $(\mathbf{\Pi}_{\rm NF}, \mathbf{V}_{\rm NF,1}, \mathbf{V}_{\rm NF,2})$ is rather straightforward, namely:

$$\mathbf{\Pi}_{\rm NF} = \begin{bmatrix} I_{H^2(\mathcal{F}_*)} & 0 \\ 0 & \omega_{\rm NF,D} \end{bmatrix} \mathbf{\Pi}_D,$$

$$\mathbf{V}_{\rm NF,j} = \begin{bmatrix} I_{H^2(\mathcal{F}_*)} & 0 \\ 0 & \omega_{\rm NF,D} \end{bmatrix} V_{D,j} \begin{bmatrix} I_{H^2(\mathcal{F}_*)} & 0 \\ 0 & \omega^*_{\rm NF,D} \end{bmatrix} \text{ for } j = 1, 2.$$

Using this correspondence combined with the result of Theorem 5.7.1 gives us the following immediate corollary.

**Corollary 5.7.2** *Let $(T_1, T_2)$ be a commuting contractive operator pair on a Hilbert space $\mathcal{H}$ and let $(\mathcal{F}_*, \Lambda_*, P_*, U_*)$, $(\mathcal{F}'_*, \Lambda'_*, P'_*, U'_*)$ be two Type I Ando tuples for $(T_1^*, T_2^*)$. Let $(\mathbf{\Pi}_{\rm NF}, \mathbf{V}_{\rm NF,1}, \mathbf{V}_{\rm NF,2})$ and $\left(\mathbf{\Pi}'_{\rm NF}, \mathbf{V}'_{\rm NF,1}, \mathbf{V}'_{\rm NF,2}\right)$ be the Sz.-Nagy–Foias model Ando lifts of $(T_1, T_2)$ corresponding to $(\mathcal{F}_*, \Lambda_*, P_*, U_*)$ and $(\mathcal{F}'_*, \Lambda'_*, P'_*, U'_*)$ as in (5.77). Then $(\mathbf{\Pi}, \mathbf{V}_{\rm NF,1}, \mathbf{V}_{\rm NF,2})$ and $(\mathbf{\Pi}', \mathbf{V}_{\rm NF,1}, \mathbf{V}_{\rm NF,2})$ are unitarily equivalent (as lifts of $(T_1, T_2)$) if and only if $(\mathcal{F}_*, \Lambda_*, P_*, U_*)$ and $(\mathcal{F}'_*, \Lambda'_*, P'_*, U'_*)$ coincide (as Ando tuples).*

**Remark 5.7.3** A well-explored research field in any subject is perhaps not without a few surprising results. The dilation theory of commuting Hilbert space operators is certainly not one. Since every single contraction operator has an isometric lifting and every commuting contractive pair has a commuting isometric lifting, one would expect that a commuting contractive triple would also have a commuting isometric lifting. Parrott's example in [74], however, proves otherwise. This radical change in dimension three onwards may be explained by the 1978 model theory for commuting isometries found in [31]. It is shown in [31] (also discussed in Chapter 3) that every pair of commuting isometries can be modeled by an orthogonal projection $P$ and a unitary $U$ acting on a (coefficient) Hilbert space $\mathcal{F}$. Here the projection $P$ and the unitary $U$ are free parameters in the sense that they are independent of each other. For tuples of commuting isometries of length $d \geq 3$, however, one needs $d$ orthogonal projections $P_1, P_2, \ldots, P_d$ and unitary operators $U_1, U_2, \ldots, U_d$ acting on a coefficient Hilbert space $\mathcal{F}$ satisfying a stringent set of conditions for the model to be commuting. See [31, Section 3] for more details on the model and the conditions required for the model to be commuting.

In [21], the authors show that, for a commuting contractive tuple $(T_1, T_2, \ldots, T_d)$ of size $d \geq 3$, it is possible to construct an analogue of the canonical commuting unitary tuple $(W_1, W_2, \ldots, W_d)$ so that, just as in the pair case, $W_D = W_1 W_2 \cdots W_d$ with $W_D$ acting on $\mathcal{Q}_{T^*}$ is the unitary as in the Douglas construction of the isometric lift of $T = T_1 T_2 \cdots T_d$ discussed in Section 2.2.2.

It is even possible to construct an analogue of the Ando tuple $(\mathcal{F}, P_j, U_j)_{j=1}^d$ and an isometric embedding

$$\Pi\colon \mathcal{H} \to \begin{bmatrix} H^2(\mathcal{F}) \\ Q_{T^*} \end{bmatrix}$$

so that the lifting relations $\Pi T_j^* = V_j^* \Pi$ hold, where

$$V_j = \begin{bmatrix} M_{P_j^\perp U_j + zP_jU_j} & 0 \\ 0 & W_j \end{bmatrix} \quad \text{for } j = 1, 2, \ldots, d.$$

However, as expected, the projections $P_j$ and the unitaries $U_j$ so constructed do not lead to a model isometric tuple $(V_1, V_2, \ldots, V_d)$ that is commutative.

There have been various other attempts at understanding the dilation theory for commuting contractive $d$-tuples of higher order $d \geq 3$; see the recent papers by Das and coworkers [25–27] and the references therein for such attempts. These papers find some classes of commuting contractive tuples satisfying certain positivity conditions that can be lifted to commuting isometric tuples.

## 5.8 Notes to Chapter 5

**1.** Let us note that the partially defined unitary in (5.29) (or in (4.5)) played a pivotal role in the original proof of Ando dilation theorem presented in Section 4.2. It played an important role in later works of Das and coworkers (see [47] and [48]) as well, where constructions of Ando isometric lifts were made in special cases, specifically, in the case where the commutative, contractive pair $(T_1, T_2)$ is bipure. Indeed, the importance of what is called an Ando tuple here was made explicit in the seminal work [48].

**2.** The construction in Example 5.5.2 of an Ando lift for a contractive matrix pair $(T_1, T_2)$ with a basis of joint eigenvectors with corresponding joint eigenvalues all contained in $\mathbb{D}^2$ follows the argument of Agler–McCarthy [5] toward the same result. Our invocation of the Nevanlinna–Pick interpolation theorem short-circuits the latter part of the argument there. Instead, Agler–McCarthy use a *lurking-isometry* argument; in the end, the content is the same since such a lurking-isometry argument is one way to prove the matrix Nevanlinna–Pick interpolation theorem (see [4, 14]). In any case, let us note that either approach gives a constructive elementary proof of the Ando lifting theorem for the case where the commuting contractive pair acts on a finite-dimensional Hilbert space with a basis of joint eigenvectors with joint eigenvalues in the open bidisk.

**3.** Let us catalogue here the open problem mentioned in Section 5.4:

*Is every Type I Ando tuple for a given contractive pair $(T_1, T_2)$ also special?*

# 6

# Schäffer-Type Model for Ando Isometric Lifts

Recall from Section 2.2.1 that the minimal isometric lift of a given contraction operator $T$ acting on a Hilbert space $\mathcal{H}$ constructed by Schäffer is given by

$$\begin{bmatrix} T & 0 \\ \mathbf{ev}^*_{0,\mathcal{D}_T} D_T & M_z^{\mathcal{D}_T} \end{bmatrix} : \begin{bmatrix} \mathcal{H} \\ H^2(\mathcal{D}_T) \end{bmatrix} \to \begin{bmatrix} \mathcal{H} \\ H^2(\mathcal{D}_T) \end{bmatrix}.$$

In this chapter, we shall construct a Schäffer-type model for minimal Ando isometric lifts of a given commuting contractive pair $(T_1, T_2)$. The key tool for this is the notion of Type II Ando tuples (see Definition 6.1.2). These are Ando tuples satisfying a different set of conditions from those that Type I Ando tuples satisfy (Definition 5.1.5). Remarkably, the special Ando tuples, constructed in Section 5.3, turn out to be Type II Ando tuples; therefore, a new proof of the Ando Dilation Theorem emerges as a by-product. This chapter may be viewed as a Schäffer parallel of Chapter 5.

## 6.1 The Construction

Let $(T_1, T_2)$ on $\mathcal{H}$ be a pair of commuting contractions on the Hilbert space $\mathcal{H}$ and let $(\Pi, V_1, V_2)$ with $\Pi\colon \mathcal{H} \to \mathcal{K}$ and $V_1, V_2$ on $\mathcal{K}$ be an Ando lift of $(T_1, T_2)$. Up to unitary equivalence we may arrange that $\mathcal{H} \subset \mathcal{K}$ and $\Pi\colon \mathcal{H} \to \mathcal{K}$ is the inclusion map. Hence, with respect to the decomposition $\mathcal{K} = \mathcal{H} \oplus (\mathcal{K} \ominus \mathcal{H})$, we have

$$(V_1, V_2) = \left( \begin{bmatrix} T_1 & 0 \\ C_1 & D_1 \end{bmatrix}, \begin{bmatrix} T_2 & 0 \\ C_2 & D_2 \end{bmatrix} \right)$$

for some operators

$$C_j\colon \mathcal{H} \to \mathcal{K} \ominus \mathcal{H}, \quad D_j\colon \mathcal{K} \ominus \mathcal{H} \to \mathcal{K} \ominus \mathcal{H} \tag{6.1}$$

for $j = 1, 2$. As $V_1$ and $V_2$ commute, we must also have that

$$\begin{bmatrix} T_1 & 0 \\ C_1 & D_1 \end{bmatrix} \begin{bmatrix} T_2 & 0 \\ C_2 & D_2 \end{bmatrix} = \begin{bmatrix} T_2 & 0 \\ C_2 & D_2 \end{bmatrix} \begin{bmatrix} T_1 & 0 \\ C_1 & D_1 \end{bmatrix},$$

leading to the matrix identity

$$V := V_1 V_2 = \begin{bmatrix} T_1 T_2 & 0 \\ C_1 T_2 + D_1 C_2 & D_1 D_2 \end{bmatrix} = \begin{bmatrix} T_2 T_1 & 0 \\ C_2 T_1 + D_2 C_1 & D_2 D_1 \end{bmatrix} = V_2 V_1, \tag{6.2}$$

which gives us the operator identities (in addition to the assumed commutativity of $(T_1, T_2)$)

$$C_1 T_2 + D_1 C_2 = C_2 T_1 + D_2 C_1, \quad D_1 D_2 = D_2 D_1. \tag{6.3}$$

From the fact that each $V_j$ $(j = 1, 2)$ is an isometry, we have

$$\begin{bmatrix} T_j^* & C_j^* \\ 0 & D_j^* \end{bmatrix} \begin{bmatrix} T_j & 0 \\ C_j & D_j \end{bmatrix} = \begin{bmatrix} I_{\mathcal{H}} & 0 \\ 0 & I_{\mathcal{K} \ominus \mathcal{H}} \end{bmatrix},$$

giving us the identities

$$C_j^* C_j = I - T_j^* T_j, \quad C_j^* D_j = 0, \quad D_j^* D_j = I_{\mathcal{K} \ominus \mathcal{H}} \quad \text{for } j = 1, 2. \tag{6.4}$$

In particular, from the last identity in (6.3) and in (6.4) we see that the pair $(D_1, D_2)$ is a commuting isometric pair on $\mathcal{K} \ominus \mathcal{H}$. In constructing the Schäffer-type model for Ando isometric lift, for reasons to become clear later, we choose to work with the BCL2-model (3.10) for the commuting isometric pair $(D_1, D_2)$ rather than with the BCL1-model (3.9) for $(V_1, V_2)$ as we did for the Douglas model. Hence by Theorem 3.1.4 there exist a Hilbert space $\mathcal{F}$, a projection $P$ and unitary $U$ acting on $\mathcal{F}$ and a pair $(Y_1, Y_2)$ of commuting unitaries acting on some Hilbert space $\mathcal{Y}$, and a unitary identification map $\tau_{\rm BCL} \colon \mathcal{K} \ominus \mathcal{H} \to \mathcal{K}_S := \begin{bmatrix} H^2(\mathcal{F}) \\ \mathcal{Y} \end{bmatrix}$ so that we have

$$\tau_{\rm BCL}\,(D_1, D_2) = \left( \begin{bmatrix} M_{(P^\perp + zP)U} & 0 \\ 0 & Y_1 \end{bmatrix}, \begin{bmatrix} M_{U^*(P + zP^\perp)} & 0 \\ 0 & Y_2 \end{bmatrix} \right) \tau_{\rm BCL}$$

and the product operator $D = V_1 V_2$ satisfying

$$\tau_{\rm BCL} D := \begin{bmatrix} M_z & 0 \\ 0 & Y \end{bmatrix}, \quad \text{where } Y = Y_1 Y_2 = Y_2 Y_1.$$

For $j = 1, 2$ let us define operators $\begin{bmatrix} C_{s1} \\ C_{u2} \end{bmatrix}, \begin{bmatrix} C_{s2} \\ C_{u2} \end{bmatrix}, \begin{bmatrix} C_s \\ C_u \end{bmatrix}$ from $\mathcal{H}$ to $\begin{bmatrix} H^2(\mathcal{F}) \\ \mathcal{Y} \end{bmatrix}$ by

$$\begin{bmatrix} C_{s1} \\ C_{u1} \end{bmatrix} = \tau_{\rm BCL} C_1, \quad \begin{bmatrix} C_{s2} \\ C_{u2} \end{bmatrix} = \tau_{\rm BCL} C_2, \quad \begin{bmatrix} C_s \\ C_u \end{bmatrix} = \tau_{\rm BCL} C,$$

where the operators $C_1$ and $C_2$ are as in (6.1) and where $C$ is given by

$C = [V_1V_2]_{21} = C_1T_2+D_1C_2$ or $C = [V_2V_1]_{21} = C_2T_1+D_2C_1$ as in (6.2),

where the subscripts $s$ and $u$ indicate the *shift* component $H^2(\mathcal{F})$ and the *unitary* component $\mathcal{Y}$, respectively. Let us denote $V = V_1V_2$ and $Y = Y_1Y_2$. Then we have

$$\begin{bmatrix} I_{\mathcal{H}} & 0 \\ 0 & \tau_{\text{BCL}} \end{bmatrix} (V_1, V_2, V) = (\mathbf{V}_{S,1}, \mathbf{V}_{S,2}, \mathbf{V}_S) \begin{bmatrix} I_{\mathcal{H}} & 0 \\ 0 & \tau_{\text{BCL}} \end{bmatrix},$$

where

$$(\mathbf{V}_{S,1}, \mathbf{V}_{S,2}, \mathbf{V}_S)$$
$$= \left( \begin{bmatrix} T_1 & 0 & 0 \\ C_{s1} & M_{(P^\perp+zP)U} & 0 \\ C_{u1} & 0 & Y_1 \end{bmatrix}, \begin{bmatrix} T_2 & 0 & 0 \\ C_{s2} & M_{U^*(P+zP^\perp)} & 0 \\ C_{u2} & 0 & Y_2 \end{bmatrix}, \begin{bmatrix} T & 0 & 0 \\ C_s & M_z & 0 \\ C_u & 0 & Y \end{bmatrix} \right). \tag{6.5}$$

The identities (6.3) lead to the equalities

$$\begin{aligned} \begin{bmatrix} C_s \\ C_u \end{bmatrix} &= \begin{bmatrix} C_{s1} \\ C_{u1} \end{bmatrix} T_2 + \begin{bmatrix} M_{P^\perp U+zPU} & 0 \\ 0 & Y_1 \end{bmatrix} \begin{bmatrix} C_{s2} \\ C_{u2} \end{bmatrix} \\ &= \begin{bmatrix} C_{s2} \\ C_{u2} \end{bmatrix} T_1 + \begin{bmatrix} M_{U^*P+zU^*P^\perp} & 0 \\ 0 & Y_2 \end{bmatrix} \begin{bmatrix} C_{s1} \\ C_{u1} \end{bmatrix}. \end{aligned} \tag{6.6}$$

Similarly, (6.4) leads to

$$\begin{aligned} I_{\mathcal{H}} &= T_1^*T_1 + C_{s1}^*C_{s1} + C_{u1}^*C_{u1} = T_2^*T_2 \\ &\quad + C_{s2}^*C_{s2} + C_{u2}^*C_{u2} = T^*T + C_s^*C_s + C_u^*C_u, \\ 0 &= C_{s1}^*M_{(P^\perp+zP)U} = C_{s2}^*M_{U^*(P+zP^\perp)} = C_s^*M_z = 0, \\ C_{u1}^*Y_1 &= C_{u2}^*Y_2 = C_u^*Y = 0. \end{aligned} \tag{6.7}$$

Multiply the first equation in (6.6) on the left by $\begin{bmatrix} M_{(P^\perp+zP)U} & 0 \\ 0 & Y_1 \end{bmatrix}^*$ to get

$$\begin{bmatrix} C_{s2} \\ C_{u2} \end{bmatrix} = \left( \begin{bmatrix} M_{P^\perp U+zPU} & 0 \\ 0 & Y_1 \end{bmatrix} \right)^* \begin{bmatrix} C_s \\ C_u \end{bmatrix} - \left( \begin{bmatrix} (M_{P^\perp U+zPU} & 0 \\ 0 & Y_1 \end{bmatrix} \right)^* \begin{bmatrix} C_{s1} \\ C_{u1} \end{bmatrix} T_2. \tag{6.8}$$

However, by taking adjoints in the second and third lines of (6.7) we see in particular that

$$\left( \begin{bmatrix} M_{P^\perp U+zPU} & 0 \\ 0 & Y_1 \end{bmatrix} \right)^* \begin{bmatrix} C_{s1} \\ C_{u1} \end{bmatrix} = 0.$$

Thus equation (6.8) simplifies to

$$\begin{bmatrix} C_{s2} \\ C_{u2} \end{bmatrix} = \left( \begin{bmatrix} M_{P^\perp U+zPU} & 0 \\ 0 & Y_1 \end{bmatrix} \right)^* \begin{bmatrix} C_s \\ C_u \end{bmatrix},$$

or, more simply,

$$C_{s2} = (M_{P^{\perp}U+zPU})^{*}\, C_s, \quad C_{u2} = Y_1^{*} C_u. \tag{6.9}$$

A similar analysis starting with the second equation in (6.6) leads to

$$\begin{bmatrix} C_{s1} \\ C_{u1} \end{bmatrix} = \left( \begin{bmatrix} M_{U^{*}P+zU^{*}P^{\perp}} & 0 \\ 0 & Y_2 \end{bmatrix} \right)^{*} \begin{bmatrix} C_s \\ C_u \end{bmatrix}$$

or, more simply,

$$C_{s1} = \left(M_{U^{*}(P+zP^{\perp})}\right)^{*} C_s, \quad C_{u1} = Y_2^{*} C_u. \tag{6.10}$$

From the second and third lines in (6.7) we see in particular that

$$M_z^{*} C_s = 0, \quad Y^{*} C_u = 0. \tag{6.11}$$

The first item in (6.11) forces $C_s$ to have the form

$$C_s = \mathbf{ev}_{0,\mathcal{F}}^{*} C_{s0}, \qquad \text{where } C_{s0} \colon \mathcal{F} \to \mathcal{F}. \tag{6.12}$$

Since $Y$ is unitary, the second item in (6.11) forces

$$C_u = 0. \tag{6.13}$$

From the second equations in (6.10) and (6.9) we get the further collapsing

$$C_{u1} = 0, \quad C_{u2} = 0. \tag{6.14}$$

From the first line in (6.7) we see in particular that $I_{\mathcal{H}} - T^{*}T = C_s^{*} C_s + C_u^{*} C_u$. Since we now have established that $C_u = 0$, this simplifies to

$$I_{\mathcal{H}} - T^{*}T = C_s^{*} C_s = C_{s0}^{*} C_{s0}.$$

This identity in turn tells us that we can factor $C_{s0}$ as

$$C_{s0} = \Lambda D_T, \tag{6.15}$$

where $\Lambda \colon \mathcal{D}_T \to \mathcal{F}$ is an isometry.

Finally, by combining the first equations in (6.9) and (6.10), together with (6.12), (6.15), and (6.14), we see that

$$\begin{aligned}
\begin{bmatrix} C_{s1} \\ C_{u1} \end{bmatrix} &= \begin{bmatrix} \left(M_{U^{*}(P+zP^{\perp})}\right)^{*} \mathbf{ev}_{0,\mathcal{F}}^{*} \Lambda D_T \\ 0 \end{bmatrix} = \begin{bmatrix} \mathbf{ev}_{0,\mathcal{F}}^{*} PU\Lambda D_T \\ 0 \end{bmatrix}, \\
\begin{bmatrix} C_{s2} \\ C_{u2} \end{bmatrix} &= \begin{bmatrix} \left(M_{(P^{\perp}+zP)U}\right)^{*} \mathbf{ev}_{0,\mathcal{F}}^{*} \Lambda D_T \\ 0 \end{bmatrix} = \begin{bmatrix} \mathbf{ev}_{0,\mathcal{F}}^{*} U^{*}P^{\perp}\Lambda D_T \\ 0 \end{bmatrix}, \\
\begin{bmatrix} C_s \\ C_u \end{bmatrix} &= \begin{bmatrix} \mathbf{ev}_{0,\mathcal{F}}^{*} \Lambda D_T \\ 0 \end{bmatrix}.
\end{aligned} \tag{6.16}$$

Let us summarize: Given a commuting contractive operator pair $(T_1, T_2)$ on $\mathcal{H}$ having an Ando isometric lift $(\Pi, V_1, V_2)$, we have now come upon a collection $(\mathcal{F}, \Lambda, U, P)$, where $\mathcal{F}$ is another Hilbert space, $\Lambda\colon \mathcal{D}_T \to \mathcal{F}$ is an isometry (where $T = T_1T_2$), $P$ and $U$ are operators on $\mathcal{F}$ with $P$ equal to a projection and $U$ equal to a unitary operator, respectively, that is, in the terminology defined in Definition 5.1.1, and the collection $(\mathcal{F}, \Lambda, P, U)$ is a *Ando tuple for* $(T_1, T_2)$. All this leads to a proof of the following result.

**Theorem 6.1.1** *Given any Ando lift* $(\Pi, V_1, V_2)$ *of a given commuting contractive operator pair* $(T_1, T_2)$*, there is an Ando tuple* $(\mathcal{F}, \Lambda, P, U)$ *for* $(T_1, T_2)$*, another Hilbert space* $\mathcal{Y}$ *along with a commuting pair* $(Y_1, Y_2)$ *of unitary operators on* $\mathcal{Y}$*, together with a unitary operator* $\boldsymbol{\tau}$ *from* $\mathcal{K}$ *onto the Schäffer model space* $\mathcal{K}_S$ *defined below, so that*

$$(\boldsymbol{\tau}\Pi, \boldsymbol{\tau}V_1\boldsymbol{\tau}^*, \boldsymbol{\tau}V_2\boldsymbol{\tau}^*, \boldsymbol{\tau}V\boldsymbol{\tau}^*) = \left(\mathbf{\Pi}_S, \mathbf{V}_{S,1}, \mathbf{V}_{S,2}, \mathbf{V}_S\right),$$

*where*

$$\mathcal{K}_S = \begin{bmatrix} \mathcal{H} \\ H^2(\mathcal{F}) \\ \mathcal{Y} \end{bmatrix}, \quad \mathbf{\Pi}_S = \begin{bmatrix} I_{\mathcal{H}} \\ 0 \\ 0 \end{bmatrix} : \mathcal{H} \to \mathcal{K}_S,$$

$$(\mathbf{V}_{S,1}, \mathbf{V}_{S,2}) = \left( \begin{bmatrix} T_1 & 0 & 0 \\ \mathbf{ev}^*_{0,\mathcal{F}} PU\Lambda D_T & M_{P^\perp U + zPU} & 0 \\ 0 & 0 & Y_1 \end{bmatrix}, \begin{bmatrix} T_2 & 0 & 0 \\ \mathbf{ev}^*_{0,\mathcal{F}} U^*P^\perp\Lambda D_T & M_{U^*P + zU^*P^\perp} & 0 \\ 0 & 0 & Y_2 \end{bmatrix} \right). \tag{6.17}$$

*Furthermore one can choose the isometry* $\Lambda$ *so that*

$$\mathbf{V}_S := \mathbf{V}_{S,1}\mathbf{V}_{S,2} = \mathbf{V}_{S,2}\mathbf{V}_{S,1} = \begin{bmatrix} T & 0 & 0 \\ \mathbf{ev}^*_{0,\mathcal{F}}\Lambda D_T & M_z^{\mathcal{F}} & 0 \\ 0 & 0 & Y \end{bmatrix}, \tag{6.18}$$

*where we set* $T := T_1T_2 = T_2T_1$, $Y := Y_1Y_2 = Y_2Y_1$.

We now investigate the converse direction; that is, given a pair of commuting contractions $(T_1, T_2)$, an Ando tuple for $(T_1, T_2)$, and a Hilbert space $\mathcal{Y}$ equipped with a commuting pair of unitary operators $(Y_1, Y_2)$, when is the pair $\left(\mathbf{V}_{S,1}, \mathbf{V}_{S,2}\right)$ as defined in (6.17) (clearly a lift of $(T_1, T_2)$ as a result of the triangular form in (6.17)) a commuting pair of isometries? It turns out that the answer to this question is negative in general: The Ando tuple must satisfy some additional conditions, which we now explore. This discussion motivates the following definition.

**Definition 6.1.2** Let $(T_1, T_2)$ be a pair of commuting contractions on a Hilbert space $\mathcal{H}$. An Ando tuple $(\mathcal{F}, \Lambda, P, U)$ for $(T_1, T_2)$ is called a *Type II Ando tuple* if we have the following.

(i) **Commutativity**

$$PU\Lambda D_T T_2 + P^{\perp}\Lambda D_T = U^* P^{\perp}\Lambda D_T T_1 + U^* PU\Lambda D_T = \Lambda D_T.$$

(ii) **Isometry**

$$D_T\Lambda^* U^* PU\Lambda D_T = D_{T_1}^2, \quad D_T\Lambda^* P^{\perp}\Lambda D_T = D_{T_2}^2.$$

We say that $(\mathcal{F}, \Lambda, P, U)$ is a *weak Type II Ando tuple* if it satisfies item (ii) and the following weak form of item (i):

(i′) $PU\Lambda D_T T_2 + P^{\perp}\Lambda D_T = U^* P^{\perp}\Lambda D_T T_1 + U^* PU\Lambda D_T$.

We say that an Ando tuple $(\mathcal{F}, \Lambda, P, U)$ is *Schäffer-minimal* if

$$\bigvee_{n_1,n_2\geq 0} \begin{bmatrix} T_1 & 0 \\ \mathbf{ev}_{0,\mathcal{F}}^* PU\Lambda D_T & M_{P^{\perp}U+zPU} \end{bmatrix}^{n_1} \begin{bmatrix} T_2 & 0 \\ \mathbf{ev}_{0,\mathcal{F}}^* U^*P^{\perp}\Lambda D_T & M_{U^*P+zU^*P^{\perp}} \end{bmatrix}^{n_2} \begin{bmatrix} \mathcal{H} \\ 0 \end{bmatrix}$$
$$= \begin{bmatrix} \mathcal{H} \\ H^2(\mathcal{F}) \end{bmatrix}.$$

A minimal Type II or weak Type II Ando tuple is a Schäffer-minimal Ando tuple.

We shall see in Section 6.4 that the class of Type II Ando tuples is strictly smaller than the class of weak Type II Ando tuples by concrete examples.

**Remark 6.1.3** A couple of remarks are in order.

(i) The point of distinction between Type II and weak Type II Ando tuples is the following. The pair

$$(V_1, V_2) = \left[\begin{bmatrix} T_1 & 0 \\ \mathbf{ev}_{0,\mathcal{F}}^* PU\Lambda D_T & M_{P^{\perp}U+zPU} \end{bmatrix}\right] \left[\begin{bmatrix} T_2 & 0 \\ \mathbf{ev}_{0,\mathcal{F}}^* U^*P^{\perp}\Lambda D_T & M_{U^*P+zU^*P^{\perp}} \end{bmatrix}\right]$$

is a commuting isometric pair acting on $\begin{bmatrix} \mathcal{H} \\ H^2(\mathcal{F}) \end{bmatrix}$ in both the case of Type II and weak Type II Ando tuples $(\mathcal{F}, \Lambda, P, U)$, and therefore is an Ando isometric lift of $(T_1, T_2)$. The difference is that only in the case where $(\mathcal{F}, \Lambda, P, U)$ is a Type II Ando tuple do we have that the product isometry $V_1V_2$ has the form

$$V_1V_2 = \begin{bmatrix} T & 0 \\ \mathbf{ev}_{0,\mathcal{F}}^* \Lambda D_T & M_z \end{bmatrix} = \begin{bmatrix} I_{\mathcal{H}} & 0 \\ 0 & I_{H^2}\otimes\Lambda \end{bmatrix} \begin{bmatrix} T & 0 \\ \mathbf{ev}_{0,\mathcal{D}_T}^* D_T & M_z \end{bmatrix}.$$

Thus $V_1V_2$, in the case of Type II Ando tuples, is the product of the isometry

$$\begin{bmatrix} I_{\mathcal{H}} & 0 \\ 0 & I_{H^2}\otimes\Lambda \end{bmatrix} : \begin{bmatrix} \mathcal{H} \\ H^2(\mathcal{D}_T) \end{bmatrix} \to \begin{bmatrix} \mathcal{H} \\ H^2(\mathcal{F}) \end{bmatrix}$$

and the Schäffer-minimal isometric lift of the product contraction $T = T_1T_2$. For this reason, Type II Ando tuples are more convenient to work with than weak Type II Ando tuples. This is demonstrated in Section 6.5.

(ii) Let us note that both notions, the *Type II Ando tuple* and the *weak Type II Ando tuple*, are *coordinate-free* in the following sense: *If* $(\mathcal{F}, \Lambda, P, U)$ *is a Type II (respectively weak Type II) Ando tuple for the commuting contractive pair* $(T_1, T_2)$, *and* $(\mathcal{F}', \Lambda', P', U')$ *is an Ando tuple for* $(T_1, T_2)$ *that coincides with* $(\mathcal{F}, \Lambda, P, U)$ *in the sense of Definition 5.1.2, then* $(\mathcal{F}', \Lambda', P', U')$ *is also a Type II (respectively weak Type II) Ando tuple for* $(T_1, T_2)$.

Now we are ready to state the complete description of a Schäffer-type model of Ando isometric lifts in terms of (weak) Type II Ando tuples.

**Theorem 6.1.4** *Let* $(T_1, T_2)$ *be a commuting contractive operator pair on a Hilbert space* $\mathcal{H}$ *and let* $(\mathcal{F}, \Lambda, P, U)$ *be a minimal weak Type II Ando tuple for* $(T_1, T_2)$. *Let* $\mathcal{K}_S$ *be the space, the isometric embedding* $\mathbf{\Pi}_S \colon \mathcal{H} \to \mathcal{K}_S$, *and* $\mathbf{V}_{S,1}, \mathbf{V}_{S,2}$ *on* $\mathcal{K}_S$ *be the operators given by*

$$\mathcal{K}_S = \begin{bmatrix} \mathcal{H} \\ H^2(\mathcal{F}) \end{bmatrix}, \quad \mathbf{\Pi}_S = \begin{bmatrix} I_{\mathcal{H}} \\ 0 \end{bmatrix} : \mathcal{H} \to \mathcal{K}_S,$$

$$(\mathbf{V}_{S,1}, \mathbf{V}_{S,2}) = \left( \begin{bmatrix} T_1 & 0 \\ \mathbf{ev}^*_{0,\mathcal{F}} PU\Lambda D_T & M_{P^\perp U + zPU} \end{bmatrix}, \begin{bmatrix} T_2 & 0 \\ \mathbf{ev}^*_{0,\mathcal{F}} U^* P^\perp \Lambda D_T & M_{U^*P + zU^*P^\perp} \end{bmatrix} \right). \tag{6.19}$$

*Then* $(\mathbf{V}_{S,1}, \mathbf{V}_{S,2})$ *is a minimal isometric lift of* $(T_1, T_2)$. *If* $(\mathcal{F}, \Lambda, P, U)$ *is a Type II Ando tuple, then in addition we have*

$$\mathbf{V}_S := \mathbf{V}_{S,1}\mathbf{V}_{S,2} = \mathbf{V}_{S,2}\mathbf{V}_{S,1} = \begin{bmatrix} T & 0 \\ \mathbf{ev}^*_{0,\mathcal{F}} \Lambda D_T & M_z^{\mathcal{F}} \end{bmatrix}. \tag{6.20}$$

*Conversely, any minimal Ando lift* $(\Pi, V_1, V_2)$ *of* $(T_1, T_2)$ *is unitarily equivalent (as a lift of* $(T_1, T_2)$*) to an Ando lift of the form* (6.19) *coming from a minimal Type II Ando tuple for* $(T_1, T_2)$.

*Proof* A matrix computation from the formulas (6.17) shows that

$$\mathbf{V}_{S,1}\mathbf{V}_{S,2} = \begin{bmatrix} T_1T_2 & 0 \\ [\mathbf{V}_{S,1}\mathbf{V}_{S,2}]_{21} & M_z^{\mathcal{F}} \end{bmatrix}, \tag{6.21}$$

where

$$\begin{aligned} [\mathbf{V}_{S,1}\mathbf{V}_{S,2}]_{21} &= \mathbf{ev}^*_{0,\mathcal{F}} PU\Lambda D_T T_2 + M_{P^\perp U + zPU}\, \mathbf{ev}^*_{0,\mathcal{F}} U^* P^\perp \Lambda D_T \\ &= \mathbf{ev}^*_{0,\mathcal{F}} PU\Lambda D_T T_2 + \mathbf{ev}^*_{0,\mathcal{F}} P^\perp \Lambda D_T \\ &= \mathbf{ev}^*_{0,\mathcal{F}} \left( PU\Lambda D_T T_2 + P^\perp \Lambda D_T \right). \end{aligned} \tag{6.22}$$

Similarly, one can compute that

$$\mathbf{V}_{S,2}\mathbf{V}_{S,1} = \begin{bmatrix} T_2T_1 & 0 \\ [\mathbf{V}_{S,2}\mathbf{V}_{S,1}]_{21} & M_z \end{bmatrix}, \tag{6.23}$$

where

$$\begin{aligned}[\mathbf{V}_{S,2}\mathbf{V}_{S,1}]_{21} &= \mathbf{ev}_{0\mathcal{F}}^* U^* P^\perp \Lambda D_T T_1 + M_{U^*P+zU^*P^\perp} \mathbf{ev}_{0,\mathcal{F}}^* PU\Lambda D_T \\ &= \mathbf{ev}_{0,\mathcal{F}}^*(U^* P^\perp \Lambda D_T T_1 + U^* PU\Lambda D_T).\end{aligned} \tag{6.24}$$

We conclude that $\mathbf{V}_{S,1}\mathbf{V}_{S,2} = \mathbf{V}_{S,2}\mathbf{V}_{S,1}$ exactly when the following system of equations holds:

$$\begin{aligned} T_1T_2 &= T_2T_1, \\ PU\Lambda D_T T_2 + P^\perp \Lambda D_T &= U^* P^\perp \Lambda D_T T_1 + U^* PU\Lambda D_T. \end{aligned} \tag{6.25}$$

The first equation is valid because of our assumption that $(T_1, T_2)$ is a commuting pair. Note that the last equation is just condition (i′) in Definition 6.1.2 (the commutativity condition).

We next argue that condition (ii) in Definition 6.1.2 implies that $\mathbf{V}_{S,1}$ and $\mathbf{V}_{S,2}$ are isometries. Since $M_{(P^\perp+zP)U}$ is an isometry, to verify that $\mathbf{V}_{S,1}$ is an isometry it suffices to check that the $(1,1)$-entry of $\mathbf{V}_{S,1}^*\mathbf{V}_{S,1}$ equals $I_{\mathcal{H}}$ and that the $(1,2)$-entry of $\mathbf{V}_{S,1}^*\mathbf{V}_{S,1}$ is equal to 0, that is, we need to show

$$T_1^*T_1 + D_T\Lambda^* U^* P\mathbf{ev}_{0,\mathcal{F}}\mathbf{ev}_{0,\mathcal{F}}^* PU\Lambda D_T = I_{\mathcal{H}}, \quad D_T\Lambda^* U^* P\mathbf{ev}_{0,\mathcal{F}} M_{(P^\perp+zP)U} = 0.$$

The first identity is an immediate consequence of the first equality in condition (ii) (the isometry condition) in Definition 6.1.2. As for the second equation, note that

$$\begin{aligned} D_T\Lambda^* U^* P\mathbf{ev}_{0,\mathcal{F}} M_{(P^\perp+zP)U} &= D_T\Lambda^* U^* P\mathbf{ev}_{0,\mathcal{F}} M_{P^\perp U} \\ &= D_T\Lambda^* U^* PP^\perp U\mathbf{ev}_{0,\mathcal{F}} = 0, \end{aligned}$$

as desired. Hence $\mathbf{V}_{S,1}$ is an isometry. Similarly, one can show that $\mathbf{V}_{S,2}$ is an isometry by making use of the second part of condition (ii) in Definition 6.1.2. Note that the isometric lift $(\mathbf{V}_{S,1}, \mathbf{V}_{S,2})$ being minimal is exactly the same as the Ando tuple $(\mathcal{F}, \Lambda, P, U)$ being Schäffer minimal (see Definition 6.1.2).

Finally, when $(\mathcal{F}, \Lambda, P, U)$ is a Type II Ando tuple, then

$$[\mathbf{V}_{S,1}\mathbf{V}_{S,2}]_{21} = [\mathbf{V}_{S,2}\mathbf{V}_{S,1}]_{21} = \mathbf{ev}_{0,\mathcal{F}}^*\Lambda D_T.$$

From (6.21) and (6.23) we see that

$$\mathbf{V}_S := \mathbf{V}_{S,1}\mathbf{V}_{S,2} = \mathbf{V}_{S,2}\mathbf{V}_{S,1} = \begin{bmatrix} T & 0 \\ \mathbf{ev}_{0,\mathcal{F}}^*\Lambda D_T & M_z^{\mathcal{F}} \end{bmatrix},$$

where we set $T = T_1T_2 = T_2T_1$. This completes the proof of the direct side of Theorem 6.1.4.

Conversely, suppose that $(\Pi, V_1, V_2)$ is a minimal Ando lift of the commuting contractive pair $(T_1, T_2)$. The discussion preceding the theorem tells us that the unitary operator

$$\tau := \begin{bmatrix} I_{\mathcal{H}} & 0 \\ 0 & \tau_{\mathrm{BCL}} \end{bmatrix} : \mathcal{K} \cong \begin{bmatrix} \mathcal{H} \\ \mathcal{K} \ominus \mathcal{H} \end{bmatrix} \to \mathcal{K}_S := \begin{bmatrix} \mathcal{H} \\ H^2(\mathcal{F}) \end{bmatrix}$$

transforms the commuting isometric pair $(V_1, V_2)$ on $\mathcal{K}$ to the pair $(\mathbf{V}_{S,1}, \mathbf{V}_{S,2})$ on $\mathcal{K}_S$ as given by (6.17) with $\mathbf{V}_S = \mathbf{V}_{S,1}\mathbf{V}_{S,2} = \mathbf{V}_{S,2}\mathbf{V}_{S,1}$ as in (6.18), where the tuple $(\mathcal{F}, \Lambda, P, U)$ is as in condition (i$'$) in Definition 6.1.2. By reversing the analysis given in the direct analysis, the fact that $\mathbf{V}_{S,1}\mathbf{V}_{S,2} = \mathbf{V}_{S,2}\mathbf{V}_{S,1}$ gives us the system of equations (6.25) that then forces the commutativity condition (i$'$) in Definition 6.1.2) to hold, and the fact that $\mathbf{V}_{S,1}$ and $\mathbf{V}_{S,2}$ are isometries, forces the isometry condition (item (ii) in Definition 6.1.2) to hold. Furthermore, the fact that by construction $\mathbf{V}_S = \mathbf{V}_{S,1}\mathbf{V}_{S,2} = \mathbf{V}_{S,2}\mathbf{V}_{S,1}$ has the form (6.20) forces the strong commutativity condition (i) in Definition 6.1.2) to hold, so in fact $(\mathcal{F}, \Lambda, P, U)$ is a Type II Ando tuple. Furthermore, the assumption that $(\mathbf{\Pi}_S, \mathbf{V}_{S,1}, \mathbf{V}_{S,2})$ is minimal forces the space $\mathcal{Y}$ to be trivial and the tuple $(\mathcal{F}, \Lambda, P, U)$ to be Schäffer minimal (see Definition 6.1.2). □

## 6.2 Ando's Dilation Theorem via Type II Ando Tuples

We first show here that, remarkably, special Ando tuples for $(T_1, T_2)$ are automatically also Type II Ando tuples.

**Theorem 6.2.1** *Suppose that $(T_1, T_2)$ is a commuting contractive operator pair on $\mathcal{H}$ and that $(\mathcal{F}_\dagger, \Lambda_\dagger, P_\dagger, U_\dagger)$ is a special Ando tuple for $(T_1, T_2)$ constructed as in Definition 5.3.2. Then $(\mathcal{F}_\dagger, \Lambda_\dagger, P_\dagger, U_\dagger)$ is, in fact, a Type II Ando tuple for $(T_1, T_2)$ (see Definition 6.1.2).*

*Proof* Let $(\mathcal{F}_\dagger, \Lambda_\dagger, P_\dagger, U_\dagger)$ be a special Ando tuple for $(T_1, T_2)$ constructed via the algorithm explained in Definition 5.3.2. We must verify that conditions (i) and (ii) in Definition 6.1.2 are satisfied. Let $h$ be an arbitrary element of $h$. The computation

$$\begin{aligned}
&P_\dagger U_\dagger \Lambda_\dagger D_T T_2 h + P_\dagger^\perp \Lambda_\dagger D_T h \\
&= P_\dagger U_\dagger \left( D_{T_1} T_2^2 h \oplus D_{T_2} T_2 h \right) + P_\dagger^\perp \left( D_{T_1} T_2 h \oplus D_{T_2} h \right) \\
&= P_\dagger \left( D_{T_1} T_2 h \oplus D_{T_2} T_1 T_2 h \right) + \left( 0 \oplus D_{T_2} h \right) = D_{T_1} T_2 h \oplus D_{T_2} h = \Lambda_\dagger D_T h,
\end{aligned}$$

and with the computation

$$
\begin{aligned}
&U_\dagger^* \left(P_\dagger^\perp \Lambda_\dagger D_T T_1 + P_\dagger U_\dagger \Lambda_\dagger D_T\right) h \\
&= U_\dagger^* \left(P_\dagger^\perp \left(D_{T_1} T_2 T_1 h \oplus D_{T_2} T_1 h\right) + P_\dagger U_\dagger \left(D_{T_1} T_2 h \oplus D_{T_2} h\right)\right) \\
&= U_\dagger^* \left(\left(0 \oplus D_{T_2} T_1 h\right) + P_\dagger \left(D_{T_1} h \oplus D_{T_2} T_1 h\right)\right) \\
&= U_\dagger^* \left(\left(0 \oplus D_{T_2} T_1 h\right) + \left(D_{T_1} h \oplus 0\right)\right) = U_\dagger^* \left(D_{T_1} h \oplus D_{T_2} T_1 h\right) = \Lambda_\dagger D_T h
\end{aligned}
$$

verify condition (i).

The following easy inner product computations for arbitrary $h, h' \in \mathcal{H}$

$$
\begin{aligned}
\left\langle D_T \Lambda_\dagger^* U_\dagger^* P_\dagger U_\dagger \Lambda_\dagger D_T h, h' \right\rangle &= \left\langle P_\dagger U_\dagger \Lambda_\dagger D_T h, P_\dagger U_\dagger \Lambda_\dagger D_T h' \right\rangle \\
&= \left\langle D_{T_1} h \oplus 0, D_{T_1} h' \oplus 0 \right\rangle = \left\langle D_{T_1}^2 h, h' \right\rangle
\end{aligned}
$$

and

$$
\begin{aligned}
\left\langle D_T \Lambda_\dagger^* P_\dagger^\perp \Lambda_\dagger D_T h, h' \right\rangle &= \left\langle P_\dagger^\perp \Lambda_\dagger D_T h, P_\dagger^\perp \Lambda_\dagger D_T h' \right\rangle \\
&= \left\langle 0 \oplus D_{T_2} h, 0 \oplus D_{T_2} h' \right\rangle = \left\langle D_{T_2}^2 h, h' \right\rangle
\end{aligned}
$$

establish condition (ii). □

*The second proof of Ando's dilation theorem* Since the definition of special Ando tuples (whether for $(T_1, T_2)$ or for $(T_1^*, T_2^*)$) is constructive, we are, in particular, assured that the class of special Ando tuples for $(T_1, T_2)$ is not empty. Furthermore, Theorem 6.2.1, combined with Theorem 6.1.4, shows us how to use special Ando tuples for $(T_1, T_2)$ to construct a class of Ando isometric lifts for $(T_1, T_2)$. In this way we arrive at a second new constructive proof of Ando's theorem [10], via the Schäffer model construction. □

A priori, it is conceivable that the unitary equivalence classes of Ando lifts arising via the Douglas model construction from special Ando tuples for $(T_1^*, T_2^*)$ are distinct from those arising via the Schäffer model construction from special Ando tuples for $(T_1, T_2)$.

## 6.3 Examples: Schäffer-Type Model of Ando Lifts

It is of interest to work out the Schäffer model for the case where the commuting contractive pair $(T_1, T_2)$ is a commuting pair of coisometries. First, we note that there is a bridge between the Type I Ando tuples and Type II Ando tuples at least for a nice class of commuting contractive pairs.

**Lemma 6.3.1** *Suppose that $(V_1, V_2)$ is a commuting isometric pair. If $(\mathcal{F}, \Lambda, P, U)$ is a Type II Ando tuple for $(V_1^*, V_2^*)$, then $(\mathcal{F}, \Lambda, P, U)$ is a Type I Ando tuple for $(V_1^*, V_2^*)$, and vice versa.*

*Proof* As usual, let us denote by $V$ the product $V_1V_2$. Suppose that $(\mathcal{F}, \Lambda, P, U)$ is a Type II Ando tuple for $(V_1^*, V_2^*)$. By Definition 6.1.2, this means we have the following operator equations:

$$PU\Lambda D_{V^*}V_2^* + P^\perp \Lambda D_{V^*} = \Lambda D_{V^*} \tag{6.26}$$
$$= U^*P^\perp \Lambda D_{V^*}V_1^* + U^*PU\Lambda D_{V^*}. \tag{6.27}$$

Surprisingly, perhaps, this will be enough to draw the conclusion; we shall not need the additional information that item (ii) of Definition 6.1.2 provides. Indeed, to show that $(\mathcal{F}, \Lambda, P, U)$ is a Type I Ando tuple for $(V_1, V_2)$, by Definition 5.1.5, we must show

$$P^\perp U\Lambda D_{V^*} + PU\Lambda D_{V^*}V^* = \Lambda D_{V^*}V_1^*, \tag{6.28}$$
$$U^*P\Lambda D_{V^*} + U^*P^\perp \Lambda D_{V^*}V^* = \Lambda D_{V^*}V_2^*. \tag{6.29}$$

By the operation $(6.26) \times V_1^*$, we get

$$\Lambda D_{V^*}V_1^* = PU\Lambda D_{V^*}V^* + P^\perp \Lambda D_{V^*}V_1^*.$$

Thus we shall have (6.28), once we show that

$$P^\perp \Lambda D_{V^*}V_1^* = P^\perp U\Lambda D_{V^*}.$$

For this we do $U \times (6.27)$ to get

$$P^\perp \Lambda D_{V^*}V_1^* = U\Lambda D_{V^*} - PU\Lambda D_{V^*} = P^\perp U\Lambda D_{V^*}.$$

Similarly, one can obtain (6.29).

For the converse, we need to show equations (6.26) and (6.27) from (6.28) and (6.29). To this end, we do $U^*P^\perp \times (6.28)$ to get

$$U^*P^\perp U\Lambda D_{V^*} = U^*P^\perp \Lambda D_{V^*}V_1^*.$$

This gives (6.27) as follows:

$$U^*P^\perp \Lambda D_{V^*}V_1^* + U^*PU\Lambda D_{V^*} = U^*P^\perp U\Lambda D_{V^*} + U^*PU\Lambda D_{V^*} = \Lambda D_{V^*}.$$

Similar computation can be done to obtain (6.26) and thereby to complete the proof. □

We now turn to the Schäffer parallel of Theorem 5.5.1.

**Theorem 6.3.2** *Let $\mathcal{F}$ be a coefficient Hilbert space and let $(P, U)$ be any pair of operators on $\mathcal{F}$ with $P$ a projection and $U$ a unitary operator. Then there is a choice of commuting pair of coisometries $(T_1, T_2)$ (with product $T = T_1T_2$ having no unitary piece) on a Hilbert space $\mathcal{H}$ and a unitary operator $\Lambda \colon \mathcal{D}_T \to \mathcal{F}$ so that $(\mathcal{F}, \Lambda, P, U)$ is special (and hence also Type II by Theorem*

*6.2.1) Ando tuple for* $(T_1, T_2)$. *In fact, one can take* $(T_1^*, T_2^*)$ *to have the BCL1-model generated by the BCL-tuple* $(\mathcal{F}, P, U)$.

*Conversely, if* $(T_1, T_2)$ *is a commuting pair of coisometries with* $T_1^* T_2^*$ *a shift, then any irreducible Type II Ando tuple* $(\mathcal{F}, \Lambda, P, U)$ *for* $(T_1, T_2)$ *has the property that* $\Lambda\colon \mathcal{D}_T \to \mathcal{F}$ *is unitary and* $(\mathcal{F}, P, U)$ *is a BCL-tuple generating a BCL1-model for the commuting isometric pair* $(T_1^*, T_2^*)$.

*Proof* We assume that we are given a projection $P$ and a unitary operator $U$ on the coefficient Hilbert space $\mathcal{F}$. View $H^2(\mathcal{F})$ as a subspace of $L^2(\mathcal{F}) = L^2 \otimes \mathcal{F}$ (where $L^2$ is the Lebesgue space of modulus-squared integrable functions with respect to normalized arc-length measure on $\mathbb{T}$) and set $H^2(\mathcal{F})^\perp = L^2(\mathcal{F}) \ominus H^2(\mathcal{F})$. Introduce the commuting unitary operators

$$(W_1, W_2) = \left(M_{P^\perp U + \zeta PU}, M_{U^* P + \zeta U^* P^\perp}\right) \quad \text{on } L^2(\mathcal{F})$$

with product

$$W := W_1 W_2 = W_2 W_1 = M_\zeta \quad \text{on } L^2(\mathcal{F}).$$

We set $(T_1, T_2)$ equal to the commuting pair of coisometries on $\mathcal{H} := H^2(\mathcal{F})^\perp$:

$$T_1 = P_\mathcal{H} W_1|_\mathcal{H}, \quad T_2 = P_\mathcal{H} W_2|_\mathcal{H}, \quad T = T_1 T_2 = T_2 T_1 = P_\mathcal{H} M_\zeta|_\mathcal{H}. \tag{6.30}$$

Let us view $(T_1, T_2)$ as a commuting pair of contractions. Clearly a choice of Ando lift for $(T_1, T_2)$ is the commuting isometric pair (in fact commuting unitary pair) $(W_1, W_2)$ acting on $\mathcal{K} = L^2(\mathcal{F}) \supset \mathcal{H} = H^2(\mathcal{F})^\perp$.

Let us compute the Schäffer model for this Ando lift of $(T_1, T_2)$. Note that

$$(W_1, W_2)|_{\mathcal{K} \ominus \mathcal{H}} = (W_1, W_2)|_{H^2(\mathcal{F})} = \left(M_{(P^\perp + zP)U}, M_{U^*(P + zP^\perp)}\right) \quad \text{on } H^2(\mathcal{F})$$

is just the BCL2-model for the commuting isometric pair associated with the BCL-tuple $(\mathcal{F}, P, U)$. Note that the defect space $\mathcal{D}_T$ and the defect operator $D_T$ are given by

$$\mathcal{D}_T = \left\{f_{-1}\zeta^{-1} \colon f_{-1} \in \mathcal{F}\right\} \quad D_T\colon \sum_{n=-\infty}^{-1} f_n \zeta^n \mapsto f_{-1}\zeta^{-1}.$$

If we define an isometry $\Lambda\colon \mathcal{D}_T \to \mathcal{F}$ by

$$\Lambda\colon f_{-1}\zeta^{-1} \mapsto f_{-1}$$

then it is a simple direct check that $W_1, W_2, W$ have exactly the form of $\mathbf{V}_{S,1}, \mathbf{V}_{S,2}, \mathbf{V}_S$ in (6.19) and (6.20), and we conclude that $(\mathcal{F}, \Lambda, P, U)$ is exactly the Ando tuple giving rise to this Schäffer model Ando lift of $(T_1, T_2)$. Hence, by the proof of the first part of Theorem 6.1.4 it follows that $(\mathcal{F}, \Lambda, P, U)$ is a Type II Ando tuple for $(T_1, T_2)$, as can also be checked directly.

Next, let us check that $(\mathcal{F}, P, U)$ gives rise to the BCL1-model for $(T_1^*, T_2^*)$. Note that

$$(T_1^*, T_2^*) = \left(M_{U^*P^\perp + \zeta^{-1}U^*P}, M_{PU+\zeta^{-1}P^\perp U}\right)\Big|_{H^2(\mathcal{F})^\perp}.$$

If we let $\omega\colon H^2(\mathcal{F})^\perp \to H^2(\mathcal{F})$ be the unitary transformation given by

$$f(\zeta) \mapsto g(z) = z^{-1} f\left(z^{-1}\right),$$

then it is easily checked that

$$\omega \circ \left(M^*_{P^\perp U+\zeta PU}, M^*_{U^*P+\zeta U^*P^\perp}\right)\Big|_{H^2(\mathcal{F})^\perp} = (M_{U^*P^\perp + zU^*P}, M_{PU+zP^\perp U}) \circ \omega,$$

that is, $(T_1^*, T_2^*)$ is unitarily equivalent to the BCL1-model generated by the BCL-tuple $(\mathcal{F}, P, U)$. That the Ando tuple $(\mathcal{F}, \Lambda, P, U)$ is a special Ando tuple for $(T_1, T_2)$ can now be seen as part of the content of Theorem 5.5.1 applied to the commuting isometric pair $(T_1^*, T_2^*)$.

For the converse direction, note that, from Definition 5.1.3, irreducibility of an Ando tuple does not depend on its type. Therefore, if $(\mathcal{F}, \Lambda, P, U)$ is an irreducible Type II Ando tuple for $(T_1^*, T_2^*)$, then by Lemma 6.3.1, $(\mathcal{F}, \Lambda, P, U)$ is an irreducible Type I Ando tuple for $(T_1^*, T_2^*)$ as well. Hence, by the converse part of Theorem 5.5.1, $\Lambda\colon \mathcal{D}_{T^*} \to \mathcal{F}$ is a unitary and $(\mathcal{F}, P, U)$ generates a BCL1- model for $(T_1^*, T_2^*)$. This is what is needed to complete the proof. □

**Remark 6.3.3** In parallel with the discussion in Remark 5.5.6 concerning the choice of BCL-model in the construction of the Douglas model for an Ando lift of $(T_1, T_2)$, here we discuss the choice of BCL-model in the construction of the Schäffer model for an Ando lift of $(T_1, T_2)$. In the representation for an Ando lift $(V_1, V_2)$ on $\mathcal{K}$ for a commuting contractive pair $(T_1, T_2)$ on $\mathcal{H} \subset \mathcal{K}$, we could have used the BCL1-model (rather than the BCL2-model) for $(V_1, V_2)|_{\mathcal{K}\ominus\mathcal{H}}$ to arrive at the unitarily equivalent Ando lift having the form $\left(\mathbf{\Pi}'_S, \mathbf{V}'_{S,1}, \mathbf{V}'_{S,2}\right)$, where

$$\left(\mathbf{V}'_{S,1}, \mathbf{V}'_{S,2}, \mathbf{V}'_S\right) = \left(\begin{bmatrix} T_1 & 0 & 0 \\ C_{s1} & M_{U^*P^\perp + zU^*P} & 0 \\ C_{u1} & 0 & Y_1 \end{bmatrix}, \begin{bmatrix} T_2 & 0 & 0 \\ C_{s2} & M_{PU+zP^\perp U} & 0 \\ C_{u2} & 0 & Y_2 \end{bmatrix}, \begin{bmatrix} T & 0 & 0 \\ C_s & M_z^{\mathcal{F}} & 0 \\ C_u & 0 & Y \end{bmatrix}\right), \tag{6.31}$$

in place of (6.5). Then the same analysis leading from (6.5) to (6.17) and (6.18) would lead us instead to

$$
\begin{aligned}
&\left(\mathbf{V}'_{S,1}, \mathbf{V}'_{S,2}\right) \\
&= \left( \begin{bmatrix} T_1 & 0 & 0 \\ \mathbf{ev}^*_{0,\mathcal{F}} U^* P \Lambda D_T & M_{U^* P^\perp + z U^* P} & 0 \\ 0 & 0 & Y_1 \end{bmatrix}, \begin{bmatrix} T_2 & 0 & 0 \\ \mathbf{ev}^*_{0,\mathcal{F}} P^\perp U \Lambda D_T & M_{PU + z P^\perp U} & 0 \\ 0 & 0 & Y_2 \end{bmatrix} \right),
\end{aligned} \tag{6.32}
$$

$$
\mathbf{V}'_S := \mathbf{V}'_{S,1}\mathbf{V}'_{S,2} = \mathbf{V}'_{S,2}\mathbf{V}'_{S,1} = \begin{bmatrix} T & 0 & 0 \\ \mathbf{ev}^*_{0,\mathcal{F}} \Lambda D_T & M^{\mathcal{F}}_z & 0 \\ 0 & 0 & Y \end{bmatrix}, \tag{6.33}
$$

where we set $T := T_1T_2 = T_2T_1$, $Y := Y_1Y_2 = Y_2Y_1$, and where $\Lambda\colon \mathcal{D}_T \to \mathcal{F}$ is an isometry; let us note that a short-cut way to see this is to make use of the flip transformation $\mathfrak{f}$ given by (3.13).

Conversely, for a collection of spaces and operators of the form (6.32) and (6.33) to be an Ando lift of $(T_1, T_2)$ requires the following additional compatibility conditions.

(a) **Strengthened commutativity condition**

$$
U^* P \Lambda D_T T_2 + U^* P^\perp U \Lambda D_T = P^\perp U \Lambda D_T T_1 + P \Lambda D_T = \Lambda D_T.
$$

(b) **Isometry condition**

$$
D_T \Lambda^* P \Lambda D_T = D^2_{T_1}, \quad D_T \Lambda^* U^* P^\perp U \Lambda D_T = D^2_{T_2}.
$$

(a′) **Commutativity condition**

$$
U^* P \Lambda D_T T_2 + U^* P^\perp U \Lambda D_T = P^\perp U \Lambda D_T T_1 + P \Lambda D_T.
$$

We note that these equations are obtained simply by replacing $(P, U)$ by $(U^*PU, U^*)$ (i.e., by applying the flip transformation $\mathfrak{f}$ (3.13)) in the conditions (i), (ii), and (i′) in Definition 6.1.2. Let us say that an Ando tuple for $(T_1, T_2)$ is a *Type* II′ *Ando tuple* for $(T_1, T_2)$ if the modified compatibility conditions (a) and (b) hold, and is a *weak Type* II′ *Ando tuple for* $(T_1, T_2)$ if conditions (a′) and (b) hold.

Just as in the Douglas model setting, the drawback of this alternative approach is that a special Ando tuple for $(T_1, T_2)$ need not be a Type II′ Ando tuple for $(T_1, T_2)$, that is, the analogue of Theorem 6.2.1 with *Type* II′ *Ando tuple* inserted in place of *Type* II *Ando tuple* fails in general. To construct an explicit example, one can proceed as follows.

Let $\mathcal{F}$ be any coefficient Hilbert space and we let $P$ be any projection and let $U$ be any unitary operator on $\mathcal{F}$. Let $(T_1, T_2)$ be the commuting coisometric pair on $H^2(\mathcal{F})^\perp$ as in Theorem 6.3.2 (see (6.30)) and define the unitary operator

$\Lambda \colon \mathcal{D}_T \to \mathcal{F}$ by $\Lambda \colon \zeta^{-1} f_{-1} \to f_{-1}$. Then we have seen in the proof of Theorem 6.3.2 that $(\mathcal{F}, \Lambda, P, U)$ is a special Ando tuple (and hence also a Type II Ando tuple) for $(T_1, T_2)$. We shall now show that, with appropriate choice of projection $P$ and unitary operator $U$ on $\mathcal{F}$, it can be the case that this $(\mathcal{F}, \Lambda, P, U)$ is not even a weak Type II′ Ando tuple for $(T_1, T_2)$.

Toward this goal, let us first compute, for $f(\zeta) = \sum_{n=-\infty}^{-1} f_n \zeta^n \in H^2(\mathcal{F})^\perp$,

$$\begin{aligned} \Lambda D_T \colon f &\mapsto f_{-1}, \\ \Lambda D_T T_1 \colon f &\mapsto \left[ M_{P^\perp U + \zeta P U} f \right]_{-1} = P^\perp U f_{-1} + P U f_{-2}, \\ \Lambda D_T T_2 \colon f &\mapsto \left[ M_{U^* P + \zeta U^* P^\perp} f \right]_{-1} = U^* P f_{-1} + U^* P^\perp f_{-2}, \end{aligned}$$

and hence we have

$$\begin{aligned} U^* P \Lambda D_T T_2 + U^* P^\perp U \Lambda D_T \colon f &\mapsto (U^* P U^* P + U^* P^\perp U) f_{-1} + U^* P U^* P^\perp f_{-2}, \\ P^\perp U \Lambda D_T T_1 + P \Lambda D_T \colon f &\mapsto (P^\perp U P^\perp U + P) f_{-1} + P^\perp U P U f_{-2}, \\ P \Lambda D_T \colon f &\mapsto P f_{-1}. \end{aligned}$$

Hence condition (a′) requires that

$$U^* P U^* P + U^* P^\perp U = P^\perp U P^\perp U + P, \quad U^* P U^* P^\perp = P^\perp U P U, \tag{6.34}$$

while condition (a) requires in addition that the common value of the first expression is $I$ and the common value of the second expression is 0. To get a counterexample to condition (a′) (and hence also to (a)), it again suffices to take $\mathcal{F} = \mathbb{C}$, $P = 1$, $U^2 \neq 1$.

It turns out that the isometry condition (b) also fails to hold in general. Indeed, one can verify that

$$D_{T_1} \colon f \mapsto (U^* P U f_{-1}) \zeta^{-1}, \quad D_{T_2} \colon f \mapsto (P^\perp f_{-1}) \zeta^{-1}, \quad D_T \colon f \mapsto f_{-1} \zeta^{-1},$$

and condition (b) requires

$$P = U^* P U, \quad U^* P^\perp U = P^\perp, \tag{6.35}$$

which is violated as soon as $P$ and $U$ do not commute, requiring $\dim \mathcal{F} \geq 2$. One can verify that $\mathcal{F} = \mathbb{C}^2$, $U = \left[\begin{smallmatrix} 0 & 1 \\ 1 & 0 \end{smallmatrix}\right]$, $P = \left[\begin{smallmatrix} 1 & 0 \\ 0 & 0 \end{smallmatrix}\right]$ violates both (6.34) and (6.35). Thus Theorem 6.2.1 fails in general when a *Type* II *Ando tuple* is replaced with a *Type* II′ *Ando tuple*.

## 6.4 Type II Ando Tuples versus Weak Type II Ando Tuples

The class of Type II Ando tuples is strictly smaller than the class of weak Type II Ando tuples, as the following result demonstrates.

**Proposition 6.4.1** *Let $T_1$ be a contraction on a Hilbert space $\mathcal{H}$, $T = T_1^2$ and let $\tau_1, \tau_2 \colon \mathcal{D}_{T_1} \to \mathcal{G}$ be two isometries. Let $\Lambda_\dagger$ be the isometry as in Definition 5.3.2, that is,*

$$\Lambda_\dagger \colon \mathcal{D}_T \to \begin{bmatrix} \mathcal{D}_{T_1} \\ \mathcal{D}_{T_1} \end{bmatrix} \quad \text{given by} \quad \Lambda_\dagger \colon D_T \mapsto \begin{bmatrix} D_{T_1}T_1 \\ D_{T_1} \end{bmatrix}.$$

*Then the Ando tuple*

$$(\mathcal{F}, \Lambda, P, U) = \left( \begin{bmatrix} \mathcal{G} \\ \mathcal{G} \end{bmatrix}, \begin{bmatrix} \tau_1 & 0 \\ 0 & \tau_2 \end{bmatrix} \Lambda_\dagger, \begin{bmatrix} I_{\mathcal{G}} & 0 \\ 0 & 0 \end{bmatrix}, \begin{bmatrix} 0 & I_{\mathcal{G}} \\ I_{\mathcal{G}} & 0 \end{bmatrix} \right) \tag{6.36}$$

*is a weak Type II Ando tuple for $(T_1, T_1)$. Furthermore, we have the following.*

(i) *The tuple* (6.36) *is a Type II Ando tuple if and only if*

$$(\tau_1 - \tau_2)D_{T_1}T_1 = 0.$$

(ii) *If $\tau_1 = \tau_2$ is unitary, then* (6.36) *is a special Type II Ando tuple.*

*Proof* For $P$ and $U$ as in (6.36), we have

$$PU = \begin{bmatrix} 0 & I \\ 0 & 0 \end{bmatrix} = U^*P^\perp \quad \text{and} \quad U^*PU = \begin{bmatrix} 0 & 0 \\ 0 & I \end{bmatrix} = P^\perp. \tag{6.37}$$

The weak commutativity condition (i′) in Definition 6.1.2 is

$$PU\Lambda D_T T_2 + P^\perp \Lambda D_T = U^*P^\perp \Lambda D_T T_1 + U^*PU\Lambda D_T.$$

This is therefore satisfied by the tuple in (6.36) because we have (6.37) and $T_1 = T_2$. The isometry condition (ii) of Definition 6.1.2 is that

$$D_T\Lambda^*U^*PU\Lambda D_T = D_{T_1}^2 \quad \text{and} \quad D_T\Lambda^*P^\perp\Lambda D_T = D_{T_2}^2,$$

which, in view of (6.37), boils down to just

$$D_T\Lambda^*P^\perp\Lambda D_T = D_{T_1}^2. \tag{6.38}$$

Since

$$P^\perp\Lambda D_T = P^\perp \begin{bmatrix} \tau_1 & 0 \\ 0 & \tau_2 \end{bmatrix} \Lambda_\dagger D_T = \begin{bmatrix} 0 & 0 \\ 0 & I \end{bmatrix} \begin{bmatrix} \tau_1 D_{T_1}T_1 \\ \tau_2 D_{T_1} \end{bmatrix} = \begin{bmatrix} 0 \\ \tau_2 D_{T_1} \end{bmatrix},$$

and $\tau_2$ is an isometry, we see that

$$D_T\Lambda^*P^\perp\Lambda D_T = (P^\perp\Lambda D_T)^*(P^\perp\Lambda D_T) = \begin{bmatrix} 0 & D_{T_1}\tau_2^* \end{bmatrix} \begin{bmatrix} 0 \\ \tau_2 D_{T_1} \end{bmatrix} = D_{T_1}^2$$

and therefore (6.38) holds. Consequently (6.36) is always a weak Type II Ando tuple for $(T_1, T_1)$.

*Proof of (i)* Note that for the tuple (6.36) to be a Type II Ando tuple, it must, in addition, satisfy

$$PU\Lambda D_T T_1 + P^\perp \Lambda D_T = \Lambda D_T \;\left(= U^* P^\perp \Lambda D_T T_1 + U^* PU\Lambda D_T\right).$$

So we compute

$$\begin{aligned} PU\Lambda D_T T_1 + P^\perp \Lambda D_T &= \begin{bmatrix} 0 & I \\ 0 & 0 \end{bmatrix} \begin{bmatrix} \tau_1 D_{T_1} T_1^2 \\ \tau_2 D_{T_1} T_1 \end{bmatrix} + \begin{bmatrix} 0 & 0 \\ 0 & I \end{bmatrix} \begin{bmatrix} \tau_1 D_{T_1} T_1 \\ \tau_2 D_{T_1} \end{bmatrix} \\ &= \begin{bmatrix} \tau_2 D_{T_1} T_1 \\ \tau_2 D_{T_1} \end{bmatrix}. \end{aligned}$$

Thus the Ando tuple (6.36) will be of Type II if and only if

$$\begin{bmatrix} \tau_2 D_{T_1} T_1 \\ \tau_2 D_{T_1} \end{bmatrix} = PU\Lambda D_T T_1 + P^\perp \Lambda D_T = \Lambda D_T = \begin{bmatrix} \tau_1 D_{T_1} T_1 \\ \tau_2 D_{T_1} \end{bmatrix},$$

which is true if and only if $(\tau_1 - \tau_2) D_{T_1} T_1 \equiv 0$. This proves (1).
*Proof of (ii)* Let us denote $\tau_1 = \tau_2 =: \tau$ and the unitary

$$\widehat{\tau} := \begin{bmatrix} \tau & 0 \\ 0 & \tau \end{bmatrix} : \begin{bmatrix} \mathcal{D}_{T_1} \\ \mathcal{D}_{T_1} \end{bmatrix} \to \begin{bmatrix} \mathcal{G} \\ \mathcal{G} \end{bmatrix}.$$

Our goal is to show that the tuple (6.36) coincides (in the sense of Definition 5.1.2) with a special Ando tuple of $(T_1, T_1)$ in its canonical form and therefore is special (see Definition 5.3.2). Since $\tau\colon \mathcal{D}_{T_1} \to \mathcal{G}$ is a unitary, we make the following simple observations:

$$\widehat{\tau}^* \begin{bmatrix} \mathcal{G} \\ \mathcal{G} \end{bmatrix} = \begin{bmatrix} \mathcal{D}_{T_1} \\ \mathcal{D}_{T_1} \end{bmatrix} = \mathcal{F}_\dagger, \quad \widehat{\tau}^* \Lambda = \Lambda_\dagger, \quad \widehat{\tau}^* \begin{bmatrix} I_{\mathcal{G}} & 0 \\ 0 & 0 \end{bmatrix} \widehat{\tau} = \begin{bmatrix} I_{\mathcal{D}_{T_1}} & 0 \\ 0 & 0 \end{bmatrix} = P_\dagger$$

and lastly
$$\widehat{\tau}^* U \widehat{\tau} = \widehat{\tau}^* \begin{bmatrix} 0 & I_{\mathcal{G}} \\ I_{\mathcal{G}} & 0 \end{bmatrix} \widehat{\tau} = \begin{bmatrix} 0 & I_{\mathcal{D}_{T_1}} \\ I_{\mathcal{D}_{T_1}} & 0 \end{bmatrix}. \tag{6.39}$$

Note that the unitary $\begin{bmatrix} 0 & I_{\mathcal{D}_{T_1}} \\ I_{\mathcal{D}_{T_1}} & 0 \end{bmatrix}$ satisfies

$$\begin{bmatrix} 0 & I_{\mathcal{D}_{T_1}} \\ I_{\mathcal{D}_{T_1}} & 0 \end{bmatrix} : \begin{bmatrix} D_{T_1} T_1 \\ D_{T_1} \end{bmatrix} \mapsto \begin{bmatrix} D_{T_1} \\ D_{T_1} T_1 \end{bmatrix},$$

and consequently the tuple

$$\left( \begin{bmatrix} \mathcal{D}_{T_1} \\ \mathcal{D}_{T_1} \end{bmatrix}, \Lambda_\dagger, \begin{bmatrix} I_{\mathcal{D}_{T_1}} & 0 \\ 0 & 0 \end{bmatrix}, \begin{bmatrix} 0 & I_{\mathcal{D}_{T_1}} \\ I_{\mathcal{D}_{T_1}} & 0 \end{bmatrix} \right)$$

is a special Ando tuple in its canonical form and it coincides with (6.36) by the observations (6.39). □

The next result shows how close a weak Type II Ando tuple is to being a Type II Ando tuple.

**Proposition 6.4.2** *Suppose that* $(\mathcal{F}, \Lambda, P, U)$ *is a weak Type II Ando tuple for the commuting contractive operator pair* $(T_1, T_2)$. *Then there is an isometry* $\widetilde{u}$ *from* Ran $\Lambda$ *into* $\mathcal{F}$ *so that condition* (i) *in Definition 6.1.2 holds in the somewhat weaker form*

(i″) $PU\Lambda D_T T_2 + P^\perp \Lambda D_T = U^* P^\perp \Lambda D_T T_1 + U^* PU\Lambda D_T = \widetilde{u} D_T.$

*Proof* Since $(\mathcal{F}, \Lambda, P, U)$ is assumed to be a weak Type II Ando tuple, the first equality above holds. We just need to establish the second equality. We compute

$$\begin{aligned}
&\left(PU\Lambda D_T T_2 + P^\perp \Lambda D_T\right)^* \left(PU\Lambda D_T T_2 + P^\perp \Lambda D_T\right)\\
&\quad = T_2^* D_T \Lambda^* U^* PU\Lambda D_T T_2 + D_T \Lambda^* P^\perp \Lambda D_T\\
&\quad = T_2^* D_{T_1}^2 T_2 + D_{T_2}^2 \quad \text{(by condition (ii) in Definition 6.1.2)}\\
&\quad = T_2^* \left(I - T_1^* T_1\right) T_2 + \left(I - T_2^* T_2\right) = I - T_2^* T_1^* T_1 T_2 = I - T^*T = D_T \Lambda^* \Lambda D_T.
\end{aligned}$$

From this it follows that there is an isometry $\widetilde{u}\colon$ Ran $\Lambda \to \mathcal{F}$ so that $PU\Lambda D_T T_2 + P^\perp \Lambda D_T = \widetilde{u} \Lambda D_T$. □

## 6.5 Classification of Schäffer Model Ando Lifts

We saw in Theorem 6.1.4 that if $(\mathcal{F}, \Lambda, P, U)$ is a weak Type II Ando tuple, then the corresponding Schäffer model is an Ando lift of $(T_1, T_2)$. To classify the unitary equivalence of two Schäffer models of an Ando lift in terms of tuple coincidence, weak Type II Ando tuples prove to be of limited utility. It turns out to be essential to work only with Type II Ando tuples, as in the following result.

**Theorem 6.5.1** *Let* $(\mathcal{F}, \Lambda, P, U)$ *and* $(\mathcal{F}', \Lambda', P', U')$ *be two Type II Ando tuples of a given commuting contractive pair* $(T_1, T_2)$ *on a Hilbert space* $\mathcal{H}$. *Let* $(\mathbf{V}_{S,1}, \mathbf{V}_{S,2})$ *and* $\left(\mathbf{V}'_{S,1}, \mathbf{V}'_{S,2}\right)$ *be the minimal Ando lifts of* $(T_1, T_2)$ *corresponding to the Type II Ando tuples* $(\mathcal{F}, \Lambda, P, U)$ *and* $(\mathcal{F}', \Lambda', P', U')$, *respectively, as in* (6.19) *and* (6.20). *Then* $(\mathbf{V}_{S,1}, \mathbf{V}_{S,2})$ *and* $\left(\mathbf{V}'_{S,1}, \mathbf{V}'_{S,2}\right)$ *are unitarily equivalent if and only if* $(\mathcal{F}, \Lambda, P, U)$ *and* $(\mathcal{F}', \Lambda', P', U')$ *coincide.*

*Proof* We first prove the sufficiency (or "if") direction. Suppose $u\colon \mathcal{F} \to \mathcal{F}'$ is a unitary such that

$$u\Lambda = \Lambda' \quad \text{and} \quad u(P, U) = (P', U')u. \tag{6.40}$$

Define the unitary

$$\tilde{u} := \begin{bmatrix} I_{\mathcal{H}} & 0 \\ 0 & I_{H^2} \otimes u \end{bmatrix} : \begin{bmatrix} \mathcal{H} \\ H^2(\mathcal{F}) \end{bmatrix} \to \begin{bmatrix} \mathcal{H} \\ H^2(\mathcal{F}') \end{bmatrix}. \tag{6.41}$$

Then keeping the equations in (6.40) in mind, we conclude from the computations

$$\begin{aligned} \tilde{u}\mathbf{V}_{S,1} &= \begin{bmatrix} I_{\mathcal{H}} & 0 \\ 0 & I_{H^2} \otimes u \end{bmatrix} \begin{bmatrix} T_1 & 0 \\ \mathbf{ev}_0^* PU\Lambda D_T & M_{P^\perp U + zPU} \end{bmatrix} \\ &= \begin{bmatrix} T_1 & 0 \\ \mathbf{ev}_0^* uPU\Lambda D_T & M_{u(P^\perp U + zPU)} \end{bmatrix} \end{aligned}$$

and

$$\begin{aligned} \mathbf{V}'_{S,1}\tilde{u} &= \begin{bmatrix} T_1 & 0 \\ \mathbf{ev}_0^* P'U'\Lambda' D_T & M_{(P'^\perp + zP')U'} \end{bmatrix} \begin{bmatrix} I_{\mathcal{H}} & 0 \\ 0 & I_{H^2} \otimes u \end{bmatrix} \\ &= \begin{bmatrix} T_1 & 0 \\ \mathbf{ev}_0^* P'U'\Lambda' D_T & M_{(P'^\perp U' + zP'U')u} \end{bmatrix} \end{aligned}$$

that $\tilde{u}\mathbf{V}_{S,1} = \mathbf{V}'_{S,1}\tilde{u}$. Similarly, one can prove that $\tilde{u}\mathbf{V}_{S,2} = \mathbf{V}'_{S,2}\tilde{u}$. Note that the proof of this direction works for weak Type II Ando tuples as well.

Conversely, suppose two Ando isometric lifts $(\mathbf{V}_{S,1}, \mathbf{V}_{S,2})$ and $\left(\mathbf{V}'_{S,1}, \mathbf{V}'_{S,2}\right)$ of $(T_1, T_2)$ corresponding to two Type II Ando tuples $(\mathcal{F}, \Lambda, P, U)$ and $(\mathcal{F}', \Lambda',$ $P', U')$, respectively, are unitarily equivalent. This means that there exists a unitary

$$\tau = \begin{bmatrix} \tau_{11} & \tau_{12} \\ \tau_{21} & \tau_{22} \end{bmatrix} : \begin{bmatrix} \mathcal{H} \\ H^2(\mathcal{F}) \end{bmatrix} \to \begin{bmatrix} \mathcal{H} \\ H^2(\mathcal{F}') \end{bmatrix}$$

such that

$$\tau(\mathbf{V}_{S,1}, \mathbf{V}_{S,2}) = \left(\mathbf{V}'_{S,1}, \mathbf{V}'_{S,2}\right)\tau, \quad \tau \begin{bmatrix} I_{\mathcal{H}} \\ 0 \end{bmatrix} = \begin{bmatrix} I_{\mathcal{H}} \\ 0 \end{bmatrix}. \tag{6.42}$$

The second equality in (6.42) implies that $\tau$ has the form

$$\tau = \begin{bmatrix} I_{\mathcal{H}} & \tau_{12} \\ 0 & \tau_{22} \end{bmatrix}.$$

As $\tau$ is unitary, this in turn forces $\tau_{12} = 0$ and $\tau_{22}\colon H^2(\mathcal{F}) \to H^2(\mathcal{F}')$ to be unitary. The first equality in (6.42) implies in particular that

$$\tau\mathbf{V}_{S,1}\mathbf{V}_{S,2} = \mathbf{V}'_{S,1}\mathbf{V}'_{S,2}\tau, \tag{6.43}$$

where

$$\mathbf{V}_{S,1}\mathbf{V}_{S,2} = \mathbf{V}_{S,2}\mathbf{V}_{S,1} = \begin{bmatrix} T & 0 \\ \mathbf{ev}_0^* \Lambda D_T & M_z \end{bmatrix},$$

and similarly for

$$\mathbf{V}'_{S,1}\mathbf{V}'_{S,2} = \mathbf{V}'_{S,2}\mathbf{V}'_{S,1} = \begin{bmatrix} T & 0 \\ \mathbf{ev}_0^*\Lambda' D_T & M_z \end{bmatrix}$$

by the assumption that $(\mathcal{F}, \Lambda, P, U)$ and $(\mathcal{F}', \Lambda', P', U')$ are both Type II Ando tuples. Hence

$$\tau \mathbf{V}_{S,1}\mathbf{V}_{S,2} = \begin{bmatrix} I_{\mathcal{H}} & 0 \\ 0 & \tau_{22} \end{bmatrix} \begin{bmatrix} T & 0 \\ \mathbf{ev}_0^*\Lambda D_T & M_z^{\mathcal{F}} \end{bmatrix} = \begin{bmatrix} T & 0 \\ \tau_{22}\mathbf{ev}_0^*\Lambda D_T & \tau_{22} M_z^{\mathcal{F}} \end{bmatrix}$$

while

$$\mathbf{V}'_{S,1}\mathbf{V}'_{S,2}\tau = \begin{bmatrix} T & 0 \\ \mathbf{ev}_0^*\Lambda' D_T & M_z^{\mathcal{F}'} \end{bmatrix} \begin{bmatrix} I_{\mathcal{H}} & 0 \\ 0 & \tau_{22} \end{bmatrix} = \begin{bmatrix} T & 0 \\ \mathbf{ev}_0^*\Lambda' D_T & M_z^{\mathcal{F}'}\tau_{22} \end{bmatrix}.$$

As a consequence of (6.43) we are led to the identity

$$\begin{bmatrix} T & 0 \\ \tau_{22}\mathbf{ev}_0^*\Lambda D_T & \tau_{22} M_z \end{bmatrix} = \begin{bmatrix} T & 0 \\ \mathbf{ev}_0^*\Lambda' D_T & M_z \tau_{22} \end{bmatrix}. \tag{6.44}$$

Equality of the $(2,2)$-entries in (6.44), combined with the fact that $\tau_{22}$ is unitary, implies that $\tau_{22}$ has the form $\tau_{22} = I_{H^2} \otimes u$ for some unitary $u\colon \mathcal{F} \to \mathcal{F}'$, from which it then follows that $\tau_{22}\mathbf{ev}_0^* = \mathbf{ev}_0^* u$. Comparison of the $(2,1)$-entries in (6.44) then gives $u\Lambda = \Lambda'$. A similar matrix computation and a comparison of the $(2,2)$-entries of $\tau(\mathbf{V}_{S,1}, \mathbf{V}_{S,2}) = \left(\mathbf{V}'_{S,1}, \mathbf{V}'_{S,2}\right)\tau$ implies

$$\left(M_{u(P^\perp U + zPU)}, M_{u(U^* P + zU^* P^\perp)}\right) = \left(M_{(P'^\perp U' + zP'U')u}, M_{(U'^* P' + zU'^* P'^\perp)u}\right),$$

which implies that

$$uP^\perp U = P'^\perp U' u, \; uPU = P'U'u, \quad uU^*P = U'^*P'u, \quad uU^*P^\perp = U'^*P'^\perp u. \tag{6.45}$$

Adding the first two identities in (6.45) gives $uU = U'u$. Use this identity in the first equation in (6.45) to get $uP^\perp = P'^\perp u$. Apply a similar argument starting with the second identity in (6.45) instead, or alternatively plug $P^\perp = I - P$, $P'^\perp = I - P'$ into $uP^\perp = P'^\perp u$, to arrive at $uP = Pu$ as well. We conclude that $(\mathcal{F}, \Lambda, P, U)$ and $(\mathcal{F}', \Lambda', P', U')$ indeed coincide. □

## 6.6 Notes to Chapter 6

Theorem 5.2.1, combined with the uniqueness assertion in Theorem 5.7.1, provides a labeling of minimal Ando lifts for a given commuting contractive pair $(T_1, T_2)$ in terms of coincidence equivalence classes of Douglas-minimal Type I Ando tuples $(\mathcal{F}, \Lambda, P, U)$ for $\left(T_1^*, T_2^*\right)$, while Theorem 6.1.4, combined with Theorem 6.5.1, accomplishes the same result but in terms of Type II Ando tuples for $(T_1, T_2)$. While, in general, minimal Ando lifts are not unitarily equivalent, coincidence-equivalence classes of Ando lifts provide a labeling for distinct unitary equivalence classes of Ando lifts. Li and Timotin [69], in parallel with the *central contractive intertwining lift* appearing in the commutant lifting theory motivated by $H^\infty$-control theory from the 1990s (see [60]), identify a unique "central" Ando lift for a given commuting contractive pair $(T_1, T_2)$. It would be of interest to identify explicitly the corresponding central choice (up to coincidence) of Type I/II Ando tuple, which corresponds to the Li–Timotin distinguished Ando lift.

# 7

# Strongly Minimal Ando Isometric Lifts and Fundamental Operators

Let $(T_1, T_2)$ be a commuting contractive pair acting on a Hilbert space $\mathcal{H}$, and let $(\Pi, V_1, V_2)$ be a minimal Ando isometric lift of $(T_1, T_2)$ acting on $\mathcal{K}$. The minimality condition means that $\mathcal{K} = \mathcal{K}_0$, where we set

$$\mathcal{K}_0 = \bigvee_{n_1, n_2 \geq 0} V_1^{n_1} V_2^{n_2} \Pi \mathcal{H}.$$

Note that whenever $(\Pi, V_1, V_2)$ is an Ando isometric lift of $(T_1, T_2)$, then $(\Pi, V_1 V_2)$ is a Sz.-Nagy–Foias isometric lift of the product contraction $T = T_1 T_2$. The condition that $(\Pi, V_1 V_2)$ is a minimal isometric lift of $T_1 T_2$ is that $\mathcal{K} = \mathcal{K}_{00}$, where we set

$$\mathcal{K}_{00} = \bigvee_{n \geq 0} V_1^n V_2^n \Pi \mathcal{H}.$$

Note that it is always the case that

$$\mathcal{K}_{00} = \bigvee_{n \geq 0} V_1^n V_2^n \Pi \mathcal{H} = \bigvee_{n_1, n_2 \geq 0:\, n_1 = n_2} V_1^{n_1} V_2^{n_2} \mathcal{H} \subset \bigvee_{n_1, n_2 \geq 0} V_1^{n_1} V_2^{n_2} \mathcal{H} = \mathcal{K}_0.$$

A natural question to ask is: When is it the case that the containment $\subset$ between the previous string of equalities is actually an equality =?

When does there exist an Ando lift $(V_1, V_2)$ of $(T_1, T_2)$ so that $\mathcal{K}_{00} = \mathcal{K}_0$?

When does there exist an Ando lift such that $\mathcal{K} = \mathcal{K}_{00}$? Lemma 3.1.19 ensures that when $(T_1, T_2)$ is a commuting isometric pair or a coisometric pair then the answer is affirmative.[1] In this chapter, we investigate this question and its correlation with the corresponding Ando tuples.

[1] This is because of the fact that the adjoint of the minimal isometric lift of a coisometry is the minimal unitary extension of the isometry.

## 7.1 Strongly Minimal Ando Lifts and Ando Tuples

We begin by formally introducing a stronger notion of minimality for Ando lifts.

**Definition 7.1.1** An Ando lift $(\Pi, V_1, V_2)$ of a commuting contractive pair $(T_1, T_2)$ is said to be *strongly minimal* if it acts on the space $\mathcal{K}_{00}$ given by

$$\mathcal{K}_{00} = \bigvee_{n\geq 0} V_1^n V_2^n \operatorname{Ran} \Pi.$$

The companion notion *strongly minimal Ando tuple* is defined as follows.

**Definition 7.1.2** An Ando tuple $(\mathcal{F}, \Lambda, P, U)$ for $(T_1, T_2)$ is said to be *strongly minimal* if the isometry $\Lambda\colon \mathcal{D}_{T_1T_2} \to \mathcal{F}$ is surjective. When the Ando tuple $(\mathcal{F}, \Lambda, P, U)$ is a Type I or Type II Ando tuple and is strongly minimal as an Ando tuple, we say that $(\mathcal{F}, \Lambda, P, U)$ is a strongly minimal Type I (respectively, strongly minimal Type II) Ando tuple.

Note that, since $\operatorname{Ran} \Lambda = \mathcal{F}$ for a strongly minimal Ando tuple $(\mathcal{F}, \Lambda, P, U)$, it is obvious that a strongly minimal Ando tuple is Douglas minimal (see Definition 5.1.3).

The next result makes the correlation between these two notions of strong minimality precise.

**Theorem 7.1.3** *Let $(T_1, T_2)$ be a commuting contractive operator pair acting on $\mathcal{H}$. Then we have the following.*

- (i) *$(T_1, T_2)$ has a strongly minimal Ando lift $(\Pi, V_1, V_2)$ if and only if there is a strongly minimal Type I Ando tuple for $(T_1^*, T_2^*)$.*
- (ii) *$(T_1, T_2)$ has a strongly minimal Ando lift $(\Pi, V_1, V_2)$ if and only if there is a strongly minimal Type II Ando tuple for $(T_1, T_2)$.*

*Thus the existence of a strongly minimal Type I Ando tuple for $(T_1^*, T_2^*)$ is equivalent to the existence of a strongly minimal Type II Ando tuple for $(T_1, T_2)$.*

*Proof of (i)* If $(\Pi, V_1, V_2)$ is a strongly minimal Ando lift of $(T_1, T_2)$, then by definition $V = V_1V_2$ is a minimal isometric lift of $T = T_1T_2$. By Theorem 5.2.1, the Ando lift $(\Pi, V_1, V_2)$ can be modeled as $(\mathbf{\Pi}_D, \mathbf{V}_{D,1}, \mathbf{V}_{D,2})$ in (5.3) for some Douglas-minimal Type I Ando tuple $(\mathcal{F}_*, \Lambda_*, P_*, U_*)$ for $(T_1^*, T_2^*)$. Consider the minimal isometric lift $(\Pi_D, V_D)$ of $T$ acting on $\mathcal{K}_D = \begin{bmatrix} H^2(\mathcal{D}_{T^*}) \\ \mathcal{Q}_{T^*} \end{bmatrix}$ as in Section 2.2.2. By the unitary equivalence of any two minimal isometric lifts of a single contraction operator (see Theorem 2.1.3), there is a unitary

$$\tau = \begin{bmatrix} \tau' & \tau_{12} \\ \tau_{21} & \tau'' \end{bmatrix} : \begin{bmatrix} H^2(\mathcal{D}_{T^*}) \\ \mathcal{Q}_{T^*} \end{bmatrix} \to \begin{bmatrix} H^2(\mathcal{F}_*) \\ \mathcal{Q}_{T^*} \end{bmatrix}$$

such that

$$\begin{bmatrix} \tau' & \tau_{12} \\ \tau_{21} & \tau'' \end{bmatrix} \begin{bmatrix} M_z^{\mathcal{D}_{T^*}} & 0 \\ 0 & W_D \end{bmatrix} = \begin{bmatrix} M_z^{\mathcal{F}_*} & 0 \\ 0 & W_D \end{bmatrix} \begin{bmatrix} \tau' & \tau_{12} \\ \tau_{21} & \tau'' \end{bmatrix}, \tag{7.1}$$

$$\begin{bmatrix} \tau' & \tau_{12} \\ \tau_{21} & \tau'' \end{bmatrix} \begin{bmatrix} \mathcal{O}_{\mathcal{D}_{T^*}, T^*} \\ \mathcal{Q}_{T^*} \end{bmatrix} = \begin{bmatrix} (I_{H^2} \otimes \Lambda_*)\, \mathcal{O}_{\mathcal{D}_{T^*}, T^*} \\ \mathcal{Q}_{T^*} \end{bmatrix}. \tag{7.2}$$

Applying part (ii) of Lemma 2.3.1 to (7.1) we conclude that

$$\tau = \begin{bmatrix} I_{H^2} \otimes \Lambda' & 0 \\ 0 & \tau'' \end{bmatrix}$$

for some unitary operators $\Lambda' : \mathcal{D}_{T^*} \to \mathcal{F}_*$ and $\tau'' : \mathcal{Q}_{T^*} \to \mathcal{Q}_{T^*}$. Equation (7.2) therefore yields

$$(I_{H^2} \otimes \Lambda')\, \mathcal{O}_{D_{T^*}, T^*} = (I_{H^2} \otimes \Lambda_*)\, \mathcal{O}_{D_{T^*}, T^*}, \tag{7.3}$$

$$\tau'' \mathcal{Q}_{T^*} = \mathcal{Q}_{T^*} \quad \tau'' W_D = W_D \tau''. \tag{7.4}$$

Equating the constant coefficients in the series forms of (7.3), we get $\Lambda' = \Lambda_*$. Consequently, the Type I Ando tuple $(\mathcal{F}_*, \Lambda_*, P_*, U_*)$ is strongly minimal.

Conversely, suppose $(\mathcal{F}_*, \Lambda_*, P_*, U_*)$ is a Type I Ando tuple of $(T_1^*, T_2^*)$ such that $\Lambda_* : \mathcal{D}_{T^*} \to \mathcal{F}_*$ is a unitary. By the forward direction of Theorem 5.2.1, the isometric operators $\mathbf{\Pi}_D, \mathbf{V}_{D,1}, \mathbf{V}_{D,2}$ given by

$$\mathbf{\Pi}_D = \begin{bmatrix} (I_{H^2} \otimes \Lambda_*)\, \mathcal{O}_{D_{T^*}, T^*} \\ \mathcal{Q}_{T^*} \end{bmatrix} : \mathcal{H} \to \begin{bmatrix} H^2(\mathcal{F}_*) \\ \mathcal{Q}_{T^*} \end{bmatrix},$$

$$(\mathbf{V}_{D,1}, \mathbf{V}_{D,2}) = \left( \begin{bmatrix} M_{U_*^*(P_*^\perp + zP_*)} & 0 \\ 0 & W_{\flat 1} \end{bmatrix}, \begin{bmatrix} M_{(P_* + zP_*^\perp)U_*} & 0 \\ 0 & W_{\flat 2} \end{bmatrix} \right) \quad \text{on} \quad \begin{bmatrix} H^2(\mathcal{F}_*) \\ \mathcal{Q}_{T^*} \end{bmatrix}$$

constitute an Ando lift of $(T_1, T_2)$. Note that the operator

$$\tau := \begin{bmatrix} I_{H^2} \otimes \Lambda_* & 0 \\ 0 & I_{\mathcal{Q}_{T^*}} \end{bmatrix} : \begin{bmatrix} H^2(\mathcal{D}_{T^*}) \\ \mathcal{Q}_{T^*} \end{bmatrix} \to \begin{bmatrix} H^2(\mathcal{F}_*) \\ \mathcal{Q}_{T^*} \end{bmatrix}$$

is unitary from $\begin{bmatrix} H^2(\mathcal{D}_{T^*}) \\ \mathcal{Q}_{T^*} \end{bmatrix}$ onto $\begin{bmatrix} H^2(\mathcal{F}_*) \\ \mathcal{Q}_{T^*} \end{bmatrix}$ since $\Lambda_*$ is unitary from $\mathcal{D}_{T^*}$ onto $\mathcal{F}_*$, and has the properties

$$\tau \Pi_D = \begin{bmatrix} I_{H^2} \otimes \Lambda_* & 0 \\ 0 & I_{\mathcal{Q}_{T^*}} \end{bmatrix} \begin{bmatrix} \mathcal{O}_{\mathcal{D}_{T^*}, T^*} \\ \mathcal{Q}_{T^*} \end{bmatrix} = \begin{bmatrix} (I_{H^2} \otimes \Lambda_*)\, \mathcal{O}_{\mathcal{D}_{T^*}, T^*} \\ \mathcal{Q}_{T^*} \end{bmatrix} = \Pi \quad \text{and}$$

$$\tau V_D = \begin{bmatrix} I_{H^2} \otimes \Lambda_* & 0 \\ 0 & I_{\mathcal{Q}_{T^*}} \end{bmatrix} \begin{bmatrix} M_z^{\mathcal{D}_{T^*}} & 0 \\ 0 & W_D \end{bmatrix} = \begin{bmatrix} M_z^{\mathcal{F}_*} & 0 \\ 0 & W_D \end{bmatrix} \begin{bmatrix} I_{H^2} \otimes \Lambda_* & 0 \\ 0 & I_{\mathcal{Q}_{T^*}} \end{bmatrix} = \mathbf{V}_D \tau,$$

where $\mathbf{V}_D = \mathbf{V}_{D,1}\mathbf{V}_{D,2}$ and $(\Pi_D, V_D)$ is the Douglas model minimal isometric lift of $T = T_1T_2$ as in Section 2.2.2. Consequently, the Ando lift $(\mathbf{\Pi}_D, \mathbf{V}_{D,1}, \mathbf{V}_{D,2})$ of $(T_1, T_2)$ is strongly minimal. This completes the proof of (i).

*Proof of (ii)* For this proof we use the Schäffer model rather than the Douglas model. Suppose that $(\Pi, V_1, V_2)$ is a strongly minimal Ando lift of $(T_1, T_2)$. Since a strongly minimal Ando lift is obviously minimal, by the converse part of Theorem 6.1.4 we know that there is a Type II Ando tuple $(\mathcal{F}, \Lambda, P, U)$ of $(T_1, T_2)$ so that the lift $(\Pi, V_1, V_2)$ is unitarily equivalent to the Schäffer model lift $(\mathbf{\Pi}_S, \mathbf{V}_{S,1}, \mathbf{V}_{S,2})$ (6.19) determined by $(\mathcal{F}, \Lambda, P, U)$

$$\mathbf{\Pi}_S = \begin{bmatrix} I_{\mathcal{H}} \\ 0 \end{bmatrix} : \mathcal{H} \to \mathcal{K}_S = \begin{bmatrix} \mathcal{H} \\ H^2(\mathcal{F}) \end{bmatrix},$$

$$\begin{aligned}(\mathbf{V}_{S,1}, \mathbf{V}_{S,2}) \\ = \left( \begin{bmatrix} T_1 & 0 \\ \mathbf{ev}^*_{0,\mathcal{F}} PU\Lambda D_T & M_{(P^\perp + zP)U} \end{bmatrix}, \begin{bmatrix} T_2 & 0 \\ \mathbf{ev}^*_{0,\mathcal{F}} U^* P^\perp \Lambda D_T & M_{U^*(P+zP^\perp)} \end{bmatrix} \right),\end{aligned}$$

and also by (6.20)

$$\mathbf{V}_S = \mathbf{V}_{S,1}\mathbf{V}_{S,2} = \begin{bmatrix} T & 0 \\ \mathbf{ev}^*_{0,\mathcal{F}} \Lambda D_T & M_z^{\mathcal{F}} \end{bmatrix} : \begin{bmatrix} \mathcal{H} \\ H^2(\mathcal{F}) \end{bmatrix} \to \begin{bmatrix} \mathcal{H} \\ H^2(\mathcal{F}) \end{bmatrix}. \tag{7.5}$$

By hypothesis, $V$ is a minimal lift of $T = T_1T_2$. On the other hand, consider the Schäffer model minimal isometric lift $(\Pi_S, V_S)$ of $T = T_1T_2$, given by

$$\Pi_S = \begin{bmatrix} I_{\mathcal{H}} \\ 0 \end{bmatrix} : \mathcal{H} \to \mathcal{K}_S := \begin{bmatrix} \mathcal{H} \\ H^2(\mathcal{D}_T) \end{bmatrix},$$

$$V_S = \begin{bmatrix} T & 0 \\ \mathbf{ev}^*_{0,\mathcal{D}_T} D_T & M_z^{\mathcal{D}_T} \end{bmatrix} \quad \text{on} \quad \begin{bmatrix} \mathcal{H} \\ H^2(\mathcal{D}_T) \end{bmatrix}, \tag{7.6}$$

as discussed in Section 2.2.1. Again, by the uniqueness of minimal isometric lifts of single contraction operators, there exists a unitary

$$\tau = \begin{bmatrix} \tau_{11} & \tau_{12} \\ \tau_{21} & \tau_{22} \end{bmatrix} : \begin{bmatrix} \mathcal{H} \\ H^2(\mathcal{D}_T) \end{bmatrix} \to \begin{bmatrix} \mathcal{H} \\ H^2(\mathcal{F}) \end{bmatrix}$$

intertwining the isometries in (7.6) and (7.5) and such that $\tau|_{\mathcal{H}} = I_{\mathcal{H}}$. Hence $\tau$ is unitary with $\tau_{11} = I_{\mathcal{H}}$ also forcing $\tau_{12} = 0$, $\tau_{21} = 0$, so now $\tau = \left[\begin{smallmatrix} I_{\mathcal{H}} & 0 \\ 0 & \tau_{22} \end{smallmatrix}\right]$ with $\tau_{22} \colon H^2(\mathcal{D}_T) \to H^2(\mathcal{F})$ unitary. The intertwining condition thus becomes

$$\begin{bmatrix} I_{\mathcal{H}} & 0 \\ 0 & \tau_{22} \end{bmatrix} \begin{bmatrix} T & 0 \\ \mathbf{ev}^*_{0,\mathcal{D}_T} D_T & M_z^{\mathcal{D}_T} \end{bmatrix} = \begin{bmatrix} T & 0 \\ \mathbf{ev}^*_{0,\mathcal{F}} \Lambda D_T & M_z^{\mathcal{F}} \end{bmatrix} \begin{bmatrix} I_{\mathcal{H}} & 0 \\ 0 & \Lambda' \end{bmatrix}. \tag{7.7}$$

Comparing the (2,1) and (2,2) entries in the previous matrices then gives

$$\tau_{22}\mathbf{ev}_{0,\mathcal{D}_T}^* D_T = \mathbf{ev}_{0,\mathcal{F}}^* \Lambda D_T, \quad \tau_{22} M_z^{\mathcal{D}_T} = M_z^{\mathcal{F}} \tau_{22}.$$

From the first equation we see that $\tau_{22}$ takes constants into constants. From the second equation we see that $\tau_{22}$ is a multiplication operator. Putting these two conditions together says that $\tau_{22}$ is multiplication by a constant, that is, $\tau_{22}$ has the form $\tau_{22} = I_{H^2} \otimes \Lambda'$ for some $\Lambda' \colon \mathcal{D}_T \to \mathcal{F}$. Further inspection of the first equation tells us that the constant is $\Lambda$: $\Lambda' = \Lambda$. Moreover, as observed earlier, $\tau_{22}$ is unitary, that is, the operator $I_{H^2} \otimes \Lambda$ is unitary. This then forces $\Lambda$ to be unitary as an operator from $\mathcal{D}_T$ to $\mathcal{F}$. This in turn means that the tuple $(\mathcal{F}, \Lambda, P, U)$ is also strongly minimal as a Type II Ando tuple.

Conversely, let $(\mathcal{F}, \Lambda, P, U)$ be a strongly minimal Type II Ando tuple of the pair $(T_1, T_2)$. By the forward direction of Theorem 6.1.4, for any Type II Ando tuple $(\mathcal{F}, \Lambda, P, U)$ of $(T_1, T_2)$, $(\mathbf{\Pi}_S, \mathbf{V}_{S,1}, \mathbf{V}_{S,2})$ given by (6.19) is an Ando lift of $(T_1, T_2)$ with $\mathbf{V}_S := \mathbf{V}_{S,1}\mathbf{V}_{S,2}$ as in (6.20). The strongly minimal property of the Ando tuple $(\mathcal{F}, \Lambda, P, U)$ means that the operator $\Lambda \colon \mathcal{D}_T \to \mathcal{F}$ is unitary, hence the operator

$$\tau := \begin{bmatrix} I_{\mathcal{H}} & 0 \\ 0 & I_{H^2} \otimes \Lambda \end{bmatrix} : \mathcal{K}_S := \begin{bmatrix} \mathcal{H} \\ H^2(\mathcal{D}_T) \end{bmatrix} \to \begin{bmatrix} \mathcal{H} \\ H^2(\mathcal{F}) \end{bmatrix} =: \mathcal{K}_S$$

is also unitary. By definition we have

$$\tau \Pi_S = \begin{bmatrix} I_{\mathcal{H}} & 0 \\ 0 & I_{H^2} \otimes \Lambda \end{bmatrix} \begin{bmatrix} I_{\mathcal{H}} \\ 0 \end{bmatrix} = \begin{bmatrix} I_{\mathcal{H}} \\ 0 \end{bmatrix} = \mathbf{\Pi}_S$$

and

$$\begin{aligned} &\tau V_S \\ &= \begin{bmatrix} I_{\mathcal{H}} & 0 \\ 0 & I_{H^2} \otimes \Lambda \end{bmatrix} \begin{bmatrix} T & 0 \\ \mathbf{ev}_{0,\mathcal{D}_T}^* D_T & M_z^{\mathcal{D}_T} \end{bmatrix} = \begin{bmatrix} T & 0 \\ \mathbf{ev}_{0,\mathcal{F}}^* \Lambda D_T & M_z^{\mathcal{F}} \end{bmatrix} \begin{bmatrix} I_{H^2} & 0 \\ 0 & I_{H^2} \otimes \Lambda \end{bmatrix} \\ &= \mathbf{V}_S \tau, \end{aligned}$$

that is, the Sz.-Nagy–Foias lift $(\mathbf{\Pi}_S, \mathbf{V}_S)$ of $T$ is unitarily equivalent (via $\tau$) to the Schäffer model minimal Sz.-Nagy–Foias lift $(\Pi_S, V_S)$ of $T$, and hence the lift $(\mathbf{\Pi}_S, \mathbf{V}_S)$ must itself also be minimal, meaning that the Ando lift $(\mathbf{\Pi}_S, \mathbf{V}_{S,1}, \mathbf{V}_{S,2})$ of $(T_1, T_2)$ is strongly minimal. This completes the proof of the theorem. □

**Remark 7.1.4** Theorem 7.1.3 and its proof provide additional details about Theorems 5.2.1 and 6.1.4. Indeed, part (i) of Theorem 7.1.3 ensures that Theorem 5.2.1 continues to hold if, in the statement of Theorem 5.2.1, we replace "minimal Type I Ando tuple of $(T_1^*, T_2^*)$" by "strongly minimal Type I Ando tuple of $(T_1, T_2)$" and "minimal Ando lift of $(T_1, T_2)$" by "strongly minimal Ando lift of $(T_1, T_2)$." Similarly, part (ii) of Theorem 7.1.3 ensures that Theorem 6.1.4 continues to hold if we replace "minimal Type II Ando tuple for $(T_1, T_2)$" by "strongly minimal Type II Ando tuple for $(T_1, T_2)$" and "minimal Ando lift of $(T_1, T_2)$" by "strongly minimal Ando lift of $(T_1, T_2)$." Let us remark here that, as in the classification problem of Schäffer models of Ando lifts (analyzed in Section 6.5), weak Type II Ando tuples do not blend well for the analysis of strong minimality, and that is why only the minimal Type II Ando tuples are considered while making the aforementioned replacements in the statement of Theorem 6.1.4.

We have noted in Theorems 5.5.1 and 6.3.2 that commuting isometric operator pairs with the product operator equal to a shift and commuting coisometric operator pairs with the product operator equal to the adjoint of a shift have strongly minimal Type I Ando tuples (respectively, strongly minimal Type II Ando tuples). The restriction that the product is a shift or adjoint of a shift is not essential: The result still holds without this restriction. On the other hand, we have seen in item (3) of Theorem 5.3.5 that both $T_1 \cdot T_2$ and $T_2 \cdot T_1$ are regular factorizations if $(T_1, T_2)$ is a commutative isometric operator pair or a commutative coisometric operator pair. The next result shows that this confluence of observations is no accident.

**Theorem 7.1.5** *A commuting contractive pair $(T_1, T_2)$ has a strongly minimal Ando lift if and only if both factorizations $T_1 \cdot T_2$ and $T_2 \cdot T_1$ are regular. In this case, there is a unique strongly minimal canonical-form special Ando tuple of $(T_1, T_2)$.*

*Proof* We note that the last statement in the theorem is simply a direct application of item (2) in Corollary 5.3.4.

Suppose that $(T_1, T_2)$ is a commuting contractive pair such that both $T_1 \cdot T_2$ and $T_2 \cdot T_1$ are regular factorizations. This means that both spaces $\mathrm{Dom}(U_0)$ and $\mathrm{Ran}(U_0)$ given by (5.28) are equal to the whole space $\mathcal{D}_{T_1} \oplus \mathcal{D}_{T_2}$. Thus both the operator $\Lambda_\dagger$ and the operator $\Lambda_\dagger^\mathfrak{f}$ given on a dense set by

$$\Lambda_\dagger^\mathfrak{f} \colon D_T h \mapsto D_{T_1} h \oplus D_{T_2} T_1 h \quad \text{for } h \in \mathcal{H}$$

define unitary operators from $\mathcal{D}_T$ onto $\mathcal{D}_{T_1} \oplus \mathcal{D}_{T_2}$, and furthermore the operator $U_0$ defined densely by (5.29) extends to define a unitary operator from $\mathcal{F}_\dagger := \mathcal{D}_{T_1} \oplus \mathcal{D}_{T_2}$ onto itself. Define a projection $P_\dagger$ on $\mathcal{D}_{T_1} \oplus \mathcal{D}_{T_2}$ by

$$P_\dagger \colon f_1 \oplus f_2 = f_1 \oplus 0 \quad \text{for } f_1 \in \mathcal{D}_{T_1},\ f_2 \in \mathcal{D}_{T_2}.$$

Then the collection

$$\left(\mathcal{F}_{\dagger} = \mathcal{D}_{T_1} \oplus \mathcal{D}_{T_2}, \Lambda_{\dagger}, P_{\dagger}, U_{\dagger} = U_0\right)$$

is a canonical-form special Ando tuple for $(T_1, T_2)$ that is also strongly minimal as an Ando tuple. By Theorem 6.2.1, this is a Type II Ando tuple for $(T_1, T_2)$. By part (ii) of Theorem 7.1.3, $(T_1, T_2)$ has a strongly minimal Ando lift.

Conversely, suppose that $(T_1, T_2)$ acting on $\mathcal{H}$ has a strongly minimal Ando lift, which we take to be of Schäffer type $(\Pi = \iota_{\mathcal{H}}, V_1, V_2)$ with $\iota_{\mathcal{H}} \colon \mathcal{H} \to \mathcal{K}$ the inclusion map and with $(V_1, V_2)$ acting on the ambient space for the minimal isometric lift of $T = T_1 T_2$:

$$\mathcal{K} = \bigvee_{n \geq 0} V_1^n V_2^n \mathcal{H}.$$

This implies that with $V_0 := V = V_1 V_2$,

$$D_{V_j^*} \mathcal{K} = D_{V_j^*} \mathcal{H} \quad \text{for } j = 0, 1, 2. \tag{7.8}$$

Indeed, if $n \geq 1$, then since $D_{V_j^*} = \left(I_{\mathcal{K}} - V_j V_j^*\right)$ we have

$$D_{V_1^*} V_1^n V_2^n = 0 = D_{V_2^*} V_2^n V_1^n \quad \text{and} \quad D_{V^*} V^n = 0.$$

Furthermore note that the map $\omega_j \colon \mathcal{D}_{T_j^*} \to \mathcal{D}_{V_j^*}$ given by

$$\omega_j \colon D_{T_j^*} h \mapsto D_{V_j^*} h \quad \text{for } h \in \mathcal{H}$$

for $j = 0, 1, 2$ (here we set $T_0 = T_1 T_2$) is isometric: Indeed, simply note that

$$\begin{aligned}
\left\| D_{V_j^*} h \right\|^2 &= \left\langle \left(I - V_j V_j^*\right) h, h \right\rangle = \|h\|^2 - \left\| V_j^* h \right\|^2 \\
&= \|h\|^2 - \left\| T_j^* h \right\|^2 \text{ (by the lifting property)} \\
&= \left\langle \left(I - T_j T_j^*\right) h, h \right\rangle = \left\| D_{T_j^*}^2 h \right\|^2 .
\end{aligned}$$

Combining this with the observation (7.8), we see that each of the maps $\omega_j$ is unitary from $\mathcal{D}_{T_j^*}$ onto $\mathcal{D}_{V_j^*}$. Focusing now just on the case $j = 1, 2$ and defining the map $\tau \colon \mathcal{D}_{T_1^*} \oplus \mathcal{D}_{T_2^*} \to \mathcal{D}_{V_1^*} \oplus \mathcal{D}_{V_2^*}$ defined densely by

$$\tau \colon D_{T_1^*} h \oplus D_{T_2^*} k \mapsto D_{V_1^*} h \oplus \mathcal{D}_{V_2^*} k,$$

we see that $\tau$ so defined is also unitary.

On the other hand, since $\left(V_1^*, V_2^*\right)$ is a commuting coisometric pair, we know by item (3) in Theorem 5.3.5 that $V_1^* \cdot V_2^*$ is a regular factorization. Hence the operator $\sigma \colon \mathcal{D}_{V^*} \to \mathcal{D}_{V_1^*} \oplus \mathcal{D}_{V_2^*}$ defined densely by

$$\sigma \colon D_{V^*} h \mapsto D_{V_1^*} V_2^* h \oplus D_{V_2^*} h$$

is unitary as well. Now it just remains to see from the definitions of the unitary operators $\omega_0$, $\tau$, and $\sigma$ that the isometry

$$\Lambda_{\dagger *}\colon D_{T^*}h \mapsto D_{T_1^*}T_2^*h \oplus D_{T_2^*}h$$

coincides with $\tau^* \circ \sigma \circ \omega_0$ on a dense set, namely, $\{D_{T^*}h\colon h \in \mathcal{H}\}$. Therefore $\Lambda_{\dagger *}$ must be unitary and hence, by definition, the factorization $T_1^* \cdot T_2^*$ is regular, or so, equivalently, is $T_2 \cdot T_1$ (see [91]). Now note that if $(T_1, T_2)$ has a strongly minimal Ando lift, then so does the pair $(T_2, T_1)$. Thus proceeding as above for the pair $(T_2, T_1)$, one can conclude that the factorization $T_1 \cdot T_2$ is regular. This completes the proof. □

Recall by Theorem 7.1.3 that if there is a strongly minimal Type I Ando tuple for $(T_1^*, T_2^*)$, then there is a strongly minimal Type II Ando tuple for $(T_1, T_2)$, and vice versa. It is not clear whether the existence of the companion notion of a strongly minimal Ando lift for the commutative contractive pair $(T_1, T_2)$ is equivalent to the existence of a strongly minimal Ando lift for the adjoint pair $(T_1^*, T_2^*)$. This can be worked out via a systematic calculation as follows.

**Theorem 7.1.6** *Let $(T_1, T_2)$ be a commuting contractive pair acting on $\mathcal{H}$. Then $(T_1, T_2)$ has a strongly minimal Ando lift if and only if $(T_1^*, T_2^*)$ has a strongly minimal Ando lift.*

*Proof* Note that the two-way implications follow by symmetry once we establish one of them. We suppose that $(T_1, T_2)$ has a strongly minimal Ando lift and prove that then so does $(T_1^*, T_2^*)$. Let $(\Pi, V_1, V_2)$ be a strongly minimal Ando lift of $(T_1, T_2)$, that is, $V_1, V_2$ act on $\mathcal{K} = \bigvee_{n \geq 0} \{V_1^n V_2^n \operatorname{Ran} \Pi\}$. By Lemma 3.1.19, $(V_1, V_2)$ has a unitary extension $(W_1, W_2)$ acting on the space

$$\widetilde{\mathcal{K}} = \bigvee_{n \in \mathbb{Z}} \{W_1^n W_2^n \operatorname{Ran} \Pi\}.$$

Let us set

$$\mathcal{K}' := \bigvee_{n \geq 0} W_1^{*n} W_2^{*n} \operatorname{Ran} \Pi = \bigvee_{n \geq 0} W^{*n} \operatorname{Ran} \Pi, \tag{7.9}$$

where we use the notation $W = W_1 W_2$. We argue that $\mathcal{K}'$ is a $(W_1^*, W_2^*)$-invariant subspace.

Let us consider the spaces

$$\mathcal{K}_+ = \mathcal{K} \ominus \operatorname{Ran} \Pi \quad \text{and} \quad \mathcal{K}_- = \widetilde{\mathcal{K}} \ominus \mathcal{K}.$$

This induces a three-fold orthogonal decomposition of the space $\widetilde{\mathcal{K}}$:

$$\widetilde{\mathcal{K}} = \mathcal{K}_- \oplus \operatorname{Ran} \Pi \oplus \mathcal{K}_+. \tag{7.10}$$

Since $(W_1, W_2, W)$ is an extension of $(V_1, V_2, V = V_1V_2)$ and $(V_1, V_2)$ is an isometric lift of $(T_1, T_2)$, we see that $\mathcal{K} = \operatorname{Ran}\Pi \oplus \mathcal{K}_+$ is invariant for $(W_1, W_2, W)$. From the facts that $(V_1, V_2)$ is the restriction of $(W_1, W_2)$ to $\mathcal{K}$ and that $(V_1, V_2)$ is a lift for $(T_1, T_2)$ on $\mathcal{K}$ (with embedding operator $\Pi$), we see that $\mathcal{K}_+$ is also invariant for $(W_1, W_2, W)$ and, furthermore, that, with respect to the three-fold orthogonal decomposition (7.10) of the space $\widetilde{\mathcal{K}}$, $(W_1, W_2, W)$ have block lower-triangular $3 \times 3$-matrix decompositions of the form

$$W = \begin{bmatrix} * & 0 & 0 \\ * & \Pi T \Pi^* & 0 \\ * & * & * \end{bmatrix}, \quad W_j = \begin{bmatrix} * & 0 & 0 \\ * & \Pi T_j \Pi^* & 0 \\ * & * & * \end{bmatrix} \quad \text{for } j = 1, 2. \tag{7.11}$$

From these lower-triangular decompositions, we see that the space $\mathcal{K}_- \oplus \operatorname{Ran}\Pi$ is a $\left(W_1^*, W_2^*, W^*\right)$-invariant subspace.

We claim next that

$$\mathcal{K}_- \oplus \operatorname{Ran}\Pi = \bigvee_{n \geq 0} W^{*n} \operatorname{Ran}\Pi =: \mathcal{K}'. \tag{7.12}$$

The containment

$$\mathcal{K}_- \oplus \operatorname{Ran}\Pi \supset \bigvee_{n \geq 0} W^{*n} \operatorname{Ran}\Pi$$

is clear since we have seen that $\mathcal{K}_- \oplus \operatorname{Ran}\Pi$ is invariant for $W^*$. But from the definitions we have

$$\widetilde{\mathcal{K}} = \bigvee_{n \in \mathbb{Z}} W^n \operatorname{Ran}\Pi, \quad \operatorname{Ran}\Pi \oplus \mathcal{K}_+ = \bigvee_{n \geq 0} W^n \operatorname{Ran}\Pi.$$

Combining these forces us to the conclusion that, in fact, we recover $\mathcal{K}_- \oplus \operatorname{Ran}\Pi \subset \widetilde{\mathcal{K}}$ as

$$\mathcal{K}_- \oplus \operatorname{Ran}\Pi = \bigvee_{n \leq 0} W^n \operatorname{Ran}\Pi = \bigvee_{n \geq 0} W^{*n} \operatorname{Ran}\Pi$$

and the claim (7.12) follows. Therefore

$$\left(V_1', V_2'\right) := \left(W_1^*, W_2^*\right)|_{\mathcal{K}'}$$

is a pair of commuting isometries. By taking adjoints of the block matrices in (7.11) we see that $\left(\Pi, V_1', V_2'\right)$ is an Ando lift of $\left(T_1^*, T_2^*\right)$ acting on the space $\mathcal{K}_- \oplus \operatorname{Ran}\Pi = \mathcal{K}'$ equal to the space for the minimal isometric lift for $T^*$, that is, $\left(\Pi, V_1', V_2'\right)$ is a strongly minimal Ando lift of $\left(T_1^*, T_2^*\right)$ as desired. □

In Section 7.2 we find more characterizations for the existence of strongly minimal Ando lifts.

## 7.2 Fundamental-Operator Pairs: Three Proofs of Existence

Associated to every commuting contractive operator pair $(T_1, T_2)$ is a contractive pair $(F_1, F_2)$ (noncommuting, in general) that plays a pivotal role in this and later chapters. In this section, we introduce these special pairs and give three proofs of their existence.

We first need a preliminary lemma.

**Lemma 7.2.1** *Let $(T_1, T_2)$ be a commuting contractive pair acting on $\mathcal{H}$ and $T = T_1 T_2$. Then a pair of operators $F_1, F_2$ on the defect space $\mathcal{D}_T$ satisfies the pair of equations*

$$T_1 - T_2^* T = D_T F_1 D_T, \quad T_2 - T_1^* T = D_T F_2 D_T \tag{7.13}$$

*if and only if $F_1, F_2$ satisfies the pair of equations*

$$D_T T_1 = F_1 D_T + F_2^* D_T T, \quad D_T T_2 = F_2 D_T + F_1^* D_T T. \tag{7.14}$$

*Furthermore, the solution of either pair* (7.13) *or* (7.14) *is unique.*

*Proof* Let us suppose that the operator pair $(F_1, F_2)$ solves (7.13). We wish to prove that the same $(F_1, F_2)$ also solves (7.14). Let us consider only the first equation in (7.14) for the moment. Since $D_T$ is an injective, bounded operator on $\mathcal{D}_T$, the solution set of the first equation in (7.14) is unaffected if we multiply the equations (7.14) on the left by $D_T$. In particular, multiplying the first equation in (7.14) on the left by $D_T$ results in

$$(I - T^* T) T_1 = D_T F_1 D_T + D_T F_2^* D_T T.$$

Now use that $(F_1, F_2)$ solves (7.13) to eliminate $F_1$ and $F_2$ and rewrite this as

$$\begin{aligned}(I - T^* T) T_1 &= (T_1 - T_2^* T) + (T_2^* - T^* T_1) T \\ &= T_1 - T^* T_1 T = (I - T^* T) T_1,\end{aligned}$$

which is just an identity in $T_1, T_2$, and $T = T_1 T_2$, and we conclude that the first equation in (7.14) holds. A similar computation (which amounts to switching the roles of the indices $(1, 2)$) verifies that the second equation in (7.14) holds. We conclude that any solution of (7.13) is also a solution of (7.14).

Conversely, suppose that $F_1, F_2 \in \mathcal{B}(\mathcal{D}_T)$ solves the system (7.14) and we wish to show that the same $F_1, F_2$ solves the system (7.13). Multiply both equations in (7.14) on the left by $D_T$ to get the pair of equations

$$(I - T^* T) T_1 = D_T F_1 D_T + D_T F_2^* D_T T, \quad (I - T^* T) T_2 = D_T F_2 D_T + D_T F_1^* D_T T. \tag{7.15}$$

Our goal is to solve this system for $D_T F_1 D_T$ and $D_T F_2 D_T$ and then arrive at the equations (7.13) written in reverse order:

$$D_T F_1 D_T = T_1 - T_2^* T, \quad D_T F_2 D_T = T_2 - T_1^* T. \tag{7.16}$$

We shall give the details only for the first equation, as the verification of the second is completely similar.

Let us take the adjoint of the second equation in (7.15) and solve for $D_T F_2^* D_T$ to get

$$D_T F_2^* D_T = T_2^*(I - T^*T) - T^* D_T F_1 D_T.$$

Plugging this back into the first equation in (7.15) then gives us

$$(I - T^*T)T_1 = D_T F_1 D_T + T_2^*(I - T^*T)T - T^* D_T F_1 D_T T. \tag{7.17}$$

We now have $D_T F_2 D_T$ eliminated and this becomes an equation for the single unknown $D_T F_1 D_T$: If we set

$$\Sigma_1 = D_T F_1 D_T, \quad Y = (I - T^*T)T_1 - T_2^*(I - T^*T)T, \tag{7.18}$$

with $\Sigma_1$ now the unknown, then (7.17) has the form

$$\Sigma_1 - T^* \Sigma_1 T = Y. \tag{7.19}$$

Rewrite this as $\Sigma_1 = T^* \Sigma_1 T + Y$, plug this expression for $\Sigma_1$ back into the right-hand side, and iterate to get

$$\Sigma_1 = T^{*N+1} \Sigma_1 T^{N+1} + \sum_{n=0}^{N} T^{*n} Y T^n \quad \text{for all } N = 0, 1, 2, \ldots. \tag{7.20}$$

If it is the case that $T^{N+1} \Sigma_1 T^{N+1}$ tends to 0 (say in the weak operator topology), then we can take limits on both sides of (7.20) to arrive at a formula for $\Sigma_1$:

$$\Sigma_1 = \sum_{N=0}^{\infty} T^{*n} Y T^n.$$

To analyze this further, let us recall the precise formulas (7.18) for what $\Sigma_1$ and $Y$ are for our case here. Thus (7.20) specializes to

$$D_T F_1 D_T = T^{*N+1} D_T F_1 D_T T^{N+1} + \sum_{n=0}^{N} T^{*n} \big( (I - T^*T)T_1 - T_2^*(I - T^*T)T \big) T^n. \tag{7.21}$$

Let us note next that, for each $h \in \mathcal{H}$, the following series is telescoping:

$$\sum_{n=0}^{N} \|D_T T^n h\|^2 = \sum_{n=0}^{N} \langle T^{*n}(I - T^*T)T^n h, h\rangle = \sum_{n=0}^{N} \left(\|T^n h\|^2 - \left\|T^{n+1} h\right\|^2\right)$$
$$= \|h\|^2 - \left\|T^{N+1} h\right\|^2.$$

As $T$ is a contraction, $\left\|T^{N+1} h\right\|^2$ is decreasing and bounded below by zero and hence convergent. In particular, we see that the series $\sum_{n=0}^{N} \|D_T T^n h\|^2$ is convergent for each $h \in \mathcal{H}$. By the $n$th term test it follows that $\lim_{n\to\infty} \|D_T T^n h\|^2 = 0$. It follows that the operator sequence $T^{*N+1} D_T F_1 D_T T^{N+1}$ certainly converges to 0 in the weak operator topology. We conclude from (7.21) that we have solved for $D_T F_1 D_T$:

$$D_T F_1 D_T = \sum_{n=0}^{\infty} T^{*n}\big((I - T^*T)T_1 - T_2^*(I - T^*T)T\big)T^n \tag{7.22}$$

with the infinite series converging in the weak operator topology.

It remains only to show that the sum of this series is actually equal to $T_1 - T_2^* T$, that is, that $F_1$ satisfies the first of equations (7.13). Let us recall from Section 4.4 that $\lim_{n\to\infty} T^{*n}T^n$ exists in the strong operator topology with the limit denoted as $Q_T^2$ (with $Q_T$ then set equal to the unique positive-semidefinite square root of $Q_T^2$) and, moreover, given that $T = T_1 \cdot T_2$ is a commuting, contractive factorization of $T$, we also have the three identities

$$Q_T^2 = T^* Q_T^2 T, \quad Q_T^2 = T_1^* Q_T^2 T_1, \quad Q_T^2 = T_2^* Q_T^2 T_2. \tag{7.23}$$

Moreover, owing to the telescoping property of the sequence of partial sums, we see that

$$\sum_{n=0}^{N} T^{*n}(I - T^*T)T^n = I - T^{*N+1}T^{N+1}$$

and hence also

$$\sum_{n=0}^{\infty} T^{*n}(I - T^*T)T^n = I - Q_T^2$$

with convergence in the strong operator topology. Let us rewrite (7.22) as

$$D_T F_1 D_T = \sum_{n=0}^{\infty} T^{*n}(I - T^*T)T_1 T^n - \sum_{n=0}^{\infty} T^{*n} T_2^*(I - T^*T)T T^n. \tag{7.24}$$

The first term on the right-hand side of (7.24) is

$$\sum_{n=0}^{\infty} T^{*n}(I - T^*T)T_1 T^n = \left(\sum_{n=0}^{\infty} T^{*n}(I - T^*T)T^n\right) T_1 = \left(I - Q_T^2\right) T_1$$

while the second term (without the minus sign) on the right-hand side is

$$\begin{aligned}\sum_{n=0}^{\infty} T^{*n}T_2^*(I-T^*T)TT^n &= T_2^*\left(\sum_{n=0}^{\infty} T^{*n}(I-T^*T)T^n\right)T \\ &= T_2^*\left(\sum_{n=0}^{\infty} T^{*n}(I-T^*T)T^n\right)T = T_2^*\left(I-Q_T^2\right)T.\end{aligned}$$

Collecting terms and recalling (7.23) then gives

$$\begin{aligned}D_T F_1 D_T &= \left(I-Q_T^2\right)T_1 - T_2^*\left(I-Q_T^2\right)T = T_1 - Q_T^2 T_1 - T_2^* T + T_2^* Q_T^2 T_2 T_1 \\ &= T_1 - Q_T^2 T_1 - T_2^* T + Q_T^2 T_1 = T_1 - T_2^* T\end{aligned}$$

as required.

It remains to show that solutions of (7.13) (or equivalently of (7.14)) are unique whenever they exist. As for (7.14), uniqueness is immediate from the fact that $F_1$ and $F_2$ are taken to be operators on $\mathcal{D}_T$ and the fact that $D_T|_{\mathcal{D}_T}$ is an injective bounded operator on $\mathcal{D}_T$ with dense range. It is possible to give a direct proof of the uniqueness of solutions of (7.14) by showing that the only solution of the homogeneous equation is the zero solution; however, this proof is rather elaborate (much like the proof of the existence of a solution for (7.14)). A much shorter proof is to note that this uniqueness follows immediately from the equivalence between solutions of (7.13) and (7.14) together with the uniqueness of solutions of (7.13) already observed. □

Solutions of (7.13) are fundamental for later developments, so we now formally give them a name. Existence of such solutions for any commuting contractive pair will be shown shortly.

**Definition 7.2.2** Given a commuting contractive operator pair $(T_1, T_2)$, the unique solution pair $(F_1, F_2)$ in $\mathcal{B}(\mathcal{D}_T)$ (where $T = T_1T_2$) of the pair of operator equations (7.13) is called the *fundamental-operator pair* of $(T_1, T_2)$.

We are now ready to prove the existence of a fundamental-operator pair for a commuting contractive operator pair $(T_1, T_2)$. We actually give three proofs, taking full advantage of the two-dimensional dilation theory.

**Theorem 7.2.3** *Let $(T_1, T_2)$ be a commuting contractive pair acting on $\mathcal{H}$ and $T = T_1T_2$. Then a fundamental-operator pair of $(T_1, T_2)$ in the sense of Definition 7.2.2 exists, that is, there exists a unique pair of contraction operators $F_1, F_2 \in \mathcal{B}(\mathcal{D}_T)$ satisfying* (7.13) *(or, equivalently, by Lemma 7.2.1,* (7.14)*).*

*Proof* We shall give three proofs of this result.

*First proof via Ando's theorem* We first consider two illustrative special cases, and then use the second special case to prove the result for the general case.

**Case 1** $(T_1, T_2) = (V_1, V_2)$ **is a commuting isometric pair** Note that in this case we have

$$V_1 - V_2^* V = V_1 - V_2^* V_2 V_1 = V_1 - V_1 = 0$$

and the same argument with the indices interchanged gives $V_2 - V_1^* V = 0$. Hence $F_1 = F_2 = 0$ on $\mathcal{D}_T$ solves (7.13) as desired.

**Case 2** $(T_1, T_2) = (V_1^*, V_2^*)$ **is a commuting coisometric pair** To handle this case, one can apply the BCL-model for the isometric pair $(V_1, V_2)$ and then directly compute. It suffices to assume that the product $V = V_1 V_2$ is a shift, since the unitary part washes out when computing the operators $V_1^* - V_2 V^*$, $V_2^* - V_1 V^*$ as well as the defect operators $D_{V^*}, D_{V_1^*}, D_{V_2^*}$.

The claim will be established once we show

$$V_1^* - V_2 V^* = D_{V^*} D_{V_2^*} V_1^* D_{V^*}, \quad V_2^* - V_1 V^* = D_{V^*} D_{V_1^*} V_2^* D_{V^*}. \tag{7.25}$$

It suffices to verify the first equation as then the second follows by interchanging the roles of the indices 1, 2. Note first that

$$V_1^* - V_2 V^* = V_1^* - V_2 V_2^* V_1^* = D_{V_2^*} V_1^*.$$

Furthermore

$$V^* D_{V_2^*} V_1^* = V_1^* V_2^* \left(I - V_2 V_2^*\right) V_1^* = 0$$

and we conclude that

$$D_{V^*} D_{V_2^*} V_1^* = D_{V_2^*} V_1^*.$$

Similarly

$$D_{V_2^*} V_1^* V V^* = D_{V_2^*} V_1^* V_1 V_2 V^* = \left(I - V_2 V_2^*\right) V_2 V^* = 0$$

and hence we also have

$$D_{V^*} D_{V_2^*} V_1^* D_{V^*} = D_{V_2^*} V_1^* (I - V^* V) = D_{V_2^*} V_1^*.$$

Putting all the pieces together, we get the first of the equations (7.25) as expected.

Additionally, let us recall that by part (4) of Theorem 3.1.7, a BCL-tuple generating a BCL1-model for $(V_1, V_2)$ is $(\mathcal{D}_{V^*}, P, U)$ with

$$P = D_{V_1^*}|_{\mathcal{D}_{V^*}}, \quad U = \left(V_1 D_{V_2^*} + D_{V_1^*} V_2^*\right)|_{\mathcal{D}_{V^*}}. \tag{7.26}$$

Since $P^\perp = V_1 D_{V_2^*} V_1^*|_{\mathcal{D}_{V^*}}$, we note that

$$F_1 = U^* P^\perp = \left(D_{V_2^*} V_1^* + V_2 D_{V_1^*}\right)\left(V_1 D_{V_2^*} V_1^*\right)|_{\mathcal{D}_{V^*}} = D_{V_2^*} V_1^*|_{\mathcal{D}_{V^*}}$$

while

$$F_2 = PU = D_{V_1^*}\left(V_1 D_{V_2^*} + D_{V_1^*} V_2^*\right)|_{\mathcal{D}_{V^*}} = D_{V_1^*} V_2^*|_{\mathcal{D}_{V^*}}$$

arriving at the formulas

$$F_1 = U^* P^{\perp}, \quad F_2 = PU \tag{7.27}$$

with $(P, U)$ as in (7.26).

**Case 3 The general case** Now let $(T_1, T_2)$ be any commuting contractive pair acting on a Hilbert space $\mathcal{H}$. By Ando's theorem, we know that $(T_1^*, T_2^*)$ has an Ando lift $(\Pi, V_1, V_2)$ with $\Pi\colon \mathcal{H} \to \mathcal{K}$ and $V_1$, $V_2$ acting on $\mathcal{K}$. For notational convenience, we may assume $\Pi = \left[\begin{smallmatrix} I_{\mathcal{H}} \\ 0 \end{smallmatrix}\right]$, that is, $\mathcal{H} \subset \mathcal{K}$. Recall that the lifting property can be reformulated as

$$V_j^*|_{\mathcal{H}} = T_j \quad \text{for } j = 1, 2, \text{ and hence also } V^*|_{\mathcal{H}} = T, \tag{7.28}$$

where we set $V = V_1 V_2 = V_2 V_1$ and $T = T_1 T_2 = T_2 T_2$.

For $h \in \mathcal{H}$ we then have

$$\|D_{V^*} h\|^2 = \langle h, h\rangle - \langle V^* h, V^* h\rangle = \langle h, h\rangle - \langle Th, Th\rangle = \|D_T h\|^2.$$

We conclude that the map $\Lambda\colon \mathcal{D}_T \to \mathcal{D}_{V^*}$ defined densely by

$$\Lambda\colon D_T h \mapsto D_{V^*} h \tag{7.29}$$

is an isometry. From this definition, we can also write

$$D_{V^*}|_{\mathcal{H}} = \Lambda D_T. \tag{7.30}$$

Taking adjoints then gives

$$P_{\mathcal{H}} D_{V^*} = D_T \Lambda^*. \tag{7.31}$$

From Case 2 above we know that there are fundamental operators $(F_1^{\mathbf{V}}, F_2^{\mathbf{V}})$ for the commuting coisometric pair $(V_1^*, V_2^*)$ for the record by (7.27) given by

$$F_1^{\mathbf{V}} = D_{V_2^*} V_1^*|_{\mathcal{D}_{V^*}}, \quad F_2^{\mathbf{V}} = D_{V_1^*} V_2^*|_{\mathcal{D}_{V^*}},$$

which by definition satisfies the equations

$$V_1^* - V_2 V^* = D_{V^*} F_1^{\mathbf{V}} D_{V^*}, \quad V_2^* - V_1 V^* = D_{V^*} F_2^{\mathbf{V}} D_{V^*}.$$

Let us compress both sides of each of these equations to the subspace $\mathcal{H} \subset \mathcal{K}$ to get

$$P_{\mathcal{H}}\left(V_1^* - V_2 V^*\right)|_{\mathcal{H}} = P_{\mathcal{H}} D_{V^*} F_1^{\mathbf{V}} D_{V^*}|_{\mathcal{H}}, \quad P_{\mathcal{H}}\left(V_2^* - V_1 V^*\right)|_{\mathcal{H}} = P_{\mathcal{H}} D_{V^*} F_2^{\mathbf{V}} D_{V^*}|_{\mathcal{H}}. \tag{7.32}$$

Making use of the identities (7.28) then leads us to

$$P_{\mathcal{H}}\left(V_1^* - V_2V^*\right)|_{\mathcal{H}} = T_1 - T_2^*T, \quad P_{\mathcal{H}}\left(V_2^* - V_1V^*\right)|_{\mathcal{H}} = T_2 - T_1^*T.$$

On the other hand, the identities (7.30) and (7.31) give us

$$P_{\mathcal{H}}D_{V^*}F_1^{\mathbf{V}}D_{V^*}|_{\mathcal{H}} = D_T\Lambda^*F_1^{\mathbf{V}}\Lambda D_T, \quad P_{\mathcal{H}}D_{V^*}F_2^{\mathbf{V}}D_{V^*}|_{\mathcal{H}} = D_T\Lambda^*F_2^{\mathbf{V}}\Lambda D_T.$$

Plugging these last two collections of identities back into (7.32) then gives us

$$T_1 - T_2^*T = D_T\Lambda^*F_1^{\mathbf{V}}\Lambda D_T, \quad T_2 - T_1^*T = D_T\Lambda^*F_2^{\mathbf{V}}\Lambda D_T$$

and we conclude that the operator pair

$$F_1 = \Lambda^*F_1^{\mathbf{V}}\Lambda, \quad F_2 = \Lambda^*F_2^{\mathbf{V}}\Lambda$$

serves as the fundamental-operator pair for the commuting, contractive pair $(T_1, T_2)$ as desired.

*Second proof via Type I Ando tuples* Here we prove the following.

> *If $(\mathcal{F}, \Lambda, P, U)$ is a Type I Ando tuple for $(T_1, T_2)$, then the fundamental operator pair $(F_1, F_2)$ for $(T_1, T_2)$ can be given by*
>
> $$F_1 = \Lambda^*P^{\perp}U\Lambda, \quad F_2 = \Lambda^*U^*P\Lambda. \tag{7.33}$$

By Lemma 7.2.1, to find the fundamental-operator pair, it suffices to find a pair of operators $F_1, F_2 \in \mathcal{B}(\mathcal{D}_T)$ so that $(F_1, F_2)$ solves (7.14) rather than (7.13). To see this, note that by Definition 5.1.5, $(\mathcal{F}, \Lambda, P, U)$ being a Type I Ando tuple for $(T_1, T_2)$ means that

$$P^{\perp}U\Lambda D_T + PU\Lambda D_T T = \Lambda D_T T_1, \quad U^*P\Lambda D_T + U^*P^{\perp}\Lambda D_T T = \Lambda D_T T_2. \tag{7.34}$$

Multiply each of these equations on the left by $\Lambda^*$ to get the equations (7.14) with $F_1, F_2$ as claimed.

*Third proof via Type II Ando tuples* As in the second proof, we actually prove the following assertion.

> *If $(\mathcal{F}, \Lambda, P, U)$ is a Type II Ando tuple for a commuting contractive pair $(T_1, T_2)$, then the fundamental operator pair $(F_1, F_2)$ for $(T_1, T_2)$ is given as in* (7.33).

In the first step of the following computation we make use of the two expressions for $\Lambda D_T$ given by condition (i′) in Definition 6.1.2:

$$
\begin{aligned}
D_T\Lambda^* P^\perp U\Lambda D_T &= (\Lambda D_T)^* P^\perp U(\Lambda D_T) \\
&= (PU\Lambda D_T T_2 + P^\perp \Lambda D_T)^* P^\perp U(U^* P^\perp \Lambda D_T T_1 + U^* PU\Lambda D_T) \\
&= \left(T_2^* D_T \Lambda^* U^* P + D_T \Lambda^* P^\perp\right) P^\perp (P^\perp \Lambda D_T T_1 + PU\Lambda D_T) \\
&= D_T \Lambda^* P^\perp \Lambda D_T T_1 \\
&= D_{T_2}^2 T_1 \quad \text{(by the second equation in Definition 6.1.2 (ii))} \\
&= T_1 - T_2^* T.
\end{aligned}
$$

Similarly, in the first step of the next computation, we use the two expressions for $\Lambda D_T$ given by condition (i′) in Definition 6.1.2 but in reverse order:

$$
\begin{aligned}
D_T\Lambda^* U^* P\Lambda D_T &= (\Lambda D_T)^* U^* P(\Lambda D_T) \\
&= (U^* P^\perp \Lambda D_T T_1 + U^* PU\Lambda D_T)^* U^* P(PU\Lambda D_T T_2 + P^\perp \Lambda D_T) \\
&= \left(T_1^* D_T \Lambda^* P^\perp + D_T \Lambda^* U^* P\right) P(PU\Lambda D_T T_2 + P^\perp \Lambda D_T) \\
&= D_T \Lambda^* U^* PU\Lambda D_T T_2 \\
&= D_{T_1}^2 T_2 \quad \text{(by the first equation in Definition 6.1.2 (ii))} \\
&= T_2 - T_1^* T.
\end{aligned}
$$

This establishes the claim and completes the proof of Theorem 7.2.3. □

## 7.3 Strongly Minimal Ando Lifts and Fundamental-Operator Pairs

The following theorem gives a characterization of the existence of a strongly minimal Ando lift for a commuting contractive pair $(T_1, T_2)$ in terms of the fundamental-operator pair $(F_1, F_2)$ for $(T_1, T_2)$.

**Theorem 7.3.1** *Let $(T_1, T_2)$ be a commuting contractive pair acting on $\mathcal{H}$. Then $(T_1, T_2)$ has a strongly minimal Ando lift if and only if its fundamental-operator pair $(F_1, F_2)$ satisfies the system of equations*

$$
F_1 F_2 = 0 = F_2 F_1, \quad F_1^* F_1 + F_2 F_2^* = I_{\mathcal{D}_T} = F_1 F_1^* + F_2^* F_2. \tag{7.35}
$$

*Proof* Suppose that $(T_1, T_2)$ has a strongly minimal Ando lift. By part (2) of Theorem 7.1.3, $(T_1, T_2)$ has a strongly minimal Type II Ando tuple, say $(\mathcal{F}, \Lambda, P, U)$. By the third proof of Theorem 7.2.3,

$$
(F_1, F_2) = (\Lambda^* P^\perp U\Lambda,\ \Lambda^* U^* P\Lambda)
$$

is the fundamental-operator pair for $(T_1, T_2)$. Since $\Lambda$ here is a unitary, it is now a matter of easy computation to check that $F_1, F_2$ satisfy equations (7.35).

Conversely, suppose $F_1, F_2$ satisfy equations (7.35). Apply Lemma 3.1.1 to $\left(F_1^*, F_2^*\right)$ to get a projection $P$ and a unitary $U$ on $\mathcal{D}_T$ such that $(F_1, F_2) = (P^\perp U, U^* P)$. Since $(F_1, F_2)$ is the fundamental-operator pair for $(T_1, T_2)$, we have

$$T_1 - T_2^* T = D_T P^\perp U D_T, \quad T_2 - T_1^* T = D_T U^* P D_T, \tag{7.36}$$

$$D_T T_1 = P^\perp U D_T + P U D_T T, \quad D_T T_2 = U^* P D_T + U^* P^\perp D_T T. \tag{7.37}$$

We now unfold these equations to show that $(\mathcal{D}_T, I_{\mathcal{D}_T}, P, U)$ is a (strongly minimal) Type II Ando tuple for $(T_1, T_2)$. Then by part (2) of Theorem 7.1.3 we will be done. Multiply both sides of the second equation in (7.37) on the left by $D_T P U$ to get

$$D_T P U D_T T_2 = D_T P D_T.$$

Combining this with the adjoint of the second equation in (7.36) then gives us

$$\begin{aligned} D_T P^\perp D_T &= D_T^2 - D_T P D_T = D_T^2 - D_T P U D_T T_2 \\ &= (I - T^* T) - \left(T_2^* - T^* T_1\right) T_2 = D_{T_2}^2. \end{aligned} \tag{7.38}$$

This is the second isometry condition in Definition 6.1.2 (here $\Lambda = I_{\mathcal{D}_T}$).

For the other isometry condition, we multiply both sides of the first equation in (7.37) on the left by $D_T U^* P^\perp$ to get

$$D_T U^* P^\perp D_T T_1 = D_T U^* P^\perp U D_T.$$

Combining this with the adjoint of the second equation in (7.36) then gives us

$$\begin{aligned} D_T U^* P U D_T &= D_T^2 - D_T U^* P^\perp U D_T = D_T^2 - D_T U^* P^\perp D_T T_1 \\ &= I - T^* T - \left(T_1^* - T^* T_2\right) T_1 = D_{T_1}^2. \end{aligned} \tag{7.39}$$

By Definition 6.1.2, it just remains to show the commutativity condition

$$D_T = P U D_T T_2 + P^\perp D_T = U^* P^\perp D_T T_1 + U^* P U D_T.$$

Since this is an operator equation on $\mathcal{D}_T$ and $D_T$ is injective on $\mathcal{D}_T$, the above will hold if and only if

$$D_T^2 = D_T P U D_T T_2 + D_T P^\perp D_T = D_T U^* P^\perp D_T T_1 + D_T U^* P U D_T$$

holds. In view of equations (7.36), (7.38) and (7.39), the above equations boil down to the operator equations

$$D_T^2 = T_2^* \left(I - T_1^* T_1\right) T_2 + \left(I - T_2^* T_2\right) = T_1^* \left(I - T_2^* T_2\right) T_1 + \left(I - T_1^* T_1\right),$$

which is true, as observed before in (5.25). Consequently, $(\mathcal{D}_T, I_{\mathcal{D}_T}, P, U)$ is a strongly minimal Type II Ando tuple for $(T_1, T_2)$ and therefore by part (2) of Theorem 7.1.3, $(T_1, T_2)$ has a strongly minimal Ando lift. This completes the proof. □

**Remark 7.3.2** We have noted in Theorems 5.5.1 and 6.3.2 that a commuting isometric pair $(V_1, V_2)$ as well as a commuting coisometric pair $(V_1^*, V_2^*)$ have a strongly minimal Ando lift. As a further exercise, we now verify how this can also be seen as an application of Theorem 7.3.1 and Lemma 3.1.1.

Indeed, we have seen in the course of the first proof of Theorem 7.2.3 that a commuting pair of isometries $(V_1, V_2)$ has a trivial fundamental-operator pair $(F_1, F_2) = (0, 0)$ acting on the zero space $\mathcal{D}_V$. This pair clearly satisfies (7.35) and therefore every commuting isometric pair has a strongly minimal Ando lift, (namely, itself).

In the first proof of Theorem 7.2.3 we have also seen that a commuting pair of coisometries $(V_1^*, V_2^*)$ has the fundamental-operator pair $(F_1, F_2)$ of the form

$$F_1 = U^* P^\perp \quad F_2 = PU. \tag{7.40}$$

Such pairs satisfy equations (7.35). Therefore by Theorem 7.3.1, every pair of coisometries also has a strongly minimal Ando lift.

In continuation of the preceding remark, let us note that Lemma 3.1.1 actually states that pairs of the form (7.40) are the only ones that satisfy equations (7.35). It is tempting to think that perhaps the only pairs that have strongly minimal Ando lifts are commuting coisometric pairs, apart from the case of commuting isometric pairs where each of the underlying defect spaces is the zero space. Below we note a family of examples, apart from the class of commuting isometries or coisometries, whose fundamental-operator pairs satisfy (7.35). Indeed, in Chapter 9, we shall see that the fundamental-operator pair is part of a set of complete unitary invariants for a commuting contractive pair; however, there are additional invariants that must be introduced to get a complete set of unitary invariants.

**Example 7.3.3** Given a Hilbert space $\mathcal{F}$, an orthogonal projection $P$, and a unitary $U$ in $\mathcal{B}(\mathcal{F})$, let us consider

$$(T_1, T_2) = (U^* P^\perp, PU).$$

Since $T = T_1 T_2 = T_2 T_1 = 0$, one can readily check that the fundamental-operator pair for $(T_1, T_2)$ is the commuting contractive pair itself. We know that such pairs satisfy the equations in (7.35) and thus have a strongly minimal Ando lift. In fact, we know by Theorem 4.5.1 that they have unique minimal Ando isometric lifts. Therefore pairs of this form are special.

Let us now put together all the characterizations for the existence of strongly minimal Ando lifts. Let us recall that the first part of Theorem 7.1.3 uses the Douglas model to obtain a criterion for $(T_1, T_2)$ to have a strongly minimal Ando lift in terms of the existence of a particular kind of Ando tuple (Type I) for $(T_1^*, T_2^*)$, while the second part uses the Schäffer model to obtain a criterion for such a strongly minimal Ando lift, but now in terms of the existence of a particular type of Ando tuple (Type II) for $(T_1, T_2)$. On the other hand, we know by Theorem 7.1.6 that the existence of a strongly minimal Ando lift for $(T_1, T_2)$ is equivalent to the existence of such a lift for $(T_1^*, T_2^*)$. Hence we may interchange $(T_1, T_2)$ with $(T_1^*, T_2^*)$ in either part of Theorem 7.1.3 and still have a valid statement.

If we enhance Theorem 7.1.3 and succeeding theorems with these observations, we arrive at the following summary of all these results.

**Theorem 7.3.4** *Let $(T_1, T_2)$ be a commuting contractive pair.*

(i) *Then the following are equivalent.*
   (a) *$(T_1, T_2)$ has a strongly minimal Ando lift.*
   (b) *There is a strongly minimal Type I Ando tuple of $(T_1^*, T_2^*)$.*
   (c) *There is a strongly minimal Type II Ando tuple of $(T_1, T_2)$.*
   (d) *Both factorizations $T = T_1 \cdot T_2$ and $T = T_2 \cdot T_1$ are regular. In this case, there is a unique strongly minimal canonical special Ando tuple of $(T_1, T_2)$.*
   (e) *If $(F_1, F_2)$ is the fundamental-operator pair for $(T_1, T_2)$, then*

$$F_1 F_2 = 0 = F_2 F_1, \quad F_1^* F_1 + F_2 F_2^* = I_{\mathcal{D}_T} = F_1 F_1^* + F_2^* F_2.$$

(ii) *If $(T_1, T_2)$ is replaced by $(T_1^*, T_2^*)$ in the previous statements, then the corresponding statements are all mutually equivalent with each other and with any of the statements in part* (i) *above.*

## 7.4 Strongly Minimal Ando Tuples and Special Ando Tuples

Let us note that by Proposition 6.4.1, one can construct an example of a weak Type II Ando tuple that is not a Type II Ando tuple and hence, by Theorem 6.2.1, is not a special Type II Ando tuple. However, as of this writing, we do not have an example of a Type I Ando tuple that is not special or of a Type II Ando tuple that is not special. The next result shows that, within the category of strongly minimal Ando tuples, any Type I or Type II Ando tuple is in fact special.

**Theorem 7.4.1** *Every strongly minimal Type I or strongly minimal Type II Ando tuple of a commuting contractive pair coincides with a special Ando tuple.*

*Proof* Let $(T_1, T_2)$ be a commuting contractive pair and let $(\mathcal{F}, \Lambda, P, U)$ be a strongly minimal Type I Ando tuple of $(T_1, T_2)$. Then by part (1) of Theorem 7.1.3 and Theorem 7.1.6, $(T_1, T_2)$ has a strongly minimal Ando lift. By Theorem 7.1.5, there is a strongly minimal special Ando tuple of $(T_1, T_2)$ that we call $(\mathcal{F}_\dagger, \Lambda_\dagger, P_\dagger, U_\dagger)$; note that $\mathcal{F}_\dagger$ here is just the space $\mathcal{D}_{T_1} \oplus \mathcal{D}_{T_2}$. By Definition 5.1.2, we will be done if we can find a unitary $\tau\colon \mathcal{F}_\dagger \to \mathcal{F}$ such that

$$\tau \cdot \Lambda_\dagger = \Lambda \quad \text{and} \quad (P, U) = (\tau P_\dagger \tau^*, \tau U_\dagger \tau^*). \tag{7.41}$$

Set $\tau := \Lambda\Lambda_\dagger^* \colon \mathcal{F}_\dagger \to \mathcal{F}$. This is a unitary because both $\Lambda_\dagger$ and $\Lambda$ are unitary. First, note that $\Lambda = \Lambda\Lambda_\dagger^* \cdot \Lambda_\dagger$. Therefore the first equation in (7.41) is achieved. Second, since a special Ando tuple is of Type I (see Theorem 5.4.1), by the second proof of Theorem 7.2.3, both the pairs

$$(\Lambda^* P^\perp U \Lambda, \Lambda^* U^* P \Lambda) \quad \text{and} \quad \left(\Lambda_\dagger^* P_\dagger^\perp U_\dagger \Lambda_\dagger, \Lambda_\dagger^* U_\dagger^* P_\dagger \Lambda_\dagger\right)$$

are the fundamental-operator pair for $(T_1, T_2)$. Since the fundamental operators are unique, we must have

$$\Lambda^* P^\perp U \Lambda = \Lambda_\dagger^* P_\dagger^\perp U_\dagger \Lambda_\dagger \quad \text{and} \quad \Lambda^* U^* P \Lambda = \Lambda_\dagger^* U_\dagger^* P_\dagger \Lambda_\dagger. \tag{7.42}$$

Adding the first of these two equations to the adjoint of the other, we get

$$\Lambda^* U \Lambda = \Lambda_\dagger^* U_\dagger \Lambda_\dagger \quad \text{or, equivalently,} \quad U = \Lambda\Lambda_\dagger^* \cdot U_\dagger \cdot \Lambda_\dagger \Lambda^*.$$

Using this expression of $U$ in the second equation of (7.42) and simplifying, we get

$$P = \Lambda\Lambda_\dagger^* \cdot P_\dagger \cdot \Lambda_\dagger \Lambda^*.$$

Therefore the second set of equations in (7.41) is also established. This shows that the strongly minimal Type I Ando tuple of $(T_1, T_2)$ coincides with the special strongly minimal Ando tuple $(\mathcal{F}_\dagger, \Lambda_\dagger, P_\dagger, U_\dagger)$ via the unitary $\Lambda\Lambda_\dagger^* \colon \mathcal{F}_\dagger \to \mathcal{F}$.

Via a similar analysis using Theorem 6.2.1 and the third proof of Theorem 7.2.3, one can show that every strongly minimal Type II Ando tuple coincides with a special strongly minimal Ando tuple. □

The following result shows how one can recover Ando tuples (up to coincidence) for a commuting contractive pair $(T_1, T_2)$ from the fundamental-operator pair $(F_1, F_2)$ for $(T_1, T_2)$, at least for the case where $(T_1, T_2)$ has a strongly minimal Ando lift.

**Proposition 7.4.2** *Let $(T_1, T_2)$ be a commuting contractive pair and $(F_1, F_2)$ be its fundamental-operator pair. Then every strongly minimal Type II Ando tuple of $(T_1, T_2)$ coincides with $(\mathcal{D}_T, I_{\mathcal{D}_T}, F_2^* F_2, F_2^* + F_1)$. The same assertion holds for a strongly minimal Type I Ando tuple of $(T_1^*, T_2^*)$ as well.*

*Proof* Let $(\mathcal{F}, \Lambda, P, U)$ be a strongly minimal Type II Ando tuple of $(T_1, T_2)$. Since $\Lambda$ is a unitary and we are interested in the coincidence envelope of strongly minimal Ando tuples, without loss of generality we can suppose that the Ando tuple is $(\Lambda^*\mathcal{F}, \Lambda^*\Lambda, \Lambda^* P\Lambda, \Lambda^* U\Lambda) = (\mathcal{D}_T, I_{\mathcal{D}_T}, P, U)$. By the third proof of Theorem 7.2.3, the fundamental operators for $(T_1, T_2)$ then are $(F_1, F_2) = (P^\perp U, U^* P)$. This readily implies that $F_1 + F_2^* = U$ and hence $P = (F_1 + F_2^*) F_2 = F_2^* F_2$. Similarly, using the second proof of Theorem 7.2.3, one can prove the assertion for a strongly minimal Type I Ando tuple of $(T_1^*, T_2^*)$. □

In Example 7.3.3 we found a commuting contractive pair that has a strongly minimal Ando lift. In view of Theorem 4.5.1, the pair in Example 7.3.3 has the additional property that every minimal Ando isometric lift is strongly minimal. Is this a coincidence?

**Problem 7.4.3** Does there exist a commuting contractive pair $(T_1, T_2)$ which has two minimal Ando isometric lifts, one of which is strongly minimal while the other is not strongly minimal?

The answer to this question is negative. This is because if a pair has a strongly minimal Ando lift, then any two minimal Ando isometric lifts are unitarily equivalent and hence any minimal Ando lift is strongly minimal (see Theorem 8.5.1). But before we can present a proof of this, we need to do a rigorous analysis of a special type of lift that we call *pseudo-commuting contractive lifts*. This is the topic of Chapter 8.

## 7.5 Notes to Chapter 7

**1.** The notion of fundamental operators has already been introduced and had an impact in the theory of symmetrized-bidisk contractions [36] and tetrablock contractions [33], as well as for commuting $d$-tuples of contractions ($d > 2$)

[21]. For a discussion of a general framework that includes the commuting-contractive pair setting here as well as the symmetrized-polydisk, tetrablock, and polydisk settings as special cases, we refer the reader to Section A.1, where the general companion-contraction setting is introduced and discussed.

**2.** In the proof of Theorem 7.2.3, we saw that the BCL-tuple $(\mathcal{F}, P, U)$ of a commuting isometric pair $(V_1, V_2)$ uniquely determines the fundamental-operator pair for $(V_1^*, V_2^*)$. This was first observed in the 2017 unpublished preprint of the second author [83].

# 8

# Pseudo-commuting Contractive Lifts

Given a commuting contractive operator pair $(T_1, T_2)$ on $\mathcal{H}$ and an Ando lift $(\mathbf{\Pi}, \mathbf{V}_1, \mathbf{V}_2)$ of $(T_1, T_2)$ on $\mathcal{K}$, as we saw in the proof of Theorem 5.2.1, it is always possible to restrict to the subspace

$$\mathcal{K}_0 := \bigvee_{n_1, n_2 \geq 0} \mathbf{V}^{n_1} \mathbf{V}^{n_2} \operatorname{Ran} \mathbf{\Pi} \subset \mathcal{K}$$

to get a minimal Ando lift $(\mathbf{\Pi}_0, \mathbf{V}_{0,1}, \mathbf{V}_{0,2})$ of $(T_1, T_2)$, where we define $\mathbf{\Pi}_0 : \mathcal{H} \to \mathcal{K}_0$ and $\mathbf{V}_{0,1}, \mathbf{V}_{0,2}$ on $\mathcal{K}_0$ via

$$\mathbf{\Pi}_0 h = \mathbf{\Pi} h \in \operatorname{Ran} \mathbf{\Pi} \subset \mathcal{K}_0 \quad \text{for } h \in \mathcal{H}, \quad \mathbf{V}_{0,1} = \mathbf{V}_1|_{\mathcal{K}_0}, \quad \mathbf{V}_{0,2} = \mathbf{V}_2|_{\mathcal{K}_0}. \tag{8.1}$$

By introducing the even smaller (in principle) subspace

$$\mathcal{K}_{00} := \bigvee_{n \geq 0} \mathbf{V}_1^n \mathbf{V}_2^n \operatorname{Ran} \mathbf{\Pi} \subset \mathcal{K}_0 \subset \mathcal{K}, \tag{8.2}$$

we can find a minimal Sz.-Nagy–Foias lift $(\mathbf{\Pi}_{00}, \mathbf{V}_{00})$ for the product contraction operator $T = T_1 T_2$ by setting $\mathbf{\Pi}_{00} : \mathcal{H} \to \mathcal{K}_{00}$ and $\mathbf{V}_{00}$ on $\mathcal{K}_{00}$ equal to

$$\mathbf{\Pi}_{00} h = \mathbf{\Pi} h \in \operatorname{Ran} \mathbf{\Pi} \subset \mathcal{K}_{00} \text{ for } h \in \mathcal{H}, \quad \mathbf{V}_{00} = \mathbf{V}_1 \mathbf{V}_2|_{\mathcal{K}_{00}}. \tag{8.3}$$

Note that while it is always the case that $\mathcal{K}_0$ is jointly invariant for $(\mathbf{V}_1, \mathbf{V}_2)$, $\mathcal{K}_{00}$ is invariant only for the product isometry $\mathbf{V}_1 \mathbf{V}_2$. However, the case where $\mathcal{K}_{00}$ is invariant for $\mathbf{V}_1$ and $\mathbf{V}_2$ individually is the special situation studied in Chapter 7, where the minimal Ando lift $(\mathbf{\Pi}_{00}, \mathbf{V}_{00,1}, \mathbf{V}_{00,2})$ acting on $\mathcal{K}_{00}$ given by

$$(\mathbf{\Pi}_{00}, \mathbf{V}_{00,1}, \mathbf{V}_{00,2}) = (\mathbf{\Pi}, \mathbf{V}_1|_{\mathcal{K}_{00}}, \mathbf{V}_2|_{\mathcal{K}_{00}})$$

is actually *strongly minimal*. Nevertheless, we show here that in the general situation it is still of interest to consider the compressions

$$\mathbf{W}_1 := P_{\mathcal{K}_{00}} \mathbf{V}_1|_{\mathcal{K}_{00}}, \quad \mathbf{W}_2 := P_{\mathcal{K}_{00}} \mathbf{V}_2|_{\mathcal{K}_{00}}$$

of $\mathbf{V}_1, \mathbf{V}_2$ to $\mathcal{K}_{00}$ even though, when this is done, the compressed pair $(\mathbf{W}_1, \mathbf{W}_2)$ on $\mathcal{K}_{00}$ may not inherit the commuting and isometric properties of the original pair $(\mathbf{V}_1, \mathbf{V}_2)$ on $\mathcal{K}$.

## 8.1 Compressed Ando Lifts versus Pseudo-commuting Contractive Lifts

It is useful to have the following more flexible definition of the *compression of an Ando lift of* $(T_1, T_2)$ *to an embedded Sz.-Nagy–Foias lift for the product contraction* $T = T_1 T_2$, which we shall refer to simply as a *minimal Sz.-Nagy–Foias compression of an Ando lift of* $(T_1, T_2)$ for short.

**Definition 8.1.1** Suppose that $(\mathbf{\Pi}, \mathbf{V}_1, \mathbf{V}_2)$ is an Ando lift of the commuting contractive operator pair $(T_1, T_2)$ on $\mathcal{H}$ with embedded minimal isometric lift $(\mathbf{\Pi}_{00}, \mathbf{V}_{00})$ of the product contraction operator $T = T_1 T_2$ given by (8.3). Suppose that $\Pi \colon \mathcal{H} \to \mathcal{K}$ is an isometric embedding and $V$ is an isometry on another Hilbert space $\mathcal{K}$ such that $(\Pi, V)$ is a minimal isometric lift of the product contraction $T = T_1 T_2$. By uniqueness of the Sz.-Nagy–Foias minimal isometric lift (see Theorem 2.1.3), there is a unitary operator $\tau \colon \mathcal{K} \to \mathcal{K}_{00}$ so that

$$\tau \Pi = \mathbf{\Pi}_{00}, \quad \tau V = \mathbf{V}_{00} \tau.$$

Let us also view $\tau$ as an isometry from $\mathcal{K}$ into $\mathcal{K}$ with final space equal to $\mathcal{K}_{00}$ depending on the context. Define operators $\Pi \colon \mathcal{H} \to \mathcal{K}$ and $\mathbb{W}_1, \mathbb{W}_2\, \mathbb{W}$ on $\mathcal{K}$ by

$$\Pi = \tau^* \mathbf{\Pi}, \quad \mathbb{W}_1 = \tau^* \mathbf{V}_1 \tau, \quad \mathbb{W}_2 = \tau^* \mathbf{V}_2 \tau, \quad \mathbb{W} = \tau^* \mathbf{V}_1 \mathbf{V}_2 \tau = \tau^* \mathbf{V}_{00} \tau = V.$$

Then we say that the collection $(\Pi, \mathbb{W}_1, \mathbb{W}_2, V)$ *is the compression of the Ando lift* $(\mathbf{\Pi}, \mathbf{V}_1, \mathbf{V}_2, \mathbf{V}_1\mathbf{V}_2)$ *of* $(T_1, T_2, T = T_1 T_2)$ *to the minimal Sz.-Nagy–Foias lift* $(\Pi, V)$ *of* $T$.

It turns out that such compressed Ando lifts to immersed minimal Sz.-Nagy–Foias lifts have an intrinsic characterization independent of any reference to having a dilation to some Ando lift. For further discussion, the following formal definitions will be useful.

**Definition 8.1.2** **1.** Suppose that $(\mathbb{W}_1, \mathbb{W}_2, \mathbb{W})$ is a triple of operators on the Hilbert space $\mathcal{K}$. We say that $(\mathbb{W}_1, \mathbb{W}_2, \mathbb{W})$ is a *pseudo-commuting contractive operator triple* if:

(i) $\mathbb{W}_1, \mathbb{W}_2$ are contractions while $\mathbb{W}$ is an isometry;
(ii) both $\mathbb{W}_1$ and $\mathbb{W}_2$ commute with $\mathbb{W}$ (but not necessarily with each other);
(iii) $\mathbb{W}_1 = \mathbb{W}_2^* \mathbb{W}$.

We shall say that $(\mathbb{W}_1, \mathbb{W}_2, \mathbb{W})$ is a *pseudo-commuting algebraic operator triple* if $(\mathbb{W}_1, \mathbb{W}_2, \mathbb{W})$ satisfies conditions (ii) and (iii) above, but condition (i) is weakened to

(i′) $\mathbb{W}$ is an isometry,

that is, if $\mathbb{W}_1$ and $\mathbb{W}_2$ are now only required to be bounded operators on $\mathcal{K}$ rather than contractions.

**2.** Suppose that $(T_1, T_2)$ is a commuting contractive operator pair on a Hilbert space $\mathcal{H}$, $\Pi\colon \mathcal{H} \to \mathcal{K}$ is an isometric embedding of $\mathcal{H}$ into $\mathcal{K}$, and that $(\mathbb{W}_1, \mathbb{W}_2, \mathbb{W})$ is a pseudo-commuting contractive operator triple on $\mathcal{K}$. We shall say that $(\Pi, \mathbb{W}_1, \mathbb{W}_2, \mathbb{W})$ is a *pseudo-commuting contractive lift* of $(T_1, T_2)$ if $(W_1, W_2, W)$ is a pseudo-commuting contractive operator triple on $\mathcal{K}$ and in addition:

(iv) $(\Pi, \mathbb{W}_1, \mathbb{W}_2, \mathbb{W})$ is a lift of $(T_1, T_2, T := T_1 T_2)$ in the sense that

$$\left(\mathbb{W}_1^*, \mathbb{W}_2^*, \mathbb{W}^*\right) \Pi = \Pi \left(T_1^*, T_2^*, T_1^* T_2^*\right)$$

(in particular, $(\Pi, \mathbb{W})$ is an isometric lift of $T = T_1 T_2$), and in addition
(v) $(\Pi, \mathbb{W})$ is a minimal isometric lift for $T = T_1 T_2$, that is,

$$\mathcal{K} = \bigvee_{n \geq 0} \mathbb{W}^n \operatorname{Ran} \Pi.$$

**Remark 8.1.3** It should be observed that, whenever $(\mathbb{W}_1, \mathbb{W}_2, \mathbb{W})$ is a pseudo-commuting contractive operator triple, in addition to condition (ii) in the definition one also has

(iii′) $\mathbb{W}_2 = \mathbb{W}_1^* \mathbb{W}$.

Indeed, from the identity $\mathbb{W}_1 = \mathbb{W}_2^* \mathbb{W}$ we get

$$\mathbb{W}^* \mathbb{W}_1 = \mathbb{W}^* \left(\mathbb{W}_2^* \mathbb{W}\right) = \left(\mathbb{W}^* \mathbb{W}_2^*\right) \mathbb{W} = \left(\mathbb{W}_2^* \mathbb{W}^*\right) \mathbb{W} = \mathbb{W}_2^* (\mathbb{W}^* \mathbb{W}) = \mathbb{W}_2^*.$$

The next result gives the promised intrinsic characterization of Ando lifts of $(T_1, T_2)$ compressed to a minimal Sz.-Nagy–Foias lift of $T = T_1 T_2$, namely, they are the same as pseudo-commuting contractive lifts of $(T_1, T_2)$ defined as above with no reference to any Ando lift of $(T_1, T_2)$.

**Theorem 8.1.4** *Suppose that* $\Pi\colon \mathcal{H} \to \mathcal{K}$ *is an isometric embedding operator,* $(\mathbb{W}_1, \mathbb{W}_2, \mathbb{W})$ *is a triple of operators on* $\mathcal{K}$*, and* $(T_1, T_2)$ *is a commuting contractive operator pair on* $\mathcal{H}$*. Then* $(\Pi, \mathbb{W}_1, \mathbb{W}_2, \mathbb{W})$ *is the compression of an Ando lift of* $(T_1, T_2)$ *to an embedded minimal Sz.-Nagy–Foias lift of* $T$ *as in Definition 8.1.1 if and only if* $(\Pi, W_1, W_2, W)$ *is a pseudo-commuting contractive lift of* $(T_1, T_2)$ *as in Definition 8.1.2.*

*Proof* We suppose first that $(\Pi, \mathbb{W}_1, \mathbb{W}_2, \mathbb{W})$ is the compression of the Ando lift of $(T_1, T_2)$ to some minimal Sz.-Nagy–Foias lift $(\Pi, V)$ of $T$. In detail, this means that there is an Ando lift $(\mathbf{\Pi}, \mathbf{V}_1, \mathbf{V}_2)$ (say $\mathbf{\Pi}\colon \mathcal{H} \to \mathcal{K}$ and $\mathbf{V}_1, \mathbf{V}_2$ are commuting isometries on $\mathcal{K}$) and a minimal isometric lift $(\Pi, V)$ of the product contraction $T = T_1T_2$ (say $\Pi\colon \mathcal{H} \to \mathcal{K}$ and $V$ is an isometry on $\mathcal{K}$), and an isometry $\tau\colon \mathcal{K} \to \mathcal{K}$ with range equal to $\mathcal{K}_{00}$ as in (8.2) so that

$$\Pi = \tau^*\mathbf{\Pi}, \quad \mathbb{W}_1 = \tau^*\mathbf{V}_1\tau, \quad \mathbb{W}_2 = \tau^*\mathbf{V}_2\tau, \quad \mathbb{W} = \tau^*\mathbf{V}_1\mathbf{V}_2\tau = V.$$

Since $\mathbf{V}_1, \mathbf{V}_2, \mathbf{V} = \mathbf{V}_1\mathbf{V}_2$ are all isometries and furthermore $\mathcal{K}_{00}$ is invariant for $\mathbf{V} := \mathbf{V}_1\mathbf{V}_2$, (i) follows.

Condition (ii) follows from the fact that $\mathbf{V}_1$ and $\mathbf{V}_2$ commute with $\mathbf{V}$ and again $\mathcal{K}_{00}$ is invariant for $\mathbf{V}$.

Since $\mathbf{V} = \mathbf{V}_1\mathbf{V}_2 = \mathbf{V}_2\mathbf{V}_1$ and $\mathbf{V}_1$ is isometric, we see that we can solve for $\mathbf{V}_1$ as $\mathbf{V}_1 = \mathbf{V}_2^*\mathbf{V}$. The formulas for $\mathbb{W}, \mathbb{W}_1, \mathbb{W}_2$ combined with the fact that $\mathcal{K}_{00}$ is invariant for $\mathbf{V}$ then lead us to condition (iii).

Since $(\mathbf{\Pi}, \mathbf{V}_1, \mathbf{V}_2)$ is a lift of $(T_1, T_2)$, we know that

$$\left(\mathbf{V}_1^*, \mathbf{V}_2^*, \mathbf{V}^*\right)\mathbf{\Pi} = \mathbf{\Pi}(T_1^*, T_2^*, T^*),$$

which is actually the same as

$$\left(\mathbf{V}_1^*, \mathbf{V}_2^*, \mathbf{V}^*\right)\mathbf{\Pi}_{00} = \mathbf{\Pi}_{00}\left(T_1^*, T_2^*, T^*\right).$$

Recalling now that $\mathbf{\Pi}_{00} = \tau\Pi$ and that $\tau\colon \mathcal{K} \to \mathcal{K}_{00}$ is unitary, this last expression becomes

$$\tau^*\left(\mathbf{V}_1^*, \mathbf{V}_2^*, \mathbf{V}^*\right)\tau\Pi = \Pi\left(T_1^*, T_2^*, T^*\right)$$

and (iv) follows.

Finally, by construction $\mathbb{W} = \tau^*\mathbf{V}_{00}\tau = V$, where by definition the compressed Ando tuple $(\Pi, V)$ is a minimal lift of $T$, from which we see that (v) holds. This completes the proof of *compressed Ando lift* $\Rightarrow$ *pseudo-commuting contractive lift*.

We postpone the proof of the converse (*pseudo-commuting contractive lift* $\Rightarrow$ *compressed Ando lift*) until after we develop the Douglas model for compressed Ando lifts in the next section (see Corollary 8.2.3). □

## 8.2 Douglas Model Pseudo-commuting Contractive Lifts

We know by Theorem 5.2.1 that minimal Ando lifts for a commuting contractive operator pair $(T_1, T_2)$ can be given up to unitary equivalence in the Douglas model form (5.3) specified by a Type I Ando tuple $(\mathcal{F}_*, \Lambda_*, P_*, U_*)$ for $(T_1^*, T_2^*)$. Identifying the embedded minimal Sz.-Nagy–Foias lift space $\mathcal{K}_{D,00}$ for the product contraction inside $\mathcal{K}_D$ and then identifying this with the Douglas model isometric lift $(\Pi_D, V_D)$ for $T$ then leads to a Douglas model for a compressed Ando lift of $(T_1, T_2)$ as follows.

**Theorem 8.2.1** *Given a commuting contractive operator pair $(T_1, T_2)$ on $\mathcal{H}$, let $\mathcal{K}_D = \begin{bmatrix} H^2(\mathcal{D}_{T^*}) \\ \mathcal{Q}_{T^*} \end{bmatrix}$ be the Douglas isometric-lift model space for $T$, let $\Pi_D \colon \mathcal{H} \to \mathcal{K}_D$ be the Douglas isometric embedding operator $\Pi_D = \begin{bmatrix} \mathcal{O}_{D_{T^*}, T^*} \\ Q_{T^*} \end{bmatrix}$, let $(G_1, G_2)$ be the fundamental-operator pair for $(T_1^*, T_2^*)$, and define operators $W_{\flat 1}, W_{\flat 2}, W_D$ on $\mathcal{Q}_{T^*}$ as in Theorem 4.4.2. Finally, define operators $\mathbb{W}_{D,1}, \mathbb{W}_{D,2}, V_D$ on $\mathcal{K}_D$ according to the formulas*

$$(\mathbb{W}_{D,1}, \mathbb{W}_{D,2}, V_D) = \left( \begin{bmatrix} M_{G_1^* + zG_2} & 0 \\ 0 & W_{\flat 1} \end{bmatrix}, \begin{bmatrix} M_{G_2^* + zG_1} & 0 \\ 0 & W_{\flat 2} \end{bmatrix}, \begin{bmatrix} M_z^{\mathcal{D}_{T^*}} & 0 \\ 0 & W_D \end{bmatrix} \right). \tag{8.4}$$

*Then $(\Pi_D, \mathbb{W}_{D,1}, \mathbb{W}_{D,2}, V_D)$ is the compression of the Douglas model Ando lift of $(T_1, T_2)$ to the embedded Douglas model Sz.-Nagy–Foias lift $(\Pi, V_D)$ and hence also is a pseudo-commuting contractive lift of $(T_1, T_2)$.*

*Conversely, with $\Pi_D, V_D$ and $\mathcal{K}_D$ as previously, suppose that $(\Pi_D, \mathbb{W}_1, \mathbb{W}_2, V_D)$ is a pseudo-commuting contractive lift of $(T_1, T_2)$ on $\mathcal{K}_D$. Then necessarily $(\mathbb{W}_1, \mathbb{W}_2) = (\mathbb{W}_{D,1}, \mathbb{W}_{D,2})$ is given as in formula* (8.4).

*Proof* Let $(\mathcal{F}_*, \Lambda_*, P_*, U_*)$ be a Type I Ando tuple for $(T_1^*, T_2^*)$. Then the Douglas model Ando lift $(\mathbf{\Pi}_D, \mathbf{V}_{D,1}, \mathbf{V}_{D,2})$ on $\mathcal{K}_D$ associated with this Ando tuple is defined as in (5.3):

$$\mathcal{K}_D = \begin{bmatrix} H^2(\mathcal{F}_*) \\ \mathcal{Q}_{T^*} \end{bmatrix},$$

$$(\mathbf{V}_{D,1}, \mathbf{V}_{D,2}) = \left( \begin{bmatrix} M_{U_*^* P_*^\perp + zU_*^* P_*} & 0 \\ 0 & W_{\flat 1} \end{bmatrix}, \begin{bmatrix} M_{P_* U_* + zP_*^\perp U_*} & 0 \\ 0 & W_{\flat 2} \end{bmatrix} \right) \quad \text{acting on } \mathcal{K}_D,$$

$$\mathbf{\Pi}_D = \begin{bmatrix} (I_{H^2} \otimes \Lambda_*)\, \mathcal{O}_{D_{T^*}, T^*} \\ Q_{T^*} \end{bmatrix} : \mathcal{H} \to \mathcal{K}_D.$$

Then, as seen in the proof of Theorem 5.2.1, the Sz.-Nagy–Foias isometric lift of $T$ embedded in the Ando lift $(\mathbf{\Pi}, \mathbf{V}_1, \mathbf{V}_2)$ of $(T_1, T_2)$ is $(\mathbf{\Pi}_{D,00}, \mathbf{V}_{D,00})$, where $\mathbf{\Pi}_{D,00}$ is the same as $\mathbf{\Pi}$ but with codomain taken to be $\mathcal{K}_{D,00} = \begin{bmatrix} H^2(\operatorname{Ran} \Lambda_*) \\ \mathcal{Q}_{T^*} \end{bmatrix}$

and where $\mathbf{V}_{D,00} = \mathbf{V}_{D1}\mathbf{V}_{D2}|_{\mathcal{K}_{D,00}}$, and, furthermore, the unique unitary operator $\tau\colon \mathcal{K}_D \to \mathcal{K}_{D,00}$ implementing the unitary equivalence between the two minimal isometric lifts $(\Pi_D, V_D)$ and $(\mathbf{\Pi}_{D,00}, \mathbf{V}_{D,00})$ of $T = T_1T_2$ is

$$\tau = \begin{bmatrix} I_{H^2} \otimes \Lambda_* & 0 \\ 0 & I_{Q_{T^*}} \end{bmatrix} : \mathcal{K}_D \to \mathcal{K}_{D,00}.$$

Hence the associated Douglas model compressed Ando lift (obtained by using the Douglas model for the the Ando lift $(T_1, T_2)$ as well as the Douglas model for the Sz.-Nagy–Foias lift of the product contraction $T = T_1T_2$) is given by

$$\begin{aligned}(\mathbb{W}_{D,1}, \mathbb{W}_{D,2}, V_D) &:= \tau^*(V_1, V_2, V_1V_2)\tau \\ &:= \left(\begin{bmatrix} M_{\Lambda_*^*(U_*^*P_*^\perp + zU_*^*P_*)\Lambda_*} & 0 \\ 0 & W_{\flat 1} \end{bmatrix}, \begin{bmatrix} M_{\Lambda_*^*(P_*U_*^* + zP_*^\perp U_*)\Lambda_*} & 0 \\ 0 & W_{\flat 2} \end{bmatrix}, \begin{bmatrix} M_z & 0 \\ 0 & W_D \end{bmatrix}\right) \\ &= \left(\begin{bmatrix} M_{G_1^* + zG_2} & 0 \\ 0 & W_{\flat 1} \end{bmatrix}, \begin{bmatrix} M_{G_2^* + zG_1} & 0 \\ 0 & W_{\flat 2} \end{bmatrix}, \begin{bmatrix} M_z & 0 \\ 0 & W_D \end{bmatrix}\right), \end{aligned} \tag{8.5}$$

where here we make use of the connection between the fundamental-operator pair of $(T_1^*, T_2^*)$ and a Type I Ando tuple $(\mathcal{F}_*, \Lambda_*, P_*, U_*)$ for $(T_1^*, T_2^*)$ (see (7.33)) coming out of the second proof of Theorem 7.2.3:

$$(G_1, G_2) = (\Lambda_*^* P_*^\perp U_* \Lambda_*, \Lambda_*^* U_*^* P_* \Lambda_*).$$

Then by definition the model triple (8.4) is a compressed Ando lift, and hence, by the part of Theorem 8.1.4 already proved, is also a pseudo-commuting contractive lift of $(T_1, T_2)$.

Conversely, suppose that the operator triple $\left(\Pi_D, \mathbb{W}_1, \mathbb{W}_2, \begin{bmatrix} M_z^{\mathcal{D}_{T^*}} & 0 \\ 0 & W_D \end{bmatrix}\right)$ is a pseudo-commuting contractive lift of $(T_1, T_2)$. We break the proof into two steps.

**Step 1 Show** *If* $\left(\mathbb{W}_1, \mathbb{W}_2, \begin{bmatrix} M_z^{\mathcal{D}_{T^*}} & 0 \\ 0 & W_D \end{bmatrix}\right)$ *is a pseudo-commuting contractive triple, then*

$$(\mathbb{W}_1, \mathbb{W}_2) = \left(\begin{bmatrix} M_{G_1^* + zG_2} & 0 \\ 0 & \widetilde{W}_2^* W_D \end{bmatrix}, \begin{bmatrix} M_{G_2^* + zG_1} & 0 \\ 0 & \widetilde{W}_2 \end{bmatrix}\right) \tag{8.6}$$

*for some operators* $G_1, G_2 \in \mathcal{B}(\mathcal{D}_{T^*})$ *such that* $\varphi_1(z) := G_1^* + zG_2$ *and* $\varphi_2(z) := G_2^* + zG_1$ *are contractive analytic functions on* $\mathbb{D}$, *and* $\widetilde{W}_2$ *is some contraction operator on* $Q_{T^*}$ *commuting with* $W_D$.

*Proof of Step 1* Assume that $\left(\mathbb{W}_1, \mathbb{W}_2, \begin{bmatrix} M_z & 0 \\ 0 & W_D \end{bmatrix}\right)$ is a pseudo-commuting contractive triple on $\mathcal{K}_D := \begin{bmatrix} H^2(\mathcal{D}_{T^*}) \\ Q_{T^*} \end{bmatrix}$. As a first step we write out $\mathbb{W}_1, \mathbb{W}_2$ as block $2 \times 2$-matrices with respect to the decomposition of $\mathcal{K}_D$ as $\begin{bmatrix} H^2(\mathcal{D}_{T^*}) \\ Q_{T^*} \end{bmatrix}$:

$$\mathbb{W}_j = \begin{bmatrix} \mathbb{W}_{j,11} & \mathbb{W}_{j,12} \\ \mathbb{W}_{j,21} & \mathbb{W}_{j,22} \end{bmatrix} \quad \text{for } j = 1, 2.$$

By axiom (ii) in Definition 8.1.2, combined with Lemma 2.3.1, we see immediately that $\mathbb{W}_{j,12} = 0$ for $j = 1, 2$ and the commutativity of $\mathbb{W}_j$ with $\left[\begin{smallmatrix} M_z^{\mathcal{D}_{T^*}} & 0 \\ 0 & W_D \end{smallmatrix}\right]$ comes down to

$$\begin{bmatrix} \mathbb{W}_{j,11} M_z^{\mathcal{D}_{T^*}} & 0 \\ \mathbb{W}_{j,21} M_z^{\mathcal{D}_{T^*}} & \mathbb{W}_{j,22} W_D \end{bmatrix} = \begin{bmatrix} M_z^{\mathcal{D}_{T^*}} \mathbb{W}_{j,11} & 0 \\ W_D \mathbb{W}_{j,21} & W_D \mathbb{W}_{j,22} \end{bmatrix}. \tag{8.7}$$

By axiom (iii) in Definition 8.1.2 we know that $\mathbb{W}_1 = \mathbb{W}_2^* \left[\begin{smallmatrix} M_z^{\mathcal{D}_{T^*}} & 0 \\ 0 & W_D \end{smallmatrix}\right]$; writing this out in detail gives

$$\begin{bmatrix} \mathbb{W}_{1,11} & 0 \\ \mathbb{W}_{1,21} & \mathbb{W}_{1,22} \end{bmatrix} = \begin{bmatrix} \mathbb{W}_{2,11}^* M_z^{\mathcal{D}_{T^*}} & \mathbb{W}_{2,21}^* W_D \\ 0 & \mathbb{W}_{2,22}^* W_D \end{bmatrix}. \tag{8.8}$$

From the $(2, 1)$ entry we see that $\mathbb{W}_{1,21} = 0$ and from the $(1, 2)$ entry we see that $\mathbb{W}_{2,21} = 0$ since $W_D$ is unitary. Thus both $\mathbb{W}_1$ and $\mathbb{W}_2$ are diagonal

$$\mathbb{W}_j = \begin{bmatrix} \mathbb{W}_{j,11} & 0 \\ 0 & \mathbb{W}_{j,22} \end{bmatrix} \quad \text{for } j = 1, 2,$$

and we see from (8.7) that

$$\mathbb{W}_{j,11} M_z^{\mathcal{D}_{T^*}} = M_z^{\mathcal{D}_{T^*}} \mathbb{W}_{j,11}, \quad \mathbb{W}_{j,22} W_D = W_D \mathbb{W}_{j,22} \quad \text{for } j = 1, 2. \tag{8.9}$$

From the first relation in (8.9), by standard Hardy space theory we conclude that $W_{j,11}$ must be a multiplication operator $M_{\varphi_j} : h(z) \mapsto \varphi_j(z)h(z)$ for a contractive analytic function $\varphi_j(z) = \sum_{k=0}^{\infty} \varphi_{j,k} z^{\ell}$ holomorphic on the unit disk $\mathbb{D}$, where the Taylor coefficients $\varphi_{j,k}$ ($j = 1, 2,\ k = 0, 1, 2, \ldots$) are operators on $\mathcal{D}_{T^*}$. From (8.8) we see that $M_{\varphi_1} = M_{\varphi_2}^* M_z^{\mathcal{D}_{T^*}}$. A Taylor series argument then shows that the pair $(\varphi_1(z), \varphi_2(z))$ must have the coupled pencil form

$$\varphi_1(z) = G_1^* + zG_2, \quad \varphi_2(z) = G_2^* + zG_1$$

for some operators $G_1, G_2$ on $\mathcal{D}_{T^*}$.

Finally, from (8.7) we see that both $\widetilde{W}_1 := W_{1,22}$ and $\widetilde{W}_2 := W_{2,22}$ commute with $W_D$. By the Fuglede–Putnam theorem, it follows that each of $\widetilde{W}_1$ and $\widetilde{W}_2$ also commutes with $W_D^*$. If we let $\widetilde{W}_2$ be any contractive operator commuting with the unitary operator $W_D$ and set $\widetilde{W}_1 = \widetilde{W}_2^* W_D$, then $\widetilde{W}_1$ automatically commutes with $W_D$ and this is the general form for a pseudo-commuting contractive triple $(W_1, W_2, W_D)$ with last component equal to the unitary operator $W_D$. This completes the proof of Step 1.

**Step 2 Show** *If* $G_1, G_2, \widetilde{W}_1 = \widetilde{W}_2^* W_D, \widetilde{W}_2$ *are as in Step* 1 *and*

$$\left(\Pi_D, \begin{bmatrix} M_{G_1^*+zG_2} & 0 \\ 0 & \widetilde{W}_1 \end{bmatrix}, \begin{bmatrix} M_{G_2^*+zG_1} & 0 \\ 0 & \widetilde{W}_2 \end{bmatrix}\right)$$

*is a lift of* $(T_1, T_2)$*, then* $(G_1, G_2)$ *is the fundamental-operator pair for* $(T_1, T_2)$ *and* $\left(\widetilde{W}_1, \widetilde{W}_2\right) = (W_{\flat 1}, W_{\flat 2})$ *is the canonical pair of unitaries on the space* $\mathcal{Q}_{T^*}$ *associated with the contractive operator pair* $(T_1, T_2)$ *as in Theorem* 4.4.2.

*Proof of Step 2* The hypothesis that $\left(\Pi_D, \left[\begin{smallmatrix} M_{G_1^*+zG_2} & 0 \\ 0 & \widetilde{W}_1 \end{smallmatrix}\right], \left[\begin{smallmatrix} M_{G_2^*+zG_1} & 0 \\ 0 & \widetilde{W}_2 \end{smallmatrix}\right]\right)$ is a lift of $(T_1, T_2)$ means that

$$\left(\left[\begin{smallmatrix} M_{G_1^*+zG_2} & 0 \\ 0 & \widetilde{W}_1 \end{smallmatrix}\right]^*, \left[\begin{smallmatrix} M_{G_2^*+zG_1} & 0 \\ 0 & \widetilde{W}_2 \end{smallmatrix}\right]^*\right) \left[\begin{smallmatrix} \mathcal{O}_{D_{T^*}, T^*} \\ \mathcal{Q}_{T^*} \end{smallmatrix}\right] = \left[\begin{smallmatrix} \mathcal{O}_{D_{T^*}, T^*} \\ \mathcal{Q}_{T^*} \end{smallmatrix}\right] (T_1^*, T_2^*),$$

which breaks apart into the set of conditions

$$M_{G_1^*+zG_2}^* \mathcal{O}_{D_{T^*}, T^*} = \mathcal{O}_{D_{T^*}, T^*} T_1^*, \quad M_{G_2^*+zG_1}^* \mathcal{O}_{D_{T^*}, T^*} \mathcal{O}_{D_{T^*}, T^*} = \mathcal{O}_{D_{T^*}, T^*} T_2^*, \tag{8.10}$$

$$\widetilde{W}_1^* \mathcal{Q}_{T^*} = \mathcal{Q}_{T^*} T_1^*, \quad \widetilde{W}_2^* \mathcal{Q}_{T^*} = \mathcal{Q}_{T^*} T_2^*. \tag{8.11}$$

Since $\mathbb{W}_1$ and $\mathbb{W}_2$ commute with $V_D$, $\widetilde{W}_1$ and $\widetilde{W}_2$ commute with $W_D$ and therefore $\widetilde{W}_1^*$ and $\widetilde{W}_2^*$ commute with $W_D$ as well, because $W_D$ is a unitary. In the following computation, we use this double commutativity of $\widetilde{W}_1, \widetilde{W}_2$ with $W_D$ and equation (8.11) to show that $W_D = \widetilde{W}_1 \widetilde{W}_2$:

$$\begin{aligned} \widetilde{W}_1^* \widetilde{W}_2^* \left(W_D^n \mathcal{Q}_{T^*} h\right) &= W_D^n \widetilde{W}_1^* \widetilde{W}_2^* \mathcal{Q}_{T^*} h = W_D^n \mathcal{Q}_{T^*} T^* h \\ &= W_D^n W_D^* \mathcal{Q}_{T^*} h = W_D^* \left(W_D^n \mathcal{Q}_{T^*} h\right). \end{aligned}$$

Thus it follows that $W_D = \widetilde{W}_1 \widetilde{W}_2$ because $\left\{W_D^n (\mathcal{Q}_{T^*} h \colon n \geq 0, h \in \mathcal{H}\right\}$ is dense in $\mathcal{Q}_{T^*}$. By Corollary 4.4.5 it now follows from (8.11) that

$$\left(\widetilde{W}_1, \widetilde{W}_2\right) = (W_{\flat 1}, W_{\flat 2})$$

as claimed. As for the first equation in (8.10), note that each side is an operator from $\mathcal{H}$ into $H^2(\mathcal{D}_{T^*})$. Applying each side to a fixed vector in $\mathcal{H}$ gives

$$\sum_{k=0}^{\infty} \left(G_1^* D_{T^*} T^{*k} + G_2 D_{T^*} T^{*k+1}\right) h z^k = \sum_{k=0}^{\infty} D_{T^*} T^{*k} T_1^* h z^k.$$

Equating Taylor coefficients and canceling off the vector $h$ gives us the system of operator equations

$$G_1^* D_{T^*} T^{*k} + G_2 D_{T^*} T^{*k+1} = D_{T^*} T^{*k} T_1^* \quad \text{for all } k = 0, 1, 2, \ldots.$$

As $T_1^*$ commutes with $T^*$, we can rewrite this with a common right factor of $T^{*k}$:

$$\left(G_1^* D_{T^*} + G_2 D_{T^*} T^*\right) T^{*k} = \left(D_{T^*} T_1^*\right) T^{*k} \quad \text{for } k = 0, 1, 2, \ldots.$$

For this to hold for all $k = 0, 1, 2, \ldots$, it is now clear that it suffices that it should hold for $k = 0$:

$$G_1^* D_{T^*} + G_2 D_{T^*} T^* = D_{T^*} T_1^*,$$

that is, $(G_1, G_2)$ is a solution of the first of equations (7.14) (with $(G_1, G_2)$ in place of $(F_1, F_2)$ and with $\left(T_1^*, T_2^*\right)$ in place of $(T_1, T_2)$. A similar analysis starting with the second of equations (8.10) leads to the equation

$$G_2^* D_{T^*} + G_1 D_{T^*} T^* = D_{T^*} T_2^*,$$

that is, $(G_1, G_2)$ also solves the second equation in (7.14) (again with $(G_1, G_2)$ in place of $(F_1, F_2)$ and $\left(T_1^*, T_2^*\right)$ in place of $(T_1, T_2)$). By Definition 7.2.2 and the uniqueness result Theorem 7.2.3, it follows that $(G_1, G_2)$ turns out to be the fundamental-operator pair for $\left(T_1^*, T_2^*\right)$ as claimed. □

**Remark 8.2.2** The proof of Theorem 8.2.1 made reference to the construction of $(\mathbb{W}_{D,1}, \mathbb{W}_{D,2}, V_D)$ as the compression (in the sense of Definition 8.1.1) of a Douglas model Ando lift of $(T_1, T_2)$ to the embedding of a Douglas model minimal isometric lift of the product contraction $T = T_1 T_2$ to conclude that the Douglas model triple $(\mathbb{W}_{D,1}, \mathbb{W}_{D,2}, V_D)$ is a pseudo-commuting contractive lift. However, it is also of interest to see if it is possible to check this directly from the formula (8.4), which a priori has no reference to the existence of an Ando lift of $(T_1, T_2)$. To get the lifting property, the proof of the direct statement uses the connection of the fundamental operator pair $(G_1, G_2)$ for $\left(T_1^*, T_2^*\right)$ with the existence of a Type I Ando tuple $(\mathcal{F}_*, \Lambda_*, P_*, U_*)$ defining a Douglas model Ando lift of $(T_1, T_2)$. However, in the proof of the converse it is shown how to get a more direct statement: The lifting property for $(\Pi_D, \mathbb{W}_{D,1}, \mathbb{W}_{D,2}, V_D)$ is associated with the fundamental-operator system of equations (7.14) (with $\left(T_1^*, T_2^*\right)$ in place of $(T_1, T_2)$ and with $(G_1, G_2)$ in place of $(F_1, F_2)$). Conditions (ii) and (iii) in Definition 8.1.2 follow by a direct check from the formulas (8.4) and we conclude that one can show directly that $(\Pi_D, \mathbb{W}_{D,1}, \mathbb{W}_{D,2}, V_D)$ is at least a *pseudo-commuting algebraic lift* (as defined in Definition 8.1.2) of $(T_1, T_2, T_1 T_2)$. The only argument to show that $\mathbb{W}_{D,1}$ and $\mathbb{W}_{D,2}$ are contraction operators to our knowledge goes through the fact that $\mathbb{W}_{D,1}$ and $\mathbb{W}_{D,2}$ are compressions of the commuting isometries $(\mathbf{V}_1, \mathbf{V}_2)$ in a Douglas model Ando lift of $(T_1, T_2)$:

$$\mathbb{W}_{D,1} := M_{G_1^*+zG_2} = \left(I_{H^2} \otimes \Lambda_*^*\right) M_{U_*^* P_*^\perp + zU_*^* P_*} \left(I_{H^2} \otimes \Lambda_*\right),$$
$$\mathbb{W}_{D,2} := M_{G_2^*+zG_1} = \left(I_{H^2} \otimes \Lambda_*^*\right) M_{P_* U_* + zP_*^\perp U_*} \left(I_{H^2} \otimes \Lambda_*\right).$$

We are now ready to complete the proof of Theorem 8.1.4.

**Corollary 8.2.3** *Suppose that $(T_1, T_2)$ is a commuting contractive operator pair $(T_1, T_2)$ on $\mathcal{H}$, and suppose that $\Pi\colon \mathcal{H} \to \mathcal{K}$ is an embedding of $\mathcal{H}$ into $\mathcal{K}$ and that $(\mathbb{W}_1, \mathbb{W}_2, \mathbb{W})$ is a triple of operators on $\mathcal{K}$. Then the following are equivalent.*

(i) $(\Pi, \mathbb{W}_1, \mathbb{W}_2, \mathbb{W})$ *is the compression of an Ando lift of $(T_1, T_2, T_1T_2)$ to an immersed minimal Sz.-Nagy–Foias lift of $T$, that is, $(\Pi, \mathbb{W})$ is a minimal Sz.-Nagy–Foias (isometric) lift of $T = T_1T_2$ on $\mathcal{K}$ and there is an Ando lift $(\mathbf{\Pi}, \mathbf{V}_1, \mathbf{V}_2)$ of $(T_1, T_2)$ on $\mathcal{K}$ together with a unitary embedding*

$$\tau\colon \mathcal{K} \to \mathcal{K}_{00} := \bigvee_{n \geq 0} \mathbf{V}_1^n \mathbf{V}_2^n \operatorname{Ran} \mathbf{\Pi} \subset \mathcal{K}$$

*so that*

$$\tau \Pi = \mathbf{\Pi}, \quad \tau^* (\mathbf{V}_1, \mathbf{V}_2, \mathbf{V}_1\mathbf{V}_2)\tau = (\mathbb{W}_1, \mathbb{W}_2, \mathbb{W}).$$

(ii) $(\Pi, \mathbb{W}_1, \mathbb{W}_2, \mathbb{W})$ *is a pseudo-commuting contractive lift of $(T_1, T_2, T_1T_2)$.*

(iii) $(\Pi, \mathbb{W}_1, \mathbb{W}_2, \mathbb{W})$ *is unitarily equivalent to the Douglas model pseudo-commutative contractive lift $\left(\Pi_D, \begin{bmatrix} M_{G_1^*+zG_2} & 0 \\ 0 & W_{\flat 1} \end{bmatrix}, \begin{bmatrix} M_{G_2^*+zG_1} & 0 \\ 0 & W_{\flat 2} \end{bmatrix}, \begin{bmatrix} M_z & 0 \\ 0 & W_D \end{bmatrix}\right)$ given by Theorem 8.2.1.*

*Furthermore, let $(\Pi, \mathbb{W}_1, \mathbb{W}_2, \mathbb{W})$ and $(\Pi', \mathbb{W}_1', \mathbb{W}_2', \mathbb{W}')$ be two pseudo-commuting contractive lifts of $(T_1, T_2, T_1T_2)$ on $\mathcal{K}$ and $\mathcal{K}'$, respectively. Let $\tau'\colon \mathcal{K} \to \mathcal{K}'$ be a unitary such that*

$$\Pi' = \tau' \Pi, \quad \mathbb{W}' \tau' = \tau' \mathbb{W}. \tag{8.12}$$

*(Such a unitary exists by Theorem 2.1.3 applied to the minimal isometric lifts $(\Pi, \mathbb{W})$ and $(\Pi', \mathbb{W}')$ of $T = T_1T_2$). Then it also follows that*

$$\mathbb{W}_1' \tau' = \tau' \mathbb{W}_1, \quad \mathbb{W}_2' \tau' = \tau' \mathbb{W}_2, \tag{8.13}$$

*that is, $\tau'$ implements a unitary equivalence of pseudo-commuting contractive lifts $(\Pi, \mathbb{W}_1, \mathbb{W}_2, \mathbb{W})$ and $\left(\Pi', \mathbb{W}_1', \mathbb{W}_2', \mathbb{W}'\right)$ of $(T_1, T_2, T_1T_2)$.*

*Proof* We first show that (i) $\Rightarrow$ (ii) $\Rightarrow$ (iii) $\Rightarrow$ (i).

**(i) $\Rightarrow$ (ii)** This follows from the part of Theorem 8.1.4 already proved earlier.

**(ii) ⇒ (iii)** Assume (ii). Then $(\Pi, \mathbb{W})$ on $\mathcal{K}$ is a minimal Sz.-Nagy–Foias lift $T = T_1T_2$. By uniqueness of the minimal isometric lift for $T$, there is a unitary $\tau\colon \mathcal{K} \to \mathcal{K}_D$ which brings the Sz.-Nagy–Foias lift $(\Pi, \mathbb{W})$ to the Douglas model Sz.-Nagy–Foias lift:

$$\tau\Pi = \Pi_D, \quad \tau\mathbb{W} = V_D\tau.$$

Since $(\mathbb{W}_1, \mathbb{W}_2, \mathbb{W})$ is a pseudo-commuting contractive triple and $\tau$ is unitary, it is easily checked that

$$\tau(\mathbb{W}_1, \mathbb{W}_2, \mathbb{W})\tau^* = (\tau\mathbb{W}_1\tau^*, \tau\mathbb{W}_2\tau^*, V_D)$$

is also a pseudo-commuting contractive triple. Similarly, since $(\Pi, \mathbb{W}_1, \mathbb{W}_2, \mathbb{W})$ is a lift of $(T_1, T_2, T_1T_2)$, it follows that $(\tau\Pi = \Pi_D, \tau\mathbb{W}_1\tau^*, \tau\mathbb{W}_2\tau^*, V_D)$ is also a lift of $(T_1, T_2, T_1T_2)$. But as a consequence of Theorem 8.2.1 we see that this then forces

$$\tau\mathbb{W}_1\tau^* = \begin{bmatrix} M_{G_1^*+zG_2} & 0 \\ 0 & W_{\flat 1} \end{bmatrix}, \quad \tau\mathbb{W}_2\tau^* = \begin{bmatrix} M_{G_2^*+zG_1} & 0 \\ 0 & W_{\flat 2} \end{bmatrix},$$

and hence $\tau$ implements a unitary equivalence of the pseudo-commuting contractive lift $(\Pi, \mathbb{W}_1, \mathbb{W}_2, \mathbb{W})$ with the Douglas model pseudo-commuting contractive lift as in (8.4)), and so (iii) follows.

**(iii) ⇒ (i)** This is part of the content of the first part of Theorem 8.4.1.

We verify the last part of the corollary as follows. We suppose that $(\Pi, \mathbb{W}_1, \mathbb{W}_2, \mathbb{W})$ are $(\Pi', \mathbb{W}_1', \mathbb{W}_2', \mathbb{W}')$ are two pseudo-commuting contractive lifts of $(T_1, T_2, T_1T_2)$ on $\mathcal{K}$ and $\mathcal{K}'$ such that (8.12) holds. Let $\tau_D : \mathcal{K} \to \mathcal{K}_D$ be a unitary identification map bringing the pseudo-commuting contractive lift $(\Pi, \mathbb{W}_1, \mathbb{W}_2, \mathbb{W})$ to the Douglas model form:

$$\tau_D\Pi = \Pi_D, \quad \tau_D\mathbb{W}_1 = \mathbb{W}_{D,1}\tau_D, \quad \tau_D\mathbb{W}_2 = \mathbb{W}_{D,2}\tau_D, \quad \tau_D\mathbb{W}_D = V_D\tau_D.$$

Set $\tau = \tau_D\tau'^* : \mathcal{K}' \to \mathcal{K}_D$. Then note that

$$\begin{aligned} \tau\Pi' &= \tau_D\Pi = \Pi_D, \\ \tau\mathbb{W}' &= \tau_D\tau'^*\mathbb{W}' = \tau_D\mathbb{W}\tau'^* = V_D\tau_D\tau'^* = V_D\tau. \end{aligned}$$

Thus

$$\left(\tau\Pi', \tau W_1'\tau^*, \tau W_2'\tau^*, \tau W'\tau^*\right) = \left(\Pi_D, \tau\mathbb{W}_1'\tau^*, \tau\mathbb{W}_2'\tau^*, V_D\right)$$

is a pseudo-commuting contractive lift of $(T_1, T_2, T_1T_2)$. By the converse statement in Theorem 8.4.1, we see that we must have

$$\tau\mathbb{W}_1'\tau^* = \mathbb{W}_{D,1}, \quad \tau\mathbb{W}_2'\tau^* = \mathbb{W}_{D,2}.$$

Thus from the definitions we have

$$\mathbb{W}_1' = \tau^* \mathbb{W}_{D,1} \tau = \tau' \tau_D^* \mathbb{W}_{D,1} \tau_D \tau'^* = \tau' \mathbb{W}_1 \tau'^*$$

and similarly

$$\mathbb{W}_2' = \tau' \mathbb{W}_2 \tau'^*,$$

and (8.13) follows as desired. □

## 8.3 Sz.-Nagy–Foias Model Pseudo-commuting Contractive Lifts

We have defined the Sz.-Nagy–Foias model minimal lift of a contraction operator $T$, as well as the Sz.-Nagy–Foias model Ando lift of a commuting contractive operator pair $(T_1, T_2)$, as a simple transformation, using the unitary operator $U_{\mathrm{NF,D}} \colon \mathcal{K}_D \to \mathcal{K}_{\Theta_T}$ as in (2.66) or $\mathbf{U}_{\mathrm{NF,D}} \colon \mathcal{K}_{\mathrm{D}} \to \mathcal{K}_{\mathrm{NF}}$ as in (5.75), of the corresponding Douglas model. The analogous procedure applies also to the construction of a Sz.-Nagy–Foias model pseudo-commuting contractive lift of a commuting contractive operator pair $(T_1, T_2)$ as follows.

**Theorem 8.3.1** *Let $(T_1, T_2)$ be a commuting contractive operator pair on a Hilbert space $\mathcal{H}$ and set $T = T_1 T_2$. Let*

$$U_{\mathrm{NF,D}} = \begin{bmatrix} I_{H^2(\mathcal{D}_{T^*})} & 0 \\ 0 & \omega_{\mathrm{NF,D}} \end{bmatrix} : \begin{bmatrix} H^2(\mathcal{D}_{T^*}) \\ \mathcal{Q}_{T^*} \end{bmatrix} \to \begin{bmatrix} H^2(\mathcal{D}_{T^*}) \\ \hline \Delta_{\Theta_T}\left(L^2(\mathcal{D}_T)\right) \end{bmatrix}$$

*be the unitary as in (2.66). Let us define operators*

$$\underline{\mathbb{W}}_{\mathrm{NF}} = (\mathbb{W}_{\mathrm{NF},1}, \mathbb{W}_{\mathrm{NF},2}, V_{\mathrm{NF}}) := U_{\mathrm{NF},D}(\mathbb{W}_{D,1}, \mathbb{W}_{D,2}, V_D) U_{\mathrm{NF},D}^*$$
$$= \left( \begin{bmatrix} M_{G_1^* + zG_2} & 0 \\ 0 & W_{\sharp 1} \end{bmatrix}, \begin{bmatrix} M_{G_2^* + zG_1} & 0 \\ 0 & W_{\sharp 2} \end{bmatrix}, \begin{bmatrix} M_z^{\mathcal{D}_{T^*}} & 0 \\ 0 & M_\zeta|_{\overline{\Delta_{\Theta_T} L^2(\mathcal{D}_T)}} \end{bmatrix} \right) \quad \text{on } \mathcal{K}_{\Theta_T},$$

$$\Pi_{\mathrm{NF}} = U_{\mathrm{NF,D}} \Pi_D := \begin{bmatrix} \mathcal{O}_{D_{T^*}, T^*} \\ \omega_{\mathrm{NF,D}} \mathcal{Q}_{T^*} \end{bmatrix} : \mathcal{H} \to \mathcal{K}_{\Theta_T}, \tag{8.14}$$

*where $\mathcal{K}_{\Theta_T} = \begin{bmatrix} H^2(\mathcal{D}_{T^*}) \\ \hline \Delta_{\Theta_T}(L^2(\mathcal{D}_T)) \end{bmatrix}$ and where $(\mathbb{W}_{D,1}, \mathbb{W}_{D,2}, V_D)$ is the Douglas model pseudo-commuting contractive operator triple as defined in (8.5) (so $(G_1, G_2)$ is the fundamental-operator pair for $(T_1^*, T_2^*)$) and the pair $(W_{\sharp 1}, W_{\sharp 2})$ is the commuting unitary operator pair as in (5.74). Then $(\Pi_{\mathrm{NF}}, W_{\mathrm{NF},1}, W_{\mathrm{NF}2}, V_{\mathrm{NF}})$ is a pseudo-commuting contractive lift of $(T_1, T_2, T)$, called the* Sz.-Nagy–Foias model pseudo-commuting contractive lift *of $(T_1, T_2)$.*

*Proof* We showed in the proof of Theorem 8.2.1 that $(\Pi_D, \mathbb{W}_{D,1}, \mathbb{W}_{D,2}, V_D)$ is a pseudo-commuting contractive lift of $(T_1, T_2)$. Since the isometry $\Pi_{\rm NF} \colon \mathcal{H} \to \mathcal{K}_{\Theta_T}$ is given by $\Pi_{\rm NF} = U_{\rm NF,D}\Pi_D$, it then follows that the collection $(\Pi_{\rm NF}, \underline{V}_{\rm NF})$ is unitarily equivalent (in the sense of lifts) to the lift $(\Pi_{\rm D}, \underline{V}_D)$ (where we set $\underline{V}_D = (\mathbb{W}_{D,1}, \mathbb{W}_{D,2}, V_D))$, and hence itself must also be a lift, of $(T_1, T_2, T_1T_2)$; in detail we have

$$\begin{aligned} \underline{\mathbb{W}}^*_{\rm NF}\Pi_{\rm NF} &= U_{\rm NF,D}\, \underline{\mathbb{W}}^*_D\, U^*_{\rm NF,D} \cdot U_{\rm NF,D}\Pi_D \\ &= U_{\rm NFD}\, \underline{\mathbb{W}}^*_D\, \Pi_D = U_{\rm NF,D}\, \Pi_D\, \underline{T}^* = \Pi_{\rm NF}\, \underline{T}^* \end{aligned}$$

(where here we set $\underline{T} = (T_1, T_2, T = T_1T_2))$ thereby verifying that $(\Pi_{\rm NF}, \underline{\mathbb{W}}_{\rm NF})$ is a lift of $\underline{T}$. Since the pseudo-commuting contractive property is invariant under unitary equivalence, it also follows that $\mathbb{W}_D$ being a pseudo-commuting contractive triple implies that $\mathbb{W}_{\rm NF}$ is also a pseudo-commuting contractive triple. □

**Remark 8.3.2** Let us observe that, since $(\Pi_{\rm NF}, \underline{\mathbb{W}}_{\rm NF})$ is related to $(\Pi_{\rm D}, \underline{\mathbb{W}}_D)$ via the innocuous change of coordinates in the second coordinate $U_{\rm NF,D}$, it is a routine exercise to see that all the results concerning $(\Pi_D, \underline{\mathbb{W}}_D)$ in Theorem 8.2.1, Remark 8.2.2, Corollary 8.2.3 hold equally well for $(\Pi_{\rm NF}, \underline{\mathbb{W}}_{\rm NF})$. In particular, the Sz.-Nagy–Foias model pseudo-commuting contractive lift $(\Pi_{\rm NF}, \underline{\mathbb{W}}_{\rm NF})$ can also be viewed as the compression of a Sz.-Nagy–Foias model Ando lift (based on some fundamental-operator pair $(G_1, G_2)$ for $(T_1^*, T_2^*)$) to an embedded Sz.-Nagy–Foias model for a minimal Sz.-Nagy–Foias lift $(\Pi_{\rm NF}, V_{\rm NF})$ of $T = T_1T_2$, thereby proving that $(\Pi_{\rm NF}, \underline{V}_{\rm NF})$ is a pseudo-commuting contractive lift of $\underline{T}$ as a consequence of the first part of the abstract Theorem 8.1.4. Alternatively, it is possible to check directly from the formulas (8.14) and identities satisfied by the fundamental-operator pair $(G_1, G_2)$ that $(\Pi_{\rm NF}, \underline{\mathbb{W}}_{\rm NF})$ is at least a pseudo-commuting algebraic lift of $\underline{T}$.

## 8.4 Schäffer Model Pseudo-commuting Contractive Lifts

The Schäffer model for a general Ando lift arising from a weak Type II Ando tuple does not appear to be sufficiently tractable for the identification of a pseudo-commuting contractive lift. We therefore restrict ourselves to Type II Ando tuples (see Definition 6.1.2).

We have seen in Theorem 6.1.4 that any minimal Ando lift for a commuting contractive pair $(T_1, T_2)$ is unitarily equivalent to the Schäffer model Ando lift associated with some Type II Ando tuple $(\mathcal{F}, \Lambda, P, U)$ for $(T_1, T_2)$ and, conversely, the Schäffer model Ando lift associated with a Type II Ando tuple for $(T_1, T_2)$ is an Ando lift for $(T_1, T_2)$. By Theorem 8.1.4 the compression of such a Schäffer model Ando lift in the sense of Definition 8.1.1 yields a pseudo-commuting contractive lift $(\mathbb{W}_1, \mathbb{W}_2, \mathbb{W})$ for $(T_1, T_2, T_1T_2)$. The next result computes such a Schäffer model compressed Ando lift for a given commuting contractive pair $(T_1, T_2)$ also using the Schäffer model for the minimal isometric lift of the product contraction operator $T = T_1T_2$.

**Theorem 8.4.1** *Given a commuting contractive operator pair $(T_1, T_2)$ on $\mathcal{H}$, let $\mathcal{K}_S = \begin{bmatrix} \mathcal{H} \\ H^2(\mathcal{D}_T) \end{bmatrix}$ be the Schäffer model isometric-lift space for the product contraction $T := T_1T_2$, let $\Pi_S \colon \mathcal{H} \to \mathcal{K}_S$ be the Schäffer model embedding operator $\Pi_S = \begin{bmatrix} I_{\mathcal{H}} \\ 0 \end{bmatrix}$, let $(F_1, F_2)$ be the fundamental-operator pair for $(T_1, T_2)$ (see Theorem 7.2.3), and define operators $\mathbb{W}_{S,1}, \mathbb{W}_{S,2}, \mathbb{W}_S$ on $\mathcal{K}_S$ according to the formula*

$$\mathbb{W}_{S,1} = \begin{bmatrix} T_1 & 0 \\ \mathbf{ev}^*_{0,\mathcal{D}_T} F_2^* D_T & M_{F_1 + zF_2^*} \end{bmatrix}, \quad \mathbb{W}_{S,2} = \begin{bmatrix} T_2 & 0 \\ \mathbf{ev}^*_{0,\mathcal{D}_T} F_1^* D_T & M_{F_2 + zF_1^*} \end{bmatrix},$$
$$\mathbb{W}_S = V_S = \begin{bmatrix} T & 0 \\ \mathbf{ev}^*_{0,\mathcal{D}_T} D_T & M_z^{\mathcal{D}_T} \end{bmatrix} \quad \text{on } \mathcal{K}_S := \begin{bmatrix} \mathcal{H} \\ H^2(\mathcal{D}_T) \end{bmatrix}. \tag{8.15}$$

*Then $(\Pi_S, \mathbb{W}_{S,1}, \mathbb{W}_{S,2}, V_S)$ is a compression of an Ando lift of $(T_1, T_2, T_1T_2)$ to an embedded minimal Sz.-Nagy–Foias lift of $T = T_1T_2$ and hence also a pseudo-commuting contractive lift of $(T_1, T_2, T_1T_2)$.*

*Conversely, if $(\Pi_S, \mathbb{W}_1, \mathbb{W}_2, V_S)$ is any pseudo-commuting contractive lift of $(T_1, T_2)$ such that*

$$(\Pi_S, V_S) = \left( \begin{bmatrix} I_{\mathcal{H}} \\ 0 \end{bmatrix} : \mathcal{H} \to \mathcal{K}_S, \begin{bmatrix} T & 0 \\ \mathbf{ev}^*_{0,\mathcal{F}} D_T & M_z^{\mathcal{D}_T} \end{bmatrix} \text{ on } \mathcal{K}_S \right) \tag{8.16}$$

*is the Schäffer model minimal isometric lift of $T$ on $\mathcal{K}_S = \begin{bmatrix} \mathcal{H} \\ H^2(\mathcal{D}_T) \end{bmatrix}$, then necessarily $(\mathbb{W}_1, \mathbb{W}_2) = (\mathbb{W}_{S,1}, \mathbb{W}_{S,2})$ are also given, as in formula* (8.15).

*Proof* Let $(\mathcal{F}, \Lambda, P, U)$ be a strong Type II Ando tuple for the commuting contractive operator pair $(T_1, T_2)$. Then the associated Schäffer model Ando lift of $(T_1, T_2)$ is given by $(\mathbf{\Pi}_S, \mathbf{V}_{S,1}, \mathbf{V}_{S,2})$, where the isometric embedding operator $\mathbf{\Pi}_S \colon \mathcal{H} \to \mathcal{K}_S$ and the isometries $\mathbf{V}_{S,1}, \mathbf{V}_{S,2}$ on $\mathcal{K}_S$ are given as in Theorem 6.1.4:

$$\mathcal{K}_S = \begin{bmatrix} \mathcal{H} \\ H^2(\mathcal{F}) \end{bmatrix}, \quad \mathbf{\Pi}_S = \begin{bmatrix} I_{\mathcal{H}} \\ 0 \end{bmatrix} : \mathcal{H} \to \mathcal{K}_S,$$

$$(\mathbf{V}_{S,1}, \mathbf{V}_{S,2}) = \left( \begin{bmatrix} T_1 & 0 \\ \mathbf{ev}^*_{0,\mathcal{F}} PU\Lambda D_T & M_{P^\perp U + zPU} \end{bmatrix}, \right.$$
$$\left. \begin{bmatrix} T_2 & 0 \\ \mathbf{ev}^*_{0,\mathcal{F}} U^* P^\perp \Lambda D_T & M_{U^*P + zU^*P^\perp} \end{bmatrix} \right),$$

$$\mathbf{V}_S = \mathbf{V}_{S,1}\mathbf{V}_{S,2} = \mathbf{V}_{S,2}\mathbf{V}_{S,1} = \begin{bmatrix} T & 0 \\ \mathbf{ev}^*_{0,\mathcal{F}} \Lambda D_T & M_z \end{bmatrix},$$

where $T := T_1T_2 = T_2T_1$ is the product contraction operator on $\mathcal{H}$. Let us next compute the space

$$\mathcal{K}_{S,00} := \bigvee_{n \geq 0} \mathbf{V}_S^n \operatorname{Ran} \mathbf{\Pi}_S.$$

By an induction argument one can see that

$$\mathbf{V}_S^n \mathbf{\Pi}_S = \begin{bmatrix} T^n \\ \sum_{j=0}^{n-1} M_z^j \mathbf{ev}^*_{0,\mathcal{F}} \Lambda D_T T^{n-1-j} \end{bmatrix},$$

where the bottom entry should be interpreted to be 0 for the case $n = 0$. By taking $n = 0$ we see that $\begin{bmatrix} \mathcal{H} \\ 0 \end{bmatrix} \subset \mathcal{K}_{S,00}$. By next taking $n = 1$ we see that

$$\bigvee_{n=0,1} \operatorname{Ran} \mathbf{V}_S^n \mathbf{\Pi}_S = \begin{bmatrix} \mathcal{H} \\ \mathbf{ev}^*_{0,\mathcal{F}} \Lambda \mathcal{D}_T \end{bmatrix} \subset \mathcal{K}_{S,00}.$$

Inductively assume that

$$\bigvee_{n=0,1,\dots,K} \operatorname{Ran} \mathbf{V}_S^n \mathbf{\Pi}_S = \begin{bmatrix} \mathcal{H} \\ \oplus_{j=0}^{K-1} \left( M_z^{\mathcal{D}_T} \right)^j \mathbf{ev}^*_{0,\mathcal{F}} \Lambda \mathcal{D}_T \end{bmatrix}.$$

It then follows that

$$\bigvee_{n=0,1,\dots,K,K+1} \operatorname{Ran} \mathbf{V}_S^n \mathbf{\Pi}_S = \begin{bmatrix} \mathcal{H} \\ \left( \oplus_{j=0}^{K-1} \left( M_z^{\mathcal{D}_T} \right)^j \mathbf{ev}^*_{0\mathcal{F}} \Lambda \mathcal{D}_T \right) \oplus \left( M_z^{\mathcal{D}_T} \right)^K \mathbf{ev}^*_{0,\mathcal{F}} \Lambda \mathcal{D}_T \end{bmatrix}$$
$$= \begin{bmatrix} \mathcal{H} \\ \oplus_{j=0}^{K} \left( M_z^{\mathcal{D}_T} \right)^j \mathbf{ev}^*_{0\mathcal{F}} \Lambda \mathcal{D}_T \end{bmatrix}.$$

Hence

$$\mathcal{K}_{S,00} = \operatorname{closure} \left( \bigcup_{K \geq 0} \begin{bmatrix} \mathcal{H} \\ \left( \oplus_{j=0}^{K-1} \left( M_z^{\mathcal{D}_T} \right)^j \mathbf{ev}^*_{0\mathcal{F}} \Lambda \mathcal{D}_T \right) \end{bmatrix} \right) = \begin{bmatrix} \mathcal{H} \\ H^2(\operatorname{Ran} \Lambda) \end{bmatrix}. \tag{8.17}$$

Let us now introduce the Schäffer model $(\Pi_S, V_S)$ for the minimal isometric lift of the product contraction operator $T := T_1T_2$. By uniqueness of minimal isometric lifts for a single contraction operator, there exists an isometry $\tau$ from $\mathcal{K}_S$ onto $\mathcal{K}_{S,00}$ so that

$$\tau\Pi_S = \mathbf{\Pi}_{S,00}, \quad \tau V_S = \mathbf{V}_{S,00}\tau,$$

where we set $\mathbf{V}_{S,00} := \mathbf{V}_S|_{\mathcal{K}_{00}}$. It is easy to check that

$$\tau := \begin{bmatrix} I_{\mathcal{H}} & 0 \\ 0 & I_{H^2} \otimes \Lambda \end{bmatrix} \tag{8.18}$$

does the job.

Putting all the pieces together, it follows by definition that the compressed Ando lift of $(T_1, T_2)$ associated with

(i) the Schäffer model Ando lift $(\mathbf{\Pi}_S, \mathbf{V}_{S,1}, \mathbf{V}_{S,2})$ determined by the Ando-tuple $(\mathcal{F}, \Lambda, P, U)$ for $(T_1, T_2)$ and
(ii) the Schäffer model minimal Sz.-Nagy–Foias lift $(\Pi_S, V_S)$ for $T = T_1T_2$

is given by

$$(\mathbb{W}_{S,1}, \mathbb{W}_{S,2}, \mathbb{W}_S) = \left(\tau^*\mathbf{V}_{S,1}\tau,\ \tau^*\mathbf{V}_{S,2}\tau,\ V_S\right) \quad \text{on } \mathcal{K}_S. \tag{8.19}$$

Thus

$$\begin{aligned} \mathbb{W}_{S,1} &= \begin{bmatrix} I_{\mathcal{H}} & 0 \\ 0 & I_{H^2} \otimes \Lambda^* \end{bmatrix} \begin{bmatrix} T_1 & 0 \\ \mathbf{ev}^*_{0,\mathcal{F}} PU\Lambda D_T & M_{P^\perp U + zPU} \end{bmatrix} \begin{bmatrix} I_{\mathcal{H}} & 0 \\ 0 & I_{H^2} \otimes \Lambda \end{bmatrix} \\ &= \begin{bmatrix} T_1 & 0 \\ \mathbf{ev}^*_{0,\mathcal{D}_T} \Lambda^* PU\Lambda D_T & M_{\Lambda^* P^\perp U\Lambda + z\Lambda^* PU\Lambda} \end{bmatrix} \\ &= \begin{bmatrix} T_1 & 0 \\ \mathbf{ev}^*_{0,\mathcal{D}_T} F_2^* D_T & M_{F_1 + zF_2^*} \end{bmatrix}, \end{aligned} \tag{8.20}$$

where $(F_1, F_2)$ is the fundamental-operator pair for the commuting contractive pair $(T_1, T_2)$. Here in the last step we used the characterization (7.33) of the fundamental-operator pair in terms of a strong Type II Ando tuple for $(T_1, T_2)$

$$F_1 = \Lambda^* P^\perp U\Lambda, \quad F_2 = \Lambda^* U^* P\Lambda,$$

coming out of the third proof of Theorem 7.2.3.

A similar computation shows that $\mathbb{W}_{S,2}$ is given by

$$\begin{bmatrix} I_{\mathcal{H}} & 0 \\ 0 & I_{H^2} \otimes \Lambda^* \end{bmatrix} \begin{bmatrix} T_2 & 0 \\ 0 & \mathbf{ev}^*_{0,\mathcal{F}} U^* P^{\perp} \Lambda D_T & M_{U^* P + z U^* P^{\perp}} \end{bmatrix} \begin{bmatrix} I_{\mathcal{H}} & 0 \\ 0 & I_{H^2} \otimes \Lambda \end{bmatrix}$$
$$= \begin{bmatrix} T_2 & 0 \\ \mathbf{ev}^*_{0,\mathcal{D}_T} \Lambda^* U^* P^{\perp} \Lambda D_T & M_{\Lambda^* U^* P \Lambda + z \Lambda^* U^* P^{\perp} \Lambda} \end{bmatrix}$$
$$= \begin{bmatrix} T_2 & 0 \\ \mathbf{ev}^*_{0,\mathcal{D}_T} F_1^* D_T & M_{F_2 + z F_1^*} \end{bmatrix}.$$

We have now verified that the formula (8.15) gives a compressed Ando lift for the commuting contractive pair $(T_1, T_2)$.

The fact that $(\Pi_S, W_{S,1}, W_{S,2}, W_S = V_S)$ is then also a pseudo-commuting contractive lift of $(T_1, T_2, T = T_1 T_2)$ follows as a consequence of the general principle *compressed Ando lift* $\Rightarrow$ *pseudo-commuting contractive lift* of $(T_1, T_2)$ verified in Theorem 8.1.4.

The converse statement of the theorem follows from the model-independent result in Corollary 8.2.3. Let us note that the proof of Corollary 8.2.3 reduces the statement to the special case where the Ando lift of $(T_1, T_2)$ and the minimal isometric lift $(\Pi, V)$ for the product contraction $T = T_1 T_2$ are both taken in Douglas model form. As an instructive exercise, we include here a direct proof of the converse statement under the assumption that the Ando lift of $(T_1, T_2)$ and the minimal isometric lift of the product contraction $T = T_1 T_2$ are both taken in Schäffer model form. This also leads to a new proof of the abstract Corollary 8.2.3: One can follow the steps in the proof of Corollary 8.2.3 but with substitution of Schäffer models for the Douglas models.

To prove the converse statement of Theorem 8.4.1 directly in the context of Schäffer models, let us suppose that $(T_1, T_2)$ is a commuting contractive operator pair on $\mathcal{H}$ and that $(\Pi_S, \mathbb{W}_1, \mathbb{W}_2, V_S)$ is a pseudo-commuting contractive triple on $\mathcal{K}_S$ such that $(\Pi_S, V_S)$ is the Schäffer model minimal isometric lift of the product contraction $T = T_1 T_2$. We wish to show that necessarily $(W_1, W_2) = (W_{S,1}, W_{S,2})$, where $W_{S,1}$ and $W_{S,2}$ are as in (8.15). As is done in the context of the Douglas model (see Remark 8.2.2), we show that the conclusion still holds without using the assumption that $W_1$ and $W_2$ are contraction operators, that is, the proof goes through with the weakened assumption that $(\Pi_S, W_1, W_2, V_S)$ is a pseudo-commuting algebraic lift of $(T_1, T_2, T = T_1 T_2)$. We break the proof up into steps.

**Step 1** Show: *$(\Pi_S, \mathbb{W}_1, \mathbb{W}_2)$ on $\mathcal{K}_S = \begin{bmatrix} \mathcal{H} \\ H^2(\mathcal{D}_T) \end{bmatrix}$ is a lift of $(T_1, T_2)$ $\Leftrightarrow$ $(\mathbb{W}_1, \mathbb{W}_2)$, written as a block $2 \times 2$-matrix with respect to the direct-sum decomposition $\mathcal{K}_S = \begin{bmatrix} \mathcal{H} \\ H^2(\mathcal{D}_T) \end{bmatrix}$ of $\mathcal{K}$, and has the lower-triangular form*

$$\mathbb{W}_i = \begin{bmatrix} T_i & 0 \\ \mathbb{W}_{i,21} & \mathbb{W}_{i,22} \end{bmatrix} \quad \textit{for } i = 1, 2. \tag{8.21}$$

*Proof of Step 1* Suppose that $\mathbb{W}_i$ has matrix representation $\mathbb{W}_i = \begin{bmatrix} \mathbb{W}_{i,11} & \mathbb{W}_{i,12} \\ \mathbb{W}_{i,21} & \mathbb{W}_{i,22} \end{bmatrix}$ for $i = 1, 2$. That $(\Pi_S, \mathbb{W}_i)$ is a lift of $T_i$ means simply that $\mathbb{W}_i^* \Pi_S = \Pi_S T_i^*$. Recalling that $\Pi_S = \begin{bmatrix} I_{\mathcal{H}} \\ 0 \end{bmatrix} : \mathcal{H} \to \mathcal{K}_S$ then gives us

$$\begin{bmatrix} \mathbb{W}_{i,11}^* & \mathbb{W}_{i,21}^* \\ \mathbb{W}_{i,12}^* & \mathbb{W}_{i,22}^* \end{bmatrix} \begin{bmatrix} I_{\mathcal{H}} \\ 0 \end{bmatrix} = \begin{bmatrix} I_{\mathcal{H}} \\ 0 \end{bmatrix} T_i^* \quad \text{for } i = 1, 2,$$

which is to say

$$\begin{bmatrix} \mathbb{W}_{i,11}^* \\ \mathbb{W}_{i,12}^* \end{bmatrix} = \begin{bmatrix} T_i^* \\ 0 \end{bmatrix},$$

and hence $\mathbb{W}_i$ has the form (8.21) for $i = 1, 2$ as claimed.

The remaining unused assumption is that $(\mathbb{W}_1, \mathbb{W}_2, V_S)$ is a pseudo-commuting algebraic operator triple. Let us spell this out using the updated information that $\mathbb{W}_i$ has the form (8.21), namely the following.

(i) $\mathbb{W}_i$ commutes with $V_S$ for $i = 1, 2$ means

$$\begin{bmatrix} T_i & 0 \\ \mathbb{W}_{i,21} & \mathbb{W}_{i,22} \end{bmatrix} \begin{bmatrix} T & 0 \\ \mathbf{ev}_{0,\mathcal{D}_T}^* D_T & M_z^{\mathcal{D}_T} \end{bmatrix} = \begin{bmatrix} T & 0 \\ \mathbf{ev}_{0,\mathcal{D}_T}^* D_T & M_z^{\mathcal{D}_T} \end{bmatrix} \begin{bmatrix} T_i & 0 \\ \mathbb{W}_{i,21} & \mathbb{W}_{i,22} \end{bmatrix},$$

which is equivalent to the system of matrix-entry identities

$$\mathbb{W}_{i,21} T + \mathbb{W}_{i,22} \mathbf{ev}_{0,\mathcal{D}_T}^* D_T = \mathbf{ev}_{0,\mathcal{D}_T}^* D_T T_i + M_z^{\mathcal{D}_T} \mathbb{W}_{i,21} \quad \text{for } i = 1, 2, \tag{8.22}$$

$$\mathbb{W}_{i,22} M_z^{\mathcal{D}_T} = M_z^{\mathcal{D}_T} \mathbb{W}_{i,22} \quad \text{for } i = 1, 2. \tag{8.23}$$

(ii) $\mathbb{W}_i = \mathbb{W}_j^* \mathbb{W}$ for $(i, j) = (1, 2)$ or $(2, 1)$ means

$$\begin{bmatrix} T_i & 0 \\ \mathbb{W}_{i,21} & \mathbb{W}_{i,22} \end{bmatrix} = \begin{bmatrix} T_j^* & \mathbb{W}_{j,21}^* \\ 0 & \mathbb{W}_{j,22}^* \end{bmatrix} \begin{bmatrix} T & 0 \\ \mathbf{ev}_{0,\mathcal{D}_T}^* D_T & M_z^{\mathcal{D}_T} \end{bmatrix}$$

for $(i, j) = (1, 2)$ or $(2, 1)$, which is equivalent to the system of matrix-entry identities

$$T_i = T_j^* T + \mathbb{W}_{j,21}^* \mathbf{ev}_{0,\mathcal{D}_T}^* D_T, \tag{8.24}$$

$$0 = \mathbb{W}_{j,21}^* M_z^{\mathcal{D}_T}, \tag{8.25}$$

$$\mathbb{W}_{i,21} = \mathbb{W}_{j,22}^* \mathbf{ev}_{0,\mathcal{D}_T}^* D_T, \tag{8.26}$$

$$\mathbb{W}_{i,22} = \mathbb{W}_{j,22}^* M_z^{\mathcal{D}_T}. \tag{8.27}$$

**Step 2** Show: *Given that* $\mathbb{W}_i$ *has the form* (8.21) *for* $i = 1, 2$, (8.23) *and* (8.27) *hold* $\Leftrightarrow$ *there are operators* $F_1, F_2$ *on* $\mathcal{D}_T$ *so that*

$$\mathbb{W}_i = \begin{bmatrix} T_i & 0 \\ \mathbb{W}_{i,21} & M_{F_i + zF_j^*} \end{bmatrix} \quad \textit{for } (i, j) = (1, 2) \textit{ or } (2, 1), \tag{8.28}$$

*that is,*

$$\mathbb{W}_{i,22} = M_{F_i + zF_j^*}$$

*for* $(i, j) = (1, 2)$ *or* $(2, 1)$.

*Proof of Step 2* The proof is a somewhat simpler version of the proof of Step 1 in the proof of the converse side of Theorem 8.2.1. Given that (8.23) holds, we see that there are analytic operator functions $\varphi_i(z)$ with values in $\mathcal{B}(\mathcal{D}_T)$ so that $\mathbb{W}_{i,22} = M_{\varphi_i(z)}$ $(i = 1, 2)$. Then condition (8.27) takes the form

$$M_{\varphi_i(z)} = M^*_{\varphi_j(z)} M_z^{\mathcal{D}_T} \quad \text{for } (i, j) = (1, 2) \text{ or } (2, 1).$$

A Taylor series argument then leads to the conclusion that $\varphi_1(z)$, $\varphi_2(z)$ have the coupled pencil form $\varphi_i(z) = F_i + zF_j^*$ for some operators $F_1, F_2$ on $\mathcal{D}_T$ for $(i, j) = (1, 2)$ or $(2, 1)$, and (8.28) follows.

Conversely, it is a direct check to see that if $\mathbb{W}_i$ is of the form (8.28), then (8.23) and (8.27) hold.

**Step 3** Show: *Given that* $\mathbb{W}_i$ *has the form* (8.28), *then* (8.26) *holds* $\Leftrightarrow$

$$\mathbb{W}_{i,21} = \mathbf{ev}^*_{0,\mathcal{D}_T} F_j^* D_T$$

*and then* $W_i$ *has the form*

$$\mathbb{W}_i = \begin{bmatrix} T_i & 0 \\ \mathbf{ev}^*_{0,\mathcal{D}_T} F_j^* D_T & M_{F_i + zF_j^*} \end{bmatrix} \quad \textit{for } (i, j) = (1, 2) \textit{ or } (2, 1). \tag{8.29}$$

*Proof of Step 3* Given that $\mathbb{W}_{i,22} = M_{G_1 + zG_j^*}$ as in (8.28), we see that the validity of (8.26) is equivalent to

$$\begin{aligned} \mathbb{W}_{i,21} &= \mathbb{W}^*_{j,22} \mathbf{ev}^*_{0,\mathcal{D}_T} D_T = M^*_{F_j + zF_i^*} \mathbf{ev}^*_{0,\mathcal{D}_T} D_T \\ &= M_{F_j^*} \mathbf{ev}^*_{0,\mathcal{D}_T} D_T + \left(M_z^{\mathcal{D}_T}\right)^* \mathbf{ev}^*_{0,\mathcal{D}_T} F_i D_T \\ &= \mathbf{ev}^*_{0,\mathcal{D}_T} F_j^* D_T, \end{aligned}$$

as claimed.

**Step 4** Show: *If* $\mathbb{W}_i$ *has the form* (8.29), *then* (8.22) *holds* $\Leftrightarrow$ $(F_1, F_2)$ *is the fundamental-operator pair for the commuting contractive pair* $(T_1, T_2)$.

*Proof of Step 4* Given that $\mathbb{W}_i$ has the form (8.29), we see that (8.26) becomes

$$\mathbf{ev}^*_{0,\mathcal{D}_T} F^*_j D_T T + M_{F_i + zF^*_j} \mathbf{ev}^*_{0,\mathcal{D}_T} D_T = \mathbf{ev}^*_{0,\mathcal{D}_T} D_T T_i + M_z^{\mathcal{D}_T} \mathbf{ev}^*_{0,\mathcal{D}_T} F^*_j D_T.$$

Applying each side of this equality of operators from $\mathcal{H}$ into $H^2(\mathcal{D}_T)$ to a fixed vector $h \in \mathcal{H}$ leads to the vector-function equality

$$\left(F^*_j D_T T + F_i D_T + zF^*_j D_T\right) h = \left(D_T T_i + zF^*_j D_T\right) h \quad \text{for all } h \in \mathcal{H}.$$

As the coefficients of $z$ are the same in each equation, this in turn amounts to the system of operator equations

$$F^*_j D_T T + F_i D_T = D_T T_i \quad \text{for } (i, j) = (1, 2) \text{ or } (2, 1).$$

This system amounts to the system (7.14) appearing in Chapter 7 for which we know that the fundamental-operator pair $(F_1, F_2)$ for $(T_1, T_2)$ is the unique solution. As the analysis is necessary and sufficient, the proof of Step 4 as well as of the converse statement of Theorem 8.4.1 is complete. □

**Remark 8.4.2** In parallel with the discussion for the Douglas model pseudo-commuting contractive lifts, let us point out that the proof of Theorem 8.4.1 makes reference to the construction of $(\mathbb{W}_{S,1}, \mathbb{W}_{S,2}, V_S)$ as a compression (in the sense of Definition 8.1.1) of an Ando lift for $(T_1, T_2)$ to conclude that $(\mathbb{W}_{S,1}, \mathbb{W}_{S,2}, V_S)$ is a pseudo-commuting contractive lift for $(T_1, T_2)$, and, as in Remark 8.2.2, it would be of interest to find a direct proof using just the formulas (8.15) for $(\mathbb{W}_{S,1}, \mathbb{W}_{S,2}, V_S)$, which shows that $(\mathbb{W}_{S,1}, \mathbb{W}_{S,2}, V_S)$ is pseudo-commuting contractive and is a lift of $(T_1, T_2, T = T_1T_2)$. As $(\mathbb{W}_{S,1}, \mathbb{W}_{S,2}, V_S)$ is of the form (8.21), we see by the analysis coming right after the proof of Step 1 in the proof of the converse statement in Theorem 8.4.1 that $(\mathbb{W}_{S,1}, \mathbb{W}_{S,2}, V_S)$ is at least a *pseudo-commuting algebraic lift* of $(T_1, T_2)$ (in the sense of Definition 8.1.2) if and only if all the conditions (8.22), (8.23), (8.24), (8.25), (8.26), and (8.27) are satisfied. The analysis there shows that $(\mathbb{W}_1, \mathbb{W}_2, V_S)$ satisfying only conditions (8.22), (8.23), (8.26), and (8.27) is already enough to force $(\mathbb{W}_1, \mathbb{W}_2, V_S) = (\mathbb{W}_{S,1}, \mathbb{W}_{S,2}, V_S)$. As this analysis is reversible, we see that in particular $(\mathbb{W}_{S,1}, \mathbb{W}_{S,2}, V_S)$ satisfies conditions (8.22), (8.23), (8.26), and (8.27). Hence, to obtain a direct proof that $(\mathbb{W}_{S,1}, \mathbb{W}_{S,2}, V_S)$ is at least a pseudo-commuting algebraic lift of $(T_1, T_2)$, it remains to show that $(\mathbb{W}_{S,1}, \mathbb{W}_{S,2}, V_S)$ satisfies (8.24) and (8.25). We can check condition (8.24) as follows. From the formula $\mathbb{W}_{S,j;21} = \mathbf{ev}^*_{0,\mathcal{D}_T} F^*_i D_T$, we see that (8.24) becomes

$$\begin{aligned} T_i &= T^*_j T + \mathbb{W}^*_{S,j;21} \mathbf{ev}^*_{0,\mathcal{D}_T} D_T && (8.30) \\ &= T^*_j T + D_T F_i D_T \quad \text{for } (i, j) = (1, 2) \text{ or } (2, 1). && (8.31) \end{aligned}$$

This amounts to the system of equations (7.13), the unique solution of which also characterizes the fundamental-operator pair $(F_1, F_2)$ for $(T_1T_2)$. In fact, we could have used (8.24) rather than (8.22) to squeeze the parameter pair $(F_1, F_2)$ to be the fundamental-operator pair in Step 4 in the proof of Theorem 8.4.1.

As for (8.25), this is an immediate consequence of $\mathbb{W}_{S,i}$ having the form $\mathbb{W}_{S,i;12} = \mathbf{ev}^*_{0,\mathcal{D}_T} F_j^* D_T$ $((i,j) = (1,2)$ or $(2,1))$ for some operators $F_1, F_2$ on $\mathcal{D}_T$. Indeed, in this case (8.25) becomes

$$\begin{aligned} 0 = \mathbb{W}^*_{j,12} M_z^{\mathcal{D}_T} &= \left(\mathbf{ev}^*_{0,\mathcal{D}_T} F_i^*\right)^* M_z^{\mathcal{D}_T} \\ &= F_i \mathbf{ev}_{0,\mathcal{D}_T} M_z^{\mathcal{D}_T} = 0, \end{aligned}$$

since any function $h \in \operatorname{Ran} M_z^{\mathcal{D}_T} \subset H^2(\mathcal{D}_T)$ has $h(0) = 0$.

As was the case for the Douglas model setting discussed in Remark 8.2.2, we are unable to verify the contractivity of $\mathbb{W}_{S,1}, \mathbb{W}_{S,2}$ without bringing in the fact that $(\mathbb{W}_{S,1}, \mathbb{W}_{S,2})$ is the compression of the commuting isometric pair $(\mathbf{V}_{S,1}, \mathbf{V}_{S,2})$ associated with a Schäffer model Ando lift $(\mathbf{\Pi}_S, \mathbf{V}_{S,1}, \mathbf{V}_{S,2})$ coming from a Type II Ando tuple $(\mathcal{F}, \Lambda, P, U)$ for $(T_1, T_2)$.

## 8.5 Strongly Minimal Ando Lifts Revisited

Recall that two minimal isometric lifts of a commuting contractive pair need not be unitary equivalent. What if we replace minimality by strong minimality? The next result shows that the existence of one strongly minimal Ando lift is a strong enough criterion to have uniqueness of any two minimal Ando lifts. In particular, this result answers the question in Problem 7.4.3.

**Theorem 8.5.1** *Let $(T_1, T_2)$ be a commuting contractive pair such that it has a strongly minimal Ando lift. Then every minimal isometric lift of $(T_1, T_2)$ is strongly minimal and, consequently, any two minimal isometric lifts of $(T_1, T_2)$ are unitarily equivalent as lifts.*

*Proof* Let $(V_1, V_2)$ acting on $\mathcal{K}$ be a minimal isometric lift of $(T_1, T_2)$ via the isometric embedding $\Pi \colon \mathcal{H} \to \mathcal{K}$. Consider the space $\mathcal{K}_{00} = \bigvee_{n \geq 0} V_1^n V_2^n \Pi \mathcal{H}$. Let $\iota \colon \mathcal{K}_{00} \to \mathcal{K}$ be the embedding of $\mathcal{K}_{00}$ into $\mathcal{K}$. Consider the compression of the Ando lift $(V_1, V_2)$ to $\mathcal{K}_{00}$:

$$(V_{1,00}, V_{2,00}, V_{00}) = \iota^*(V_1, V_2, V_1V_2)\iota.$$

By Theorem 8.1.4, $(V_{1,00}, V_{2,00}, V_{00})$ is a pseudo-commuting contractive lift of $(T_1, T_2)$. Let $(W_1, W_2)$ be a strongly minimal Ando lift of $(T_1, T_2)$ acting on $\mathcal{W} = \bigvee_{n \geq 0} W_1^n W_2^n \mathcal{H}$. For simplicity, we take the embedding to be the inclusion map. It is routine to see that the triple $(W_1, W_2, W_1W_2)$ satisfies all the conditions for a pseudo-commuting contractive lift. By Corollary 8.2.3, the compression $(V_{1,00}, V_{2,00})$ is unitarily equivalent to the Ando lift $(W_1, W_2)$, which in turn implies that $(V_{1,00}, V_{2,00})$ must be an Ando lift of $(T_1, T_2)$. By minimality of the Ando lift $(V_1, V_2)$, we must have $\mathcal{K} = \mathcal{K}_{00}$, showing that $(V_1, V_2)$ is strongly minimal. Since strongly minimal Ando lifts are all pseudo-commuting contractive lifts, and since the latter class are all unique up to unitary equivalence, any two minimal isometric lifts of $(T_1, T_2)$ must be unitarily equivalent. □

## 8.6 Notes to Chapter 8

The notion of a pseudo-commuting contractive lift makes sense in the general companion-contraction setting introduced in the Appendix (Chapter A), in particular Sections A.1.2 and A.1.3. The examples explored here can be seen as formal generalizations of the present setting (commuting operator pairs having the bidisk $\mathbb{D}^2$ as a spectral set), but with the bidisk $\mathbb{D}^2 \subset \mathbb{C}^2$ replaced by a more general domain $X \subset \mathbb{C}^d$, specifically with $X$ equal to (i) the symmetrized bidisk, (ii) the tetrablock, (iii) the polydisk, and (iv) the symmetrized polydisk. In these examples it can happen that the the dilation theorem in the sense of Arveson (see Section A.1.1) fails in general, yet there is still a Sz.-Nagy–Foias functional model for $X$-contractive commuting operator tuples $(T_1, \ldots, T_d)$ based on the existence of a pseudo-commutative contractive lift. For further details and details on additional references, we invite the reader to consult the Appendix (Chapter A), specifically Section A.1.

# 9

# Functional Model and Invariants for Commuting Contractive Pairs

We have seen in the preceding chapters that there is a lack of uniqueness (up to unitary equivalence of lifts) in general for Ando lifts $(\mathbf{\Pi}, \mathbf{V}_1, \mathbf{V}_2)$ of a given contractive commuting operator pair $(T_1, T_2)$, but there is uniqueness for a pseudo-commuting contractive lift $(\Pi, \mathbb{W}_1, \mathbb{W}_2, \mathbb{W})$ of $(T_1, T_2, T = T_1T_2)$, which has embedded in it a minimal Sz.-Nagy–Foias isometric lift $(\Pi, V = \mathbb{W})$ of the single contraction operator $T$. When we use the Sz.-Nagy–Foias model version $(\Pi_{\rm NF}, V_{\rm NF})$ of the minimal isometric lift for $T$, expressed in terms of the characteristic function $\Theta_T$ for $T$, together with the Sz.-Nagy–Foias model $(\Pi_{\rm NF}, \mathbb{W}_{{\rm NF},1}, \mathbb{W}_{{\rm NF},2}, V_{\rm NF})$ for the pseudo-commuting contractive lift of $(T_1, T_2, T)$ as in (8.14), we arrive at a functional model

$$(T_1, T_2, T) \underset{\rm u}{\cong} P_{\mathcal{H}_{\Theta_T}} (\mathbb{W}_{{\rm NF},1}, \mathbb{W}_{{\rm NF},2}, V_{\rm NF})|_{\mathcal{H}_{\Theta_T}}$$

for the commuting triple $(T_1, T_2, T)$ itself. Conversely, there is a notion of an *admissible triple* $\Xi$ for the case where no commuting contractive pair $(T_1, T_2)$ is initially specified, from which one can build a commuting contractive operator pair $(T_{\Xi,1}, T_{\Xi,2})$ on a functional model space $\mathcal{H}_\Theta$, which in turn has a characteristic triple $\Xi_{T_{\Xi,1}, T_{\Xi,2}}$ that can be shown to coincide with the original admissible triple, giving a complete parallel with the Sz.-Nagy–Foias theory outlined in Remark 2.2.7 (where one now has *characteristic triple* in place of *characteristic function* and *admissible triple* in place of *purely contractive analytic function*). Perhaps as is to be expected, however, there are some compatibility conditions in the definition of admissible triple that may be difficult to check in practice. In the following sections we spell out the details. The first order of business is to understand precisely the notion of *completely nonunitary* for the commuting contractive operator pair case.

## 9.1 Characteristic Triples and Functional Models

The following definition makes precise the notion of *characteristic triple* for a commuting contractive operator pair $(T_1, T_2)$.

**Definition 9.1.1** Let $(T_1, T_2)$ be a commuting contractive operator pair on $\mathcal{H}$. Let us introduce the following objects.

(i) $(G_1, G_2)$ = the *fundamental-operator pair* for $(T_1^*, T_2^*)$ as in Definition 7.2.2.

(ii) $(W_{\sharp 1}, W_{\sharp 2})$ = the commuting unitary operator pair canonically associated with $(T_1, T_2)$ as in (5.74).

(iii) $\Theta_T$ = *the characteristic operator function* (2.50) for the product contraction operator $T = T_1T_2$. Here we set $\underline{T}$ to the operator triple $\underline{T} = (T_1, T_2, T = T_1T_2)$.

Then the triple $\Xi_{\underline{T}} := \left((G_1, G_2), \left(W_{\sharp 1}, W_{\sharp 2}\right), \Theta_T\right)$ is called the *characteristic triple* for $(T_1, T_2)$.

Note that the set of components $\left((G_1, G_2), \left(W_{\sharp 1}, W_{\sharp 2}\right), \Theta_T\right)$ is all the information needed to write down the Sz.-Nagy–Foias model pseudo-commuting contractive lift $(\Pi_{\rm NF}, W_{{\rm NF},1}, W_{{\rm NF},2}, V_{\rm NF})$ of $(T_1, T_2, T = T_1T_2)$ acting on the space $\mathcal{K}_{\Theta_T}$ as in Theorem 8.3.1.

We now present the bivariate analogue of the result discussed in Remark 2.2.7 and reviewed in the preceding paragraphs.

**Theorem 9.1.2** *Let $(T_1, T_2)$ be a commuting contractive operator pair and let its characteristic triple be*

$$\Xi_{\underline{T}} = ((G_1, G_2), \left(W_{\sharp 1}, W_{\sharp 2}\right), \Theta_T).$$

*Then the Sz.-Nagy–Foias model space*

$$\mathcal{H}_{\Theta_T} = \begin{bmatrix} H^2(\mathcal{D}_{T^*}) \\ \Delta_{\Theta_T} \cdot L^2(\mathcal{D}_T) \end{bmatrix} \ominus \begin{bmatrix} \Theta_T \\ \Delta_{\Theta_T} \end{bmatrix} \cdot H^2(\mathcal{D}_T) \tag{9.1}$$

*is coinvariant under*

$$\left(\begin{bmatrix} M_{G_1^*+zG_2} & 0 \\ 0 & W_{\sharp 1} \end{bmatrix}, \begin{bmatrix} M_{G_2^*+zG_1} & 0 \\ 0 & W_{\sharp 2} \end{bmatrix}, \begin{bmatrix} M_z & 0 \\ 0 & M_\zeta|_{\overline{\Delta_{\Theta_T}(L^2(\mathcal{D}_T))}} \end{bmatrix}\right),$$

*and $\underline{T} = (T_1, T_2, T_1T_2)$ is unitarily equivalent to*

$$P_{\mathcal{H}_{\Theta_T}} \left(\begin{bmatrix} M_{G_1^*+zG_2} & 0 \\ 0 & W_{\sharp 1} \end{bmatrix}, \begin{bmatrix} M_{G_2^*+zG_1} & 0 \\ 0 & W_{\sharp 2} \end{bmatrix}, \begin{bmatrix} M_z & 0 \\ 0 & M_\zeta|_{\overline{\Delta_{\Theta_T}(L^2(\mathcal{D}_T))}} \end{bmatrix}\right)\Bigg|_{\mathcal{H}_{\Theta_T}} \tag{9.2}$$

*via the unitary operator* $\Pi_{\rm NF,0}\colon \mathcal{H} \to \mathcal{H}_{\Theta_T}$ *given by*

$$\Pi_{\rm NF,0}\colon h \mapsto \Pi_{\rm NF} h \in \mathcal{H}_{\Theta_T} \quad \textit{for } h \in \mathcal{H}. \tag{9.3}$$

*Proof* Let us denote by $\underline{\mathbb{W}}_{\rm NF}$ the operator triple

$$\begin{aligned} \underline{\mathbb{W}}_{\rm NF} &:= (\mathbb{W}_{\rm NF,1}, \mathbb{W}_{\rm NF,2}, V_{\rm NF}) \\ &:= \left( \begin{bmatrix} M_{G_1^* + zG_2} & 0 \\ 0 & W_{\sharp 1} \end{bmatrix}, \begin{bmatrix} M_{G_2^* + zG_1} & 0 \\ 0 & W_{\sharp 2} \end{bmatrix}, \begin{bmatrix} M_z & 0 \\ 0 & M_\zeta|_{\overline{\Delta_{\Theta_T}(L^2(\mathcal{D}_T))}} \end{bmatrix} \right) \end{aligned} \tag{9.4}$$

acting on $\mathcal{K}_{\Theta_T} := \begin{bmatrix} H^2(\mathcal{D}_{T^*}) \\ \overline{\Delta_{\Theta_T} L^2(\mathcal{D}_T)} \end{bmatrix}$ and set $\Pi_{\rm NF} = \begin{bmatrix} \mathcal{O}_{D_{T^*}, T^*} \\ \omega_{\rm NF,D} \mathcal{Q}_{T^*} \end{bmatrix}$ equal to the Sz.-Nagy–Foias embedding operator from $\mathcal{H}$ into $\mathcal{K}_{\Theta_T}$ with range equal to $\mathcal{H}_{\Theta_T}$. By Theorem 8.3.1 we know that $(\Pi_{\rm NF}, \mathbb{W}_{\rm NF,1}, \mathbb{W}_{\rm NF,2}, V_{\rm NF})$ is a lift of $(T_1, T_2, T = T_1 T_2)$, that is,

$$\left( \mathbb{W}_{\rm NF,1}^*, \mathbb{W}_{\rm NF,2}^*, V_{\rm NF}^* \right) \Pi_{\rm NF} = \Pi_{\rm NF} \left( T_1^*, T_2^*, T^* = T_1^* T_2^* \right), \tag{9.5}$$

where (as noted in Remark 2.2.7) $\operatorname{Ran} \Pi_{\rm NF} = \mathcal{H}_{\Theta_T}$. This shows immediately that $\mathcal{H}_{\Theta_T}$ is coinvariant under $\underline{\mathbb{W}}_{\rm NF}$. Apply $\Pi_{\rm NF}^*$ on the left to both sides of (9.5), use that $\Pi_{\rm NF}$ is an isometry $\left(\Pi_{\rm NF}^* \Pi_{\rm NF} = I_{\mathcal{H}}\right)$, and then take adjoints to arrive at

$$\Pi_{\rm NF}^* (W_{\rm NF,1}, W_{\rm NF,2}, V_{\rm NF}) \Pi_{\rm NF} = (T_1, T_2, T_1 T_2).$$

Then use the connection (9.3) between $\Pi_{\rm NF}$ and $\Pi_{\rm NF,0}$ to reinterpret this last identity as

$$\Pi_{\rm NF,0}^* \left( P_{\mathcal{H}_{\Theta_T}} (W_{\rm NF,1}, W_{\rm NF,2}, V_{\rm NF})|_{\mathcal{H}_{\Theta_T}} \right) \Pi_{\rm NF,0} = (T_1, T_2, T_1 T_2).$$

This last equality is the statement that the model operator triple (9.2)

$$P_{\mathcal{H}_{\Theta_T}} (\mathbb{W}_{\rm NF,1}, \mathbb{W}_{\rm NF,2}, V_{\rm NF})|_{\mathcal{H}_{\Theta_T}}$$

is unitarily equivalent via $\Pi_{\rm NF,0}$ to the original contractive operator triple $(T_1, T_2, T_1 T_2)$ as claimed, and the theorem follows. □

## 9.2 Canonical Decomposition for Pairs of Commuting Contractions

Our eventual goal is to prove that characteristic triples form a complete unitary invariant for commuting contractive pairs $(T_1, T_2)$ with the condition that $T = T_1 T_2$ is a completely nonunitary (c.n.u.) contraction. The goal of this section is to argue that this c.n.u. assumption on $T = T_1 T_2$ can be discarded without any substantive loss of generality due to the existence of a *canonical decomposition* for any commuting contractive pair $(T_1, T_2)$. This is analogous to the single-variable phenomenon that every single contraction has a decomposition as the direct sum of a unitary operator and a c.n.u. contraction operator. We shall need the following lemma.

**Lemma 9.2.1** *Let $A$ be a bounded operator on a Hilbert space such that $\omega A$ has negative-semidefinite real part for all $\omega$ on the unit circle:*

$$\operatorname{Re}(\omega A) := \omega A + (\omega A)^* \preceq 0 \quad \textit{for all } \omega \in \mathbb{T}.$$

*Then $A = 0$.*

*Proof* The hypothesis means that the operator-valued function $R(\omega) := \omega A + \overline{\omega} A^*$ satisfies $R(\omega) \preceq 0$ for every $\omega \in \mathbb{T}$. Note that $R(-\omega) = -R(\omega)$ for every $\omega \in \mathbb{T}$ and hence

$$R(\omega) := \omega A + \overline{\omega} A^* = 0 \quad \text{for all } \omega \in \mathbb{T},$$

which readily implies that $A = \frac{1}{2}(R(1) - iR(i)) = 0$. □

**Theorem 9.2.2** *For every pair $(T_1, T_2)$ of commuting contractions on a Hilbert space $\mathcal{H}$ there corresponds a decomposition of $\mathcal{H}$ into the orthogonal sum of two subspaces reducing for both $T_1$ and $T_2$, say $\mathcal{H} = \mathcal{H}_u \oplus \mathcal{H}_c$, such that, with notation*

$$(T_{1u}, T_{2u}) = (T_1, T_2)|_{\mathcal{H}_u} \quad \textit{and} \quad (T_{1c}, T_{2c}) = (T_1, T_2)|_{\mathcal{H}_c}, \tag{9.6}$$

*we have that $T_u = T_{1u}T_{2u}$ is a unitary and $T_c = T_{1c}T_{2c}$ is a c.n.u. contraction. Moreover, then $T_u \oplus T_c$ with respect to $\mathcal{H} = \mathcal{H}_u \oplus \mathcal{H}_c$ is the Sz.-Nagy–Foias canonical decomposition for the contraction operator $T = T_1T_2$.*

*Proof* Let $(T_1, T_2)$ be a commuting contractive operator pair on a Hilbert space $\mathcal{H}$ such that $(F_1, F_2)$ is the fundamental-operator pair for $(T_1^*, T_2^*)$. By Definition 7.2.2 (combined with Lemma 7.2.1), on one hand $(F_1, F_2)$ is characterized as the unique solution of the pair of operator equations

$$T_i - T_j^* T = D_T F_{\sharp i} D_T, \quad \text{where } (i, j) = (1, 2) \text{ or } (2, 1), \tag{9.7}$$

but on the other hand, as a consequence of the second proof of Theorem 7.2.3, can also be expressed directly in terms of a Type I Ando tuple $(\mathcal{F}, \Lambda, P, U)$ for $(T_1^*, T_2^*)$ as

$$(F_1, F_2) = \Lambda^*(P^\perp U, U^* P)\Lambda. \tag{9.8}$$

Since both $F_1$ and $F_2$ are contractions, we have for every $\omega$ and $\zeta$ in $\mathbb{T}$

$$I_{\mathcal{D}_T} - \operatorname{Re}(\omega F_1) \succeq 0 \quad \text{and} \quad I_{\mathcal{D}_T} - \operatorname{Re}(\zeta F_2) \succeq 0. \tag{9.9}$$

Adding these two inequalities then gives

$$2I_{\mathcal{D}_T} - \operatorname{Re}(\omega F_1 + \zeta F_2) \succeq 0, \qquad \text{for all } \omega, \zeta \in \mathbb{T}. \tag{9.10}$$

As (9.10) is an operator inequality on the space $\mathcal{D}_T$, we may multiply by $D_T$ on the left and on the right to get an equivalent inequality, namely

$$2D_T^2 - \operatorname{Re}(\omega D_T F_1 D_T + \zeta D_T F_2 D_T) \succeq 0 \quad \text{for all } \omega, \zeta \in \mathbb{T}.$$

By (9.7) this is the same as

$$2D_T^2 - \operatorname{Re}\left(\omega\left(T_1 - T_2^* T\right)\right) - \operatorname{Re}\left(\zeta\left(T_2 - T_1^* T\right)\right) \succeq 0, \qquad \text{for all } \omega, \zeta \in \mathbb{T}. \tag{9.11}$$

Let

$$T = \begin{bmatrix} T_u & 0 \\ 0 & T_c \end{bmatrix} : \mathcal{H}_u \oplus \mathcal{H}_c \to \mathcal{H}_u \oplus \mathcal{H}_c \tag{9.12}$$

be the canonical decomposition of $T$ into unitary piece $T_u$ and completely nonunitary piece $T_c$. It remains to show that $T_1$ and $T_2$ are also block-diagonal with respect to this decomposition. To get started, we consider the $2 \times 2$-matrix representation of each $T_j$ with respect to the decomposition $\mathcal{H} = \mathcal{H}_u \oplus \mathcal{H}_c$:

$$T_j = \begin{bmatrix} A_j & B_j \\ C_j & D_j \end{bmatrix} : \mathcal{H}_u \oplus \mathcal{H}_c \to \mathcal{H}_u \oplus \mathcal{H}_c \quad \text{for } j = 1, 2. \tag{9.13}$$

Next, apply (9.11) to obtain that

$$\begin{aligned} \begin{bmatrix} 0 & 0 \\ 0 & 2D_{T_c}^2 \end{bmatrix} &- \operatorname{Re}\left(\omega \begin{bmatrix} A_1 - A_2^* T_u & B_1 - C_2^* T_c \\ C_1 - B_2^* T_u & D_1 - D_2^* T_c \end{bmatrix}\right) \\ &- \operatorname{Re}\left(\zeta \begin{bmatrix} A_2 - A_1^* T_u & B_2 - C_1^* T_c \\ C_2 - B_1^* T_u & D_2 - D_1^* T_c \end{bmatrix}\right) \succeq 0 \quad \text{for all } \omega, \zeta \in \mathbb{T}. \end{aligned} \tag{9.14}$$

In particular, the $(1, 1)$-entry in this inequality works out as

$$\mathbb{P}_{11}(\omega, \zeta) := \operatorname{Re}\left(\omega\left(A_1 - A_2^* T_u\right)\right) + \operatorname{Re}\left(\zeta\left(A_2 - A_1^* T_u\right)\right) \preceq 0, \quad \text{for all } \omega, \zeta \in \mathbb{T}. \tag{9.15}$$

This in turn implies that

$$\begin{aligned} \mathbb{P}_{11}(\omega, 1) + \mathbb{P}_{11}(\omega, -1) &= 2 \operatorname{Re}\left(\omega\left(A_1 - A_2^* T_u\right)\right) \preceq 0 \quad \text{and} \\ \mathbb{P}_{11}(1, \zeta) + \mathbb{P}_{11}(-1, \zeta) &= 2 \operatorname{Re}\left(\zeta\left(A_2 - A_1^* T_u\right)\right) \preceq 0. \end{aligned}$$

Now we apply Lemma 9.2.1 to conclude that

$$A_1 = A_2^* T_u, \quad A_2 = A_1^* T_u. \tag{9.16}$$

This shows that the $(1, 1)$-entry of the matrix on the left-hand side of (9.14) is zero. Since the matrix is positive-semidefinite, the $(1, 2)$-entry (and hence also the $(2, 1)$-entry) is also zero, that is, for all $\omega, \zeta \in \mathbb{T}$,

$$\begin{aligned}\mathbb{P}_{12}(\omega, \zeta) &:= \omega\left(B_1 - C_2^*T_c\right) + \bar{\omega}\left(C_1^* - T_u^*B_2\right) + \zeta\left(B_2 - C_1^*T_c\right) + \bar{\zeta}\left(C_2^* - T_u^*B_1\right)\\ &= 0.\end{aligned}$$

In particular, we then get that

$$\mathbb{P}(\omega) := \mathbb{P}_{12}(\omega, 1) + \mathbb{P}_{12}(\omega, -1) = 2\omega\left(B_1 - C_2^*T_c\right) + 2\bar{\omega}\left(C_1^* - T_u^*B_2\right) = 0$$

for every $\omega \in \mathbb{T}$. This implies the first two of the following equations while the last two are obtained similarly:

$$B_1 = C_2^*T_c, \quad C_1^* = T_u^*B_2, \quad B_2 = C_1^*T_c, \quad C_2^* = T_u^*B_1. \tag{9.17}$$

Now from the commutativity of $T_j$ with $T$ we have the following for $j = 1, 2$:

$$A_jT_u = T_uA_j, \quad B_jT_c = T_uB_j, \quad C_jT_u = T_cC_j, \quad T_cD_j = D_jT_c. \tag{9.18}$$

Let us note that commutativity of $(T_1, T_2)$ has been used in the beginning of the proof, namely, $F_1$ and $F_2$ are contractions because of the commutativity of $T_1$ and $T_2$.

For $(i, j) = (1, 2)$ or $(2, 1)$ we obtain by using the second and third equation in (9.17) and the third equation in (9.18) that

$$B_j^*T_u^2 = C_iT_u = T_cC_i = T_cB_j^*T_u, \tag{9.19}$$

which implies that $B_j^*T_u = T_cB_j^*$, for $j = 1, 2$. Using this and the second equality in (9.18) we obtain

$$T_cT_c^*B_j^* = T_cB_j^*T_u^* = B_j^* = B_j^*T_u^*T_u = T_c^*B_j^*T_u = T_c^*T_cB_j^*,$$

which implies that $T_c$ is unitary on $\overline{\operatorname{Ran}} B_j^*$, for every $j = 1, 2$. Since $T_c$ is completely nonunitary, $B_j = 0$ for each $j = 1, 2$. Similarly one can show that $C_j = 0$, for each $j = 1, 2$. This completes the proof. □

## 9.3 Characteristic Triple as a Complete Unitary Invariant

As already discussed in Remark 2.2.7, it was proved by Sz.-Nagy–Foias (see [92, Chapter VI]) that the characteristic function $\Theta_T$ for a c.n.u. contraction $T$ is a complete unitary invariant. This means that two c.n.u. contractions $T$ and $T'$ are unitarily equivalent if and only if their characteristic functions *coincide* in the sense that there exist unitary operators $u\colon \mathcal{D}_T \to \mathcal{D}_{T'}$ and $u_*\colon \mathcal{D}_{T^*} \to \mathcal{D}_{T'^*}$ such that the following diagram commutes for every $z \in \mathbb{D}$:

$$\begin{array}{ccc} \mathcal{D}_T & \xrightarrow{\Theta_T(z)} & \mathcal{D}_{T^*} \\ u\downarrow & & \downarrow u_* \\ \mathcal{D}_{T'} & \xrightarrow[\Theta_{T'}(z)]{} & \mathcal{D}_{T'^*} \end{array}. \tag{9.20}$$

Theorem 9.3.3 shows that such a result holds for characteristic triples of pairs of commuting contractions also. First, we define a notion of coincidence for such a triple.

Let us recall (see Remark 2.2.7 and also [92] for complete details) that a contractive analytic function $(\mathcal{D}, \mathcal{D}_*, \Theta)$ is a $\mathcal{B}(\mathcal{D}, \mathcal{D}_*)$-valued analytic function on $\mathbb{D}$ such that

$$\|\Theta(z)\| \le 1 \quad \text{for all } z \in \mathbb{D}.$$

Such a function is called *purely contractive* if $\Theta(0)$ does not preserve the norm of any nonzero vector, that is,

$$\|\Theta(0)\xi\|_{\mathcal{D}_*} < \|\xi\|_{\mathcal{D}} \quad \text{for all nonzero } \xi \in \mathcal{D}. \tag{9.21}$$

We note that a Sz.-Nagy–Foias characteristic function $\Theta_T$ is always purely contractive (see [92, Section VI.1]), and that it is always the case that a general contractive analytic function $(\mathcal{D}, \mathcal{D}_*, \Theta)$ has a block-diagonal decomposition $\Theta = \Theta' \oplus \Theta^0$, where $(\mathcal{D}', \mathcal{D}'_*, \Theta')$ is a unitary constant function and $(\mathcal{D}^0, \mathcal{D}^0_*, \Theta^0)$ is purely contractive. A key easily checked property of this decomposition is the following.

**Observation 9.3.1** The model space associated with a contractive analytic function $\Theta$ is defined to be

$$\mathcal{H}_\Theta := \begin{bmatrix} H^2(\mathcal{D}_*) \\ \overline{\Delta_\Theta L^2(\mathcal{D})} \end{bmatrix} \ominus \begin{bmatrix} \Theta \\ \Delta_\Theta \end{bmatrix} H^2(\mathcal{D}),$$

where here we set

$$\Delta_\Theta(\zeta) = (I_{\mathcal{D}} - \Theta(\zeta)^*\Theta(\zeta))^{\frac{1}{2}}.$$

We define the model operator associated with the contractive analytic function $\Theta$ to be

$$T_\Theta = P_{\mathcal{H}_\Theta} \begin{bmatrix} M_z & 0 \\ 0 & M_\zeta \end{bmatrix}\Big|_{\mathcal{H}_\Theta}.$$

The observation here is that the model operator $T_\Theta$ remains exactly the same (after some natural identification of respective coefficient spaces) when $\Theta$ is replaced by its purely contractive part $\Theta^0$. Thus only purely contractive analytic functions are relevant when discussing Sz.-Nagy–Foias functional models.

In view of Observation 9.3.1, for the moment we consider only purely contractive analytic functions.

**Definition 9.3.2** Let $(\mathcal{D}, \mathcal{D}_*, \Theta)$, $(\mathcal{D}', \mathcal{D}'_*, \Theta')$ be two purely contractive analytic functions, let $(G_1, G_2)$ on $\mathcal{D}_*$ and $(G'_1, G'_2)$ on $\mathcal{D}'_*$ be two pairs of contraction operators, and let $(W_1, W_2)$ on $\overline{\Delta_\Theta L^2(\mathcal{D})}$ and $(W'_1, W'_2)$ on $\overline{\Delta_{\Theta'} L^2(\mathcal{D}')}$ be two pairs of commuting unitary operators with each having a product equal to $M_\zeta$ on the respective space. We say that the two triples $((G_1, G_2), (W_1, W_2), \Theta)$ and $((G'_1, G'_2), (W'_1, W'_2), \Theta')$ *coincide* if:

(i) $(\mathcal{D}, \mathcal{D}_*, \Theta)$ and $(\mathcal{D}', \mathcal{D}'_*, \Theta')$ coincide, that is, there exist unitary operators $u\colon \mathcal{D} \to \mathcal{D}'$ and $u_*\colon \mathcal{D}_* \to \mathcal{D}'_*$ such that

$$u_*\Theta(z) = \Theta'(z)u \qquad \text{for all } z \in \mathbb{D},$$

that is, the diagram (9.20) commutes with $\Theta$ and $\Theta'$ in place of $\Theta_T$ and $\Theta_{T'}$, respectively;

(ii) the unitary operators $u, u_*$ also have the intertwining properties

$$\begin{cases} (G'_1, G'_2) = u_*(G_1, G_2)u_*^* = (u_*G_1u_*^*, u_*G_2u_*^*) \quad \text{and} \\ (W'_1, W'_2) = \omega_u(W_1, W_2)\omega_u^* = \left(\omega_u W_1\omega_u^*, \omega_u W_2\omega_u^*\right), \end{cases} \tag{9.22}$$

where $\omega_u\colon \overline{\Delta_\Theta L^2(\mathcal{D})} \to \overline{\Delta_{\Theta'} L^2(\mathcal{D}')}$ is the unitary map induced by $u$ defined by

$$\omega_u := (I_{L^2} \otimes u)\,|_{\overline{\Delta_\Theta L^2(\mathcal{D})}}. \tag{9.23}$$

**Theorem 9.3.3** *Let $(T_1, T_2)$ and $\left(T'_1, T'_2\right)$ be two pairs of commuting contractions such that their products $T = T_1T_2$ and $T' = T'_1T'_2$ are c.n.u. contractions. Then $(T_1, T_2)$ and $\left(T'_1, T'_2\right)$ are unitarily equivalent if and only if their characteristic triples coincide.*

*Proof* Let $(T_1, T_2)$ on $\mathcal{H}$ and $\left(T'_1, T'_2\right)$ on $\mathcal{H}'$ be unitarily equivalent via the unitary operator $U\colon \mathcal{H} \to \mathcal{H}'$. Let

$$((G_1, G_2), \left(W_{\sharp 1}, W_{\sharp 2}\right), \Theta_T), \quad \left((G'_1, G'_2), \left(W'_{\sharp 1}, W'_{\sharp 2}\right), \Theta_{T'}\right)$$

be their respective characteristic triples. It is easy to see that $UD_T = D_{T'}U$ and $UD_{T^*} = D_{T'^*}U$ and that the unitaries

$$u := U|_{\mathcal{D}_T}\colon \mathcal{D}_T \to \mathcal{D}_{T'} \quad \text{and} \quad u_* := U|_{\mathcal{D}_{T^*}}\colon \mathcal{D}_{T^*} \to \mathcal{D}_{T'^*} \tag{9.24}$$

have the following property:

$$u_*\Theta_T(z) = \Theta_{T'}(z)u \quad \text{for all } z \in \mathbb{D}. \tag{9.25}$$

Hence $\Theta_T$ and $\Theta_{T'}$ coincide.

We next show that the unitary $u_*$ also implements a unitary equivalence of $(G_1, G_2)$ with $(G_1', G_2')$ as follows. Note first that since by definition $(G_1, G_2)$ and $(G_1', G_2')$ are the fundamental-operator pairs for $(T_1^*, T_2^*)$ and $(T_1'^*, T_2'^*)$ respectively, by definition we have:

$$T_i^* - T_jT^* = D_{T^*}G_iD_{T^*}, \quad T_i'^* - T_j'T'^* = D_{T'^*}G_i'D_{T'^*} \quad \text{for } (i,j) = (1,2) \text{ or } (2,1).$$

It then follows that

$$u_*(G_1, G_2) = (G_1', G_2')\, u_*. \tag{9.26}$$

We have seen in Theorem 8.3.1 that $(\Pi_{\mathrm{NF}}, \underline{\mathbb{W}}_{\mathrm{NF}})$ is a pseudo-commuting contractive lift of $\underline{T} = (T_1, T_2, T_1T_2)$ acting on the space $\mathcal{K}_{\Theta_T}$. Let $(\Pi_{\mathrm{NF}}', \underline{\mathbb{W}'}_{\mathrm{NF}})$ be the corresponding pseudo-commuting contractive lift of $\underline{T'} = (T_1', T_2', T_1'T_2')$ acting on the space $\mathcal{K}_{\Theta_{T'}}$. Let $\Pi''$ denote the isometry

$$\Pi'' := \begin{bmatrix} I_{H^2}\otimes u_*^* & 0 \\ 0 & \omega_u^* \end{bmatrix} \Pi_{\mathrm{NF}}'U \colon \mathcal{H} \to \begin{bmatrix} H^2(\mathcal{D}_{T^*}) \\ \overline{\Delta_{\Theta_T}L^2(\mathcal{D}_T)} \end{bmatrix}, \tag{9.27}$$

where $\omega_u \colon \overline{\Delta_{\Theta_T}L^2(\mathcal{D}_T)} \to \overline{\Delta_{\Theta_{T'}}L^2(\mathcal{D}_{T'})}$ is the unitary $\omega_u = (I_{L^2}\otimes u)|_{\overline{\Delta_{\Theta_T}L^2(\mathcal{D}_T)}}$. We observe that $(\Pi'', \mathcal{K}_{\mathrm{NF}}, \underline{\mathbb{W}''})$ is also a pseudo-commuting contractive lift of $(T_1, T_2, T)$, where $(W_1'', W_2'') = \omega_u^*\left(W_{\sharp 1}', W_{\sharp 2}'\right)\omega_u$ and

$$\underline{\mathbb{W}''} := \left(\begin{bmatrix} M_{G_1^*+zG_2} & 0 \\ 0 & W_1'' \end{bmatrix}, \begin{bmatrix} M_{G_2^*+zG_1} & 0 \\ 0 & W_2'' \end{bmatrix}, \begin{bmatrix} M_z & 0 \\ 0 & M_\zeta|_{\overline{\Delta_{\Theta_T}(L^2(\mathcal{D}_T))}} \end{bmatrix}\right).$$

By making use of (9.26), we see that, for $(i,j) = (1,2)$ or $(2,1)$, we have

$$\begin{aligned}
\Pi''T_i^* &= \begin{bmatrix} I_{H^2}\otimes u_*^* & 0 \\ 0 & \omega_u^* \end{bmatrix} \Pi_{\mathrm{NF}}'T_i'^*U \quad \text{(by (9.27))} \\
&= \begin{bmatrix} I_{H^2}\otimes u_*^* & 0 \\ 0 & \omega_u^* \end{bmatrix} \begin{bmatrix} M^*_{G_i'^*+zG_j'} & 0 \\ 0 & W_{\sharp i}'^{\,*} \end{bmatrix} \Pi_{\mathrm{NF}}'U \\
&= \begin{bmatrix} M_{G_i^*+zG_j} & 0 \\ 0 & W_i''^* \end{bmatrix} \begin{bmatrix} I_{H^2}\otimes u_*^* & 0 \\ 0 & \omega_u^* \end{bmatrix} \Pi_{\mathrm{NF}}'U \quad \text{(by (9.26))}.
\end{aligned}$$

Now, since the last entry of $\underline{\mathbb{W}''}$ is the same as that of $\underline{\mathbb{W}}_{\mathrm{NF}}$, applying Corollary 8.2.3, we get $\underline{\mathbb{W}''} = \underline{\mathbb{W}}_{\mathrm{NF}}$ and we conclude that

$$(W_1'', W_2'') = \omega_u^*\left(W_{\sharp 1}', W_{\sharp 2}'\right)\omega_u = (W_{\sharp 1}, W_{\sharp 2}).$$

This, together with equations (9.25) and (9.26), establishes the first part of the theorem.

Conversely, let $((G_1,G_2),(W_{\sharp 1},W_{\sharp 2}),\Theta_T)$ and $\left((G_1',G_2'),\left(W_{\sharp 1}',W_{\sharp 2}'\right),\Theta_{T'}\right)$ be the characteristic triples of $(T_1,T_2)$ and $(T_1',T_2')$ respectively, and suppose the respective characteristic triples coincide. Thus there exist unitaries $u\colon \mathcal{D}_T \to \mathcal{D}_{T'}$ and $u_*\colon \mathcal{D}_{T^*} \to \mathcal{D}_{T'^*}$ such that part (i) and part (ii) in Definition 9.3.2 hold. Let $\omega_u$ be the unitary induced by $u$ as defined in (9.23). Then it is easy to see that the unitary

$$\begin{bmatrix} I_{H^2}\otimes u_* & 0 \\ 0 & \omega_u \end{bmatrix} : \begin{bmatrix} H^2(\mathcal{D}_{T^*}) \\ \overline{\Delta_{\Theta_T}L^2(\mathcal{D}_T)} \end{bmatrix} \to \begin{bmatrix} H^2(\mathcal{D}_{T'^*}) \\ \overline{\Delta_{\Theta_{T'}}L^2(\mathcal{D}_{T'})} \end{bmatrix} \tag{9.28}$$

intertwines

$$\underline{\mathbb{W}} = \left(\begin{bmatrix} M_{G_1^*+zG_2} & 0 \\ 0 & W_{\sharp 1} \end{bmatrix}, \begin{bmatrix} M_{G_2^*+zG_1} & 0 \\ 0 & W_{\sharp 2} \end{bmatrix}, \begin{bmatrix} M_z^{\mathcal{D}_{T^*}} & 0 \\ 0 & M_\zeta^{\mathcal{D}_T}|_{\overline{\Delta_{\Theta_T}(L^2(\mathcal{D}_T))}} \end{bmatrix}\right)$$

with

$$\underline{\mathbb{W}'} = \left(\begin{bmatrix} M_{G_1'^*+zG_2'} & 0 \\ 0 & W_{\sharp 1}' \end{bmatrix}, \begin{bmatrix} M_{G_2'^*+zG_1'} & 0 \\ 0 & W_{\sharp 2}' \end{bmatrix}, \begin{bmatrix} M_z^{\mathcal{D}_{T'^*}} & 0 \\ 0 & M_\zeta|_{\overline{\Delta_{\Theta_{T'}}(L^2(\mathcal{D}_{T'}))}} \end{bmatrix}\right).$$

Also, the unitary in (9.28) clearly takes the space $\begin{bmatrix} \Theta_T \\ \Delta_{\Theta_T} \end{bmatrix} H^2(\mathcal{D}_T)$ onto the space $\begin{bmatrix} \Theta_{T'} \\ \Delta_{\Theta_{T'}} \end{bmatrix} H^2(\mathcal{D}_{T'})$ and hence

$$\begin{bmatrix} H^2(\mathcal{D}_{T^*}) \\ \overline{\Delta_{\Theta_T}L^2(\mathcal{D}_T)} \end{bmatrix} \ominus \begin{bmatrix} \Theta_T \\ \Delta_{\Theta_T} \end{bmatrix} H^2(\mathcal{D}_T) \quad \text{onto} \begin{bmatrix} H^2(\mathcal{D}_{T'^*}) \\ \overline{\Delta_{\Theta_{T'}}L^2(\mathcal{D}_{T'})} \end{bmatrix} \ominus \begin{bmatrix} \Theta_{T'} \\ \Delta_{\Theta_{T'}} \end{bmatrix} H^2(\mathcal{D}_{T'}).$$

This implies that the functional models for $(T_1,T_2)$ and $(T_1',T_2')$ as in (9.2) are unitarily equivalent and hence by (9.2) the pairs $(T_1,T_2)$ and $(T_1',T_2')$ are unitarily equivalent also. □

## 9.4 Admissible Triples

In this section, we consider general contractive analytic functions $(\mathcal{D},\mathcal{D}_*,\Theta)$ and do not insist that $\Theta$ is also pure. We start with a contractive analytic function $(\mathcal{D},\mathcal{D}_*,\Theta)$, a commuting unitary operator pair $(W_1,W_2)$ on $\overline{\Delta_\Theta L^2(\mathcal{D})}$, and a pair of contraction operators $(G_1,G_2)$ on $\mathcal{D}_*$, and we seek to investigate when the triple $((G_1,G_2),(W_1,W_2),\Theta)$ gives rise to a commuting contractive operator pair $(T_1,T_2)$ such that $T=T_1T_2$ is completely nonunitary (c.n.u.).

If $((G_1,G_2),(W_1,W_2),\Theta)$ is equal to the characteristic triple $((G_1,G_2),(W_{\sharp 1},W_{\sharp 2}),\Theta_T)$ for a commuting contractive operator pair $(T_1,T_2)$, it is easy to check from the fact that $(\Pi_{\rm NF},\mathbb{W}_{{\rm NF},1},\mathbb{W}_{{\rm NF},2},V_{\rm NF})$ is a pseudo-commuting contractive lift of $(T_1,T_2)$ that the characteristic triple $((G_1,G_2),(W_1,W_2),\Theta)$

in particular satisfies the set of *admissibility conditions* listed in Definition 9.4.1 for the *admissibility conditions* for such a triple.

**Definition 9.4.1 Admissibility conditions** For $(i, j) = (1, 2)$ or $(2, 1)$ we have the following.

(i) $M_{G_i^*+zG_j} \oplus W_i$ is a contraction.
(ii) $W_1W_2 = W_2W_1 = M_\zeta|_{\overline{\Delta_\Theta L^2(\mathcal{D})}}$.
(iii) The space $\mathcal{Q}_\Theta := \{\Theta f \oplus \Delta_\Theta f : f \in H^2(\mathcal{D})\}$ is jointly invariant under $\left(M_{G_1^*+zG_2} \oplus W_1, M_{G_2^*+zG_1} \oplus W_2, M_z \oplus M_\zeta|_{\overline{\Delta_\Theta L^2(\mathcal{D})}}\right)$.
(iv) With $\mathcal{K}_\Theta := H^2(\mathcal{D}_*) \oplus \overline{\Delta_\Theta L^2(\mathcal{D})}$ and $\mathcal{H}_\Theta := \mathcal{K}_\Theta \ominus \mathcal{Q}_\Theta$ we have

$$\left(M^*_{G_i^*+zG_j} \oplus W_i{}^*\right)\left(M^*_{G_j^*+zG_i} \oplus W_j{}^*\right)\Big|_{\mathcal{H}_\Theta} = \left(M_z^* \oplus M_\zeta^*|_{\overline{\Delta_\Theta L^2(\mathcal{D})}}\right)\Big|_{\mathcal{H}_\Theta}.$$

In particular, since condition (iv) holds for both $(i, j) = (1, 2)$ and $(i, j) = (2, 1)$, we see that $T_1^* := \left(M^*_{G_1^*+zG_2} \oplus W_1{}^*\right)\Big|_{\mathcal{H}_\Theta}$ *commutes with* $T_2^* := \left(M^*_{G_2^*+zG_1} \oplus W_2{}^*\right)\Big|_{\mathcal{H}_\Theta}$.

This motivates the following definition.

**Definition 9.4.2** Let $(\mathcal{D}, \mathcal{D}_*, \Theta)$ be a contractive analytic function, let $(G_1, G_2)$ be a contractive operator pair on $\mathcal{D}_*$, and let $(W_1, W_2)$ be a pair of commuting unitaries on $\overline{\Delta_\Theta L^2(\mathcal{D})}$. We say that the triple $\Xi = ((G_1, G_2), (W_1, W_2), \Theta)$ is *admissible* if it satisfies the admissibility conditions (i)–(iv) in Definition 9.4.1. We then say that the triple

$$\begin{aligned}\underline{\mathbf{T}}_\Xi &:= (\mathbf{T}_1, \mathbf{T}_2, \mathbf{T}_1\mathbf{T}_2)_\Xi \\ &:= P_{\mathcal{H}_\Theta}\left(M_{G_1^*+zG_2} \oplus W_1, M_{G_2^*+zG_1} \oplus W_2, M_z \oplus M_\zeta|_{\overline{\Delta_\Theta(L^2(\mathcal{D}))}}\right)\Big|_{\mathcal{H}_\Theta}\end{aligned} \tag{9.29}$$

is the functional model associated with the admissible triple $((G_1, G_2), (W_1, W_2), \Theta)$.

Let us also say that the admissible triple $((G_1, G_2), (W_1, W_2), \Theta)$ is *pure* if its last component $\Theta$ is a purely contractive analytic function.

Then we have the following analogue of Observation 9.3.1 for the Sz.-Nagy–Foias model.

**Proposition 9.4.3** *We suppose that* $((G_1, G_2), (W_1, W_2), \Theta)$ *is an admissible triple and that* $\Theta$ *has a (possibly nontrivial) decomposition* $\Theta = \Theta' \oplus \Theta^0$ *with* $(\mathcal{D}', \mathcal{D}'_*, \Theta')$ *a unitary constant function and* $\Theta^0$ *a purely contractive analytic function. Then there is an admissible triple of the form* $\left((G_1^0, G_2^0), (W_1^0, W_2^0), \Theta^0\right)$ *so that the functional model for* $((G_1, G_2), (W_1, W_2), \Theta)$ *is unitarily equivalent to the functional model for* $\left((G_1^0, G_2^0), (W_1^0, W_2^0), \Theta^0\right)$.

We shall refer to $\left((G_1^0, G_2^0), (W_1^0, W_2^0), \Theta^0\right)$ as the *pure part* of the admissible triple $((G_1, G_2), (W_1, W_2), \Theta)$.

*Proof* We suppose that $((G_1, G_2), (W_1, W_2), \Theta)$ is an admissible triple and that $\Theta$ has a (possibly nontrivial) decomposition $\Theta = \Theta' \oplus \Theta^0$ with $(\mathcal{D}', \mathcal{D}'_*, \Theta')$ a unitary constant function and $(\mathcal{D}^0, \mathcal{D}^0_*, \Theta^0)$ a purely contractive analytic function. Let

$$\mathcal{H}_\Theta = H^2(\mathcal{D}_*) \oplus \overline{\Delta_\Theta L^2(\mathcal{D})} \ominus \left\{\Theta f \oplus \Delta_\Theta f : f \in H^2(\mathcal{D})\right\}$$

be the Sz.-Nagy–Foias functional model space associated with $\Theta$ (and hence also the functional model space associated with the admissible triple $((G_1, G_2), (W_1, W_2), \Theta)$), and let

$$\begin{aligned}(\mathbf{T}_1^*, \mathbf{T}_2^*, \mathbf{T}^*) \\ = \left(\left(M_{G_1^* + zG_2} \oplus W_1\right)^*, \left(M_{G_2^* + zG_1} \oplus W_2\right)^*, \left(M_z \oplus M_\zeta|_{\overline{\Delta_\Theta L^2(\mathcal{D})}}\right)^*\right)\Big|_{\mathcal{H}(\Theta)}\end{aligned}$$

be the associated functional-model triple of contraction operators (with $\mathbf{T} = \mathbf{T}_1\mathbf{T}_2$). As a result of [92, Theorem VI.3.1], we know that $\mathbf{T}$ is c.n.u. with characteristic function $\Theta_\mathbf{T}$ coinciding with $\Theta^0$. Thus the characteristic triple for $(\mathbf{T}_1^*, \mathbf{T}_2^*, \mathbf{T}^*)$ has the form

$$\widetilde{\Xi} := \left(\left(\widetilde{G}_1, \widetilde{G}_2\right), \left(\widetilde{W}_1, \widetilde{W}_2\right), \Theta_\mathbf{T}\right),$$

and by Theorem 9.1.2 it follows that $(\mathbf{T}_1, \mathbf{T}_2, \mathbf{T})$ is unitarily equivalent to the model operators associated with $\widetilde{\Xi}$. As already noted, $\Theta_\mathbf{T}$ coincides with $\Theta^0$; hence there are unitary operators $u : \mathcal{D}_T \to \mathcal{D}^0$ and $u_* : \mathcal{D}_{T^*} \to \mathcal{D}^0_*$ so that

$$\Theta^0(z)u = u_* \Theta_T(z) \qquad \text{for all } z \in \mathbb{D}.$$

Define operators $G_1^0$, $G_2^0$ on $\mathcal{D}^0_*$ and $W_1^0$, $W_2^0$ on $\overline{\Delta_{\Theta^0} L^2(\mathcal{D}^0)}$ by

$$G_i^0 = u_* \widetilde{G}_i u_*^*, \quad W_i^0 = (u \otimes I_{L^2}) \widetilde{W}_i (u^* \otimes I_{L^2})|_{\overline{\Delta_{\Theta^0} L^2(\mathcal{D}^0)}}$$

for $i = 1, 2$. Then by construction the triple

$$\Xi^0 = \left(\left(G_1^0, G_2^0\right), \left(W_1^0, W_2^0\right), \Theta^0\right)$$

coincides with $\widetilde{\Xi}$ and hence is also admissible. Then by Theorem 9.3.3 the commuting contractive pair $(\mathbf{T}_1, \mathbf{T}_2)$ is also unitarily equivalent to the functional-model commuting contractive pair associated with the admissible triple $\Xi^0$. This completes the proof of Proposition 9.4.3. □

**Proposition 9.4.4** *Let* $\Xi = ((G_1, G_2), (W_1, W_2), \Theta)$ *be a pure admissible triple and let* $\underline{T}_\Xi$ *be the functional model associated with* $\Xi$*. Then the model triple of operators*

$$\underline{\mathbb{W}} = \left(M_{G_1^*+zG_2} \oplus W_1, \quad M_{G_2^*+zG_1} \oplus W_2, \quad M_z \oplus M_\zeta|_{\overline{\Delta_\Theta(L^2(\mathcal{D}))}}\right)$$

*on $\mathcal{K}_\Theta$ is a pseudo-commuting contractive lift of $\underline{\mathbf{T}}_\Xi$ with the inclusion map $i\colon \mathcal{H}_\Theta \to \mathcal{K}_\Theta$ as the associated isometric embedding operator.*

*Proof* By the analysis of Case 1 in the proof of Theorem 8.2.1 we see that the content of the admissibility conditions (i) and (ii) in Definition 9.4.1 is that the triple $\underline{\mathbb{W}} = (\mathbb{W}_1, \mathbb{W}_2, \mathbb{W})$ given by

$$\mathbb{W}_1 = M_{G_1^*+zG_2} \oplus W_1, \quad \mathbb{W}_2 = M_{G_2^*+zG_1} \oplus W_2, \quad \mathbb{W} = M_z^{\mathcal{D}_*} \oplus M_\zeta|_{\overline{\Delta_\Theta L^2(\mathcal{D})}},$$

all acting on $H^2(\mathcal{D}_*) \oplus \overline{\Delta_\Theta L^2(\mathcal{D})}$, is the general form for a pseudo-commuting contractive triple with third component specified to be $\mathbb{W} = M_z^{\mathcal{D}_*} \oplus M_\zeta|_{\overline{\Delta_\Theta L^2(\mathcal{D})}}$. Thus, it follows that $\underline{\mathbb{W}}$ is a pseudo-commuting contractive triple. Condition (iii) in Definition 9.4.1 is equivalent to saying that $\mathcal{H}_\Theta := \mathcal{K}_\Theta \ominus \mathcal{Q}_\Theta$ is jointly invariant for the adjoint $\underline{\mathbb{W}}^* = \left(\mathbb{W}_1^*, \mathbb{W}_2^*, \mathbb{W}^*\right)$ and we can define $\underline{\mathbf{T}}^*$ on $\mathcal{H}(\Theta)$ by

$$\underline{\mathbf{T}}^* = \underline{\mathbb{W}}^*|_{\mathcal{H}(\Theta)} =: \left(\mathbf{T}_1^*, \mathbf{T}_2^*, \mathbf{T}^*\right). \tag{9.30}$$

As a consequence of admissibility condition (iv) in Definition 9.4.1 we see that $\mathbf{T}_1^*$ and $\mathbf{T}_2^*$ commute, and the product operator $\mathbf{T}_1^*\mathbf{T}_2^*$ is equal to $M_z^* \oplus M_\zeta^*|_{\overline{\Delta_\Theta L^2(\mathcal{D})}} = \mathbf{T}^*$, and we conclude that $\underline{\mathbf{T}}$ is equal to the functional model operator triple (9.29) associated with the admissible triple $\Xi$. Next note that equation (9.30) is just the statement that $(i, \underline{\mathbb{W}})$ is a lift of $\mathbf{T}_\Xi$ on $\mathcal{H}(\Theta)$, where the isometric embedding operator $i\colon \mathcal{H}_\Theta \to \mathcal{K}_\Theta$ is just the inclusion map. Finally, the fact that $(i, \mathbb{W})$ is a minimal lift of $\mathbf{T}$ is part of the assertion of Theorem VI.3.1 in [92]. We can now conclude that $(i, \mathbb{W}_1, \mathbb{W}_2, \mathbb{W})$ is a pseudo-commuting lift of $(\mathbf{T}_1, \mathbf{T}_2, \mathbf{T} = \mathbf{T}_1\mathbf{T}_2)$ in the sense of Definition 8.1.2. This completes the proof. □

For $\Theta$ a purely contractive analytic function, we have the following result.

**Theorem 9.4.5** *Let $(\mathcal{D}, \mathcal{D}_*, \Theta)$ be a purely contractive analytic function, let $(G_1, G_2)$ on $\mathcal{D}_*$ be a pair of contractions, and let $(W_1, W_2)$ on $\overline{\Delta_\Theta L^2(\mathcal{D})}$ be a pair of commuting unitaries such that their product $W_1 W_2$ is equal to $M_\zeta|_{\overline{\Delta_\Theta L^2(\mathcal{D})}}$. Then $((G_1, G_2), (W_1, W_2), \Theta)$ is admissible if and only if it coincides with the characteristic triple of some commuting contractive pair $(T_1, T_2)$ with product operator $T = T_1 T_2$ equal to a c.n.u. contraction. In fact, the triple $((G_1, G_2), (W_1, W_2), \Theta)$ coincides with the characteristic triple of its functional model as defined in (9.29).*

*Proof* We have already observed that the characteristic triple of a pair $(T_1, T_2)$ of commuting contractions with $T = T_1T_2$ being a c.n.u. contraction is indeed a pure admissible triple (since characteristic functions $\Theta_T$ are necessarily purely contractive analytic functions).

Conversely suppose that $\Xi = ((G_1, G_2), (W_1, W_2), \Theta)$ is a pure admissible triple. This means that the pair $(\mathbf{T}_1, \mathbf{T}_2)$ defined on

$$\mathcal{H}_\Theta := \left(H^2(\mathcal{D}_*) \oplus \overline{\Delta_\Theta L^2(\mathcal{D})}\right) \ominus \left\{\Theta f \oplus \Delta_\Theta f \colon f \in H^2(\mathcal{D})\right\}$$

by

$$(\mathbf{T}_1, \mathbf{T}_2) := P_{\mathcal{H}_\Theta}\left(M_{G_1^*+zG_2} \oplus W_1, M_{G_2^*+zG_1} \oplus W_2\right)\Big|_{\mathcal{H}_\Theta}$$

is a commuting pair of contractions with product operator given by

$$\mathbf{T} := \mathbf{T}_1\mathbf{T}_2 = P_{\mathcal{H}_\Theta}\left(M_z \oplus M_\zeta|_{\overline{\Delta_\Theta(L^2(\mathcal{D}))}}\right)\Big|_{\mathcal{H}_\Theta}. \tag{9.31}$$

By the Sz.-Nagy–Foias model theory for a single contraction operator $T$ (see [92, Theorem VI.3.1]), we conclude that $\mathbf{T}$ is a c.n.u. contraction. We claim that the triple $((G_1, G_2), (W_1, W_2), \Theta)$ coincides with the characteristic triple for $(\mathbf{T}_1, \mathbf{T}_2)$, which we assume to be $\left((G_1', G_2'), (W_1', W_2'), \Theta_{\mathbf{T}}\right)$. Since $\Theta$ is a purely contractive analytic function, by (9.31) and Theorem VI.3.1 in [92], we conclude that $\Theta$ coincides with $\Theta_{\mathbf{T}}$. By definition this means that there exist unitaries $u\colon \mathcal{D} \to \mathcal{D}_{\mathbf{T}}$ and $u_*\colon \mathcal{D}_* \to \mathcal{D}_{\mathbf{T}^*}$ such that $\Theta_{\mathbf{T}}u = u_*\Theta$. Then the unitary operator $u_* \oplus \omega_u$ takes the space $\mathcal{K}_\Theta := H^2(\mathcal{D}_*) \oplus \overline{\Delta_\Theta L^2(\mathcal{D})}$ onto the space $\mathcal{K}_{\Theta_{\mathbf{T}}} := H^2(\mathcal{D}_{\mathbf{T}^*}) \oplus \overline{\Delta_{\Theta_{\mathbf{T}}} L^2(\mathcal{D}_{\mathbf{T}})}$ and also

$$u_* \oplus \omega_u \colon \left\{\Theta f \oplus \Delta_\Theta f \colon f \in H^2(\mathcal{D})\right\} \underset{\text{onto}}{\to} \left\{\Theta_{\mathbf{T}} g \oplus \Delta_{\Theta_{\mathbf{T}}} g \colon g \in H^2(\mathcal{D}_{\mathbf{T}})\right\}.$$

We can therefore conclude that furthermore $u_* \oplus \omega_u$ maps $\mathcal{H}_\Theta$ onto $\mathcal{H}_{\Theta_{\mathbf{T}}}$, where

$$\begin{aligned}\mathcal{H}_\Theta &:= \mathcal{K}_\Theta \ominus \left\{\Theta f \oplus \Delta_\Theta f \colon f \in H^2(\mathcal{D})\right\},\\ \mathcal{H}_{\Theta_{\mathbf{T}}} &:= \mathcal{K}_{\Theta_{\mathbf{T}}} \ominus \left\{\Theta_{\mathbf{T}} g + \Delta_{\Theta_{\mathbf{T}}} g \colon g \in H^2(\mathcal{D}_{\mathbf{T}}\right\}.\end{aligned}$$

Denote by $\tau$ the restriction of $u_* \oplus \omega_u$ to $\mathcal{H}_\Theta$. Then we have the following commuting diagram, where $i$ and $i'$ are the inclusion maps:

$$\begin{array}{ccc} \mathcal{H}_\Theta & \xrightarrow{\ i\ } & \mathcal{K}_\Theta \\ \tau\big\downarrow & & \big\downarrow u_*\oplus\omega_u. \\ \mathcal{H}_{\mathrm{NF}} & \xrightarrow[i']{} & \mathcal{K}_{\mathrm{NF}} \end{array}$$

By Proposition 9.4.4, we know that $(i, \underline{\mathbb{W}})$ is a pseudo-commuting contractive lift of $\underline{\mathbf{T}}_\Xi$ on $\mathcal{K}_\Theta$, where

$$\underline{\mathbb{W}} = \left(M_{G_1^*+zG_2} \oplus W_1, M_{G_2^*+zG_1} \oplus W_2, M_z \oplus M_\zeta|_{\overline{\Delta_\Theta(L^2(\mathcal{D}))}}\right).$$

Theorem 8.3.1 and the diagram above shows that $(i' \circ \tau, \underline{\mathbb{W}'})$ is also a pseudo-commuting contractive lift of $\underline{\mathbf{T}}_{\Xi}$, on $\mathcal{K}_{\Theta_{\mathbf{T}}}$, where

$$\underline{\mathbb{W}'} = \left( M_{G_1'^* + zG_2'} \oplus W_1', M_{G_2'^* + zG_1'} \oplus W_2', M_z \oplus M_\zeta|_{\overline{\Delta_{\mathbf{T}}(L^2(\mathcal{D}_{\mathbf{T}}))}} \right).$$

Now by the uniqueness result in Corollary 8.2.3, there exists a unitary $U: \mathcal{K}_\Theta \to \mathcal{K}_{\Theta_{\mathbf{T}}}$ such that $U\underline{\mathbb{W}} = \underline{\mathbb{W}'}U$ and $U \circ i = i' \circ \tau$. Since the last entries of $\underline{\mathbb{W}}$ and $\underline{\mathbb{W}'}$ are the minimal isometric dilations of $T = T_1T_2$, such a unitary is, in fact, unique. By the earlier commuting diagram we see that $u_* \oplus \omega_u$ is one such unitary. Therefore we get

$$(u_* \oplus \omega_u)\underline{W} = \underline{W'}(u_* \oplus \omega_u).$$

Consequently $((G_1, G_2), (W_1, W_2), \Theta)$ coincides with $\big((G_1', G_2'), (W_1', W_2'), \Theta_{\mathbf{T}}\big)$ and the theorem follows. □

Let us mention that the results of this and Section 9.3 can be stated more succinctly in the language of category theory, as follows.

**Proposition 9.4.6** *Define the following categories.*

*(i) Let $\mathfrak{C}_1$ be the category of all commuting pairs of contraction operators $\mathbf{T} = (T_1, T_2)$, where we set $T = T_1 \cdot T_2 = T_2 \cdot T_1$ and we assume that $T$ is c.n.u.*

*(ii) Let $\mathfrak{C}_2$ be the category of all purely contractive admissible triples $\Xi = ((G_1, G_2), (W_1, W_2), \Theta)$.*

*Define functors $\mathfrak{f}: \mathfrak{C}_1 \to \mathfrak{C}_2$ and $\mathfrak{g}: \mathfrak{C}_2 \to \mathfrak{C}_1$ by*

$\mathfrak{f}: \mathbf{T} \mapsto \Xi_{\mathbf{T}} =$ *characteristic triple for* $\mathbf{T}$,

$\mathfrak{g}: \Xi \mapsto \mathbf{T}_\Xi =$ *functional-model commuting contractive pair associated with $\Xi$ as in* (9.29).

*Then, for $\mathbf{T}, \mathbf{T}' \in \mathfrak{C}_1$ and $\Xi, \Xi' \in \mathfrak{C}_2$, we have*

(i) $\mathbf{T} \underset{\mathrm{u}}{\cong} \mathbf{T}' \Leftrightarrow \mathfrak{f}(\mathbf{T}) \underset{\mathrm{c}}{\cong} \mathfrak{f}(\mathbf{T}')$,

(ii) $\Xi \underset{\mathrm{c}}{\cong} \Xi' \Leftrightarrow \mathfrak{g}(\Xi) \underset{\mathrm{u}}{\cong} \mathfrak{g}(\Xi')$,

(iii) $\mathfrak{g} \circ \mathfrak{f}(\mathbf{T}) \underset{\mathrm{u}}{\cong} \mathbf{T}$,

(iv) $\mathfrak{f} \circ \mathfrak{g}(\Xi) \underset{\mathrm{c}}{\cong} (\Xi)$,

*where $\underset{\mathrm{u}}{\cong}$ denotes* unitary equivalence of operator tuples *and $\underset{\mathrm{c}}{\cong}$ denotes* coincidence of admissible triples.

Finally, we now show how it is possible to apply the theory of admissible triples to the following factorization problem.

**Commuting contractive factorization problem** *Given a c.n.u. contraction operator* $T$*, find all commuting contractive operator pairs* $(T_1, T_2)$ *which generate a commuting contractive factorization of* $T$: $T = T_1T_2 = T_2T_1$.

A closely related problem can be formulated as follows.

**Admissible-triple completion problem** *Given a pure contractive analytic function* $(\mathcal{D}, \mathcal{D}_*, \Theta)$*, find all pairs* $((G_1, G_2), (W_1, W_2))$*, where*

(i) $(G_1, G_2)$ *is a contractive operator pair on* $\mathcal{D}_*$,
(ii) $(W_1, W_2)$ *is a unitary operator pair on* $\overline{\Delta_\Theta L^2(\mathcal{D})}$,

*so that* $((G_1, G_2), (W_1, W_2), \Theta)$ *is an admissible triple.*

Then we have the following result.

**Corollary 9.4.7** *Let* $T$ *be a c.n.u. contraction operator on* $\mathcal{H}$*. Then solutions* $(T_1, T_2)$ *of the commuting contractive factorization problem for* $T$ *are in one-to-one correspondence with solutions of the admissible-triple completion problem for* $\Theta_T$*, that is, solutions of the commuting contractive factorization problem for* $T$ *are in one-to-one correspondence with pairs* $((G_1, G_2), \ (W_1, W_2))$ *of operators on* $\mathcal{D}_{T^*}$ *and on* $\overline{\Delta_{\Theta_T} L^2(\mathcal{D})}$ *respectively, such that* $((G_1, G_2), (W_1, W_2), \Theta_T)$ *is an admissible triple.*

*Proof* If $(T_1, T_2)$ is a solution of the commuting contractive factorization problem for $T$, then from the definitions we see that the characteristic triple of $(T_1, T_2)$ solves the admissible-triple completion problem for $\Theta_T$.

Conversely, suppose that $((G_1, G_2), (W_1, W_2))$ solves the admissible-triple completion problem for $\Theta_T$. Then the model operator triple $\underline{\mathbf{T}}_\Xi$ for

$$\Xi = ((G_1, G_2), (W_1, W_2), \Theta_T)$$

(see (9.29)) generates a commuting contractive operator pair $(\mathbf{T}_1, \mathbf{T}_2)$ with product $\mathbf{T}_1\mathbf{T}_2 = \mathbf{T}_2\mathbf{T}_1$ equal to

$$\mathbf{T} := P_{\mathcal{H}(\Theta_T)} \left( M_z^{\mathcal{D}_{T^*}} \oplus M_\zeta |_{\overline{\Delta_{\Theta_T} L^2(\mathcal{D}_T)}} \right) |_{\mathcal{H}(\Theta_T)},$$

which is the Sz.-Nagy–Foias functional-model operator for $T$, and hence is unitarily equivalent to $T$. □

**Example 9.4.8** As an illustrative example of the previous result, let $(\mathbb{C}^L, \Theta)$ be a rational $L \times L$-matrix inner function; thus $\Theta$ has a Blaschke–Potapov product representation (see [77])

$$\Theta(z) = \prod_{j=1}^{J} U_j \begin{bmatrix} \frac{\lambda_j - z}{1-\overline{\lambda_j} z} I_{p'_j} & 0 \\ 0 & I_{p_j} \end{bmatrix} U_j^*, \tag{9.32}$$

where $\lambda_j \in \mathbb{D}$, $U_j$ are $L \times L$-unitary matrices and nonnegative integers $p_j, p'_j$ are such that $p_j + p'_j = L$ for every $j = 1, 2, \ldots, J$. Then the model space $\mathfrak{H}(\Theta)$ collapses to $\mathfrak{H}(\Theta) = H^2(\mathbb{C}^L) \ominus \Theta H^2(\mathbb{C}^L)$ and is finite-dimensional. As a mildly simplifying assumption, let us suppose that $\mathfrak{H}(\Theta)$ has a basis consisting of vector-valued kernel functions

$$\mathfrak{B} = \bigvee \{d_{m,n} k_{w_m} : 1 \le m \le M,\ 1 \le n \le n_m\}. \tag{9.33}$$

The relation between the representation (9.32) for $\Theta$ versus the condition that $\mathfrak{B}$ (9.33) be a basis for $H^2(\mathbb{C}^L) \ominus \Theta \cdot H^2(\mathbb{C}^L)$ is not easily tractable. Let us point out that there are various realization formulas $\Theta(z) = A + zB(I - zD)^{-1}C$ for the rational $L \times L$-matrix function $\Theta$ (uniquely determined up to a right unitary constant factor) so that $\mathfrak{H}(\Theta)$ has $\mathfrak{B}$ (9.33) as a basis (see [17, Theorem 7.4.2 and 7.6.3]). As we shall not have need of such formulas, we shall now proceed with only making use of the assumption that $\mathfrak{H}(\Theta) = \bigvee \mathfrak{B}$. For ease of notation, in the following discussion let us set $\mathbb{C}^L = \mathcal{D}$.

Since $\Theta$ is inner, the second component $(W_1, W_2)$ of any admissible triple solving the admissible-triple completion problem for $\Theta$ is vacuous. So any solution of the admissible-triple completion problem for $\Theta$ consists simply of two matrices $G_1, G_2$ considered as operators on $\mathcal{D} = \mathbb{C}^L$. Given such a pair of matrices, let us define two matrix pencils:

$$\varphi_1(z) = G_1^* + zG_2, \quad \varphi_2(z) = G_2^* + zG_1. \tag{9.34}$$

Then $(G_1, G_2)$ solves the admissible-triple completion problem for $\Theta$ if and only if:

(i) the multiplication operators $M_{\varphi_1}$ and $M_{\varphi_2}$ are contractions on $H^2(\mathcal{D})$, that is, $\|\varphi_i(z)\| \le 1$ for all $z \in \mathbb{D}$ and $i = 1, 2$;
(ii) the space $\mathcal{H}_\Theta$ is jointly invariant for $\left(M_{\varphi_1}^*, M_{\varphi_2}^*\right)$;
(iii) $M_{\varphi_1}^* M_{\varphi_2}^*|_{\mathcal{H}_\Theta} = M_{\varphi_2}^* M_{\varphi_1}^*|_{\mathcal{H}_\Theta} = \left(M_z^{\mathcal{D}}\right)^*|_{\mathcal{H}_\Theta}$.

This reduction is in a sense already a solution of the commuting contractive factorization problem, as formulated next.

**Corollary 9.4.9** *Suppose that $T$ is a pure contraction, so the characteristic function of $T$ is a pure inner function $\Theta_T$. Suppose that $T = T_1T_2 = T_2T_1$ is a commutative contractive factorization of $T$. Let $\iota = O_{D_{T^*},T^*} : \mathcal{H} \to H^2(\mathcal{D}_{T^*}) \ominus \Theta_T H^2(\mathcal{D}_{T^*})$ be the unitary map from $\mathcal{H}$ onto the Sz.-Nagy–Foias model space $\mathcal{H}_{\Theta_T} = H^2(\mathcal{D}_{T^*}) \ominus \Theta_T \cdot H^2(\mathcal{D}_{T^*})$. Then there are operators $G_1$ and $G_2$ on*

*$\mathcal{D}_{T^*}$ so that, if we let $\varphi_1(z)$ and $\varphi_2(z)$ be the operator pencils as in* (9.34), *$(M_{\varphi_1}, M_{\varphi_2})$ satisfy conditions (i), (ii), and (iii) above, and*

$$T_i = \iota^* M_{\varphi_i} \iota \quad \textit{for } i = 1, 2.$$

This is essentially the solution of the commutative factorization problem presented by Das–Sarkar–Sarkar in [48]; our result contains more precise information on the coupled form of the two pencils $\varphi_1(z)$ and $\varphi_2(z)$. Nevertheless, this paper was an inspiration in the initial stages of putting together the present book.

We next seek more information on the existence of a pair $(G_1, G_2)$ for some particular cases. Note that if we can find a solution $(G_1, G_2)$ of conditions (ii) and (iii), we get a solution of condition (i) simply by a rescaling $(G_1, G_2) \mapsto (\mu G_1, \mu G_2)$ for a scalar $\mu > 0$ sufficiently small, so it suffices to consider conditions (ii) and (iii).

Let us note the action of $M_{\varphi_i}^*$ on a general kernel function:

$$M_{\varphi_i}^* \colon d\, k_w \mapsto \left(G_i + \overline{w} G_j^*\right) d\, k_w.$$

Let us introduce for $1 \le m \le M$ the subspace

$$\mathcal{S}_m = \bigvee \{d_{m,n} \colon 1 \le n \le n_m\}.$$

Then condition (ii) amounts to the collection of invariant-subspace conditions:

(ii′) $G_i + \overline{w}_m G_j^* \colon \mathcal{S}_m \to \mathcal{S}_m$ for $1 \le m \le M$ for $(i, j) = (1, 2)$ and $(2, 1)$.

Let us note next that for a general kernel function $d\, k_w$ we have, for $(i, j) = (1, 2)$ or $(2, 1)$,

$$\begin{aligned} M_{\varphi_i}^* M_{\varphi_j}^* (d\, k_w) &= M_{\varphi_i}^* (G_j + \overline{w} G_i^*) d k_w \\ &= G_i (G_j + \overline{w} G_i^*) d k_w + \overline{w} G_j^* (G_j + \overline{w} G_i^*) d k_w \\ &= \left(G_i G_j + \overline{w} \left(G_i G_i^* + G_j^* G_j\right) + \overline{w}^2 G_j^* G_i^*\right) d k_w \end{aligned}$$

and hence condition (iii) boils down to the following condition.

(iii′) For $1 \le m \le M$, for all $d \in \mathcal{S}_m$ we have

$$\begin{aligned} &\left(G_1 G_2 + \overline{w}_m \left(G_1 G_1^* + G_2^* G_2\right) + \overline{w}_m^2 G_2^* G_1^*\right) d \\ &= \left(G_2 G_1 + \overline{w}_m \left(G_2 G_2^* + G_1^* G_1\right) + \overline{w}_m^2 G_1^* G_2^*\right) d \\ &= \overline{w}_m d. \end{aligned}$$

It appears to be difficult to sort out how to solve these conditions (ii′) and (iii′) in general. Therefore we resort to discussing a couple of more tractable special cases where a complete solution can be found.

**Special case 1 $\mathcal{D}_* = \mathbb{C}$, $M > 1$ and** $w_1 = 0$ In this case, condition (ii′) is automatic as $\mathcal{S}_m$ is the whole space $\mathbb{C}$ and any solution $(G_1, G_2)$ of the admissible-triple completion problem has the form $(G_1, G_2) = (g_1, g_2)$, where $g_1, g_2$ are complex numbers. We shall show the following.

**Claim** *The only solution of the admissible-triple completion problem for this special case is that either* $(g_1, g_2)$ *or* $(g_2, g_1)$ *is in the set* $\mathbb{T} \times \{0\}$.

Indeed, note that condition (ii′) now assumes the form

$$g_1g_2 + \overline{w}_m \left(|g_1|^2 + |g_2|^2\right) + \overline{w}_m^2 \overline{g}_2\overline{g}_1 = g_2g_1 + \overline{w} \left(|g_2|^2 + |g_1|^2\right) + \overline{w}_m^2 \overline{g_1g_2} = \overline{w}_m \tag{9.35}$$

for $m = 1, \ldots, M$. In particular, if we apply criterion (9.35) to the case $m = 1$, where by hypothesis $w_1 = 0$, we get

$$g_1g_2 = 0$$

and hence at least one of $g_1$ and $g_2$ is equal to 0. If both $g_1 = g_2 = 0$ and we apply (9.35) to the case where $w_m \neq 0$, we get $0 = \overline{w}_m$, leading to a contradiction. Hence exactly one of $g_1$, $g_2$ is zero and the other is nonzero; say $g_1 \neq 0$ and $g_2 = 0$. Apply (9.35) again to the case of any $\alpha$ where $w_m \neq 0$ to get

$$\overline{w}_m |g_1|^2 = \overline{w}_m,$$

leading to the conclusion that $|g_1| = 1$. Similarly, if we assume that $g_1 = 0$, we are forced to the solution $(g_1, g_2) = (0, \omega)$ for some $\omega \in \mathbb{T}$. Conversely, one can easily check that $(g_1, g_2) = (\omega, 0)$ or $(g_1, g_2) = (0, \omega)$ for some unimodular $\omega \in \mathbb{T}$ leads to a solution of (9.35) for all $m = 1, \ldots, M$, and the claim follows.

As a corollary, we get a solution of the associated commuting contractive factorization problem: *Suppose that $T$ is a contraction operator on a finite-dimensional Hilbert space $\mathcal{H}$ with* $\dim \mathcal{H} \geq 2$ *having defect rank* $\operatorname{rank}(I - TT^*)$ *equal to* 1 *and having distinct simple eigenvalues all in the open unit disk, one of which is* 0, *and suppose that $T_1$ and $T_2$ are commuting contractive operators on $\mathcal{H}$ such that $T = T_1T_2$. Then there is unimodular $\omega \in \mathbb{T}$ so that*

$$\text{either } (T_1, T_2) = (\omega T, \overline{\omega} I) \quad \text{or} \quad (T_1, T_2) = (\overline{\omega} I, \omega T).$$

**Special case 2 $\Theta = (\{0\}, \mathcal{D}_*, 0)$** We consider next the limiting special case where we replace the finite point-set $(w_1, \ldots, w_M)$ with the whole unit disk $\mathbb{D}$, and the associated coefficient space $\mathcal{S}_m$ (now indexed as $\mathcal{S}_{w_m}$) with the whole coefficient space $\mathcal{D}_*$. Then the span of the kernels $\{dk_w : d \in \mathcal{D}_*, w \in \mathbb{D}\}$ is the whole space $H^2(\mathcal{D}_*)$, which can be viewed as $\mathcal{H}_\Theta$, where $\Theta$ is the zero function from $\{0\}$ into $\mathcal{D}_*$ (and hence can now be viewed as a nonsquare inner function). Then we seek a solution of the admissible-triple completion problem for the case where we have the nonsquare inner function $(\{0\}, \mathcal{D}_*, \Theta = 0)$. Note again that the third component $(W_1, W_2)$ of an admissible triple solving the admissible-triple completion problem for the case $\Theta = (\{0\}, \mathcal{D}_*, 0)$ is vacuous, since in this case the input space is trivial: $\mathcal{D} = 0$. Thus we wish to find an operator pair $(G_1, G_2)$ on the coefficient Hilbert space $\mathcal{D}_*$ so that conditions (i), (ii), and (iii) hold on $\mathcal{H}_\Theta$, which is now equal to all of $H^2(\mathcal{D}_*)$. As the kernel function $\{dk_w : d \in \mathcal{D}_*, w \in \mathbb{D}\}$ has span equal to all of $H^2(\mathcal{D}_*) = \mathcal{H}_\Theta$, we can again formulate the problem just as was done above, but now with the point-set $\{w_1, w_2, \dots\}$ equal to all of $\mathbb{D}$ and with the associated vector spaces $\mathcal{S}_w$ equal to the whole space $\mathcal{D}_*$ for each $w \in \mathbb{D}$. This leads us to conditions (i), (ii), and (iii) having to hold. Again, condition (i) can be handled by a rescaling of $(G_1, G_2)$, and conditions (ii) and (iii) become conditions (ii′) and (iii′). Condition (ii′) is again automatic since $\mathcal{S}_w = \mathcal{D}_*$ for all $w \in \mathbb{D}$. Then condition (iii′) must hold for all $w_m \in \mathbb{D}$. This leads to the system of quadratic operator polynomial equations in the variable $\overline{w} \in \mathbb{D}$:

$$\begin{aligned} &G_1G_2 + \overline{w}\left(G_1G_1^* + G_2^*G_2\right) + \overline{w}^2 G_2^*G_1^* \\ &= G_2G_1 + \overline{w}\left(G_2G_2^* + G_1^*G_1\right) + \overline{w}^2 G_1^*G_2^* \\ &= \overline{w} I_{\mathcal{D}_*}. \end{aligned}$$

As this must hold for all $w \in \mathbb{D}$, we are led to the system of operator equations

$$\begin{aligned} &G_1G_2 = G_2G_1 = 0, \\ &G_1G_1^* + G_2^*G_2 = G_2G_2^* + G_1^*G_1 = I_{\mathcal{D}_*}, \\ &G_2^*G_1^* = G_1^*G_2^* = 0. \end{aligned}$$

Note that the third line of these equations is just the adjoint version of the first line and hence can be dropped. Next observe that this slimmed-down system of operator equations is exactly the same as the equations in (3.4), (3.5), and (3.6) (with $(G_1, G_2)$ replacing the $(E_1, E_2)$ appearing there). By Lemma 3.1.1 we know that any solution has the form

$$(G_1, G_2) = (U^*P^\perp, PU),$$

where $U$ is a unitary operator and $P$ is a projection operator on $\mathcal{D}_*$. As a corollary we get what we can call the Berger–Coburn–Lebow solution of the associated commuting contractive factorization problem: If $(T_1, T_2)$ is a commuting contractive operator pair on $H^2(\mathcal{D}_*)$ solving the factorization problem

$$M_z^{\mathcal{D}_*} = T_1T_2 = T_2T_1,$$

then there is a projection $P$ and unitary $U$ on $\mathcal{D}_*$ so that

$$T_1 = M_{P^\perp U + zPU}, \quad T_2 = M_{U^*P + zU^*P^\perp}.$$

In the special case where $\mathcal{D}_* = \mathbb{C}$, the possibilities for projection operators $P$ and unitary operators $U$ are limited to

$$P = 0 \text{ or } P = 1, \quad U = \omega \in \mathbb{T}.$$

Then solutions of the associated commuting contractive factorization problem find a commuting contractive operator pair $(T_1, T_2)$ on $H^2$ so that $M_z = T_1T_2 = T_2T_2$ has the same trivial form as was the case for Special Case 1 discussed earlier: There is an $\omega \in \mathbb{T}$ so that

$$(T_1, T_2) = (\omega I_{H^2}, \overline{\omega} M_z) \quad \text{or} \quad (T_1, T_2) = (\omega M_z, \overline{\omega} I_{H^2}).$$

## 9.5 Application: Characterization of Joint Invariant Subspaces

In this section we characterize invariant subspaces for pairs $(T_1, T_2)$ of commuting contractions such that $T = T_1T_2$ is a c.n.u. contraction. Sz.-Nagy and Foias characterized how invariant subspaces for c.n.u. contractions arise in the functional model (see [92, Chapter VII]). They showed that invariant subspaces of a c.n.u. contraction $T$ are in one-to-one correspondence with regular factorizations of the characteristic function of $T$. Recall that, given contractive analytic functions $(\mathcal{D}, \mathcal{D}_*, \Theta)$, $(\mathcal{D}, \mathcal{F}, \Theta')$, and $(\mathcal{F}, \mathcal{D}_*, \Theta'')$ such that there is a factorization, $\Theta = \Theta''\Theta'$, the factorization is said to be *regular* if the contractive operator factorization $\Theta(\zeta) = \Theta''(\zeta)\Theta'(\zeta)$ is regular in the sense discussed in Remark 5.3.3. This means that the map $Z\colon \overline{\Delta_\Theta L^2(\mathcal{D})} \to \overline{\Delta_{\Theta''} L^2(\mathcal{F})} \oplus \overline{\Delta_{\Theta'} L^2(\mathcal{D})}$ defined densely by

$$Z\colon \Delta_\Theta(\zeta)h(\zeta) \mapsto \Delta_{\Theta''}(\zeta)\Theta'(\zeta)h(\zeta) \oplus \Delta_{\Theta'}(\zeta)h(\zeta), \tag{9.36}$$

which is necessarily an isometry (the pointwise version of the map (9.36) given by equation (5.24), is actually surjective and hence unitary. A minor complication in the theory is that the factors in a regular factorization of a purely contractive analytic function need not again be purely contractive. We now recall their result as we shall have use of it later in this chapter.

**Theorem 9.5.1** (Sz.-Nagy–Foias) *Let $(\mathcal{D}, \mathcal{D}_*, \Theta)$ be a purely contractive analytic function and let* $\mathbf{T}$ *be the contraction on*

$$\mathbb{H} := \mathcal{H}_\Theta = H^2(\mathcal{D}_*) \oplus \overline{\Delta_\Theta L^2(\mathcal{D})} \ominus \{\Theta f \oplus \Delta_\Theta f : f \in H^2(\mathcal{D})\}$$

*defined by*

$$\mathbf{T} = P_{\mathbb{H}}\big(M_z \oplus M_\zeta\big)|_{\mathbb{H}}.$$

*A subspace* $\mathbb{H}'$ *of* $\mathbb{H}$ *is invariant under* $\mathbf{T}$ *if and only if there exist contractive analytic functions* $(\mathcal{D}, \mathcal{F}, \Theta')$, $(\mathcal{F}, \mathcal{D}_*, \Theta'')$ *such that*

$$\Theta = \Theta''\Theta'$$

*is a regular factorization, and with the unitary* $Z$ *as in* (9.36) *we have*

$$\mathbb{H}' = \left\{\Theta'' f \oplus Z^{-1}(\Delta_{\Theta''} f \oplus g) \colon\ f \in H^2(\mathcal{F}), g \in \overline{\Delta_{\Theta'} L^2(\mathcal{D})}\right\} \ominus \left\{\Theta h \oplus \Delta_\Theta h \colon\ h \in H^2(\mathcal{D})\right\} \tag{9.37}$$

*and*

$$\mathbb{H}'' := \mathbb{H} \ominus \mathbb{H}' = H^2(\mathcal{D}_*) \oplus Z^{-1}\left(\overline{\Delta_{\Theta''} L^2(\mathcal{F})} \oplus \{0\}\right) \ominus \left\{\Theta'' f \oplus Z^{-1}(\Delta_{\Theta''} f \oplus 0) \colon\ f \in H^2(\mathcal{F})\right\}. \tag{9.38}$$

*Moreover, the characteristic function of* $\mathbf{T}|_{\mathbb{H}'}$ *coincides with the purely contractive part of* $\Theta'$*, and the characteristic function of* $P_{\mathbb{H}''}\mathbf{T}|_{\mathbb{H}''}$ *coincides with the purely contractive part of* $\Theta''$.

**Remark 9.5.2** For future reference, let us mention the special case of Theorem 9.5.1, where the input coefficient space $\mathcal{D}$ is taken to be the zero space and the contractive analytic function $\Theta$ is just the zero function $\Theta(z) = 0 \colon \{0\} \to \mathcal{D}_*$. In this case the only contractive analytic factorizations $\Theta = \Theta''\Theta'$ have $\Theta''$ equal to an arbitrary contractive analytic function $(\mathcal{F}, \mathcal{D}_*, \Theta'')$ and $\Theta'$ equal to a zero function $(\{0\}, \mathcal{F}, \Theta')$ $(\Theta'(z) = 0 \colon \{0\} \to \mathcal{F})$. The only such factorizations that are regular are those for which $\Theta''$ is an inner function. In this case, in the context of Theorem 9.5.1, $\mathbb{H} = H^2(\mathcal{D}_*)$, $\mathbf{T} = M_z$ on $H^2(\mathcal{D}_*)$, and the invariant subspace corresponding to the regular factorization $0 = \Theta'' \cdot 0$ is $\mathbb{H}' = \Theta'' H^2(\mathcal{F})$. Thus Theorem 9.5.1 can be thought of as a generalization of the Beurling–Lax theorem.

Let $T$ be a c.n.u. contraction such that $T = T_1 T_2$ for a pair $(T_1, T_2)$ of commuting contractions. It is natural that one would need more conditions than (9.37) and (9.38) for an invariant subspace of $T$ to be jointly invariant under $(T_1, T_2)$.

**Theorem 9.5.3** *Let* $(\mathcal{D}, \mathcal{D}_*, \Theta)$ *be a purely contractive analytic function and let the triple* $((G_1, G_2), (W_1, W_2), \Theta)$ *be admissible. Define the pair* $(\mathbf{T_1}, \mathbf{T_2})$ *of commuting contractions on*

$$\mathbb{H} = H^2(\mathcal{D}_*) \oplus \overline{\Delta_\Theta L^2(\mathcal{D})} \ominus \left\{\Theta f \oplus \Delta_\Theta f \colon\ f \in H^2(\mathcal{D})\right\} \tag{9.39}$$

*by*

$$(\mathbf{T_1}, \mathbf{T_2}) = P_{\mathbb{H}} \left( M_{G_1^* + zG_2} \oplus W_1, M_{G_2^* + zG_1} \oplus W_2 \right)\Big|_{\mathbb{H}}. \tag{9.40}$$

*A subspace* $\mathbb{H}'$ *of* $\mathbb{H}$ *is jointly invariant under* $(\mathbf{T_1}, \mathbf{T_2})$ *if and only if there exist a Hilbert space* $\mathcal{F}$*, two contractions* $G_1'$*,* $G_2'$ *in* $\mathcal{B}(\mathcal{F})$*, two contractive analytic functions* $(\mathcal{D}, \mathcal{F}, \Theta')$*,* $(\mathcal{F}, \mathcal{D}_*, \Theta'')$ *such that*

$$\Theta = \Theta''\Theta'$$

*is a regular factorization, a pair* $(W_1', W_2')$ *of unitary operators on* $\overline{\Delta_{\Theta'} L^2(\mathcal{D})}$ *with the property*

$$W_1'W_2' = W_2'W_1' = M_\zeta|_{\overline{\Delta_{\Theta'} L^2(\mathcal{D})}}, \tag{9.41}$$

*and also, with* $Z$ *the pointwise unitary operator as in (9.36),*

$$\begin{aligned} \mathbb{H}' = &\left\{ \Theta'' f \oplus Z^{-1}(\Delta_{\Theta''} f \oplus g) \colon\ f \in H^2(\mathcal{F}), g \in \overline{\Delta_{\Theta'} L^2(\mathcal{D})} \right\} \\ &\ominus \left\{ \Theta h \oplus \Delta_\Theta h \colon\ h \in H^2(\mathcal{D}) \right\}, \end{aligned} \tag{9.42}$$

$$\begin{aligned} \mathbb{H}'' := \mathbb{H} \ominus \mathbb{H}' = &H^2(\mathcal{D}_*) \oplus Z^{-1} \left( \overline{\Delta_{\Theta''} L^2(\mathcal{F})} \oplus \{0\} \right) \\ &\ominus \left\{ \Theta'' f \oplus Z^{-1}(\Delta_{\Theta''} f \oplus 0) \colon\ f \in H^2(\mathcal{F}) \right\}, \end{aligned} \tag{9.43}$$

*and for every* $f \in H^2(\mathcal{F})$ *and* $g \in \overline{\Delta_{\Theta'} L^2(\mathcal{D})}$*,*

$$\begin{bmatrix} M_{G_i^* + zG_j} & 0 \\ 0 & W_i \end{bmatrix} \begin{bmatrix} \Theta'' f \\ Z^{-1}(\Delta_{\Theta''} f \oplus g) \end{bmatrix} = \begin{bmatrix} \Theta'' M_{G_i' + zG_j'} f \\ Z^{-1} \left( \Delta_{\Theta''} M_{G_i' + zG_j'} f \oplus W_i' g \right) \end{bmatrix}, \tag{9.44}$$

*where* $(i, j) = (1, 2), (2, 1)$.

*Proof* We first prove the easier part – the proof of sufficiency. Suppose that

$$((G_1, G_2), (W_1, W_2), \Theta)$$

is a purely contractive admissible triple (i.e., $(\mathcal{D}, \mathcal{D}_*, \Theta)$ is a purely contractive analytic function) such that $\Theta$ has a regular factorization $\Theta = \Theta''\Theta'$ with $(\mathcal{D}, \mathcal{F}, \Theta')$ and $(\mathcal{F}, \mathcal{D}_*, \Theta'')$ contractive analytic functions. We suppose also that $G_1'$ and $G_2'$ are contraction operators on $\mathcal{F}$, and $W_1'$, $W_2'$ are unitary operators on $\overline{\Delta_{\Theta'} L^2(\mathcal{D})}$ so that (9.41) and (9.44) hold. Then we have all the ingredients to define $\mathbb{H}'$ and $\mathbb{H}''$ as in (9.42) and (9.43). Note next that $\mathbb{H}'$ is indeed a subspace of $\mathbb{H}$. We wish to show that the space $\mathbb{H}'$ given in (9.42) is jointly invariant under the pair $(\mathbf{T_1}, \mathbf{T_2})$ defined in (9.40). Firstly, it is easy to see that $\mathbb{H}'$ is a subspace of $\mathbb{H}$. Since the operator

$$\mathcal{I}: H^2(\mathcal{F}) \oplus \overline{\Delta_{\Theta'} L^2(\mathcal{D})} \to H^2(\mathcal{D}_*) \oplus \overline{\Delta_{\Theta} L^2(\mathcal{D})}, \qquad f \oplus g \mapsto \Theta'' f \oplus Z^{-1}(\Delta_{\Theta''} f \oplus g) \tag{9.45}$$

is an isometry, the space

$$\left\{\Theta'' f \oplus Z^{-1}(\Delta_{\Theta''} f \oplus g) : f \in H^2(\mathcal{F}) \text{ and } g \in \overline{\Delta_{\Theta'} L^2(\mathcal{D})}\right\}$$

is closed and by (9.44) we see that it is jointly invariant under

$$\left(M_{G_1^* + zG_2} \oplus W_1, M_{G_2^* + zG_1} \oplus W_2, M_z \oplus M_\zeta\right).$$

We also see that

$$\text{Ran}\, \mathcal{I} = \left(H^2(\mathcal{D}_*) \oplus \overline{\Delta_\Theta L^2(\mathcal{D})}\right) \ominus (\mathbb{H} \ominus \mathbb{H}').$$

Now the sufficiency follows from the definition of $(\mathbf{T_1}, \mathbf{T_2})$ and from the general fact that if $V$ is an operator on $\mathcal{K}$ containing $\mathcal{H}$, $V(\mathcal{K} \ominus \mathcal{H}) \subset \mathcal{K} \ominus \mathcal{H}$, and $V^*|_{\mathcal{H}} = T^*$, then for a subspace $\mathcal{H}'$ of $\mathcal{H}$,

$$V(\mathcal{K} \ominus (\mathcal{H} \ominus \mathcal{H}')) \subseteq (\mathcal{K} \ominus (\mathcal{H} \ominus \mathcal{H}')) \quad \text{if and only if } T(\mathcal{H}') \subseteq \mathcal{H}'.$$

Now we show that the conditions are necessary. The first step of the proof is an application of Theorem 9.5.1. Indeed, if $\mathbb{H}' \subset \mathbb{H}$ is jointly invariant under $(\mathbf{T_1}, \mathbf{T_2})$, then it is also invariant under the product $\mathbf{T_1 T_2}$ and by definition of admissibility,

$$\mathbf{T} = \mathbf{T_1 T_2} = \mathbf{T_2 T_1} = P_{\mathbb{H}}(M_z \oplus M_\zeta)|_{\mathbb{H}}.$$

Hence, by Theorem 9.5.1, there exist two contractive analytic functions

$$(\mathcal{D}, \mathcal{F}, \Theta'), \quad (\mathcal{F}, \mathcal{D}_*, \Theta'')$$

such that $\Theta = \Theta''\Theta'$ is a regular factorization and the spaces $\mathbb{H}'$ and $\mathbb{H}''$ are realized as in (9.42) and (9.43), respectively. It only remains to produce contraction operators $G_1'$, $G_2'$ on $\mathcal{F}$ and unitary operators $W_1'$, $W_2'$ on $\overline{\Delta_{\Theta'} L^2(\mathcal{D})}$ so that conditions (9.41) and (9.44) hold. Note that, once we have found $G_1'$, $G_2'$, $W_1'$, and $W_2'$, verification of (9.44) breaks up into three linear pieces, where $(i, j) = (1, 2)$ or $(2, 1)$:

$$M_{G_i^* + zG_j} \Theta'' f = \Theta'' M_{G_i' + zG_j'} f \quad \text{for all } f \in H^2(\mathcal{F}), \tag{9.46}$$

$$W_i Z^{-1}(\Delta_{\Theta''} f \oplus 0) = Z^{-1}\left(\Delta_{\Theta''} M_{G_i' + zG_j'} f \oplus 0\right) \quad \text{for all } f \in H^2(\mathcal{F}), \tag{9.47}$$

$$W_i Z^{-1}(0 \oplus g) = Z^{-1}(0 \oplus W_i' g) \quad \text{for all } g \in \overline{\Delta_{\Theta'} L^2(\mathcal{D})}. \tag{9.48}$$

As a first step, we define operators $X_i$ on $H^2(\mathcal{F})$ and $W_i'$ on $\overline{\Delta_{\Theta'}L^2(\mathcal{D})}$, for $i = 1, 2$, such that for every $f \in H^2(\mathcal{F})$ and $g \in \overline{\Delta_{\Theta'}L^2(\mathcal{D})}$,

$$\begin{aligned}\begin{bmatrix} M_{G_i^*+zG_j} & 0 \\ 0 & W_i \end{bmatrix} \mathcal{I} \begin{bmatrix} f \\ g \end{bmatrix} &= \begin{bmatrix} M_{G_i^*+zG_j} & 0 \\ 0 & W_i \end{bmatrix} \begin{bmatrix} \Theta'' f \\ Z^{-1}(\Delta_{\Theta''} f \oplus g) \end{bmatrix} \\ &= \begin{bmatrix} \Theta'' X_i f \\ Z^{-1}\left(\Delta_{\Theta''} X_i f \oplus W_i' g\right) \end{bmatrix} = \mathcal{I} \begin{bmatrix} X_i & 0 \\ 0 & W_i' \end{bmatrix} \begin{bmatrix} f \\ g \end{bmatrix},\end{aligned} \tag{9.49}$$

where $\mathcal{I}$ is the isometry as defined in (9.45). The operators $X_1, X_2$ and $W_1', W_2'$ are well defined because the operator $\mathcal{I}$ is an isometry. Indeed, it follows that $X_1, X_2$ and $W_1', W_2'$ are contractions. Since the unitary $Z$ commutes with $M_\zeta$, it is easy to see from the definition of $\mathcal{I}$ that it has the following intertwining property

$$\mathcal{I}\left(M_z \oplus M_\zeta|_{\overline{\Delta_{\Theta'}L^2(\mathcal{D})}}\right) = \left(M_z \oplus M_\zeta|_{\overline{\Delta_{\Theta}L^2(\mathcal{D})}}\right)\mathcal{I}. \tag{9.50}$$

From the intertwining properties (9.50) and (9.49) of $\mathcal{I}$, we get for $i = 1, 2$,

$$(X_i \oplus W_i')\left(M_z \oplus M_\zeta|_{\overline{\Delta_{\Theta'}L^2(\mathcal{D})}}\right) = \left(M_z \oplus M_\zeta|_{\overline{\Delta_{\Theta'}L^2(\mathcal{D})}}\right)(X_i \oplus W_i'),$$

which implies that $(X_1, X_2) = \left(M_{\varphi_1}, M_{\varphi_2}\right)$, for some $\varphi_1$ and $\varphi_2$ in $L^\infty(\mathcal{B}(\mathcal{F}))$. We next show that $\varphi_1$ and $\varphi_2$ are actually linear pencils. Toward this end, notice from (9.49) that

$$\begin{bmatrix} M_{\varphi_1} & 0 \\ 0 & W_1' \end{bmatrix} = \mathcal{I}^* \begin{bmatrix} M_{G_1^*+zG_2} & 0 \\ 0 & W_1 \end{bmatrix} \mathcal{I}, \tag{9.51}$$

$$\begin{bmatrix} M_{\varphi_2} & 0 \\ 0 & W_2' \end{bmatrix} = \mathcal{I}^* \begin{bmatrix} M_{G_2^*+zG_1} & 0 \\ 0 & W_2 \end{bmatrix} \mathcal{I}. \tag{9.52}$$

Now multiplying (9.51) on the left by $M_z^* \oplus M_\zeta^*|_{\overline{\Delta_{\Theta'}L^2(\mathcal{D})}}$, then using the intertwining property (9.50) of $\mathcal{I}$, and then remembering that $(W_1, W_2)$ is a commuting pair of unitaries such that $W_1 W_2 = M_\zeta^*|_{\overline{\Delta_\Theta L^2(\mathcal{D})}}$, we get

$$\begin{bmatrix} M_z^* & 0 \\ 0 & M_\zeta^*|_{\overline{\Delta_{\Theta'}L^2(\mathcal{D})}} \end{bmatrix} \begin{bmatrix} M_{\varphi_1} & 0 \\ 0 & W_1' \end{bmatrix} = \mathcal{I}^* \begin{bmatrix} M^*_{G_2^*+zG_1} & 0 \\ 0 & W_2^* \end{bmatrix} \mathcal{I} = \begin{bmatrix} M^*_{\varphi_2} & 0 \\ 0 & W_2'^* \end{bmatrix}.$$

Consequently, $M_{\varphi_2} = M^*_{\varphi_1} M_z$. A similar argument as above yields $M_{\varphi_1} = M^*_{\varphi_2} M_z$. Considering these two relations and the power-series expansions of $\varphi_1$ and $\varphi_2$, we get

$$\varphi_1(z) = G_1'^* + zG_2' \quad \text{and} \quad \varphi_2(z) = G_2'^* + zG_1' \tag{9.53}$$

for some $G_1', G_2' \in \mathcal{B}(\mathcal{F})$. The fact that $M_{\varphi_1}$ (and $M_{\varphi_2}$) is a contraction implies that $G_1'$ and $G_2'$ are contractions too. Recalling (9.49) and the substitution

$(X_1, X_2) = (M_{\varphi_1}, M_{\varphi_2})$, where $\varphi_1$ and $\varphi_2$ are given by (9.53), we see that we have established (9.46) with the choice of $G'_1$, $G'_2$ as in (9.53).

Next note that the bottom component of (9.49) gives us

$$W_i Z^{-1}(\Delta_{\Theta''} f \oplus g) = Z^{-1}\left(\Delta_{\Theta''} X_i f \oplus W'_i g\right) \tag{9.54}$$

for all $f \in H^2(\mathcal{F})$, $g \in \overline{\Delta_{\Theta'} L^2(\mathcal{D})}$, and $i = 1, 2$. In particular, setting $g = 0$ and recalling that $X_i = M_{\varphi_i} = M_{G_i'^* + zG'_j}$, we get

$$W_i Z^{-1}(\Delta_{\Theta''} f \oplus 0) = Z^{-1}\left(\Delta_{\Theta''} M_{G_i'^* + zG'_j} f \oplus 0\right), \tag{9.55}$$

thereby verifying (9.47). We next consider (9.54) with $f = 0$ and $g$ equal to a general element of $\overline{\Delta_{\Theta'} L^2(\mathcal{D})}$ to get

$$W_i Z^{-1}(0 \oplus g) = Z^{-1}\left(0 \oplus W'_i g\right), \tag{9.56}$$

thereby verifying (9.48) and hence also completing the proof of (9.44).

It remains to show that $\left(W'_1, W'_2\right)$ is a commuting pair of unitary operators satisfying condition (9.41). Toward this goal, let us rewrite (9.56) in the form

$$Z W_i Z^{-1}(0 \oplus g) = 0 \oplus W'_i g, \tag{9.57}$$

which implies that, for $i = 1, 2$,

$$Z W_i Z^{-1}\left(\{0\} \oplus \overline{\Delta_{\Theta'} L^2(\mathcal{D})}\right) \subseteq \left(\{0\} \oplus \overline{\Delta_{\Theta'} L^2(\mathcal{D})}\right).$$

On the other hand, using (9.55) and noting that $M_\zeta|_{\overline{\Delta_\Theta L^2(\mathcal{D})}}$ commutes with $Z, W_1, W_2$, and $\Delta_{\Theta''}$, we get for every $f \in H^2(\mathcal{F})$ and $n \geq 0$,

$$Z W_i Z^{-1}\left(\Delta_{\Theta''} e^{-int} f \oplus 0\right) = \left(\Delta_{\Theta''} e^{-int} X_i f \oplus 0\right),$$

which implies that $Z W_i Z^{-1}\left(\overline{\Delta_\Theta L^2(\mathcal{D})} \oplus \{0\}\right) \subseteq \left(\overline{\Delta_{\Theta'} L^2(\mathcal{D})} \oplus \{0\}\right)$, for $i = 1, 2$. We conclude that $Z^{-1}\left(\{0\} \oplus \overline{\Delta_{\Theta'} L^2(\mathcal{D})}\right)$ is a reducing subspace for the pair of unitaries $(W_1, W_2)$ and hence $(W_1, W_2)|_{Z^{-1}(\{0\} \oplus \overline{\Delta_{\Theta'} L^2(\mathcal{D})})}$ is a pair of commuting unitary operators. The intertwining (9.57) shows that the pair $\left(W'_1, W'_2\right)$ on $\overline{\Delta_{\Delta'} L^2(\mathcal{D})}$ is jointly unitarily equivalent to the commuting unitary pair $(W_1, W_2)|_{Z^{-1}(\{0\} \oplus \overline{\Delta'_\Theta} L^2(\mathcal{D}))}$ and hence is itself a commuting unitary pair. Furthermore, since $W_1 W_2 = M_\zeta$ in particular on $Z^{-1}\left(\{0\} \oplus \overline{\Delta_{\Theta'} L^2(\mathcal{S})}\right)$ and $M_\zeta$ commutes past $Z$ and $Z^{-1}$, we conclude that condition (9.41) holds as well. This completes the proof of the necessary part. □

As we see from the last part of the statement of Theorem 9.5.3, Sz.-Nagy and Foias went on to prove that, under the conditions of Theorem 9.5.1, the characteristic functions of $\mathbf{T}|_{\mathbb{H}'}$ and $P_{\mathbb{H} \ominus \mathbb{H}'} \mathbf{T}|_{\mathbb{H} \ominus \mathbb{H}'}$ coincide with the purely contractive parts of $\Theta'$ and $\Theta''$, respectively. We next find an analogous result (at least for the first part of this statement) for pairs of commuting contractions. The strategy of the proof is the same as that of Sz.-Nagy–Foias, namely: application of model theory.

**Theorem 9.5.4** *Under the conditions of Theorem 9.5.3, let $\mathbb{H}'$ be a joint invariant subspace of $\mathbb{H}$ induced by the regular factorization $\Theta = \Theta''\Theta'$. Then with the notations as in Theorem 9.5.3, the triple $((G_1', G_2'), (W_1', W_2'), \Theta')$ is admissible and its purely contractive part coincides with the characteristic triple for $(\mathbf{T}_1, \mathbf{T}_2)|_{\mathbb{H}'}$.*

*Proof* With the isometry $\mathcal{I}$ as in (9.45), define a unitary $U := \mathcal{I}^*|_{\mathrm{Ran}\,\mathcal{I}}$. Therefore

$$U\colon \left\{\Theta'' f \oplus Z^{-1}(\Delta_{\Theta''} f \oplus g) : f \in H^2(\mathcal{F}),\ g \in \overline{\Delta_{\Theta'} L^2(\mathcal{D})}\right\}$$
$$\to H^2(\mathcal{F}) \oplus \overline{\Delta_{\Theta'} L^2(\mathcal{D})},$$
$$U\colon \Theta'' f \oplus Z^{-1}(\Delta_{\Theta''} f \oplus g) \mapsto f \oplus g. \tag{9.58}$$

For every $g \in H^2(\mathcal{D})$,

$$U(\Theta g \oplus \Delta_\Theta g) = U(\Theta''\Theta' g \oplus Z^{-1}(\Delta_{\Theta''}\Theta' g \oplus \Delta_{\Theta'} g)) = \Theta' g \oplus \Delta_{\Theta'} g, \tag{9.59}$$

which implies that $U$ takes $\mathbb{H}'$ as given in (9.42) onto the Hilbert space

$$\mathfrak{H}' := H^2(\mathcal{F}) \oplus \overline{\Delta_{\Theta'} L^2(\mathcal{D})} \ominus \left\{\Theta' g \oplus \Delta_{\Theta'} g : g \in H^2(\mathcal{D})\right\}. \tag{9.60}$$

The basis of the proof is the following unitary equivalences:

$$U\left(M_{G_i^* + zG_j} \oplus W_i\right) U^* = M_{G_i'^* + zG_j'} \oplus W_i' \quad \text{for } (i, j) = (1, 2), (2, 1), \tag{9.61}$$
$$U(M_z \oplus M_\zeta)U^* = (M_z \oplus M_\zeta). \tag{9.62}$$

To verify (9.61) and (9.62), proceed as follows. Since $M_\zeta$ commutes with $Z$ and $\Delta_{\Theta''}$, (9.62) follows easily. We establish equation (9.61) only for $(i, j) = (1, 2)$ and omit the proof for the other case because it is similar. For $f \in H^2(\mathcal{F})$ and $g \in \overline{\Delta_{\Theta'} L^2(\mathcal{D})}$,

$$U\left(M_{G_1^* + zG_2} \oplus W_1\right)U^*(f \oplus g) = U\left(M_{G_1^* + zG_2} \oplus W_1\right)\left(\Theta'' f \oplus Z^{-1}(\Delta_{\Theta''} f \oplus g)\right)$$
$$= U\left(\Theta'' M_{G_1'^* + zG_2'} f \oplus Z^{-1}\left(\Delta_{\Theta''} M_{G_1'^* + zG_2'} f \oplus W_1' g\right)\right) \quad \text{[by (9.44)]}$$
$$= M_{G_1'^* + zG_2'} f \oplus W_1' g,$$

and (9.61) also follows.

We now show that the triple $((G_1', G_2'), (W_1', W_2'), \Theta')$ is admissible. Recall that, in the course of the proof of Theorem 9.5.3, we saw that both $G_1'$ and $G_2'$

are contractions and that $(W_1', W_2')$ is a pair of commuting unitaries satisfying (9.41). From (9.61) we see that, for every $f \in H^2(\mathcal{F})$ and $(i, j) = (1,2), (2,1)$,

$$\begin{aligned}\left(M_{G_i'^*+zG_j'} \oplus W_i'\right)(\Theta' f \oplus \Delta_{\Theta'} f) &= U\left(M_{G_i^*+zG_j} \oplus W_i\right)U^*(\Theta' f \oplus \Delta_{\Theta'} f)\\ &= U\left(M_{G_i^*+zG_j} \oplus W_i\right)(\Theta''\Theta' f \oplus Z^{-1}(\Delta_{\Theta''}\Theta' f \oplus \Delta_{\Theta'} f))\\ &= U\left(M_{G_i^*+zG_j} \oplus W_i\right)(\Theta f \oplus \Delta_\Theta f).\end{aligned}$$

From the admissibility of $((G_1, G_2), (W_1, W_2), \Theta)$, we know that each of the contraction operators $\left(M_{G_i^*+zG_j} \oplus W_i\right)$ takes the space $\{\Theta f \oplus \Delta_\Theta f : f \in H^2(\mathcal{D})\}$ into itself. Therefore from the last term of the above computation and (9.59), we see that, for each $(i, j) = (1,2), (2,1)$,

$$\left(M_{G_i'^*+zG_j'} \oplus W_i'\right)(\{\Theta' f \oplus \Delta_{\Theta'} f : f \in H^2(\mathcal{D})\}) \subset \{\Theta' f \oplus \Delta_{\Theta'} f : f \in H^2(\mathcal{D})\}.$$

From (9.61) it is also clear that for each $(i, j) = (1,2), (2,1)$, the operators $\left(M_{G_i'^*+zG_j'} \oplus W_i'\right)$ are contractions and that with $\mathfrak{H}'$ as in (9.60)

$$\begin{aligned}&\left(M_{G_i'^*+zG_j'} \oplus W_i'\right)^*\left(M_{G_j'^*+zG_i'} \oplus W_j'\right)^*|_{\mathfrak{H}'}\\ &= U\left(M_{G_i^*+zG_j} \oplus W_i\right)^*\left(M_{G_j^*+zG_i} \oplus W_j\right)^*|_{\mathbb{H}}\\ &= U(M_z \oplus M_\zeta)|_{\mathbb{H}} = U(M_z \oplus M_\zeta)U^*|_{\mathfrak{H}'} = (M_z \oplus M_\zeta)|_{\mathfrak{H}'} \quad \text{[by (9.62)]}.\end{aligned}$$

This completes the proof of admissibility of $((G_1', G_2'), (W_1', W_2'), \Theta')$.

And finally to prove the last part we first observe that

$$(\mathbf{T}_1, \mathbf{T}_2) = P_{\mathbb{H}'}\left(M_{G_1^*+zG_2} \oplus W_1, M_{G_2^*+zG_1} \oplus W_2\right)|_{\mathbb{H}'}.$$

Now from equations (9.61) and (9.62) again and from the fact that $U(\mathbb{H}') = \mathfrak{H}'$ (hence $UP_{\mathbb{H}'} = P_{\mathfrak{H}'}U$), we conclude that

$$(\mathbf{T}_1, \mathbf{T}_2, \mathbf{T}_1\mathbf{T}_2)|_{\mathbb{H}'} = P_{\mathbb{H}'}\left(M_{G_1^*+zG_2} \oplus W_1, M_{G_2^*+zG_1} \oplus W_2, M_z \oplus M_\zeta\right)|_{\mathbb{H}'}$$

is unitarily equivalent to the functional model associated to $((G_1', G_2'), (W_1', W_2'), \Theta')$, that is,

$$P_{\mathfrak{H}'}\left(M_{G_1^*+zG_2} \oplus W_1, M_{G_2^*+zG_1} \oplus W_2, M_z \oplus M_\zeta\right)|_{\mathfrak{H}'}$$

via the unitary $U|_{\mathbb{H}'} : \mathbb{H}' \to \mathfrak{H}'$. Therefore an application of Theorem 9.3.3, Theorem 9.4.5, and Proposition 9.4.3 completes the proof. □

When the purely contractive analytic function $(\mathcal{D}, \mathcal{D}_*, \Theta)$ is inner, the results above are much simpler, as in the following statement.

**Theorem 9.5.5** *Let $(\mathcal{D}, \mathcal{D}_*, \Theta)$ be an inner function and let $((G_1, G_2), \Theta)$ be an admissible pair. Define the pair $(\mathbf{T}_1, \mathbf{T}_2)$ of commuting contractions on*

$$\mathbb{H} = H^2(\mathcal{D}_*) \ominus \{\Theta f : f \in H^2(\mathcal{D})\} \tag{9.63}$$

*by*

$$(\mathbf{T}_1, \mathbf{T}_2) = P_{\mathbb{H}} \left( M_{G_1^* + zG_2}, M_{G_2^* + zG_1} \right) |_{\mathbb{H}}. \tag{9.64}$$

*A subspace $\mathbb{H}'$ of $\mathbb{H}$ is jointly invariant under $(\mathbf{T}_1, \mathbf{T}_2)$ if and only if there exist two inner functions $(\mathcal{D}, \mathcal{F}, \Theta')$, $(\mathcal{F}, \mathcal{D}_*, \Theta'')$ such that*

$$\Theta = \Theta''\Theta'$$

*is a regular factorization,*

$$\begin{aligned} \mathbb{H}' &= \{\Theta'' f : f \in H^2(\mathcal{F})\} \ominus \{\Theta h : h \in H^2(\mathcal{D})\}, \\ \mathbb{H}'' &:= \mathbb{H} \ominus \mathbb{H}' = H^2(\mathcal{D}_*) \ominus \{\Theta'' f : f \in H^2(\mathcal{F})\}, \end{aligned} \tag{9.65}$$

*and two contractions $G_1', G_2'$ in $\mathcal{B}(\mathcal{F})$ such that*

$$M_{G_i^* + zG_j} M_{\Theta''} = M_{\Theta''} M_{G_i'^* + zG_j'}. \tag{9.66}$$

*Moreover, the pair $((G_1', G_2'), \Theta')$ coincides with the characteristic pair for $(T_1, T_2)|_{\mathcal{H}'}$.*

Another interesting simplification is the case where the input space $\mathcal{D}$ is the zero space and the purely contractive analytic function $(\{0\}, \mathcal{D}_*, \Theta)$ is necessarily the zero function $\Theta(z) = 0 : \{0\} \to \mathcal{D}_*$ for all $z \in \mathbb{D}$. If $\Theta = \Theta''\Theta'$ is any factorization into contractive analytic functions $(\mathcal{F}, \mathcal{D}_*, \Theta'')$, $(\{0\}, \mathcal{F}, \Theta')$, then $\Theta'$ is forced to be the zero function. One can then show that the factorization $0 = \Theta'' \cdot 0$ is regular exactly when $\Theta''$ is inner, so $I - \Theta''(\zeta)^*\Theta''(\zeta) = I_{\mathcal{F}}$ for a.e. $\zeta \in \mathbb{T}$. Then Theorem 9.5.3 simplifies to the following form; in view of Remark 9.5.2, this result can be viewed as a bivariate version of the Beurling–Lax theorem.

**Theorem 9.5.6** *Suppose that $(\mathcal{D}_*, P, U)$ is a BCL-tuple. Define a pair of commuting isometries $(\mathbf{T}_1, \mathbf{T}_2)$ on $\mathbb{H} = H^2(\mathcal{D}_*)$ as the associated BCL1-model pair of commuting isometries*

$$(\mathbf{T}_1, \mathbf{T}_2) = \left( M_{U^*(P^\perp + zP)}, M_{(P + zP^\perp)U} \right).$$

*A subspace $\mathbb{H}'$ of $\mathbb{H}$ is jointly invariant under $(\mathbf{T}_1, \mathbf{T}_2)$ if and only if there exist an inner function $(\mathcal{F}, \mathcal{D}_*, \Theta'')$ and another BCL-tuple $(\mathcal{F}, P', U')$ so that*

$$\mathbb{H}' = \Theta'' H^2(\mathcal{F}), \quad \mathbb{H}'' = H^2(\mathcal{D}_*) \ominus \mathbb{H}' = H^2(\mathcal{D}_*) \ominus \Theta'' H^2(\mathcal{F}), \tag{9.67}$$

*and lastly,*

$$M_{U^*(P^\perp + zP)} M_{\Theta''} = M_{\Theta''} M_{U'^*(P'^\perp + zP')}, \quad M_{(P+zP^\perp)U} M_{\Theta''} = M_{\Theta''} M_{(P'+zP'^\perp)U'}.$$

A particular choice of BCL-tuple in Theorem 9.5.6 is the choice $(\mathcal{D}_*, P, U) = (\ell^2_{\mathbb{Z}}, P_{[1,\infty)}, \mathbf{S})$ as in Example 3.1.20, so that the resulting $(\mathbf{T}_1, \mathbf{T}_2)$ is just the BCL1-model for the bidisk shift pair $(M_{z_1}, M_{z_2})$ on $H^2_{\mathbb{D}^2}$. In this model the operator $M_{z_1 z_2}$ on $H^2_{\mathbb{D}^2}$ becomes the operator $M_z$ on $H^2(\ell^2_{\mathbb{Z}})$. The standard single-variable Beurling–Lax theorem tells us that invariant subspaces for $M_{z_1 z_2}$ on the Hardy space of the bidisk are then in one-to-one correspondence with inner functions of the form $(\mathcal{F}, \ell^2_{\mathbb{Z}}, \Theta'')$; the latter object is not easy to classify since the target coefficient space $\ell^2_{\mathbb{Z}}$ is infinite dimensional. Joint invariant subspaces for $(M_{z_1}, M_{z_2})$ on $H^2_{\mathbb{D}^2}$ then correspond to such inner $\Theta''$ such that in addition $\Theta'' \cdot H^2(\ell^2_{\mathbb{Z}})$ is jointly invariant for $(M_{\mathbf{S}^*(P^\perp + zP)}, M_{(P+zP^\perp)\mathbf{S}})$ (where $P = P_{\ell^2_{[1,\infty)}}$). The trade-off between the bidisk setting versus the BCL-setting is: In the bidisk setting one has scalar-valued functions at the cost of the functions being of two variables, while in the BCL-setting one has single-variable functions but with values in the infinite-dimensional space $(\ell^2_{\mathbb{Z}})$. Characterizing joint invariant subspaces in either setting appears to be rather intractable. There has been a great deal of work on the problem, leading to deeper appreciation of how complicated the structure is in the bidisk setting: A small sample of such work is [41–43, 56, 85, 96].

## 9.6 Notes to Chapter 9

**1.** Theorem 9.2.2 can be seen to follow as a special case of the canonical decomposition for tetrablock contractions obtained in [73]; one simply needs to apply the tetrablock result to the special example $(T_1, T_2, T_1T_2)$ of tetrablock contraction, for a commuting contractive pair $(T_1, T_2)$. See [21, Proposition 3.5] or [83, Lemma 32] to see that $(T_1, T_2, T_1T_2)$ is a tetrablock contraction, whenever $(T_1, T_2)$ is a commuting contractive pair. Here we present a more elementary direct proof for the special setting of commuting pairs of contractions. A direct analogue of Theorem 9.2.2 is true for a commuting tuple of contractions as well; see [21, Theorem 7.6].

**2.** A Sz.-Nagy–Foias canonical functional model holds for $X$-contractive commutative operator tuples for the various domains associated with the companion-contraction formalism discussed in Section A.1; here, Theorem 9.1.2 corresponds to the particular case $X = \overline{\mathbb{D}}^2$. For further detail and references we refer to Sections A.1.2 and A.1.3.

**3.** We note that an analogue of Theorem 9.5.6 appears in the context of model theory for a commuting pair of operators $(S, P)$ having the symmetrized bidisk as a spectral set (i.e., a $\Gamma$-contraction) – see Sarkar [80, Theorem 3.3] – and for a triple of commuting operators $(A, B, P)$ having the tetrablock as a spectral set (i.e., a tetrablock contraction) – see [82, Theorem 3.1].

# Appendix

# Appendix More General Domains and Open Problems

## A.1 More General Domains

There has been a considerable amount of activity on the dilation theory and operator model theory for settings different from, but having strong analogies with, the unit disk and the bidisk settings treated here. There is a general framework going back to Arveson [12] that can be summarized as follows.

### A.1.1 The Rational Dilation Problem

Suppose that we are given a compact set $X$ in $d$-dimensional Euclidean space $\mathbb{C}^d$ and we are interested in studying the Taylor operator functional calculus associated with the set $X$. More precisely, suppose $\mathbf{T} = (T_1, \dots, T_d)$ is a commuting $d$-tuple of operators on a Hilbert space with its Taylor joint spectrum contained in $X$. Suppose $r(z) = p(z)/q(z)$ is a $d$-variable rational function with $(p, q)$ a coprime pair of polynomials in $d$ variables such that $r$ is holomorphic in a neighborhood of $X$, that is, $q$ does not vanish in $X$. Then the operator $r(\mathbf{T}) = \frac{p}{q}(\mathbf{T})$ is well-defined as an operator on $\mathcal{H}$. If it is the case that

$$\|r(\mathbf{T})\| \leq \sup\{|r(z)| : z \in X\} \tag{A.1}$$

for all rational $r$ holomorphic in a neighborhood of $X$, we say that $\mathbf{T}$ has $X$ as a *spectral set*. A convenient alternative terminology is to say that $\mathbf{T}$ is an $X$-*contraction*. Suppose next that the rational function $r$ is replaced by an $N \times N$-matrix rational function $R(z) = \left[r_{ij}(z)\right]_{1 \leq i,j \leq N}$, where each $r_{ij}(z) = \frac{p_{ij}(z)}{q_{ij}(z)}$ is a scalar rational function holomorphic in a neighborhood of $X$. Then we may define $R(\mathbf{T}) = [r_{ij}(\mathbf{T})] \in \mathcal{B}\left(\mathcal{H}^N\right)$ and replace (A.1) by

$$\|R(\mathbf{T})\| \leq \sup\{\|R(z)\|_{\mathcal{B}(\mathbb{C}^N)} : z \in X\}. \tag{A.2}$$

If (A.2) holds for all matrix rational $R(z)$ holomorphic in a neighborhood of $X$, we then say that $X$ is a *complete spectral set* for $\mathbf{T}$. If $X$ is a complete spectral set for $\mathbf{T}$, it then follows by a theorem of Arveson that there is an $X$-unitary dilation $\mathbf{U}$ of $\mathbf{T}$, that is, there exists a $d$-tuple $\mathbf{U} = (U_1, \ldots, U_d)$ of commuting normal operators having Taylor joint spectrum contained in the *distinguished* (or *essential*) *boundary* $\partial_e X$ of $X$ acting on a Hilbert space $\widetilde{\mathcal{K}}$ containing $\mathcal{H}$ so that

$$r(\mathbf{T}) = P_{\mathcal{H}} r(\mathbf{U})|_{\mathcal{H}} \tag{A.3}$$

for all rational functions $r$ holomorphic in a neighborhood of $X$. Conversely, because of the spectral theory for commuting $d$-tuples of normal operators, it is clear from this last identity that any $\mathbf{T}$ represented in this way by a commuting $d$-tuple of normal operators with Taylor joint spectrum contained in $\partial_e X$ will have $X$ as a complete spectral set. When (A.3) happens we say that $\mathbf{T}$ is dilatable to an $X$-unitary $d$-tuple $\mathbf{U}$. We say that this $X$-unitary dilation is *minimal* if it is the case that

$$\bigvee \{r(\mathbf{U})\mathcal{H} \colon r \in C(\partial_e X)\} = \widetilde{\mathcal{K}}. \tag{A.4}$$

Let us say that a $d$-tuple $\mathbf{V} = (V_1, \ldots, V_d)$ on the Hilbert space $\mathcal{K}$ is $X$-*isometric* if it can be extended to an $X$-unitary $d$-tuple $\mathbf{U} = (U_1, \ldots, U_d)$ on a Hilbert space $\widetilde{\mathcal{K}}$ containing $\mathcal{K}$. We say that the $X$-isometric tuple $\mathbf{V} = (V_1, \ldots, V_d)$ on $\mathcal{K}$ is an $X$-*isometric lift* of $\mathbf{T} = (T_1, \ldots, T_d)$ on $\mathcal{H}$ if $\mathcal{K} \supset \mathcal{H}$ and $\mathbf{V}^*$ extends $\mathbf{T}^*$: $\mathbf{V}^*|_{\mathcal{H}} = \mathbf{T}^*$. Minimality for such a lift means that

$$\bigvee \{r(\mathbf{V}) \colon r \text{ rational and holomorphic on } X\} = \mathcal{K}.$$

Any $X$-unitary dilation $\mathbf{U}$ can be restricted to a minimal $X$-isometric lift $\mathbf{V}$ and, conversely, any minimal $X$-isometric lift $\mathbf{V}$ can be extended to a minimal $X$-unitary dilation $\mathbf{U}$. As we have seen for the case $X = \mathbb{D}$ and $X = \mathbb{D}^2$, it is often more convenient not to insist that $\mathcal{H}$ is a subspace of $\mathcal{K}$ but to allow $\mathcal{H}$ to be identified with a subspace of $\mathcal{K}$ via an isometric embedding $\mathbf{\Pi} \colon \mathcal{H} \to \mathcal{K}$. Then we say that $(\mathbf{\Pi}, \mathbf{V})$ is a $X$-isometric lift of $\mathbf{T}$ if $\mathbf{\Pi} \colon \mathcal{H} \to \mathcal{K}$ is isometric and $\mathbf{V}^* \mathbf{\Pi} = \mathbf{\Pi} \mathbf{T}^*$. Then the *rational dilation problem* or simply the *dilation problem* for $X$ is:

*Is it always the case that an $X$-contraction* $\mathbf{T}$ *has an $X$-isometric lift* $\mathbf{V}$*?*

### A.1.2 The Companion Contraction

For the case when $X$ is the unit disk $\mathbb{D}$, a $\mathbb{D}$-contraction amounts to a contraction operator $D$, a $\mathbb{D}$-isometry amounts to an isometry in the usual sense, and the dilation problem for $\mathbb{D}$ always has an affirmative solution, which leads to a functional model for contraction operators, as reviewed in Chapter 2. For the case $X = \mathbb{D}^2$, Ando's theorem assures us that the lifting question always has a positive solution but a minimal such lift is not unique (up to lift unitary equivalence) except in special circumstances, as demonstrated in various concrete examples (see Examples 4.1.4, 5.5.2, and 5.5.3) and uniqueness results (see Theorems 4.5.1 and 8.5.1). What is true is that, associated with any such $\mathbb{D}^2$-contraction $\mathbf{T} = (T_1, T_2)$ is a *companion-contraction operator* $T$, namely the *product contraction* $T = T_1T_2 = T_2T_1$. As we have seen in Chapter 8, any $\mathbb{D}^2$-isometric lift (say on the space $\boldsymbol{\mathcal{K}}$) has embedded inside it a copy of the minimal isometric (Sz.-Nagy–Foias) lift of the companion contraction $T$ on $\mathcal{K} \subset \boldsymbol{\mathcal{K}}$, and compression of $(\mathbf{V}_1, \mathbf{V}_2, \mathbf{V}_1\mathbf{V}_2)$ down to $\mathcal{K}$ leads to another lift $(\mathbb{W}_1, \mathbb{W}_2, \mathbb{W})$ of $(T_1, T_2, T)$ (with $\mathbb{W}$ being the minimal isometric lift of $T$), which in general is no longer commuting or $\mathbb{D}^2$-isometric, but rather is characterized as having the property of being a *pseudo-commuting contractive lift* of $(T_1, T_2, T)$. As we saw in Chapter 9, it is this lift which, when put in functional-model form, leads to a Sz.-Nagy–Foias-like functional model for the $\mathbb{D}^2$-contraction $(T_1, T_2)$.

This program has now been analyzed for a number of specific cases $X$: See, for example, the summary in the introduction of [20]. Here we wish to point out specific examples where there is this additional ingredient of a companion-contraction operator $T$ with its own Sz.-Nagy–Foias model dilation and functional model, which leads to a more complete analogue of the Sz.-Nagy–Foias model theory for an $X$-contraction $\mathbf{T}$.

For simplicity, in the following specific cases we discuss only the case where the companion contraction $T$ is pure (i.e., $T^{*n} \to 0$ strongly); removal of this assumption is usually possible via an appropriate adaptation of Theorem 4.4.2 to the particular $X$-setting. These examples all have the following common features for analysis of an $X$-contraction $\mathbf{T}$.

(i) One can identify a *companion contraction* $T$ associated with $\mathbf{T}$.

(ii) There is a uniquely determined notion of a *fundamental-operator collection* (a single operator, an operator pair, or an operator tuple, depending on the choice of set $X$) determined as the unique solution of some system of operator equations determined by $\mathbf{T}$.

(iii) By making use of the fundamental-operator collection $\mathbf{G}$ of $\mathbf{T}^*$ one can explicitly construct a collection of operators $(\underline{\mathbb{W}}, \mathbb{W}_c)$ on the Sz.-Nagy–Foias model space $\mathcal{K}_{\Theta_T}$ that is a lift of $(\mathbf{T}, T)$ (with $\mathbb{W}_c = V$ equal to the minimal isometric lift of $T$). Furthermore, this lift of $\mathbf{T}$ can be characterized intrinsically with the appropriate notion of an *$X$-pseudo-commuting contractive lift* of $\mathbf{T}$.

(iv) The compression $T_{\Xi_T} := P_{\mathcal{H}_{\Theta_T}} \mathbb{W}|_{\mathcal{H}_{\Theta_T}}$ gives a Sz.-Nagy–Foias-like functional model for the original $X$-contraction $\mathbf{T}$ determined by its *characteristic pair* $\Xi_{\mathbf{T}} = (\mathbf{G}, \Theta_T)$, a complete unitary invariant for $\mathbf{T}$. One can turn the process around and construct a functional model $T_\Xi$ from an $X$-admissible pair $(\mathbf{G}, \Theta)$ (which in general is not as explicit as the characteristic pair of $\mathbf{T}$).

### A.1.3 Concrete Examples of Companion Contractions

We discuss the following choices of sets (in addition to the disk and the bidisk already discussed) where this program has had at least a modest degree of success. Let us mention that the original motivation for the recent study of two of these sets (the symmetrized bidisk $\overline{\mathbb{D}}^2_s$ and the tetrablock $\mathbb{E}$) came from the $\mu$-synthesis problem in robust control theory (see, e.g., [57, 58]); more on these connections can be found in [21, Section 2.2].

**1. $X$ = the symmetrized bidisk $\overline{\mathbb{D}}^2_s$** We define the symmetrized bidisk $\overline{\mathbb{D}}^2_s$ to be the compact set

$$\overline{\mathbb{D}}^2_s = \left\{(z_1 + z_2, z_1 z_2) : (z_1, z_2) \in \overline{\mathbb{D}}^2\right\}.$$

A $\overline{\mathbb{D}}^2_s$-contraction has the form $(S, P)$ (satisfying $\|S\| \leq 2$, $\|P\| \leq 1$ together with some additional conditions). It is known that the isometric lifting problem is always solvable (see Agler–Young [6]). A model for $\overline{\mathbb{D}}^2_s$-contractions was obtained in Agler–Young [8]). The papers [36, 38] identify the key role of the fundamental operator $F$ of $(S, P)$ defined as the unique solution of the equation

$$S - S^* P = D_P F D_P.$$

In this case one takes the companion contraction $T$ of the $\overline{\mathbb{D}}^2_s$-contraction $(S, P)$ to be the second component $T = P$. Remarkably, in this case there is a minimal isometric lift of $(S, P)$, which is actually strongly minimal in the sense analogous to Definition 7.1.1), so there is no need for the introduction of a distinct notion of a $\overline{\mathbb{D}}^2_s$- pseudo-commuting contractive lift. The Sz.-Nagy–Foias model version of this minimal lift is given by

$$(V_1, V_2) = \left(M_{G^* + zG}, M_z^{\mathcal{D}_{P^*}}\right),$$

where $G$ is the fundamental operator for the $\overline{\mathbb{D}}^2_s$-contraction $(S^*, P^*)$. The Sz.-Nagy–Foias model representation for $(S, P)$ is then given by

$$(S, P) \underset{\mathrm{u}}{\cong} P_{\mathcal{H}_{\Theta_P}} \left(M_{G^* + Gz}, M_z^{\mathcal{D}_{P^*}}\right)\Big|_{\mathcal{H}_{\Theta_P}}$$

and is determined by its characteristic pair $(G, \Theta_P)$ (here we are assuming that $P$ is pure), where $G$ is the fundamental operator for $(S^*, P^*)$, where $\Theta_P$ is the characteristic function for $P$, and where $\mathcal{H}_{\Theta_P} = H^2(\mathcal{D}_{P^*}) \ominus \Theta_P \cdot H^2(\mathcal{D}_P)$ is the Sz.-Nagy–Foias model space associated with the inner function $\Theta_P$.

**2.** $X$ **= the tetrablock** $\mathbb{E}$ The tetrablock $\mathbb{E}$ was introduced by Abouhajar–White–Young [1] with follow-up work on operator theory aspects by Bhattacharyya [33]. The tetrablock $\mathbb{E}$ can be defined as

$$\begin{aligned}\mathbb{E} = \big\{(x_1, x_2, x_3) \in \mathbb{C}^3 : \exists A = \left[\begin{smallmatrix} a_{11} & a_{12} \\ a_{21} & a_{22} \end{smallmatrix}\right] \in \mathbb{C}^{2\times 2} \\ \text{with } \|A\| \le 1 \text{ and } x_1 = a_{11},\ x_2 = a_{22},\ x_3 = \det A\big\}.\end{aligned}$$

For this case, as of this writing it is still unknown whether an $\mathbb{E}$-isometric lift of an $\mathbb{E}$-contraction always exists (see [20, 34]). It is easy to see that $\mathbb{E} \subset \overline{\mathbb{D}}^3$, from which it follows that each coordinate $A, B, T$ of an $\mathbb{E}$-contraction $(A, B, T)$ is a contraction operator. In particular, we may consider the third component $T$ of $(A, B, T)$ as the *companion contraction* for the $\mathbb{E}$-contraction $(A, B, T)$. The paper [33] identified the *fundamental-operator pair* for the $\mathbb{E}$-contraction as the unique solution $(F_1, F_2)$ of the pair of equations

$$A - B^*T = D_T F_1 D_F, \quad B - A^*T = D_T F_2 D_T. \tag{A.5}$$

As explained in [20], if $(T_1, T_2)$ is a commuting contractive pair, then $(A, B, T) = (T_1, T_2, T_1T_2)$ is a tetrablock contraction. Comparing (A.5) with (7.13), we see that the fundamental-operator pair for the commuting contractive pair $(T_1, T_2)$ is the same as the Fundamental-Operator pair for the associated $\mathbb{E}$-contraction $(T_1, T_2, T_1T_2)$. The papers [33, 37] identify an additional commutativity condition, which must be satisfied by the fundamental-operator pair of $(A, B, T)$, which is necessary and sufficient for what we call here a *strongly minimal* $\mathbb{E}$*-isometric lift* of $(A, B, T)$, that is, a $\mathbb{E}$-isometric lift $(V_1, V_2, V)$ of $(A, B, T)$ so that the third component is a *minimal* isometric lift of the contraction operator $T$; uniqueness for such a lift when it exists is also established. However, as shown by the second author in [82] for the case where $T$ is pure and by the authors in [22] for the general case, the fundamental-operator pair for the adjoint $\mathbb{E}$-contraction $(A^*, B^*, T^*)$ can be used to construct an $\mathbb{E}$-pseudo-commuting algebraic lift of $(A, B, T)$, which leads to a Sz.-Nagy–Foias functional model for $(A, B, T)$ itself. However, it is not known whether this pseudo-commuting algebraic lift is contractive, leading to additional difficulties in solving the inverse problem; that is, in finding an explicit characterization of which candidate tuples $((G_1, G_2), (W_1, W_2), (\mathcal{D}, \mathcal{D}_*, \Theta))$ "coincide" with a characteristic triple $((G_1, G_2), (W_1, W_2, W), \Theta_T)$, that is $(G_1, G_2)$ equal to the fundamental-operator pair for $(A^*, B^*, T^*)$, $(W_1, W_2, W)$ equal to an $\mathbb{E}$-unitary triplet on $\Delta_\Theta L^2(\mathcal{D})$ for some c.n.u. $\mathbb{E}$-contraction $(A, B, T)$.

**3. $X$ = the polydisk $\overline{\mathbb{D}}^d$** In this case it is known that, in general, a commuting isometric operator tuple lift $\underline{V} = (V_1, \ldots, V_d)$ of a commuting contractive tuple $\underline{T} = (T_1, \ldots, T_d)$ can fail to exist (see [94]). However, it can be shown that any commuting contractive $d$-tuple $\underline{T}$ can be lifted to a BCL-model consisting of operators of the form

$$\left( \begin{bmatrix} M_{U_1 P_1^\perp + zU_1 P_1} & 0 \\ 0 & W_1 \end{bmatrix}, \ldots, \begin{bmatrix} M_{U_d P_d^\perp + zU_d P_d} & 0 \\ 0 & W_d \end{bmatrix} \right)$$

acting on a space of the form $\begin{bmatrix} H^2(\mathcal{F}) \\ \mathcal{E} \end{bmatrix}$ for a coefficient Hilbert space $\mathcal{F}$ and a Hilbert space $\mathcal{E}$, where $P_1, \ldots, P_d$ are projection operators and $U_1, \ldots, U_d$ are unitary operators on $\mathcal{F}$, while $W_1, \ldots, W_d$ are commuting unitary operators on $\mathcal{E}$. Each individual operator $\begin{bmatrix} M_{U_j P_j + zU_j P_j} & 0 \\ 0 & W_j \end{bmatrix}$ $(j = 1, \ldots, d)$ is isometric but when $d \geq 3$ there is no useful tractable characterization as to when it is the case that such a collection is commuting. Independent of all these observations, one can bootstrap off the tetrablock theory to define a $2d$-tuple $\{F_{j1}, F_{j2} \colon j = 1, \ldots, d\}$ of fundamental operators defined as the unique solution of a system of operator equations leading to the construction of a $d$-variable version of a pseudo-commuting algebraic lift $\underline{\mathbb{W}} = (\mathbb{W}_1, \ldots, \mathbb{W}_d, V)$ of $(\underline{T}, T = T_1 \cdots T_d)$ with the property that $V$ is a version of the minimal isometric lift of the companion contraction operator $T = T_1 \cdots T_d$. One can use the uniqueness property of the fundamental-operator tuple to see that this lift in turn is the compression of any BCL-model isometric lift of $\underline{T}$, which then implies that the pseudo-contractive lift is actually contractive (each component is a contraction operator rather than an operator with numerical radius at most 1). One can then embed all this in the Sz.-Nagy–Foias model space $\mathcal{K}_{\Theta_T}$ for the minimal isometric lift of the companion contraction operator $T = T_1 \cdots T_d$. Compression of the model version of the pseudo-commuting contractive lift $\underline{\mathbb{W}}$ to $\mathcal{H}_{\Theta_T}$ then leads to a functional model for the original commuting $d$-tuple $\underline{T}$. Thus, as was the case for $X = \mathbb{E}$ (the tetrablock), for the case $X = \mathbb{D}^d$ (the polydisk) we pick up a functional model for $X$-contractions via compression of pseudo-contractive lifts acting on the ambient space for a minimal Sz.-Nagy–Foias lift of the associated companion contraction operator $T$, even though the existence of a (commuting) $X$-isometric lift is either unknown (for the tetrablock case) or is known to fail (for the polydisk case).

**4. $X$ = the symmetrized polydisk $\overline{\mathbb{D}}_s^d$** It has only recently come to our attention that Bisai–Pal in [40] have explored these ideas in the setting of the symmetrized polydisk $\overline{\mathbb{D}}_s^d$. This example combines the symmetrized bidisk setting with the commutative polydisk setting.

We expect that more can, and will, be said about the unresolved issues appearing in the previous discussion. It would be of interest to identify additional sets $X$ for which one can attach in a canonical way a *companion contraction* $T$ and then use the Sz.-Nagy–Foias model theory for $T$ to say something about the dilation and model theory for $X$-contractions.

## A.2 Open Problems

There remain a number of open problems concerning a pair of commuting contraction operators. In this section, we compile some of the more intriguing ones for the entertainment of curious minds

### A.2.1 Some Special Commuting Contractive Pairs

Let us recall that a contractive operator $T$ is called *pure* if $T^{*n} \to 0$ strongly as $n \to \infty$. This class comes up as the starting point for the de Branges–Rovnyak/Sz.-Nagy–Foias operator model theory due to the following well-known result: *Every pure contraction $T$ acting on a Hilbert space $\mathcal{H}$ lifts to the shift operator $M_z^{\mathcal{D}_{T^*}}$ on $H^2(\mathcal{D}_{T^*})$*. A nice history of this and closely related results is given in the introduction in the paper of Agler [3] where this particular result is attributed to Sz.-Nagy–Foias [88] and de Branges–Rovnyak [51]. We observed this result in Remark 2.1.5 for the coordinate-free setting and it can also be seen from the elaboration made on the Douglas model of isometric lifts for single contractive operators in Section 2.2.2. More fundamentally, this so-called *purity* property has a stability interpretation in discrete-time mathematical systems theory and more generally in dynamical systems: given a discrete-time autonomous linear system described by the update equation with initial conditions

$$x(n+1) = Ax(n), \quad x(0) = x_0,$$

for $n = 0, 1, 2, \ldots$, where $A$ is a Hilbert space operator. In this case it is easy to solve the recursion relation giving us the explicit formula for $x(n)$:

$$x(n) = A^n x_0.$$

The system is said to be *stable* if $x(n) \to 0$ in norm, independent of the choice of initial condition $x_0$. We see that this is the case exactly when the operator $A$ is *strongly stable* in the sense that $\lim_{n\to\infty} \|A^n x_0\| = 0$ for each $x_0 \in \mathcal{H}$, that is, exactly when the operator $T := A^*$ is *pure* in the operator-theory sense specified previously.

In this book, we have produced a bivariate analogue of this phenomenon: Note that in Proposition 4.3.5 we have proved that if $(T_1, T_2)$ is a pure commuting contractive operator pair (see Definition 4.3.2), that is, if $T = T_1T_2$ is pure, then every minimal Ando isometric lift $(V_1, V_2)$ of $(T_1, T_2)$ is pure, namely, $V = V_1V_2$ is a shift operator. Thus, we have proved the following.

**Theorem A.2.1** *Every minimal Ando isometric lift $(V_1, V_2)$ of a pure commuting contractive pair $(T_1, T_2)$ is also pure.*

In the bivariate setting, however, there could be more than one way of defining a pure pair. Indeed, from Definition 4.3.2 again, recall the following stronger notion of a pure pair: A *bi-pure* commuting contractive pair $(T_1, T_2)$ is such that both $T_1$ and $T_2$ are pure, that is, we have $T_j^{*n} \to 0$ for $j = 1, 2$. A natural question is as follows.

**Problem A.2.2** Does a bipure commuting contractive pair have at least one bipure Ando isometric lift?

A stronger question and a direct parallel of Theorem A.2.1 is as follows.

**Problem A.2.3** Does a bipure commuting contractive pair have every minimal Ando isometric lift bipure?

At this stage we do not have enough data to make a conjecture one way or the other. However, there have been some developments in the affirmative direction in [50] (also in [81]). Let us say that a commuting contractive operator pair $(T_1, T_2)$ acting on $\mathcal{H}$ is *conditionally pure* if, for every $h \in \mathcal{H}$,

$$\lim_{n\to\infty} \left\| T_i^* T_j^{*n} h \right\| = \lim_{n\to\infty} \left\| T_j^{*n} h \right\| \quad \text{for } (i, j) = (1, 2) \text{ and } (2, 1). \tag{A.6}$$

It can easily be checked that this set of conditions is equivalent to

$$\lim_{n\to\infty} \|D_{T_i^*} T_j^{*n} h\| = 0 \quad \text{for } (i, j) = (1, 2) \text{ and } (2, 1) \tag{A.7}$$

for every $h \in \mathcal{H}$. Note that the first set of conditions (A.6) does not require that the commuting pair be contractive but the second set of conditions (A.7) does requires that the commuting pair be contractive. While a bipure pair is conditionally pure, the converse is not true. Indeed, for any contractive operator $T$, the pair $(T, T)$ is conditionally pure, as can easily be seen from the first set of conditions (A.7). This pair need not be bipure unless $T$ is pure.

What are the commuting isometric pairs that are conditionally pure? These are precisely the bipure commuting isometric pairs. To see this, let us first prove the following preliminary lemma.

**Lemma A.2.4** *Pure commuting contractive pairs that are also conditionally pure are bipure.*

*Proof* Suppose that $(T_1, T_2)$ is a conditionally pure pair such that $T = T_1T_2$ is pure. Pick $\epsilon > 0$ and $h \in \mathcal{H}$. Let $N > 0$ be such that $\|T^{*N}h\| < \epsilon$. Then we apply (A.6) to see that

$$\lim_{n\to\infty} \left\|T_j^{*n}h\right\| = \lim_{n\to\infty} \left\|T_i^{*(n-N)}T^N h\right\| \leq \lim_{n\to\infty} \left\|T^{*(n-N)}h\right\| \leq \epsilon.$$

This completes the proof. □

Let us note next that a commuting unitary pair is never conditionally pure. Therefore, in view of the Berger–Coburn–Lebow decomposition theorem 3.1.4, the only commuting isometric pairs that are conditionally pure are the pure commuting isometric pairs. But by the lemma above, such pairs must be bipure.

The following is proved in [50].

**Theorem A.2.5** *If* $(T_1, T_2)$ *is a conditionally pure commuting contractive pair and* $\dim \mathcal{D}_{T_j^*} < \infty$ *for* $j = 1, 2$, *then the pure part of every minimal isometric lift of* $(T_1, T_2)$ *is conditionally pure or equivalently (in view of the discussion above) bipure.*

As a consequence of this result, it therefore follows that Problem A.2.2 has an affirmative answer when the commutative contractive pair $(T_1, T_2)$ is such that $\dim \mathcal{D}_{T_j^*} < \infty$ for $j = 1, 2$. A natural question at this stage is as follows.

**Problem A.2.6** Is Theorem A.2.5 true without the finite-dimensionality assumption?

### A.2.2 The Halmos Dilation

**Unitary Halmos Dilation to Sz.-Nagy–Foias Unitary Power Dilations**
Throughout this book we have defined the operator $\mathcal{U}$ on a Hilbert space $\mathcal{K} \supset \mathcal{H}$ to be a *unitary dilation* of the contraction operator $T$ on $\mathcal{H}$ if it is the case that $\mathcal{U}$ is unitary and

$$P_{\mathcal{H}}\mathcal{U}^n|_{\mathcal{H}} = T^n \quad \text{for } n = 1, 2, \ldots. \tag{A.8}$$

An earlier idea introduced by Halmos in his 1950 paper [64] is to insist only that condition (A.8) holds for the case $n = 1$, that is,

$$P_{\mathcal{H}}\mathcal{U}|_{\mathcal{H}} = T. \tag{A.9}$$

If we write $\mathcal{U}$ as a $2 \times 2$-block matrix with respect to the decomposition of $\mathcal{K}$ as $\mathcal{K} = (\mathcal{K} \ominus \mathcal{H}) \oplus \mathcal{H}$, this just means that $\mathcal{U}$ has the form

$$\mathcal{U} = \begin{bmatrix} * & * \\ * & T \end{bmatrix} \tag{A.10}$$

(where the $*$ are arbitrary operators of the correct conformal type). Such dilations are often called *Halmos dilations*; when one wants to emphasize that $\mathcal{U}$ has the additional properties in (A.8), the term *unitary power dilation* has been used.

For $T$ any contraction operator on $\mathcal{H}$, a Halmos dilation having minimality properties, which we shall not discuss in detail here, is

$$U_T = \begin{bmatrix} -T^* & D_T \\ D_{T^*} & T \end{bmatrix}\Big|_{\left[\begin{smallmatrix} \mathcal{D}_{T^*} \\ \mathcal{H} \end{smallmatrix}\right]} : \begin{bmatrix} \mathcal{D}_{T^*} \\ \mathcal{H} \end{bmatrix} \to \begin{bmatrix} \mathcal{D}_T \\ \mathcal{H} \end{bmatrix}. \tag{A.11}$$

We note that $U_{T^*}$ comes up in Remark 3.3.5; $U_{T^*}$ given by

$$U_{T^*} = \begin{bmatrix} -T & D_{T^*} \\ D_T & T^* \end{bmatrix}\Big|_{\left[\begin{smallmatrix} \mathcal{D}_T \\ \mathcal{H} \end{smallmatrix}\right]} : \begin{bmatrix} \mathcal{D}_T \\ \mathcal{H} \end{bmatrix} \to \begin{bmatrix} \mathcal{D}_{T^*} \\ \mathcal{H} \end{bmatrix}$$

is the unitary colligation matrix inducing the Sz.-Nagy–Foias characteristic function for the contraction operator $T$:

$$\Theta_T(z) = T_{U_{T^*}}(z) := \left(-T + zD_{T^*}(I - zT^*)^{-1}D_T\right)|_{\mathcal{D}_T} : \mathcal{D}_T \to \mathcal{D}_{T^*}.$$

If we are not concerned with minimality issues, we can equally well consider the alternative Halmos dilation of $T$ given by

$$U'_T = \begin{bmatrix} -T^* & D_T \\ D_{T^*} & T \end{bmatrix} : \begin{bmatrix} \mathcal{H} \\ \mathcal{H} \end{bmatrix} \to \begin{bmatrix} \mathcal{H} \\ \mathcal{H} \end{bmatrix}. \tag{A.12}$$

We can embed a slight adjustment of the Halmos dilations $U_T$ and $U'_T$ to produce a unitary power dilation $\widetilde{\mathcal{U}}_T$ of $T$ as follows. We consider the adjustments of $U_T$ and $U'_T$ obtained by reversing the order of the columns to arrive at

$$\widetilde{U}_T := U_T \begin{bmatrix} 0 & I_{\mathcal{H}} \\ I_{\mathcal{H}} & 0 \end{bmatrix} = \begin{bmatrix} D_T & -T^* \\ T & D_{T^*} \end{bmatrix} : \begin{bmatrix} \mathcal{H} \\ \mathcal{D}_{T^*} \end{bmatrix} \to \begin{bmatrix} \mathcal{D}_T \\ \mathcal{H} \end{bmatrix}, \tag{A.13}$$

$$\widetilde{U}'_T = U'_T \begin{bmatrix} 0 & I_{\mathcal{D}_{T^*}} \\ I_{\mathcal{H}} & 0 \end{bmatrix} = \begin{bmatrix} D_T & -T^* \\ T & D_{T^*} \end{bmatrix}\Big|_{\left[\begin{smallmatrix} \mathcal{H} \\ \mathcal{D}_{T^*} \end{smallmatrix}\right]} : \begin{bmatrix} \mathcal{H} \\ \mathcal{H} \end{bmatrix} \to \begin{bmatrix} \mathcal{H} \\ \mathcal{H} \end{bmatrix}. \tag{A.14}$$

Then we can embed $\widetilde{U}_T$ and $\widetilde{U}'_T$ into a bi-infinite matrices $\widetilde{\mathcal{U}}_T$ and $\widetilde{\mathcal{U}}'_T$ as follows:

$$\widetilde{\mathcal{U}}_T = \begin{bmatrix} \ddots & & & & & & \\ & I_{\mathcal{D}_T} & & & & & \\ & & I_{\mathcal{D}_T} & & & & \\ & & & D_T & -T^* & & & \\ & & & \boxed{T} & D_{T^*} & & & \\ & & & & & I_{\mathcal{D}_T^*} & & \\ & & & & & & I_{\mathcal{D}_T^*} & \\ & & & & & & & \ddots \end{bmatrix} \quad \text{acting on} \begin{bmatrix} \ell^2_{\mathbb{Z}_-}(\mathcal{D}_T) \\ \mathcal{H} \\ \ell^2_{\mathbb{Z}_+}(\mathcal{D}_{T^*}) \end{bmatrix}, \tag{A.15}$$

$$\widetilde{\mathcal{U}}'_T = \begin{bmatrix} \ddots & & & & & & \\ & I_{\mathcal{H}} & & & & & \\ & & I_{\mathcal{H}} & & & & \\ & & & D_T & -T^* & & & \\ & & & \boxed{T} & D_{T^*} & & & \\ & & & & & I_{\mathcal{H}} & & \\ & & & & & & I_{\mathcal{H}} & \\ & & & & & & & \ddots \end{bmatrix} \quad \text{acting on} \begin{bmatrix} \ell^2_{\mathbb{Z}_-}(\mathcal{H}) \\ \mathcal{H} \\ \ell^2_{\mathbb{Z}_+}(\mathcal{H}) \end{bmatrix}. \tag{A.16}$$

Here, the boxed entry denotes the (00)-entries of the above two-sided infinite block matrices. If we denote the matrix entries as $\left[\widetilde{\mathcal{U}}_T\right]_{ij}$ and $\left[\widetilde{\mathcal{U}}'_T\right]_{ij}$ for $-\infty < i, j \leq \infty$ and in the notation we let $\mathbb{Z}_- = \{\dots, -3, -2, -1\}$ and $\mathbb{Z}_+ = \{1, 2, 3, \dots)$, then we see that the adjusted Halmos dilations appear in (A.13) and (A.14) as

$$\widetilde{U}_T = \left[\widetilde{\mathcal{U}}_T\right]_{ij;-1\leq i,j\leq 0},$$
$$\widetilde{U}'_T = \left[\widetilde{\mathcal{U}}'_T\right]_{ij:\ -1\leq i,j\leq 0}.$$

It turns out that $\widetilde{\mathcal{U}}_T$ is a minimal unitary (power) dilation of $T$ while $\widetilde{\mathcal{U}}'_T$ is a (not necessarily minimal) unitary (power) dilation of $T$. Both of these are discussed in some detail in [92, Section I.5]. Our development here leads to the general principle: *In general Halmos dilations are not unitary power dilations; however, one can construct unitary power dilations by embedding adjusted Halmos dilations into carefully constructed two-sided infinite block matrices.*

**Unitary $N$-dilations to Unitary Power Dilations**

Recall the original geometric (nonmatricial) definition (A.9) characterizing when a unitary operator $U$ on a Hilbert space $\mathcal{K} \supset \mathcal{H}$ is a Halmos dilation of a contraction operator $T$ on $\mathcal{H}$: $U = P_{\mathcal{H}}U|_{\mathcal{H}}$. According to the next definition, this is just the special case $N = 1$ of the following notion of an $N$-dilation of $T$.

**Definition A.2.7** Given a contractive operator $T$ acting on a Hilbert space $\mathcal{H}$, an integer $N \geq 1$, and an operator $U$ acting on $\mathcal{K} \supset \mathcal{H}$, we say that $U$ is an *$N$-dilation* of $T$ if it is the case that

$$p(T) = P_{\mathcal{H}}p(U)|_{\mathcal{H}} \tag{A.17}$$

for all polynomial $p$ of degree less than or equal to $N$. We shall be primarily interested in unitary $N$-dilations.

Note that if we reverse our matricial conventions in Section A.2.2 (which were convenient when discussing unitary realizations of characteristic functions) and write $\mathcal{K} \underset{\mathrm{u}}{\cong} \mathcal{H} \ominus (\mathcal{K} \ominus \mathcal{H})$ rather than $\mathcal{K} \underset{\mathrm{u}}{\cong} (\mathcal{K} \ominus \mathcal{H}) \oplus \mathcal{H}$, then the block-matrix criterion for $U$ to be a Halmos dilation of $T$ is that $U$ has the matrix representation

$$U = \begin{bmatrix} T & * \\ * & * \end{bmatrix}$$

(rather than (A.10)). If we use this convention, then the formula for $U'_T$ (A.12) becomes

$$U'_T = \begin{bmatrix} T & D_{T^*} \\ D_T & -T^* \end{bmatrix} : \begin{bmatrix} \mathcal{H} \\ \mathcal{H} \end{bmatrix} \to \begin{bmatrix} \mathcal{H} \\ \mathcal{H} \end{bmatrix}. \tag{A.18}$$

It is possible to "inflate" the unitary 1-dilation in (A.18) to obtain an $N$-dilation for every $N \geq 1$. Consider the $(N+1) \times (N+1)$-block operator matrix

$$U'_{T,N} := \begin{bmatrix} T & & & & D_{T^*} \\ D_T & & & & -T^* \\ & I_{\mathcal{H}} & & & \\ & & \ddots & & \\ & & & I_{\mathcal{H}} & 0 \end{bmatrix} \tag{A.19}$$

acting on $(N+1)$ copies of $\mathcal{H}$ denoted by

$$\mathcal{H}_N := \mathcal{H} \oplus \cdots \oplus \mathcal{H}.$$

Egerváry [61] (see also [78, 87]) constructed this block matrix and showed that it is a unitary $N$-dilation for $T$.

We next show that we can use unitary $N$-dilations to obtain a proof of the von Neumann inequality as well as a proof of the Sz.-Nagy dilation theorem. It is interesting to note that the mere existence of unitary $N$-dilations is used to prove the von Neumann inequality as follows.

**Theorem A.2.8** (von Neumann inequality) *If $T$ is a contractive operator acting on a Hilbert space $\mathcal{H}$ and $p$ is a single-variable polynomial, then*

$$\|p(T)\| \leq \sup_{\mathbb{T}} |p(\cdot)|.$$

*Proof* Suppose that $\deg p = N$. If $U_N$ is a unitary $N$-dilation for $T$, then by definition (A.17) we have $p(T) = P_{\mathcal{H}} p(U_N)|_{\mathcal{H}}$ and so

$$\|p(T)\| = \|P_{\mathcal{H}} p(U_N)|_{\mathcal{H}}\| \leq \|p(U_N)\|| = \sup_{\sigma(U_N)} |p(\cdot)| \leq \sup_{\mathbb{T}} |p(\cdot)|.$$

To obtain the inequality here we used the fact that $U_N$ is a unitary, giving $p(U_N)$ a normal operator with the spectrum being

$$\{p(z) \colon z \in \sigma(U_N)\} \subset \{p(z) \colon z \in \mathbb{T}\},$$

and that the norm of a normal operator is the same as its spectral radius. □

We next demonstrate how the explicit form of the unitary $N$-dilations leads to a proof of the Sz.-Nagy dilation theorem.

**Theorem A.2.9** *If $T$ is a contractive operator acting on a Hilbert space $\mathcal{H}$, then there is an isometric lift $V$ of $T$ acting on a Hilbert space $\mathcal{K} \supset \mathcal{H}$, that is, $V^*|_{\mathcal{H}} = T^*$.*

*Proof* The first step is to extend each unitary $N$-dilation $U'_{T,N}$ (A.19) acting on $\mathcal{H}_N$ to a countable direct sum of copies of $\mathcal{H}$ or, in other words, to the space $\ell^2 \otimes \mathcal{H}$. The extension can be done in several ways (see Remark A.2.10 below); in particular, the following trivial extension serves our purpose here:

$$\widetilde{U}'_{T,N} = \begin{bmatrix} U'_{T,N} & 0 \\ 0 & 0 \end{bmatrix} : \begin{bmatrix} \mathcal{H}_N \\ \ell^2 \otimes \mathcal{H} \end{bmatrix} \to \begin{bmatrix} \mathcal{H}_N \\ \ell^2 \otimes \mathcal{H} \end{bmatrix}.$$

We show that the sequence $\left\{\widetilde{U}'_{T,N}\right\}$ converge strongly. To that end, first let us note that, for $M > N \geq 1$, we have

$$\widetilde{U}'_{T,M} - \widetilde{U}'_{T,N} = \begin{bmatrix} 0 & & & -D_{T^*} & \cdots & 0 & D_{T^*} & \cdots \\ & \ddots & & T^* & \cdots & 0 & -T^* & \cdots \\ & & \ddots & \vdots & & & & \\ & & & 0 & & & & \\ & & & I_{\mathcal{H}} & \ddots & & & \\ & & & & \ddots & 0 & & \\ & & & & & I_{\mathcal{H}} & 0 & \\ & & & & & & & \ddots \end{bmatrix}.$$

The convention here is that all unspecified entries in the above matrix are equal to the zero operator on $\mathcal{H}$; the first nonzero column is the $(N+1)$th column, the last nonzero column is the $(M+1)$th column; the last nonzero row is the $(M+1)$th row, and the identity operators are repeated $M-N$ times. Therefore, for $f = \{h_n\}_{n\geq 0}$,

$$\left(\widetilde{U}'_{T,M} - \widetilde{U}'_{T,N}\right) f = \begin{bmatrix} -D_{T^*}h_N + D_{T^*}h_M \\ T^*h_N - T^*h_M \\ \vdots \\ h_N \\ \vdots \\ h_M \\ 0 \\ \vdots \end{bmatrix}.$$

Since $\|D_{T^*}h\|^2 + \|T^*h\|^2 = \|h\|^2$ for every $h \in \mathcal{H}$, from the above computation of $\left(\widetilde{U}'_{T,M} - \widetilde{U}'_{T,N}\right) f$ we see that

$$\begin{aligned}\left\|\left(\widetilde{U}'_{T,M} - \widetilde{U}'_{T,N}\right) f\right\|^2 &= \|h_M - h_N\|^2 + \|h_N\|^2 + \cdots + \|h_M\|^2 \\ &= \|h_M - h_N\|^2 + s_M - s_{N-1},\end{aligned}$$

where

$$s_L = \sum_{j=0}^{L} \|h_j\|^2.$$

Since $s_L$ is a convergent sequence (converging to $\|f\|^2$), $s_M - s_{N-1} \to 0$ as $M, N \to \infty$. Also note that since $\|h_N\| \to 0$ as $N \to \infty$, $\|h_M - h_N\| \to 0$

as $M, N \to \infty$ by the triangle inequality. Consequently, $\left\{\widetilde{U}'_{T,N} f\right\}$ is a Cauchy sequence in $\ell^2 \otimes \mathcal{H}$. Set

$$Vf = \lim \widetilde{U}'_{T,N} f \quad \text{for every } f \in \mathcal{H}.$$

Since $\left\|\widetilde{U}'_{T,N} f\right\| = \|f\|$ for each $f \in \mathcal{H}$,

$$0 \leq \left|\|f\| - \|Vf\|\right| = \left|\|U'_{T,N} f\| - \|Vf\|\right| \leq \|U'_{T,N} f - Vf\| \to 0$$

as $N \to \infty$ and therefore it follows that $V$ is an isometry.

Finally, let us set the embedding $\iota \colon \mathcal{H} \to \ell^2 \otimes \mathcal{H}$ as the inclusion map

$$\iota \colon h \mapsto (h, 0, \cdots) \quad \text{for every } h \in \mathcal{H}.$$

Then we see that, for $h \in \mathcal{H}$, and $f = \{h_n\}_{n \geq 0} \in \ell^2 \otimes \mathcal{H}$, we have

$$\begin{aligned} \langle V^* \iota h, f \rangle = \langle \iota h, Vf \rangle &= \lim_{N \to \infty} \left\langle \iota h, \widetilde{U}'_{T,N} f \right\rangle = \lim_{N \to \infty} \langle h, Th_0 + D_{T^*} h_N \rangle \\ &= \langle h, Th_0 \rangle = \langle T^* h, \iota^* f \rangle = \langle \iota T^* h, f \rangle. \end{aligned}$$

In the above computation, we used the fact that $h_N \to 0$ in $\mathcal{H}$. The conclusion of the computation is that $V^* \iota = \iota T^*$, namely, that $V$ is a lift of $T$. □

**Remark A.2.10** It is worthwhile to remark that there are many admissible choices of the extension $\widetilde{U}'_{T,N}$ of $U'_{T,N}$. In particular, for any bounded operator $X \in \ell^2 \otimes \mathcal{H}$, one could choose

$$\widetilde{U}'_{T,N} = \begin{bmatrix} U'_{T,N} & 0 \\ 0 & X \end{bmatrix}.$$

Of particular interest is the extension

$$\widetilde{U}'_{T,N} = \begin{bmatrix} U'_{T,N} & 0 \\ C & M_z \end{bmatrix},$$

where $C \colon \mathcal{H}_N \to \ell^2 \otimes \mathcal{H}$ is the operator given by

$$C \begin{bmatrix} h_0 \\ h_1 \\ \vdots \\ h_N \end{bmatrix} = \begin{bmatrix} h_N \\ 0 \\ \vdots \end{bmatrix}.$$

The significance of this choice is that the strong limit of $\widetilde{U}_N$, in this case, is equal to a nonminimal variant of the isometric lift of $T$ constructed by Schäffer:

$$\begin{bmatrix} T & & & & \\ D_T & & & & \\ & I_{\mathcal{H}} & & & \\ & & I_{\mathcal{H}} & & \\ & & & \ddots & \end{bmatrix}.$$

The computation to check all these details is routine; we leave it to the readers.

In this book, although we are able to find two explicit constructions of the Ando isometric lift of a commuting contractive pair, an analogue of an Ando 1-dilation, or more generally an Ando $N$-dilation, remains unknown.

**Problem A.2.11** Given a commuting contractive pair $(T_1, T_2)$ acting on a Hilbert space $\mathcal{H}$ and an integer $N \geq 1$, find a pair of commuting unitary operators $(U_{N,1}, U_{N,2})$ acting on a Hilbert space $\mathcal{H}_N \supset \mathcal{H}$ so that for every polynomial $p$ of bidegree $(m, n)$ with $m + n \leq N$, we have

$$p(T_1, T_2) = P_{\mathcal{H}} p(U_1, U_2)|_{\mathcal{H}}.$$

Here the correspondence $N \mapsto \mathcal{H}_N$ should be such that, when $\dim \mathcal{H} < \infty$, we must have $\dim \mathcal{H}_N < \infty$ for every $N \geq 1$.

Note that without the additional dimension constraint on the spaces $\mathcal{H}_N$, the problem above is already answered by Ando's dilation theorem.

Let us recall a related result of McCarthy and Shalit found in [70].

**Theorem A.2.12** (See Theorem 1.2 in [70].) *Let $(T_1, T_2, \ldots, T_d)$ be a $d$-tuple of commuting contractive matrices acting on $\mathcal{H}$, where* $\dim \mathcal{H} = L < \infty$. *Then the following are equivalent.*

(i) *$(T_1, T_2, \ldots, T_d)$ has a commuting unitary dilation $(U_1, U_2, \ldots, U_d)$ acting on $\mathcal{K}$, that is, $\mathcal{K} \supset \mathcal{H}$ and*

$$f(T_1, T_2, \ldots, T_d) = P_{\mathcal{H}} f(U_1, U_2, \ldots, U_d)|_{\mathcal{H}} \tag{A.20}$$

*for every polynomial $f$ in $d$ variables.*

(ii) *For every $N \geq 1$, there exists a unitary $N$-dilation $(U_1, U_2, \ldots, U_d)$ acting on $\mathcal{H}_N$, where* $\dim \mathcal{H}_N < \infty$, *that is, for every polynomial $f$ in $d$ variables of $d$-degree $(k_1, k_2, \ldots, k_d)$ such that $k_1 + k_2 + \cdots + k_d \leq N$ we have* (A.20).

*Moreover, the space $\mathcal{H}_N$ in item (ii) above can be chosen so that*

$$\dim \mathcal{H}_N = L^2(L+1)\frac{(N+d)!}{N!d!} + L.$$

Thus in view of the above theorem by McCarthy and Shalit, it is known, at least for commuting contractive matrix pairs, that unitary $N$-dilations acting on a finite-dimensional Hilbert space exist. An explicit form of the space $\mathcal{H}_N$ and the unitary $N$-dilation operators, however, remains unknown.

### A.2.3 Incomplete Polydisk Dilation Theory

After Sz.-Nagy published his paper [86] on the dilation of a single contraction operator, followed up by a more systematic study of the theme in his book with Foias [92], many mathematicians became interested in the possible generalizations of the theory in both single and multivariable settings, giving rise to the celebrated rational dilation problem. The rational dilation problem was discussed briefly in Section A.1. While the rational dilation problem has an affirmative answer for both the disk $\overline{\mathbb{D}}$ and the bidisk $\overline{\mathbb{D}}^2$ cases, it has a negative answer for all the polydisks $\overline{\mathbb{D}}^d$ with $d \geq 3$. The first concrete example is due to Parrott, who, in fact, produced in [74] a commuting contractive triple $(T_1, T_2, T_3)$ acting on a Hilbert space $\mathcal{H}$ that has the tridisk $\overline{\mathbb{D}}^3$ as a spectral set, yet the commuting contractive triple does not even have a commuting unitary triple $(U_1, U_2, U_3)$, so that $P_{\mathcal{H}} U_j|_{\mathcal{H}} = T_j$ for each $j = 1, 2, 3$.

As it turns out, when $d \geq 3$, a commuting contractive $d$-tuple $(T_1, \ldots, T_d)$ need not have $\overline{\mathbb{D}}^d$ as a spectral set. The first proof was due to Varopoulos [94] without any concrete examples; however, many concrete examples followed later. Concrete examples of smallest possible size were given by Holbrook [66], who found a $4 \times 4$ commuting triple of contractive matrices that does not have the tridisk $\overline{\mathbb{D}}^3$ as a spectral set. See also the earlier works by Varopoulos and Kaijser (see addendum to [94]) and Crabb and Davie [45] establishing the same negative result by giving concrete commuting contractive triples of respective matrix sizes $5 \times 5$ and $8 \times 8$. In a parallel development, Choi and Davidson [44] found an example of a quadruple of $3 \times 3$ commuting contractive matrices that has $\overline{\mathbb{D}}^4$ as a spectral set but not as a complete spectral set, thereby showing by means of a concrete example that the rational dilation problem has a negative answer for $\overline{\mathbb{D}}^4$. Then came an affirmative answer: Knese [67] showed, by using a result of Kosiński [68], that any $d$-tuple of commuting contractive $3 \times 3$-matrices has $\overline{\mathbb{D}}^d$ as a spectral set. This prompted him to make the following conjecture, which still remains open.

**Problem A.2.13** Show that every commuting triple of contractive $3 \times 3$-matrices lifts to a commuting triple of isometries.

### A.2.4 A Possible Parallel Theory

In Chapters 5 and 6, we give an explicit model and thereby classify all Ando isometric lifts of a given commuting contractive pair with the BCL-model for a commuting isometric pair (Theorem 3.1.4) as a key tool. The attentive reader may have noticed that a parallel theory using the BDF-model (Theorem 3.2.2) is missing in this exposition. In other words, the theory developed in this monograph takes the product contraction $T = T_1T_2$ as the companion contraction. This was inspired by the BCL-model theory for commuting isometries, where the product isometry $V = V_1V_2$ played the role of a companion contraction. In contrast, the BDF-model theory, thoroughly discussed in Chapter 3, develops with one of the components (either $T_1$ or $T_2$) of a commuting isometric pair $(T_1, T_2)$ taken as the companion contraction. What is missing at this point is a parallel dilation/model theory for commuting contractive pairs $(T_1, T_2)$ with $T_1$ (or $T_2$) as the companion contraction. The first objective is the following.

**Problem A.2.14** Given a commuting contractive pair $(T_1, T_2)$ acting on a Hilbert space $\mathcal{H}$, construct an Ando isometric lift in the BDF1-model (respectively, BDF2-model) form as in (3.64) (respectively, (3.65)) or simply in the form

$$\left(\begin{bmatrix} M_z & 0 \\ 0 & W \end{bmatrix}, \begin{bmatrix} M_\Theta & 0 \\ C & E \end{bmatrix}\right) \quad \text{on} \quad \begin{bmatrix} H^2(\mathcal{F}) \\ \mathcal{G} \end{bmatrix}$$

via some embedding of $\mathcal{H}$, namely $\Pi\colon \mathcal{H} \to \begin{bmatrix} H^2(\mathcal{F}) \\ \mathcal{G} \end{bmatrix}$.

In the special case when the commuting contractive pair $(T_1, T_2)$ is such that $T_1$ is pure and $\dim \mathcal{D}_{T_j^*} < \infty$ for both $j = 1, 2$, it was shown in [47] that if we take $\Theta$ to be the $\mathcal{B}(\mathcal{D}_{T_1^*})$-valued transfer function

$$\Theta(z) = A + zB(I - zD)^{-1}C$$

corresponding to the unitary colligation

$$\begin{bmatrix} A & B \\ C & D \end{bmatrix} : \begin{bmatrix} \mathcal{D}_{T_1^*} \\ \mathcal{D}_{T_2^*} \end{bmatrix} \to \begin{bmatrix} \mathcal{D}_{T_1^*} \\ \mathcal{D}_{T_2^*} \end{bmatrix}$$

that extends the norm-preserving assignment

$$\begin{bmatrix} D_{T_1^*}T_2^* \\ D_{T_2^*} \end{bmatrix} h \mapsto \begin{bmatrix} D_{T_1^*} \\ D_{T_2^*}T_1^* \end{bmatrix} h \quad \text{for all } h \in \mathcal{H},$$

then

$$(M_z, M_\Theta) \quad \text{acting on } H^2\left(\mathcal{D}_{T_1^*}\right)$$

is an isometric lift of $(T_1, T_2)$ via the isometric embedding given by

$$\Pi\colon h \mapsto D_{T_1^*}\left(I - zT_1^*\right)^{-1} h.$$

The general case appears to be challenging at this moment.

# References


[1] A. A. Abouhajar, M. C. White and N. J. Young, *A Schwarz lemma for a domain related to $\mu$-synthesis*, J. Geom. Anal. **17** (2007), 717–750.

[2] V. M. Adamjan and D. Z. Arov, *Unitary couplings of semi-unitary operators* (Russian), Mat. Issled. **1** (1966), 3–64. [English translation: Amer. Math Soc. Transl. Ser. 2, **95** (1970), p. 95.]

[3] J. Agler, *The Arveson extension theorem and coanalytic models*, Integr. Equ. Oper. Theory **5** (1982), 608–631.

[4] J. Agler and J. E. McCarthy, *Pick Interpolation and Hilbert Function Spaces*, Graduate Studies in Mathematics **44**, Amer. Math. Soc., Providence, RI, 2002.

[5] J. Agler and J. McCarthy, *Distinguished varieties*, Acta. Math. **194** (2005), 133–153.

[6] J. Agler and N. J. Young, *A commutant lifting theorem for a domain in $\mathbb{C}^2$ and spectral interpolation*, J. Funct. Anal. **161** (1999), no. 2, 452–477.

[7] J. Agler and N. J. Young, *Operators having the symmetrized bidisc as a spectral set*, Proc. Edinburgh Math. Soc. (2) **43** (2000), no. 1, 195–210.

[8] J. Agler and N. J. Young, *A model theory for $\Gamma$-contractions*, J. Oper. Theory **49** (2003), 45–60.

[9] C.-G. Ambrozie, M. Engliš and V. Müller, *Operator tuples and analytic models over general domains in $\mathbb{C}^n$*, J. Oper. Theory **47** (2002), no. 2, 287–302.

[10] T. Ando, *On a pair of commuting contractions*, Acta Sci. Math. (Szeged) **24** (1963), 88–90.

[11] D. Z. Arov and L. Z. Grossman, *Scattering matrices in the theory of unitary extension of isometric operators*, Math. Nachr. **157** (1992), 105–123.

[12] W. B. Arveson, *Subalgebras of $C^*$-algebra*. II, Acta Math. **128** (1972), 271–308.

[13] J. A. Ball, *Factorization and model theory for contraction operators with unitary part*, Mem. Amer. Math. Soc. **13** (1978), no. 198.

[14] J. A. Ball, Linear systems, operator model theory and scattering: multivariable generalizations, in *Operator Theory and its Applications (Winnipeg, MB, 1998)*, pp. 151–178, Fields Inst. Commun. **25**, Amer. Math. Soc., Providence, RI, 2000.

[15] J. A. Ball and V. Bolotnikov, de Branges–Rovnyak spaces: basics and theory, in *Operator Theory Vol.1* (ed. D. Alpay), pp. 631–679, Springer, Basel, 2015.

[16] J. A. Ball and N. Cohen, de Branges–Rovnyak operator models and systems theory: a survey, in *Topics in Matrix and Operator Theory*, pp. 103–136. Oper. Theory Adv. Appl. **50**, Birkhäuser, Basel, 1991.

[17] J. A. Ball, I. Gohberg and L. Rodman, *Interpolation theory of rational matrix functions*, Oper. Theory Adv. Appl. **45** Birkhäuser-Verlag, Basel, 1990.

[18] J. A. Ball and T. L. Kriete, *Operator-valued Nevanlinna–Pick kernels and the functional models for contraction operators*, Integr. Equ. Oper. Theory **10** (1987) no. 1, 17–61.

[19] J. A. Ball, C. Sadosky and V. Vinnikov, *Scattering systems with several evolutions and multidimensional input/state/output systems*, Integr. Equ. Oper. Theory **52** (2005), 323–393.

[20] J. A. Ball and H. Sau, *Rational dilation of tetrablock contractions revisited*, J. Funct. Anal. **278** (2020), 108275.

[21] J. A. Ball and H. Sau, Functional models for commuting Hilbert-space contractions, in *Operator Theory, Operator Algebras and Their Interactions with Geometry and Topology: Ronald G. Douglas Memorial Volume*, pp. 11–54, Oper. Theory Adv. Appl. **278**, Birkhäuser, Cham, 2020.

[22] J. A. Ball and H. Sau, *Dilation theory and functional models for tetrablock contractions*, Complex Anal. Oper. Theory **17** (2023), no. 2, paper no. 25 (40 pages).

[23] J. A. Ball and H. Sau, Commuting and doubly commuting pairs of isometries, in *Operator Theory, Related Fields, and Applications IWOTA 2023*, pp. 15–67, Oper. Theory Adv. Appl. **307**, Birkhäuser, Cham, 2025.

[24] J. A. Ball and V. Vinnikov, Hardy spaces on a finite bordered Riemann surface, multivariable operator model theory and Fourier analysis along a unimodular curve, in *Systems, Approximation, Singular Integral Operators, and Related Topics*, pp. 37–56, Oper. Theory Adv. Appl. **129**, Birkhäuser, Basel, 2001.

[25] S. Barik and B. K. Das, *Isometric dilations of commuting contractions and Brehmer positivity*, Complex Anal. Oper. Theory **16** (2022), 69.

[26] S. Barik, B. K. Das, K. J. Haria and J. Sarkar, *Isometric dilations and von Neumann inequality for a class of tuples in the polydisc*, Trans. Amer. Math. Soc. **372** (2019), no. 2, 1429–1450.

[27] S. Barik, B. K. Das and J. Sarkar, *Isometric dilations and von Neumann inequality for finite rank commuting contractions*, Bull. Sci. Math. **165** (2020), 102915 (25 pages).

[28] H. Bercovici, R. G. Douglas and C. Foias, *On the classification of multi-isometries*, Acta Sci. Math. (Szeged) **72** (2006), 639–661.

[29] H. Bercovici, R. G. Douglas and C. Foias, Bi-isometries commutant lifting, in *Characteristic Functions, Scattering Functions, and Transfer Functions*, Oper. Theory Adv. Appl. **197**, pp. 51–76, Birkhäuser-Verlag, Basel, 2010.

[30] H. Bercovici, R. G. Douglas and C. Foias, Canonical models for bi-isometries, in *A Panorama of Modern Operator Theory and Related Topics (The Israel Gohberg Memorial Volume)*, Oper. Theory Adv. Appl. **218,** Birkhäuser, Basel, 2012.

[31] C. A. Berger, L. A. Coburn and A. Lebow, *Representation and index theory for $C^*$-algebras generated by commuting isometries*, J. Funct. Anal. **27** (1978), no. 1, 51–99.

[32] R. Bhat and T. Bhattacharyya, *Dilations, Completely Positive Maps and Geometry*, Hindustan Book Agency, New Delhi, 2023.

[33] T. Bhattacharyya, *The tetrablock as a spectral set*, Indiana Univ. Math. J. **63** (2014), 1601–1629.

[34] T. Bhattacharyya and M. Bhowmik, *A note on the dilation of a certain family of tetrablock contractions*, Oper. Theory Adv. Appl., **295** Birkhäuser/Springer, Cham, 2024, pp. 115–132.

[35] T. Bhattacharyya, P. Kumar and H. Sau, *Distinguished varieties through the Berger–Coburn–Lebow theorem*, Analysis and PDE **15** (2022), no. 5, 477–506.

[36] T. Bhattacharyya, S. Pal and S. Shyam Roy, *Dilations of* $\Gamma$*-contractions by solving operator equations*, Adv. Math. **230** (2012), 577–606.

[37] T. Bhattacharyya and H. Sau, *Explicit and unique construction of tetrablock unitary dilation in a certain case*, Complex Anal. Oper. Theory **10** (2016), no. 4, 749–768.

[38] T. Bhattacharyya and H. Sau, $\Gamma$*-unitaries, dilation and a natural example*, Publ. Res. Inst. Math. Sci. **53** (2017), 261–285.

[39] T. Bhattacharyya and H. Sau, *Holomorphic functions on the symmetrized bidisk – realization, interpolation and extension*, J. Funct. Anal. **274** (2018), 504–524.

[40] B. Bisai and S. Pal, *A Nagy–Foias program for C.N.U.* $\Gamma_n$*-contractions*, Complex Anal. Oper. Theory **18** (2024), paper no. 122 (29 pages).

[41] Z. Burdak, *On the model and invariant subspaces for pairs of commuting isometries*, Integr. Equ. Oper. Theory **91** (2019), 23 pages.

[42] Z. Burdak, M. Kosiek, P. Pagacz and M. Słociński, *On the commuting isometries*, Linear Algebra Appl. **516** (2017), 167–185.

[43] Z. Burdak, M. Kosiek, and M. Słociński, *Compatible pairs of commuting isometries*, Linear Algebra Appl. **479** (2015), 216–259.

[44] M.-D. Choi and K. R. Davidson, *A* $3 \times 3$ *dilation counterexample*, Bull. London Math. Soc. **45** (2013), 511–519.

[45] M. Crabb and A. Davie, *Von Neumann's inequality for Hilbert space operators*, Bull. London Math. Soc. **7** (1975), 49–50.

[46] R. E. Curto and F.-H. Vasilescu, *Standard operator models in the polydisc*, Indiana Univ. Math. J. **42** (1993), no. 3, 791–810.

[47] B. K. Das and J. Sarkar, *Ando dilations, von Neumann inequality, and distinguished varieties*, J. Funct. Anal. **272** (2017), no. 5, 2114–2131.

[48] B. K. Das, S. Sarkar and J. Sarkar, *Factorizations of contractions*, Adv. Math. **322** (2017), 186–200.

[49] B. K. Das and H. Sau, *Pure inner functions, distinguished varieties and toral algebraic commutatifve contractive pairs*, Proc. Amer. Math. Soc. **152** (2024), no. 3, 1067–1081.

[50] B. K. Das and H. Sau, *Algebraic varieties as a complete spectral set*, preprint, 2024.

[51] L. de Branges and J. Rovnyak, Canonical models in quantum scattering theory, in *Perturbation Theory and its Applications in Quantum Mechanics* (Proc. Adv. Sem. Math. Res. Center, U.S. Army, Theoret. Chem. Inst., University of Wisconsin, Madison, WI, 1965), pp. 295–392, Wiley, New York, 1966.

[52] L. de Branges and J. Rovnyak, *Square Summable Power Series*, Holt, Rinehart and Winston, New York–Toronto, Ont.–London, 1966.

[53] R. G. Douglas, *On majorization, factorization and range inclusion of operators on Hilbert space*, Proc. Amer. Math. Soc. **17** (1966), 413–415.

[54] R. G. Douglas, *Structure theory for operators, I.*, J. Reine Angew. Math. **232** (1968) 180–193.
[55] R. G. Douglas, P. S. Muhly and C. Pearcy, *Lifting commuting operators*, Michigan Math. J. **15** (1968), 385–395.
[56] R. G. Douglas and K. Yan, *On the rigidity of Hardy submodules*, Integr. Equ. Oper. Theory **13** (1990), 350–363.
[57] J. C. Doyle and A. Packard, *The complex structured singular value*, Automatica **29** (1993) no. 1, 71–109.
[58] G. E. Dullerud and F. Paganini, *A course in robust conrol theory: a convex approach*, Texts in Applied Mathematics **36**, Springer, New York, 2000.
[59] E. Durszt, *Contractions as restricted shifts*, Acta Sci. Math. **48** (1985), 129–134.
[60] H. Dym and I. Gohberg, *A maximum entropy principle for contractive intertwining dilations*, J. Funct. Anal. **65** (1986), 83–125.
[61] E. Egerváry, *On the contractive linear transformations of n-dimensional vector space*, Acta Sci. Math. (Szeged) **15** (1954), 178–182.
[62] C. Foias and A. E. Frazho, *The Commutant Lifting Approach to Interpolation Problems*, Operator Theory: Advances and Applications **44**, Birkhäuser Verlag, Basel, 1990.
[63] D. Gaşpar and P. Gaşpar, *Wold decompositions and the unitary model for bi-isometries*, Integr. Equ. Oper. Theory **49** (2004), no. 4, 419–433.
[64] P. R. Halmos, *Normal dilations and extensions of operators*, Summa Brasil. Math. **2** (1950), 125–134.
[65] J. W. Helton, *The characteristic function of operator theory and electrical network realization*, Ind. U. Math. J. **22** (1972), no. 5, 403–414.
[66] J. A. Holbrook, *Schur norms and the multivariate von Neumann inequality*, Recent Advances in Operator Theory and Related Topics (Szeged), 1999, Operator Theory: Advances and Applications **127**, Birkäuser, Basel 2001, pp. 375–386.
[67] G. Knese, *The von Neumann inequality for* $3 \times 3$ *matrices*, Bull. London Math. Soc. **48** (2016), 53–57.
[68] L. Kosiński, *Three-point Nevanlinna Pick problem in the polydisc*, Proc. London Math. Soc. **111** (2015), 887–910.
[69] W. S. Li and D. Timotin, *The central Ando dilation and related orthogonality properties*, J. Funct. Anal. **154** (1998), 1–16.
[70] J. E. McCarthy, and O. M. Shalit, *Unitary N-dilations for tuples of commuting matrices*, Proc. Amer. Math. Soc. **141** (2013), 563–571.
[71] N. K. Nikolskii and V. Vasyunin, Notes on two function models, in *The Bieberbach Conjecture* (West Lafayette, IN, 1985), pp. 113–141, Math. Surveys Monogr. **21**, Amer. Math. Soc., Providence, RI, 1986.
[72] N. K. Nikolskii and V. Vasyunin, A unified approach to function models and the transcription problem, in *The Gohberg Anniversary Collection, Vol. II* (Calgary, AB, 1988), pp. 405–434, Oper. Theory Adv. Appl. **41**, Birkhäuser, Basel, 1989.
[73] S. Pal, *Canonical decomposition of a tetrablock contraction and operator model*, J. Math. Anal. Appl. **438** (2016), 274–284.
[74] S. Parrott, *Unitary dilations for commuting contractions*, Pacific J. Math., **34** (1970), 481–490.
[75] S. Parrott, *On a quotient norm and the Sz.-Nagy–Foiaş lifting theorem, J. Funct. Anal.* **30** (1978), no. 3, 311–328.

[76] D. Popovici, *A Wold-type decomposition for commuting isometric piars*, Proc. Amer. Math. Soc. **132** (2004), no. 8, 2303–2314.

[77] V. P. Potapov, *The multiplicative structure of J-contractive matrix functions*, Trudy Moskov. Mat. Obšč. **4** (1955), 125–236.

[78] F. Riesz and B. Sz.-Nagy, *Functional Analysis*, Dover, New York, 1990 [unabridged republication of the work first published by Frederick Ungar Publishing Co., New York (1955)].

[79] D. Sarason, *Sub-Hardy Hilbert Spaces in the Unit Disk*, University of Arkansas Lecture Notes in the Mathematical Sciences **10**, John Wiley & Sons, New York, 1994.

[80] J. Sarkar, *Operator theory on symmetrized bidisc*, Ind. U. Math. J. **64** (2015), no. 3, 847–873.

[81] S. Sarkar, *Pairs of commuting pure contractions and isometric dilation*, J. Oper. Theory, **92** (2024) 49–76.

[82] H. Sau, *A note on tetrablock contractions*, New York J. Math. **21** (2015), 1347–1369.

[83] H. Sau, *Ando dilations for a pair of commuting contractions: two explicit constructions and functional models*, arXiv:1710.11368 [math.FA].

[84] J. J. Schäffer, *On unitary dilations of contractions*, Proc. Amer. Math. Soc. **6** (1955), 322. MR 16,934c.

[85] M. Seto and R. Yang, *Inner sequence based invariant subspaces in* $H^2(\mathbb{D}^2)$, Proc. Amer. Math. Soc. **135** (2017), no. 8, 2519–2526.

[86] B. Sz.-Nagy, *Sur les contractions de l'espace de Hilbert*, Acta Sci. Math. **15** (1953), 87–92.

[87] B. Sz.-Nagy, *Extensions of Linear Transformations in Hilbert Space Which Extend Beyond This Space*, Frederick Ungar Publishing Co., New York (1960); reprinted as an Appendix of [78].

[88] B. Sz.-Nagy and C. Foias, *Sur les contractions de l'espace de Hilbert. VIII* Acta Sci. Math. **25** (1964), 38–71.

[89] B. Sz.-Nagy and C. Foias, *Dilatation des commutants d'opérateurs* (French), C. R. Acad. Sci. Paris Ser. A–B **266** (1968), A493–A495.

[90] B. Sz.-Nagy and C. Foias, *On the structure of intertwining operators*, Acta Sci. Math. (Szeged) **35** (1973), 225–254.

[91] B. Sz.-Nagy and C. Foias, *Regular factorizations of contractions*, Proc. Amer. Math. Soc. 43(1974), 91–93.

[92] B. Sz.-Nagy, C. Foias, H. Bercovici and L. Kérchy, *Harmonic Analysis of Operators on Hilbert Space*, revised and enlarged edition, Universitext, Springer, New York, 2010.

[93] D. Timotin, *Regular dilations and models for multicontractions*, Indiana Univ. Math. J. **47** (1993), no. 3, 671–684.

[94] N. Th. Varopoulos, *On an inequality of von Neumann and an application of the metric theory of tensor products to operators theory*, J. Funct. Anal. **16** (1974), 83–100.

[95] H. Wold, *A Study in the Analysis of Stationary Time Series*, 2nd edition, Stockholm, 1954.

[96] R. Yang, *A brief survey of operator theory in* $H^2(\mathbb{D}^2)$, Lecture Notes for 2018 Summer School Dalian University of Technology, India, available at arXiv:1810.00133v4.

# Index

For EU product safety concerns, contact us at Calle de José Abascal, 56–1°, 28003 Madrid, Spain or eugpsr@cambridge.org.

www.ingramcontent.com/pod-product-compliance
Ingram Content Group UK Ltd.
Pitfield, Milton Keynes, MK11 3LW, UK
UKHW020219290926
15470UKWH00001B/285
* 9 7 8 1 0 0 9 6 8 7 2 1 8 *